**Progress in Systems and Control Theory**
Volume 12

# Systems, Models and Feedback: Theory and Applications

*Proceedings of a U.S.-Italy Workshop in honor of Professor Antonio Ruberti, Capri, 15-17, June 1992*

A. Isidori
T. J. Tarn
Editors

Springer Science+Business Media, LLC 1992

Alberto Isidori
Department of Computer
and Systems Science
Università di Roma "La Sapienza"
00184 Rome, Italy

Tzyh-Jong Tarn
Department of Systems
Sciences and Mathematics
Washington University
St. Louis, MO 63130

**Library of Congress Cataloging-in-Publication Data**

Systems, models, and feedback : theory and applications : proceedings of a U.S.-Italy workshop in honor of Professor Antonio Ruberti, Capri, 15-17 June 1992 / edited by Alberto Isidori, Tzuh-Jon;g Tarn.
p. cm. (Progress in systems and control theory ; v. 12)
ISBN 978-1-4757-2206-2 ISBN 978-1-4757-2204-8 (eBook)
DOI 10.1007/978-1-4757-2204-8
1. System theory--Congresses. 2. Nonlinear control theory--Congresses. 3. Feedback control systems--Congresses. 4. Ruberti, Antonio. I. Isidori, Alberto. II. Tarn, Tzyh-Jong, 1937- III. Ruberti, Antonio. IV. Series.
Q295.S9564 1992 92-3321
003--dc20 CIP

Printed on acid-free paper.


Originally published by Birkhäuser Boston in 1992
Softcover reprint of the hardcover 1st edition 1992



ISBN 978-1-4757-2206-2

Camera-ready text prepared by the authors.

9 8 7 6 5 4 3 2 1

*This book is dedicated to*
*Professor Antonio Ruberti*
*on the occasion*
*of his 65th birthday*

# TABLE OF CONTENTS

# Preface

It is a great honor and privilege to have this opportunity of celebrating the 65th birthday of Professor Antonio Ruberti by holding an International Conference on *Systems, Models and Feedback.*

The conference, and this volume which contains its proceedings, is a tribute to Professor Ruberti in acknowledgement of his major contributions to System Theory, at a time in which this area was emerging and consolidating as an independent discipline, his role as a leader of the Italian academic community, his activity in promoting and fostering close scientific relations between Italian and U.S. scholars in Systems and Control.

The format of this conference is inspired by a series of seminars initiated exactly twenty years ago under the direction of Professor Ruberti, in Italy, and Professor R. R. Mohler, in the U.S. By bringing together many authoritative talents from both countries, these seminars were instrumental in promoting the expansion of System Theory in new areas, notably that of Nonlinear Control, and were the key to successful scientific careers for many of the younger attendants.

The papers contained in this volume are divided into three groups. The first group is a set of papers dealing with advanced applications of Systems and Control concepts. The second and third groups deal with selected topics on the conference theme, i.e., Systems, Models and Feedback. In order to facilitate the reading, we deemed it convenient to organize these papers into two separate groups: one group includes contributions in the broader area of *Systems Theory* and the other collects those papers whose common focus is the design of *Feedback* control systems.

We wish to thank all the contributors to this volume for their outstanding work, which altogether provides a remarkable picture of how powerfully System and Control Theory has developed in recent years. We also would like to thank the funding agencies that have supported organization of the conference: the Consiglio Nazionale delle Ricerce, the National Science Foundation and the Universitá di Roma "La Sapienza".

A. Isidori, T. J. Tarn

# Robust Binary Control of Robotic Systems

C. BONIVENTO, C. MELCHIORRI, and R. ZANASI

**Abstract**

**In the area of robust control, both the well-known Variable Structure approach and its more recent generalization referred as Binary Control Theory present interesting methodological properties but also some applicative inconveniences. Among major drawbacks, practical implementations of these control algorithms show undesirable chattering behaviour - for Variable Structure System - or continuous oscillations - for Binary Control - on the system variables. In this paper, experimental results obtained applying a new control technique to the control of a robotic system are reported and discussed. The main advantage of this technique, based on a Sliding Mode regulator with an additional switching integral action, is that an arbitrary reduction of the chattering phenomenon can be achieved maintaining the robustness properties of the Sliding Mode approach. The results are encouraging and comparatively interesting.**

## 1. Introduction

During the last decade, noticeable interest has arisen concerning the development of non-linear, robust control techniques, see [1]-[3], and their application in robotics. Main reasons for this interest are the unmodeled dynamics, such as friction, joint elasticity, mechanical flexibility, and the highly non-linear, coupled dynamical effects usually present in a robotic structure. Among the others, the Variable Structure System (VSS), [4], [5], is one of the most promising approaches for its robustness characteristics, simplicity of implementation, and compensation of non-linearities of the controlled system. Details about some applications of this technique to the control of manipulators may be found in [6]-[9]. On the other hand, the VSS approach has also some drawbacks such us chattering and high control activity. Except for particular cases, such as motor control using switching electronic drives, these phenomena are highly undesirable because they can excite high-frequency unmodeled dynamics and cause fast deterioration of the system.

In the 1980's a general non-linear feedback design approach for stabilization of uncertain systems, the *Binary Control* method, was proposed by S.V. Emelyanov and his co-workers, [10]-[12]. This approach aims to be considered as a general framework in which several well-known techniques of robust stabilization may be framed. In this stream, new types of non-linear feedbacks have been proposed, see for example [10] and [12]. In many cases these approaches solve the stabilization problem in a more feasible way than well-known VSS techniques, due to the fact that the resulting control signal is continuous and not switching.

In this paper, the application of a new technique, framed in the Binary Control context, to the control of a robotic mechanism is presented. The adopted control method, discussed in detail in [13], [14] and [15], has some interesting robustness properties. It is based on a standard Sliding Mode regulator, with an additional switching integral action which tracks unknown disturbances. This allows an arbitrary reduction of the amplitude of the switching term of the control signal, maintaining the robustness properties shown by the Sliding Mode approach.

The paper is organized as follows. In Section 2 some basic concepts about the Binary Control theory are given. The adopted control structure is reported in Section 3, while in Section 4 some results of the experimental activity are presented and discussed. Section 5 concludes with comments and remarks about the current and future activity.

## 2. Basic concepts of the Binary Control Theory.

The general control structure of the Binary Control is shown in Fig. 1. In this scheme, $P$ is the plant to be controlled, $y_d(t)$ is the reference signal, $\varphi(\mathrm{x}, t)$ is an unknown parametric disturbance, $R_u$, $R_\mu$, $R_\rho$ and $R_\nu$ are operators defining four different control loops with specific stabilization properties:

- $R_u$ acts directly on the state coordinates of the system and defines a feedback of *coordinate* type;

- $R_\mu$ changes the properties of $R_u$ using the information on both the state variables and the desired dynamics $\sigma = 0$. It is called a feedback of *operator* type;

- $R_\rho$, through $\delta_0$ and $\mu$, slightly relaxes the desired dynamics ($\sigma \rightarrow \bar{\sigma} = \sigma - \delta_0 \mu(R_u \circ \mathrm{x})$), so that the time-varying surface $\bar{\sigma} = 0$ is reached with a global smoother system behaviour. This is the so-called Operator Feedback (OF);

- $R_\nu$ (Operator Coordinate Feedback: OCF) defines the external control loop in order to guarantee that the closed loop system behaviour is asymptotically independent from the uncertain parameters of system $P$.

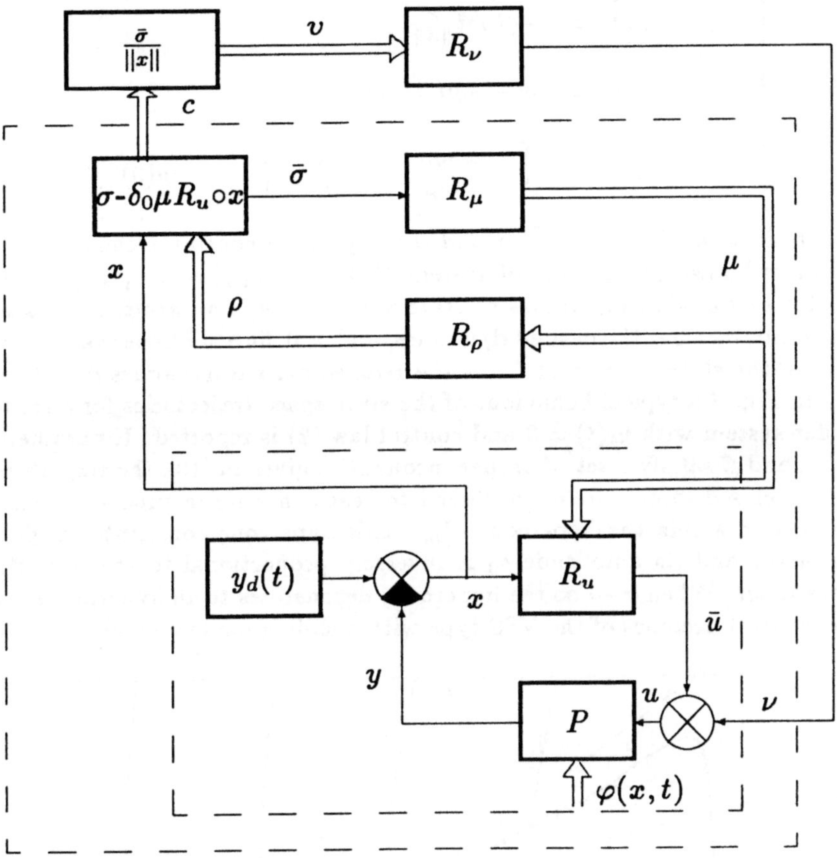


Figure 1: General Binary Control structure.

The general expression of the feedback law as proposed in [10] is the following:

$$\begin{cases} u = R_\mu \circ R_u \circ \mathbf{x} + R_\nu \circ (\mathbf{x}, \bar{\sigma}) \\ \bar{\sigma} = R_\rho \circ (\mathbf{x}, \sigma) \end{cases} \tag{1}$$

where $A \circ B$ means that the operator $A$ affects the behaviour of $B$. With different choices for $R_u$, $R_\mu$, $R_\rho$ and $R_\nu$, it is possible to obtain well-known control schemes, such as VSS, Bang-Bang, Adaptive, and so on, or design new ones. In [10] some choices for these operators have been presented and discussed. One of the structures showing good robustness properties when the external reference signal is zero ($y_d = 0$) is the following:

$$\left\{\begin{array}{rcl} u & = & R_\mu \cdot R_u \circ \mathbf{x} + R_\nu \circ (\mathbf{x}, \bar{\sigma}) \\ R_u \circ \mathbf{x} & = & \bar{\mathbf{k}}^T |\mathbf{x}| \\ R_\mu & = & \mu(t) \\ R_\nu \circ (\mathbf{x}, \bar{\sigma}) & = & -\beta \, sat \frac{\bar{\sigma}}{\|\mathbf{x}\|} \\ \bar{\sigma} & = & \sigma - \delta_0 \mu (R_u \circ \mathbf{x}) \\ \dot{\mu} & = & \left\{ \begin{array}{ll} -\alpha \, sgn \, \bar{\sigma} & |\mu| \leq 1 \\ -\omega \mu & |\mu| > 1 \end{array} \right. \qquad |\mu(0)| \leq 1 \end{array}\right. \tag{2}$$

where $\bar{\mathbf{k}} = [k_1, k_2, \ldots, k_n]^T$, $\alpha$ and $\beta$ are positive constants chosen on the basis of known parameters of system $P$, $|\mathbf{x}| = [|x_1|, |x_2|, \ldots, |x_n|]^T$ is the vector of the absolute values of state variables, $\omega$ is an arbitrary positive constant, $\sigma = 0$ is the desired dynamics, and $\delta_0$ defines an hyperconic region $G_{\delta_0}$ in the state space containing the relaxed desired dynamics $\bar{\sigma} = 0$.

In Fig. 2 a typical behaviour of the state-space trajectories for a second order system with $y_d(t) = 0$ and control law (2) is reported. If parameters $\bar{\mathbf{k}}$, $\alpha$, and $\beta$ satisfy a set of proper inequalities given in [10], the trajectories of the closed loop system are forced to reach in a finite time and remain thereafter within the hypercone $G_{\delta_0}$. This hypercone constitutes a global attractor and its amplitude $\delta_0$ is inversely proportional to the $\alpha$ control parameter. When $\alpha \to \infty$ the hypercone degenerates to an hyperplane, and the control becomes of the VSS type with a sliding mode system dynamics.

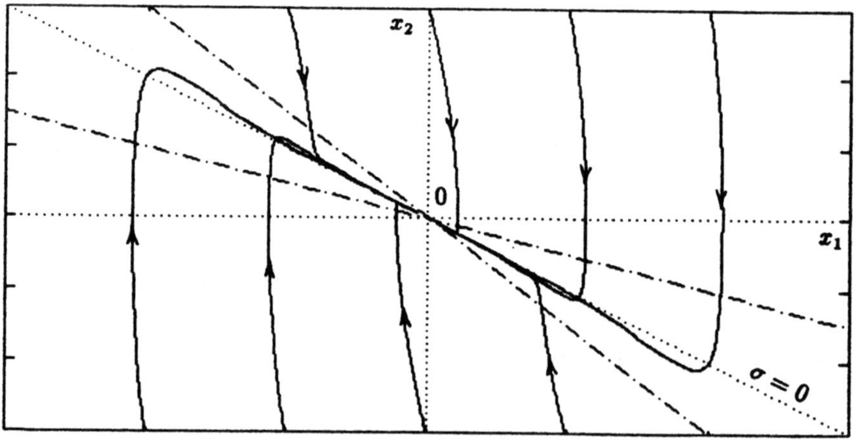


Figure 2: Typical behaviour of the trajectories for a second order system when $y_d(t) = 0$.

When the reference signal is not zero ($y_d(t) \neq 0$), a typical modification of the term $(R_u \circ \mathbf{x})$ in (2) is the following:

$$R_u \circ \mathbf{x} = \bar{\mathbf{k}}^T |\mathbf{x}| + k, \qquad k > |\psi(t)| \quad \forall t \in \Re_+ \tag{3}$$

where $\psi(t)$ is a function of the exogenous disturbances (reference signal, external loads, and so on) and their time-derivatives acting on the system. Using control law (3), the system $P$ can be stabilized with a steady-state error approaching zero only if the parameter $\alpha$ approaches infinity. Again, in this limit case the Binary Control structure becomes a VSS one, and the control signal $u$ becomes discontinuous.

Due to the presence of unmodeled dynamics or small delays in $P$, practical implementations of VSS regulators are often affected by large chattering, while, in the same conditions, the use of Binary regulators generates only continuous oscillations on the control signal $u$. Both the chattering and the continuous oscillations are undesirable disturbances when the external reference $y_d(t)$ is not zero. A possible solution to these problems has been firstly presented in [13] and [14]. The basic idea is the introduction, in the control law (2), of a switching integral part that estimates the external disturbance acting on the system. In this way, it is possible to reduce $k$ and therefore the amplitude of the switching term in the control signal. This control structure is summarized in the next Section.

## 3. The Adopted Control Structure

The control scheme implemented for the experimental activity reported in this paper is shown in Fig. 3. It consists of a velocity control loop, based on the Binary Control approach, with a standard proportional control in the external position loop. In the following, a brief description of this scheme and of its main features is given.

*Inner Loop: the Switching Integral Action*

As previously mentioned, in many situations it is desirable to have a small value for parameter $k$ in eq. (3), even in presence of large values of $|\psi|$, since it is the main cause of the chattering phenomenon. To cope with this problem, the Binary Control law (2) is modified as follows:

$$\left\{\begin{array}{rcl} u & = & R_\mu \cdot R_u \circ \mathbf{x} + R_\nu \circ (\mathbf{x}, \bar{\sigma}) + \lambda\bar{\sigma} \\ R_u \circ \mathbf{x} & = & \bar{\mathbf{k}}^T|\mathbf{x}| + k \\ R_\mu & = & \mu(t) \\ R_\nu \circ (\mathbf{x}, \bar{\sigma}) & = & \nu(t) \\ \bar{\sigma} & = & \sigma - \delta_0 \mu(R_u \circ \mathbf{x}) \\ \dot{\mu} & = & \left\{\begin{array}{ll} -\alpha\, sgn\, \bar{\sigma} & |\mu| \leq 1 \\ -\omega\mu & |\mu| > 1 \end{array}\right. \qquad |\mu(0)| \leq 1 \\ \dot{\nu} & = & \left\{\begin{array}{ll} -h\, sgn\, \bar{\sigma} & |\nu| \leq \nu_{max} \\ -\omega\mu & |\nu| > \nu_{max} \end{array}\right. \qquad |\nu(0)| \leq \nu_{max} \end{array}\right. \tag{4}$$

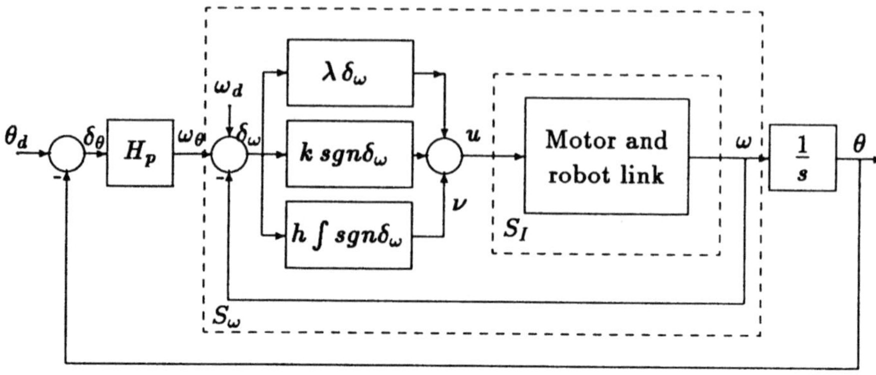


Figure 3: The new control scheme used in the experiments.

The main modifications consist of the terms $\lambda\bar{\sigma}$ and $R_\nu \circ (\mathbf{x}, \bar{\sigma})$. Due to the presence of the integral action $\nu(t)$, which acts as an observer of the disturbances, the value of $k$ in eq. (3) can be noticeably reduced. The proportional term $\lambda\bar{\sigma}$ helps in the stabilization of the system in presence of large state errors. In [14] a detailed discussion of this algorithm when $\alpha \to \infty$, i.e. in the limit case of a switching hyperplane, has been given. Here, only the basic ideas are reported.

Let us consider the uncertain system:

$$\dot{\mathbf{x}} = \mathbf{A}\mathbf{x} + \mathbf{B}(\bar{\mathbf{u}} + \varphi + \psi), \tag{5}$$

where $\mathbf{x} \in \Re^n$ is the state, $\bar{\mathbf{u}} \in \Re^m$ is the control, $\mathbf{A} \in \Re^{n\times n}$, $\mathbf{B} \in \Re^{n\times m}$ are the known parameters of the system, and $\psi = \psi(t)$ and $\varphi = \varphi(\mathbf{x}, t)$ are uncertain vector functions satisfying the conditions of existence and uniqueness of the solutions of (5).

Assuming the pair $(\mathbf{A}, \mathbf{B})$ controllable, $\mathbf{B}$ full-column rank and, without losing generality, $m = 1$, with a proper state-transformation $\mathbf{z} = \mathbf{T}\mathbf{x}$ and the linear state-feedback $\bar{\mathbf{u}} = (-\mathbf{A}_{21}\mathbf{z}_1 - \mathbf{A}_{22}z_2 + u)$ it is always possible to obtain the following equivalent representation:

$$\begin{cases} \dot{\mathbf{z}}_1 &= \mathbf{A}_{11}\mathbf{z}_1 + \mathbf{A}_{12}z_2 \\ \dot{z}_2 &= u + \varphi + \psi \end{cases} \tag{6}$$

in which $\mathbf{z}_1 \in \Re^{(n-1)}$, $z_2 \in \Re$, and matrices $\mathbf{A}_{ij}$ are a proper decomposition of matrix $\mathbf{TAT}^{-1}$, where $\mathbf{A}_{11}$ has an arbitrary preassigned spectrum. Then, the system is stabilized with the desired dynamics if control low (4) guarantees $z_2 = 0$ despite the presence of the disturbances. The control problem of the system (5) is therefore reconducted to the control of the first-order system

$$\dot{z}_2 = u + \varphi + \psi. \tag{7}$$

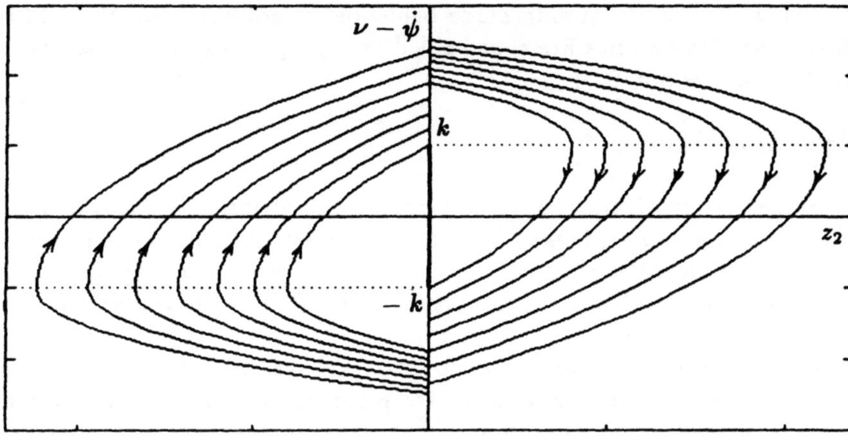


Figure 4: Typical state-space trajectories in the worst case.

The control law (4) in this case reduces to

$$\begin{cases} u &= -\lambda z_2 - (\Phi + k)\, sgn\, z_2 + \nu \\ \dot{\nu} &= -h\, sgn\, z_2. \end{cases} \tag{8}$$

where $\lambda$ is a proper non negative constant. The following results hold.

Given the system (7) with the control law (8), and disturbances $\varphi$ and $\psi$ such that:

$$\begin{array}{rclcl} |\varphi(z_2,t)| &<& \Phi(z_2) & & \Phi(0)=0, \\ |\psi(0)| &<& \Delta_0 & & |\dot{\psi}(t)| < \Delta_1, \end{array} \tag{9}$$

where $\Phi(z_2)$ is a known function, and $\Delta_0$, $\Delta_1$ are known positive constants. Then, the closed loop system is globally stable, with $z_2 = 0$ after a finite time $t^* > 0$

- in case of $\lambda = 0$, if for arbitrarily small $k > 0$ the parameter $h$ satisfies the inequality

$$h \geq \Delta_1\left(\frac{\Delta_0}{2k} + \frac{k}{2\Delta_0}\right),$$

- in case of $\lambda \neq 0$, if

$$h > \Delta_1$$

and

$$\lambda \geq \frac{\Delta_1^2}{2k(h + \sqrt{h^2 - \Delta_1^2})}.$$

Typical state-space trajectories in the worst case are shown in Fig. 4. Note that a sliding mode is obtained only in the region $[-k, k]$, which is reached in a finite time $t^*$ due to the combined actions of the integral ($\nu$) and switching ($(\Phi + k)\, sgn\, z_2$) terms.

When $\alpha < \infty$ the sliding mode behaviour is not obtained, but still the state space trajectories are constrained in a proper region close to the desired dynamics $\sigma = 0$.

*Outer Loop*

Since the algorithm previously presented is intrinsically based on a first order dynamic model, it may be profitably adopted in the velocity control loop for the actuation of a robotic system, as shown in Fig. 3. In this way, the specific robustness properties of the Binary Control scheme may be conveniently exploited.

In the considered applicative context, the desired set-points are specified both as position and velocity trajectories, and therefore an external position control loop has been added. Since the performances, in terms of steady-state errors and insensitivity to external disturbances and parametric variations, are ensured by the inner loop, the outer feedback may be chosen in the simplest possible way in order to guarantee desired stability margins. In fact, in the experimental implementation a simple proportional action $H_p$ has been used.

## 4. Application to the WAM system

The control algorithm described in the previous sections has been tested both from a simulative and an experimental point of view. The laboratory activity has been carried out on the WAM (Whole-Arm Manipulation) system [16] at the Artificial Intelligence Laboratory, MIT.

The WAM system consists of a 4 degree-of-freedom manipulator, with an independent link used as master for teleoperation tasks. All the five joints of the WAM system are cable-driven, with transmission ratios varying between 1:13 and 1:30. The motors are Moog 303-003 brushless motors, whose main specifications are reported in Table I. The experiments presented in this paper refer to the master link. In this case, the transmission ratio is 1:13.12, the cable length is 1.0 m, and the weight of the link is 0.5 Kg approximately.

| Torque constant | Voltage constant | Rotor inertia |
|---|---|---|
| 0.27 Nm/A | 0.27 V/rad/s | 41 $10^{-6}$ Kg $m^2$ |
| Mech. time const. | Electr. time const. | Resistance |
| 1.0 ms | 1.72 ms | 1.8 $\Omega$ |

Table I. Main specifications of the Moog Motor 303-003.

A three-level control architecture is devoted to the WAM system. At the lower level, each motor is independently controlled by a PWM Moog controller, which provides the upper levels with position, velocity and current information. Note that a 5 KHz PWM servo is present for each motor in order to control the phase currents. The velocity or torque set points for the five Moog controllers are generated by the second control level, implemented on a multiprocessor VME system. Up to six 68020 CPU's are used for the trajectory generation and the control of the system. Adaptive techniques are used at this level [17]. Finally, the VME is interfaced by network to a Sun-3 workstation, used as development system and user interface in real time.

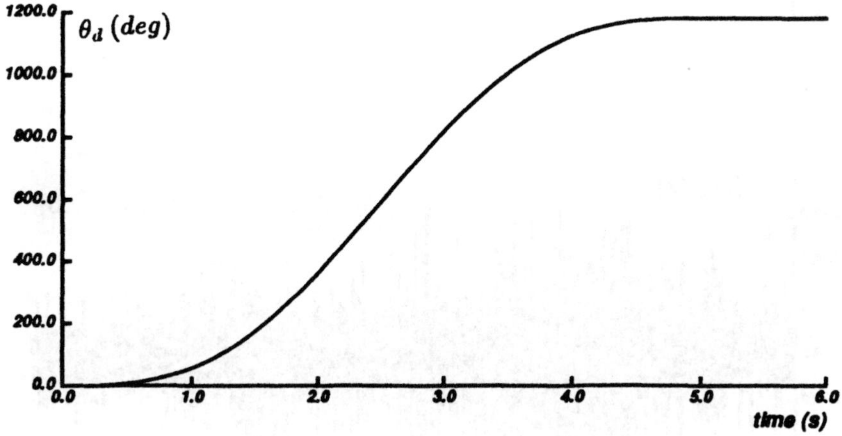


Figure 5: The desired trajectories for the reported experiments.

The reported results refer to the implementation of the control technique presented in Section 3. The desired position trajectory is shown in Fig. 5, i.e. a rotation of $90^o$ with duration 5 s: the initial position of the link is vertical, while at the end it is horizontal. Due to the characteristics of the actuation system, more suitable for high speed motions - i.e. the motors and the controllers are designed for velocities up to 5000 rpm - this is not a particularly favourable trajectory. The experiments refer to the use in different working conditions of both the new controller and a standard PD for a comparison. Note that all the position/velocity data here reported refer to the motor shaft.

Fig. 6-7 report the basic case, i.e. the position/velocity control of the master link without external disturbances acting on the system. In particular, Fig. 6.a,b show the position and velocity errors, and Fig. 6.c the integrator output. In Fig. 7 the position error for the same experiment with a linear PD controller is reported. The PD controller is obtained from the Binary one with the choice $h = k = 0$, i.e. zeroing the VSS and integral switching terms. Note the larger errors during the transient, and the presence of a steady-state offset due to the gravity.

Fig. 8-9 show the same experiment when a position-dependent external torque acts on the system. Fig. 8.a reports the resulting position error, and

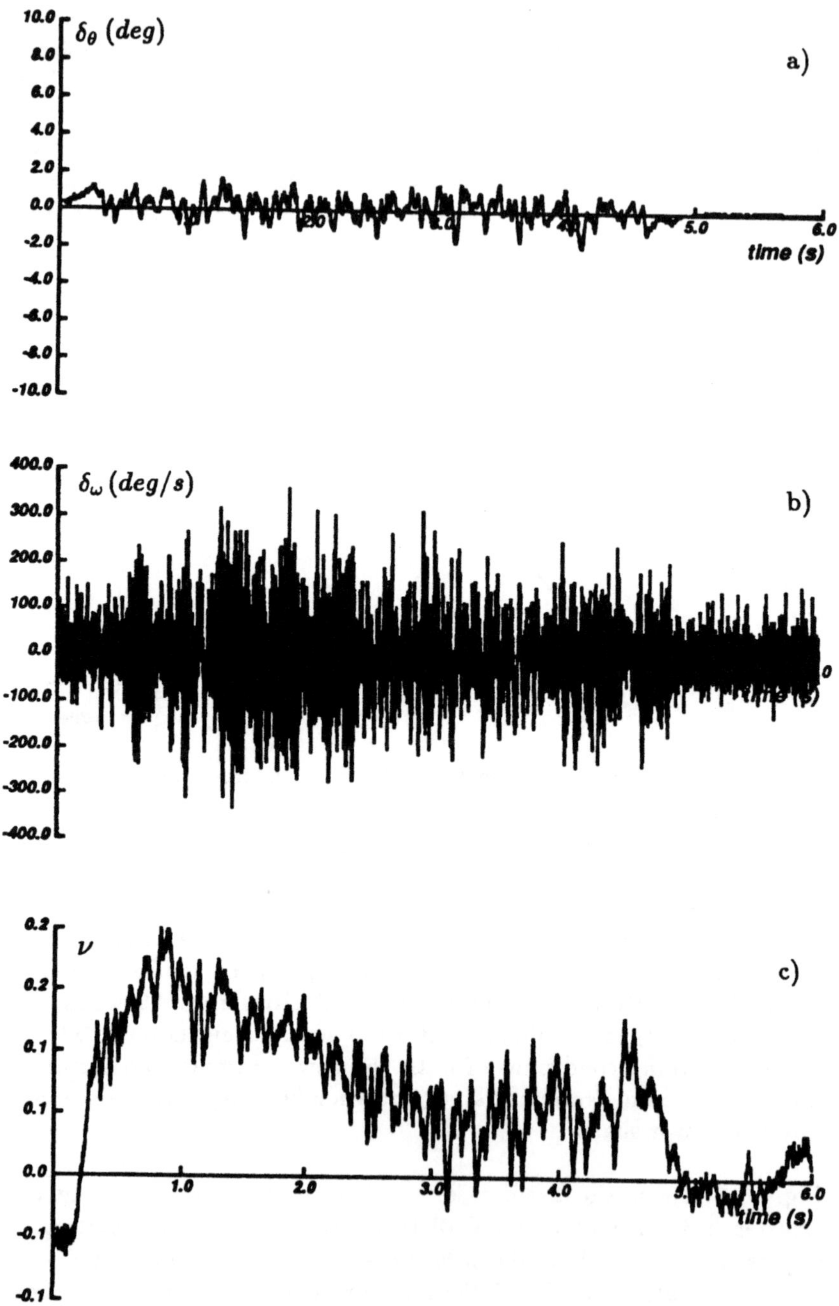


Figure 6: Basic case. Binary Control scheme: a) position error $\delta_\theta$, b) velocity error $\delta_\omega$, c) integral action $\nu$.

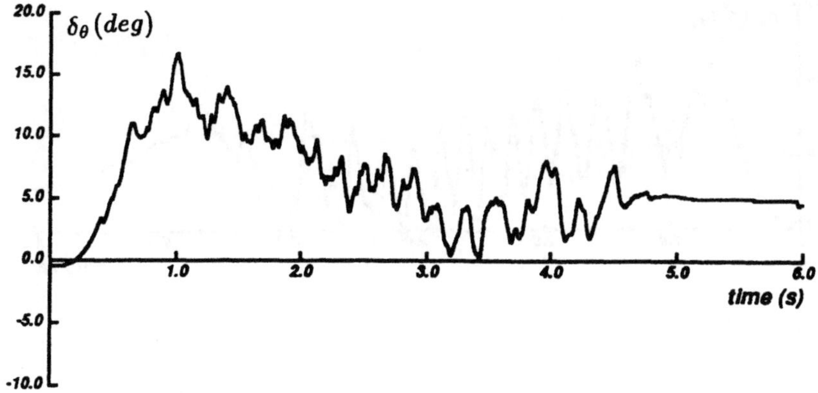


Figure 7: Basic case. PD scheme: position error $\delta_\theta$.

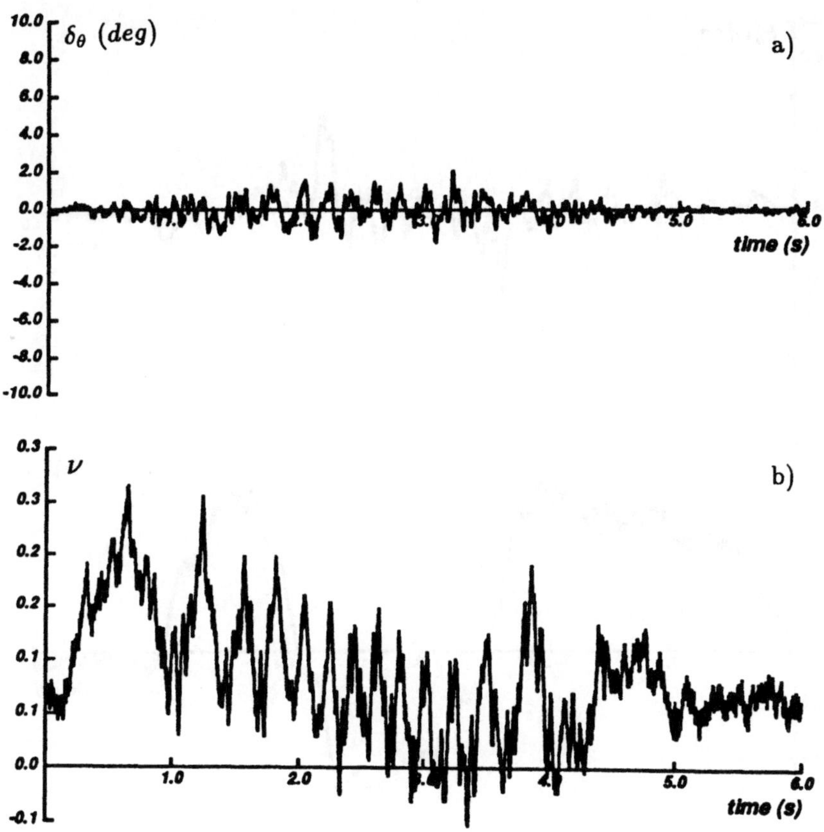


Figure 8: Trajectory with a ripple torque. Binary Control: a) position error $\delta_\theta$, b)integral action $\nu$.

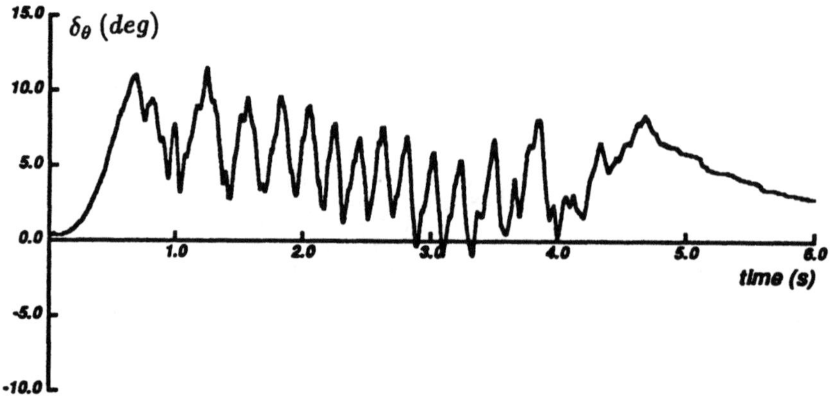


Figure 9: Trajectory with a ripple torque. PD: position error $\delta_\theta$.

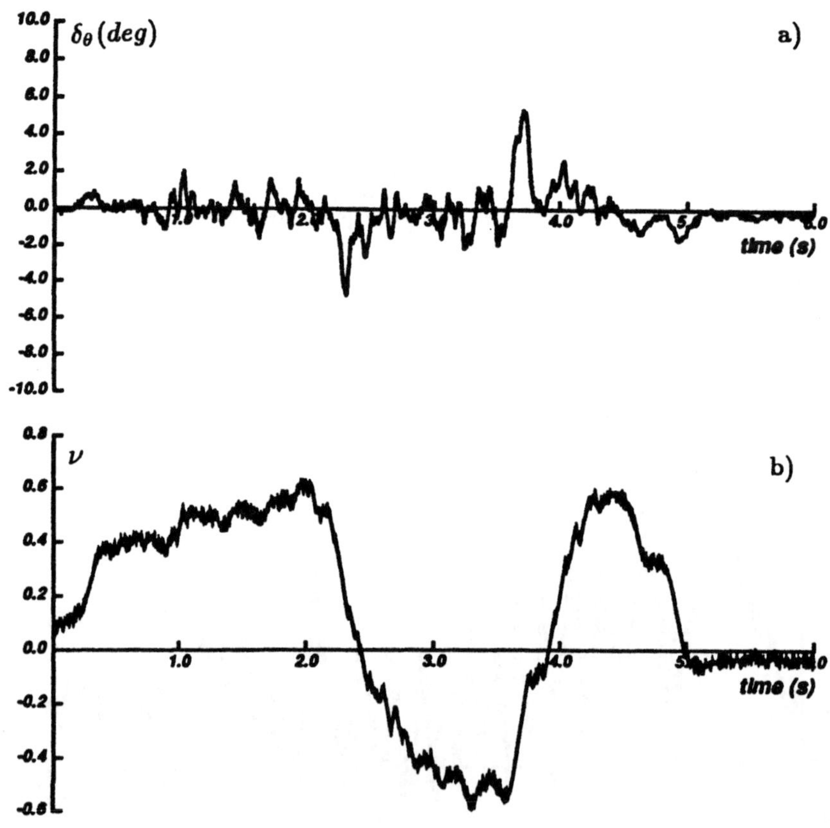


Figure 10: Trajectory with an external torque disturbance. Binary Control: a) position error $\delta_\theta$, b) integral action $\nu$.

Fig. 8.b the output of the integrator. It is possible to see how the integral action tracks the disturbances and generates the correct control action. In Fig. 9 the position error with the PD controller for the same experiment is shown. Note how in this case the torque ripple strongly affects the position during the motion.

The last reported experiment refers to the case in which an external time-varying torque is applied to the link. Only the results of the Binary scheme are given. In particular, in Fig. 10.a, the position error is shown, while the output of the integrator is reported in Fig. 10.b. Note that the integral term follows the applied torque disturbance, and that the sliding condition is momentarily lost at $t = 2.2$ and $t = 3.7$ s, i.e. when the time-derivative value of the external torque violates the condition of eq. (9).

## 5. Conclusions

In this paper, the application to a robotic system of a robust control technique, framed in the context of the Binary Theory, has been presented. With this technique, the robustness and insensitivity to external disturbances typical of Variable Structure Systems may be obtained along with the possibility of an arbitrary reduction of the associated chattering phenomenon. This is achieved with a standard Sliding Mode control with an additional integral switching action which behaves as an observer for parametric and exogenous disturbances.

In principle, the presence of non-linear time-varying complex dynamics and exogenous disturbances in robotic applications, along with a simplicity of implementation in terms of computational effort, makes the adopted technique particularly attractive.

In fact, the experimental data obtained on the WAM robotic system, here discussed, seems to confirm this expectations and encourage further work in both theoretical and application-oriented directions. In particular, research activity for the UB Hand project [18], [19] is currently envisaged.

## Acknowledgments

The experimental activity described in this paper has been carried out at the Artificial Intelligence Laboratory, MIT, with the support of the University Research Initiative Program under Office of Naval Research contract N00014-86-K-0685, NASA contract No. NAG-9-319. Support for C. Melchiorri and R. Zanasi as Visiting Scientists at MIT has been provided by Progetto Finalizzato Robotica of C.N.R., Italy, under contract No. 89.00493.PF.67.3. Finally, the authors are grateful to J.K. Salisbury and J.J. Slotine for stimulating discussions.

## REFERENCES

[1] A. Isidori, *Nonlinear Control Systems*, Springer Verlag, Berlin, 1989.

[2] K.J. Åström, B. Wittemark, *Adaptive Control*, Addison Wesley, Reading, MA, 1989.

[3] B.A. Francis, *A Course in $H_\infty$ Control Theory*, Springer-Verlag, Berlin, 1987.

[4] S.V. Emelyanov, *Variable Structure Control System*, Nauka, Moscow, 1967.

[5] V.I. Utkin, *Variable Structure Systems with Sliding Modes*, IEEE Trans. Automatic Control, Vol. 22, 212-222, 1977.

[6] K.K.D. Young, *Controller Design for Manipulators Using Theory of Variable Structure Systems*, IEEE Trans. on System, Man, and Cybernetics, Vol. 8, No. 2, 101-109, Feb. 1987.

[7] H. Hashimoto, K. Maruyama, F. Harashima, *A Microprocessor-Based Robot Manipulator Control with Sliding Mode*, IEEE Trans. on Industrial Electronics, Vol. IE-34, No. 1, Feb. 1987.

[8] J.-J.E. Slotine, S.S. Sastry, *Tracking Control of Non-Linear Systems Using Sliding Surfaces, with Application to Robot Manipulator*, Int. J. Control, Vol. 38, No. 2, 465-492, 1983.

[9] M. Foresti, C. Melchiorri, A. Tonielli, *Application of Sliding Mode Techniques to the Control of a Robotic Hand*, XI IFAC World Congress, Tallinn, URSS, August 1990.

[10] Emelyanov S.V., *Binary Control Systems*, IRIMS series, Vol.1, Moscow, 1984.

[11] Emelyanov S. V., Korovin S. K., *Development of Feedback Types and their Application to the Design of Closed Loop Dynamical Systems*, Problems of Control and Information Theory, Vol. 10, No. 3, pp. 161-174, 1981.

[12] Emelyanov S.V., S.K. Korovin and B.V. Ulanov, *Quasicontinuos Control Functions in non Stationary Binary Dynamic Systems*, IRIMS series, Vol.4, Moscow, 1985.

[13] A. Nersisian, R. Zanasi, *Binary Control Approach for Robust Stabilization and Tracking of Uncertain Dynamical Systems*, University of Bologna, Report No. GRA-91-02, Bologna, January 1991.

[14] A. Nersisian, R. Zanasi, *A Modified Variable Structure Control Algorithm for Stabilization of Uncertain Dynamical Systems*, submitted to Int. J. of Robust and Nonlinear Control, July 1991.

[15] C. Bonivento, A. Nersisian, A. Tonielli, and R. Zanasi, *Robust Control Design Combining Binary and Variable Structure Techniques*, submitted to the Nonlinear Control System Design Symposium, NOCOLS'92, Bordeaux, June 1992.

[16] J.K. Salisbury, W. Townsend, B. Eberman, and D. Di Pietro, *Preliminary Design of a Whole-Arm Manipulation System (WAMS)*, Proc. IEEE Int. Conf. on Robotics and Automation, 1988.

[17] G. Niemeyer, J.-J.E. Slotine, *Adaptive Cartesian Control of Redundant Manipulators*, Proc. American Control Conference, ACC'90, San Diego, CA, 1990.

[18] C. Bonivento, E. Faldella, G. Vassura, *The U.B. Robotic Hand Project: Current State and Future Developments*, Proc. 5th Int. Conf. on Advanced Robotics, '91 ICAR, Pisa, Italy, June 1991.

[19] C. Melchiorri, G. Vassura, *Mechanical and Control Issues for Integration of an Arm-Hand Robotic System*, 2nd. Int. Symp. on Experimental Robotics, ISER'91, Toulouse, June 1991.

The authors are with:

DEIS, Dipartimento di Elettronica Informatica e Sistemistica
University of Bologna
Viale Risorgimento, 2
40136 Bologna, Italy

# Trajectory Tracking in Flexible Robot Arms

A. DE LUCA F. NICOLÒ G. ULIVI

**Abstract.** *We consider the problem of controlling the end-effector motion of flexible robot arms. Main difficulties arise both from the typical nonlinearities of robot dynamics and from the non-minimum phase nature of the flexible arm when tip position is chosen as output. This bars the straight application of inversion control solving the trajectory tracking problem in robots with rigid links. Three alternative strategies are presented: inversion control based on a suitable minimum phase output, nonlinear regulation of the tip output, and iterative learning control. The control laws are discussed from the point of view of system requirements and complexity, while their performance is compared by simulation on a simple but significative example. Finally, we report on the experimental activity in controlling flexible robots carried out in the Robotics Laboratory at DIS.*

## 1. Introduction

The adoption of lightweight materials and the development of very large manipulators have recently given an impulse to the modeling and control of robotic systems with flexible elements. From the application point of view, robots with flexible links are found in space, as in the Space Shuttle manipulator [1], and on earth, e.g. in automated crane devices for advanced building costruction. Flexibility plays a major role whenever long and dextrous arms are needed to have access to hostile environments, such as in nuclear sites for maintenance, in underground waste deposits for inspection, or in deep sea for exploration.

Robotic tasks require two types of motion: point-to-point moves for repositioning the end-effector and path following. However, it is always convenient to specify an interpolating trajectory so to guarantee smooth transition between initial and final configurations, impose an explicit time constraint on the overall motion, and eventually simplify the synthesis of the control gains. Moreover, when the task involves contact with the environment, the accuracy in executing the planned motion becomes quite crucial. As a result, trajectory tracking is considered to be the reference problem for evaluating advanced robotic control law.

In the case of robots with fully rigid links, straightforward solutions are offered by inversion control schemes, which use nonlinear state feedback to compensate for couplings and nonlinearities of the system equation [2].

In the presence of link flexibility, both the trajectory specification and the control design should take care also of the relevant induced vibrations in the robot arm. On one hand, fast trajectories are usually associated to large deformations that may excessively stress the mechanical structure and result in long oscillation settling times at the final point. Conversely, slow motions that do not excite the characteristic eigenfrequencies of the arm would limit the operational performance. As for the controller, two main objectives are of interest: exact or asymptotic reproduction of a conveniently defined output trajectory and stabilization of the arm flexible dynamics.

When inversion control is applied to a flexible robot arm for tracking an end-effector trajectory, serious instability problems occur. This phenomenon emerges already in a linear setting. In fact, the nature of the mapping between input torque and tip location of each link is non-minimum phase, i.e. the transfer function contains right-half plane zeros [3]. Sometimes this corresponds to the practical observation of an 'opposite' output motion in response to a given torque input command. Since inversion control tends to impose a prescribed linear input-output behavior with no zeros, this suggests that cancellation with unstable poles is attempted. In a nonlinear setting, the generalization of the concept of zero dynamics allows to draw analogous conclusions about instability due to cancellations [4]. Nonlinear systems with unstable zero dynamics are often called *non-minimum phase* systems, similarly to the linear case.

Recently, some control alternatives have been proposed that achieve accurate tracking in flexible robots, complying with the non-minimum phase characteristics of the end-effector output. These can be classified as follows.

- *Inversion control for an associated minimum phase output* [5–8]. This involves the definition of a suitable output for which inversion control yields a stable closed-loop system.
- *Output regulation of the end-effector output* [9–11]. A bounded evolution of the whole state can be associated to any given bounded reference trajectory and used as a reference for a state feedback stabilizing law, without involving cancellations. This reference state trajectory produces exactly the desired time profile at the system output.
- *Iterative learning control of repetitive trajectories* [12,13]. Provided that repeated trials are allowed, it is possible to learn the input command that imposes the desired output trajectory, processing error data at previous trials with a limited knowledge of the system dynamics.

The purpose of this paper is to present a comparison of the above control methodologies. As a matter of fact, several interesting relationships exist among the different approaches: the application to a physical example, as the flexible robot arm, will help to enlighten them.

The first two classes of methods lead essentially to model-based controllers and will thus perform best when a good dynamic model of the robot arm is available. The third class uses instead a nominal model only for an efficient selection of the learning functions. Several approaches exist for modeling articulated flexible structures, ranging from finite- to infinite-dimensional descriptions and from linear to nonlinear dynamics [14,15].

Due to the overall complexity of modeling and controlling flexible manipulators, the analysis and the synthesis stages asks for an adequate experimental verification. To this aim, a two-link planar robot with a very flexible forearm has been designed and built in our Laboratory [16,17], and the robotic system is now available for control experiments. Such an apparatus allows to evaluate theoretical control findings, with respect to several real world aspects which are often neglected at the level of control design. However, it should be emphasized that the end-effector trajectory tracking in a flexible robot arm can be viewed as a prototype control problem for the larger class of nonlinear systems with unstable zero dynamics.

The paper is organized as follows. Dynamic modeling issues are summarized in Section 2. The three approaches to trajectory control based on inversion, regulation, and learning are presented in Section 3, directly for the case of multi-link flexible robot arms. In Section 4 a more specific comparison is performed on the basis of the most simple arm structure that inherits the non-minimum phase characteristic. Preliminary experimental results are then reported, obtained on the flexible manipulator available at DIS. Finally, on-going research activities and future developments are outlined.

## 2. Dynamic modeling of flexible robot arms

Accurate and complete dynamic models are mandatory for simulation purposes, where efficient numerical formats are more convenient. On the other hand, when control design is of concern, the availability of the system equations in an explicit symbolic form allows to trade off between the complexity of a model-based derivation and the expected control performance [18].

In robots with flexible links, deflection is always distributed in nature. This supports the usual argument that infinite-dimensional dynamic equations describe at best the dynamic behavior of flexible manipulators [19]. Typically, each link is represented in the limit as an Euler beam with proper boundary conditions at the two ends [20]. When small deformations are assumed, an eigenvalue problem has to be solved for each link or, more correctly, for the entire system. However, for all but the most simple structures, it is quite difficult to solve for the time evolution of the arm deformation. In particular, time-varying boundary conditions appear whenever the robot is constituted by two or more flexible links [21]. As a result, several approximations have been introduced, all leading to the derivation of finite-dimensional dynamic models. In describing the arm flexibility, separability in time and space is a common hypothesis, e.g. for the bending deformation at time $t$ of a point $x$ along the $i$th link one has

$$p_i(x,t) = \sum_{i=1}^{n_i} \phi_{ij}(x)\delta_{ij}(t). \tag{1}$$

The $n_i$ spatial components $\phi_{ij}(x)$ satisfy geometric and/or dynamic conditions, while time dependency is resumed within the generalized coordinates $\delta_{ij}(t)$.

Once the arm kinematics is properly described in terms of a set of $N$ rigid variables (usually associated to the robot joints) and of $N_e = \sum_{i=1}^{N} n_i$ deformation variables, the dynamic equations of motion follows in a standard fashion using a Lagrangian approach [22]. Consider a robot arm with an open kinematic chain structure, rotational joints and flexible links. The coupling of $N$ rigid motion equations with $N_e$ flexible ones will result in a nonlinear dynamic model of the general form

$$\begin{bmatrix} B_{11}(\theta,\delta) & B_{12}(\theta,\delta) \\ B_{12}^{\mathrm{T}}(\theta,\delta) & B_{22}(\theta,\delta) \end{bmatrix} \begin{bmatrix} \ddot{\theta} \\ \ddot{\delta} \end{bmatrix} + \begin{bmatrix} n_1(\theta,\delta,\dot{\theta},\dot{\delta}) \\ n_2(\theta,\delta,\dot{\theta},\dot{\delta}) \end{bmatrix} + \begin{bmatrix} D_1\dot{\theta} \\ K\delta + D_2\dot{\delta} \end{bmatrix} = \begin{bmatrix} u \\ 0 \end{bmatrix}, \tag{2}$$

where $\theta$ is the vector of rigid joint variables and $\delta$ collects the generalized coordinates associated to deformations. $B_{ij}$ are blocks of the positive definite inertia matrix $B$, partitioned according to the rigid and flexible components; vectors $n_i$ contain the Coriolis, centrifugal and gravity terms. The diagonal matrix $D_1$ represents joint viscous friction, while the positive definite, symmetric (and typically diagonal) matrices $K$ and $D_2$ are, respectively, the modal stiffness and the structural damping of the arm links. Accordingly, the input torques $u$ appear only in the first set of $N$ equations.

The model structure (2) holds for any finite-dimensional approximation of distributed flexibility, as long as the assumed modes of deformations satisfy the geometric *clamped* boundary conditions at the base of each link [14]. However, simplifications are obtained for the blocks in (2) whenever use is made of deformation modes $\phi_{ij}(x)$ that automatically satisfy orthonormal conditions. Moreover, relevant reductions follow from evaluating robot kinetic energy only in correspondence to the undeformed arm configuration ($\delta = 0$). Other couplings between rigid body motion and flexible deformations are also often neglected.

We point out that consistent dynamic modeling choices retain the same physical feature, relevant for control purposes: in the presence of uniform mass distribution for each link, all flexible robot models share the non-minimum phase property of the end-effector output.

## 3. Control strategies for trajectory tracking

We present here the basic steps involved in the synthesis of three trajectory control laws for general flexible manipulators. The common problem is to determine an input torque which leads to the reproduction of a desired smooth end-effector trajectory. This can be restated as finding ways to extend the well-known computed torque method [2] to flexible robot arms.

In the following, some simplifying assumptions are introduced: *(i)* for each link, deflections are limited to the plane of rigid motion; *(ii)* gravity effects are not included in the analysis. As a result, the terms $n_i$ in (2) are purely quadratic in the velocity $(\dot{\theta}, \dot{\delta})$ and any undeformed rest configuration will be an equilibrium state for the robot arm. Typical examples of this situation are the motion in a horizontal plane and robotic operations in space.

### 3.1 Joint and link-point based inversion control

Let the output of the system be defined in a parametric form as

$$y = \theta + \lambda C\delta, \qquad \lambda \in [0, 1]. \tag{3}$$

When $\lambda = 0$, (3) is just the set of rigid joint variables. More in general, the component $y_i$ is $\theta_i$ modified by a linear combination of the variables $\delta_{ij}$ relative to the $i$th link. When $\lambda = 1$, each output is an angle that points from the joint to the end of the associated flexible link.

Under the assumption that the so-called decoupling matrix is nonsingular, the inversion algorithm requires to differentiate each output component until the input appears explicitly and then solve for the input $u$ [23]. Define the inverse of the inertia matrix $H = B^{-1}$ and denote by $H_{ij}$ the blocks of its partitioned form. The generic *relative degree* for system (2) with (3) is uniformly equal to two. Then, dropping dependencies,

$$\begin{aligned}\ddot{y} = \ddot{\theta} + \lambda C\ddot{\delta} &= (H_{11} + \lambda C H_{12}^{\mathrm{T}})(u - n_1 - D_1\dot{\theta}) \\ &\quad - (H_{12} + \lambda C H_{22})(n_2 + K\delta + D_2\dot{\delta}).\end{aligned} \tag{4}$$

Setting $\ddot{y} = v$ and solving for $u$ yields, after some manipulations, the nonlinear feedback controller

$$\begin{aligned}u &= n_1 + D_1\dot{\theta} + B_{11}v \\ &\quad - (B_{12} - \lambda B_{11}C)(B_{22} - \lambda B_{12}^{\mathrm{T}}C)^{-1}(B_{12}^{\mathrm{T}}v + n_2 + K\delta + D_2\dot{\delta}).\end{aligned} \tag{5}$$

The closed-loop dynamics

$$\ddot{y} = v \tag{6a}$$

$$\ddot{\delta} = (B_{22} - \lambda B_{12}^{\mathrm{T}}C)^{-1}(B_{12}^{\mathrm{T}}v + n_2 + K\delta + D_2\dot{\delta}) \tag{6b}$$

should be asymptotically stable. The external input $v$ is designed as a linear stabilizing feedback for (6a), while the choice of the output parameter $\lambda$ will affect directly the stability of (6b).

A simplification occurs when $\lambda = 0$, i.e. when a joint-based inversion strategy is pursued. In this case, the control law becomes

$$u = n_1 + D_1\dot{\theta} - B_{12}B_{22}^{-1}(n_2 + K\delta + D_2\dot{\delta}) + (B_{11} - B_{12}B_{22}^{-1}B_{12}^{\mathrm{T}})v, \tag{7}$$

and the closed-loop unobservable dynamics is

$$\ddot{\delta} = -B_{22}^{-1}(B_{12}^{\mathrm{T}}v + n_2 + K\delta + D_2\dot{\delta}). \tag{8}$$

It can be shown that the zero dynamics associated to (8) (i.e. when $y \equiv 0$) is exponentially stable [24]. Thus, inversion at the joint level is always feasible. By continuity, there exists an interval $[0, \lambda_0]$ for $\lambda$ that guarantees closed-loop stability while providing exact tracking of the associated output trajectory. Indeed, if $\lambda_0 < 1$ end-effector exact tracking will never be realized. However, the closer is the output to the tip of each link, the better the end-effector will approximate the same trajectory.

### 3.2 Nonlinear regulation of end-effector motion

In this case the output is chosen directly at the end of each flexible link, i.e. setting $\lambda = 1$ in (3) or

$$y = \theta + C\delta. \tag{9}$$

Under the hypothesis of small link deformation, this output is one-to-one related to the cartesian position and orientation of the end-effector through the standard direct kinematics of the arm.

Following [25], the design of a regulator requires the computation of a reference *state* trajectory associated to the *output* evolution, together with a nominal *input* needed to keep the state along its reference trajectory. The output reference trajectory is generated by a dynamic autonomous exosystem that can be chosen, without loss of generality, as a linear system in observable canonical form with

$$Y_d = \{y_{d,i}, \dot{y}_{d,i}, \ddot{y}_{d,i}, \ldots, y_{d,i}^{r_i-1};\ i = 1, \ldots, N\} \tag{10}$$

taken as its state. Here, $r_i$ is the smoothness degree of the $i$th output reference trajectory.

Using (9), the robot dynamics (2) can be rewritten in terms of the new coordinates $(y, \delta)$ as

$$\begin{aligned} &B_{11}(y - C\delta, \delta)\ddot{y} + \left[B_{12}(y - C\delta, \delta) - B_{11}(y - C\delta, \delta)C\right]\ddot{\delta} \\ &\qquad + n_1(y - C\delta, \delta, \dot{y} - C\dot{\delta}, \dot{\delta}) + D_1(\dot{y} - C\dot{\delta}) = u, \end{aligned} \tag{11a}$$

$$\begin{aligned} &B_{12}^{\mathrm{T}}(y - C\delta, \delta)\ddot{y} + \left[B_{22}(y - C\delta, \delta) - B_{12}^{\mathrm{T}}(y - C\delta, \delta)C\right]\ddot{\delta} \\ &\qquad + n_2(y - C\delta, \delta, \dot{y} - C\dot{\delta}, \dot{\delta}) + K\delta + D_2\dot{\delta} = 0. \end{aligned} \tag{11b}$$

The reference state evolution is then equivalently specified in terms of $y_d$ and $\delta_d$, and of their derivatives $\dot{y}_d$ and $\dot{\delta}_d$. In particular, it is easy to see that the problem reduces to determining only the $N_e$ functions $\delta_d = \pi(Y_d)$, with $\dot{\delta}_d = (\partial\pi/\partial Y_d)\dot{Y}_d$ as a direct consequence.

Thus, the vector function $\pi(Y_d)$ should satisfy equation (11b), evaluated along the reference evolution of the output

$$\begin{aligned} B_{12}^{\mathrm{T}}(y_d, \pi(Y_d))\ddot{y}_d + \left[B_{22}(y_d, \pi(Y_d)) - B_{12}^{\mathrm{T}}(y_d, \pi(Y_d))C\right]\ddot{\pi}(Y_d, \dot{Y}_d, \ddot{Y}_d) \\ + n_2(y_d, \pi(Y_d), \dot{y}_d, \dot{\pi}(Y_d, \dot{Y}_d)) + K\pi(Y_d) + D_2\dot{\pi}(Y_d, \dot{Y}_d) = 0. \end{aligned} \tag{12}$$

This equation is independent from the applied torque and should be considered as a dynamic constraint for $\pi(Y_d)$. Being (12) a *nonlinear time-varying* differential equation, it is almost impossible to determine a bounded solution in closed form. Note that this would be equivalent to finding proper initial values for $\pi(Y_d)$ and $\dot{\pi}(Y_d, \dot{Y}_d)$ at time $t = 0$ such that forward integration of (12) yields a bounded evolution.

A feasible approach is to build an approximate solution $\widehat{\pi}(Y_d)$ by using basis elements which are bounded functions of their arguments, e.g. polynomials in $Y_d$ [10]. As long as each component $y_{d,i}(t)$ of the desired trajectory and its derivatives up to the $(r_i - 1)$ order are bounded, the approximation $\widehat{\pi}(Y_d)$ is necessarily a bounded function of time. The problem of determining the constant coefficients in the expansion is then solved through a recursive procedure based on the polynomial identity principle. Once a solution $\widehat{\pi}(Y_d) \approx \pi(Y_d)$ is obtained up to any desired accuracy, backsubstitution of the reference deformation $\delta_d$, of the desired output trajectory $y_d$, and of their time derivatives into (11a) will give the nominal feedforward term $u_d = \gamma(Y_d, \dot{Y}_d, \ddot{Y}_d)$ of the regulation law.

If the initial state is not on the reference trajectory, a stabilizing term should be added in order to drive the state towards this solution and only *asymptotic* output reproduction can be guaranteed. Since the system linearization is controllable, this can be accomplished —at least locally— using a simple linear state error feedback characterized by a matrix $F$. The final regulation law takes on the form

$$u = \gamma(Y_d, \dot{Y}_d, \ddot{Y}_d) + \widehat{F}\begin{bmatrix} y_d - y \\ \dot{y}_d - \dot{y} \\ \delta_d - \delta \\ \dot{\delta}_d - \dot{\delta} \end{bmatrix} = \gamma(Y_d, \dot{Y}_d, \ddot{Y}_d) + F\begin{bmatrix} \theta_d - \theta \\ \dot{\theta}_d - \dot{\theta} \\ \delta_d - \delta \\ \dot{\delta}_d - \dot{\delta} \end{bmatrix}. \tag{13}$$

If a particular globally stabilizing law exists, then convergence to the reference state behavior associated to the desired output trajectory can be achieved from *any* initial state. For robot arms with flexible links it is known that, in the absence of gravity, proportional-derivative (PD) joint feedback is enough for stabilizing any arm configuration [8,26]. As a result, a simplified matrix $F$ can be used in (13) and the overall regulator for the nonlinear system can be implemented using only *partial state* feedback.

### 3.3 Iterative learning of end-effector motion

Suppose that the task requires to execute the desired end-effector trajectory several times. We may then acquire from experiments (or from simulations using an accurate dynamic model) the input torque needed for accurate reproduction of that particular trajectory. At each trial, a feedforward command is applied to the robot and output error data are stored for subsequent off-line processing. The next feedforward term is computed by combining the previous command ('memory') with the error at the current trial ('update'). Provided that the robot arm is stabilized, a contraction mapping argument can be used to prove convergence of the learning process.

It has been shown [27] that for rigid robot arms one can design the learning algorithm based only on approximate information about the system dynamics. In the presence of flexibility, additional filtering of high-frequency signal components is needed to guarantee convergence. The price to pay is that tracking accuracy is preserved only within a bandwidth of interest (i.e. for $f < f_0$) [28]. However, this operation prevents other unmodeled effects from destroying learning convergence.

In the synthesis of a learning controller a linearized model can be used, e.g. obtained by expanding (2) around a static configuration $\theta = \theta_0$, $\delta = 0$. In order to better shape the system characteristics, a preliminary linear feedback is applied. Following the same arguments used in (13), a PD joint feedback law is sufficient and the closed-loop linearization yields

$$(14) \qquad B(\theta_0, 0)\begin{bmatrix} \Delta\ddot{\theta} \\ \Delta\ddot{\delta} \end{bmatrix} + \begin{bmatrix} F_P\Delta\theta + (D_1 + F_D)\Delta\dot{\theta} \\ K\Delta\delta + D_2\Delta\dot{\delta} \end{bmatrix} = \begin{bmatrix} v \\ 0 \end{bmatrix},$$

with obvious meaning of the variations $\Delta$. Taking only the diagonal elements $w_{ii}(s)$ of the matrix transfer function associated to (14) and to the tip output (9), a decentralized synthesis can be performed in the frequency domain. Let $v_k$ be the feedforward command used at the $k$trial ($v_1$ usually contains all a priori information). The next feedforward is derived as

$$(15) \qquad v_{k+1,i}(s) = \alpha_i(s)e_{k,i}(s) + \beta_i(s)v_{k,i}(s), \qquad i = 1, \ldots, N,$$

where $e_{k,i} = y_{d,i} - y_{k,i}$ is the $i$th joint error at the $k$th trial, and $\alpha_i$ and $\beta_i$ are the two learning filters to be designed. For rigid manipulators, it has been shown that the learning process is convergent provided that [28]

$$(16) \qquad \left|\beta_i(s) - \alpha_i(s)w_{ii}(s)\right| < 1, \qquad i = 1, \ldots, N.$$

Although a similar argument has not yet been proven for the flexible case, this simple design approach can be attempted and shown to converge by simulation. Note that noncausal filtering is allowed in (15), as well as anticipative shifting of signals in time, since the whole processing is performed off-line.

The applied control law at the $k$th trial is then

$$u = v_k + F \begin{bmatrix} \theta_d - \theta \\ \dot{\theta}_d - \dot{\theta} \end{bmatrix}. \tag{17}$$

Also in this case, as for regulation, the stabilizing feedback used in (17) is not closed around the output for which trajectory tracking is desired. However, the tip tracking error is taken into account here, being fed back 'off-line' through the learning procedure. Accordingly, control effort will move gradually from feedback to feedforward along iterations.

### 3.4 Discussion of the methods

It is interesting to summarize assumptions and properties of the above three tracking control methods. Table 1 collects various aspects which should help in understanding system requirements and ease of implementation.

| | Inversion control | Nonlinear regulation | Learning control |
|---|---|---|---|
| End-effector tracking | approximate | asymptotic | exact ($f < f_0$) |
| Feedback type | nonlinear | linear | linear |
| Partial state feedback | no | yes | yes |
| Trajectory known a priori | no | yes | yes |
| Repeated trials | no | no | yes |
| Anticipative action | no | no | yes |
| Model accuracy | high | high | low |

Tab. 1 – Comparison of tracking controllers for flexible robot arms

The following additional remarks are in order.

- An inversion control approach may give stable results even when applied at the end-effector level, provided that the flexible arm is correctly initialized. This would mean to 'preload' the link deflection (and its velocity) at time $t = 0$ in a way that depends on the desired tip trajectory.
- When applying the final feedforward computed by learning, the system state at time $t = 0$ will be in the right 'preloaded' conditions, as a result

of the anticipative action of the input torque. Thus, learning iterations may be seen as a computational scheme for obtaining the reference state evolution that produces exact output tracking. In practice, some error will be left due to the frequency cut-off at $f = f_0$ (see Table 1).

- At steady operation, e.g. with a sinusoidal trajectory, the feedforward terms of the regulation and the learning control schemes will coincide.

# 4. Numerical results

## 4.1 Comparative simulations

In order to compare inversion, regulation, and learning control, we consider a one-link planar arm of 1 m and 0.2 kg, modeling flexibility by just one elastic spring ($k = 10$ Nm/rad, corresponding to an eigenfrequency of $\approx$ 5 Hz) located halfway along the link. Large deflections are allowed, resulting in two second-order weakly nonlinear dynamic equations that can be found in [5]. Damping coefficients (see (2)) are $d_1 = d_2 = 0.01$ Nm/rad. This simple model is still representative since the end-effector angular position is a non-minimum phase output.

The desired output trajectory has a sinusoidal velocity profile and moves 90° in 1 second, with zero initial and final velocity. Simulations were run in nominal conditions for 1.8 seconds, using a 2nd order Runge-Kutta integration routine. In all cases the robot initial state is zero (undeformed and at rest).

The results obtained with inversion control for an output point at $\lambda = 0.3$ of the distal half link are shown in Figs. 1–3. Note that in this case $\lambda_0 = 2/3$. Although satisfactory tip tracking is obtained (maximum error is 0.45 deg), the error is persistent while the control effort is highly oscillatory. Note that the small error drift after the trajectory completion is due to the absence of a stabilizing feedback, i.e. $v = \ddot{y}_d$ is used in (6a).

Nonlinear regulation results are shown in Figs. 4–6, using a feedback matrix $F$ in (13) which sets the poles of the closed-loop system linearization at $(-20, -20, -30, -30)$. After a transient phase, which vanishes in 0.2 seconds, perfect tracking is obtained in this case. The second transient error at the final trajectory time ($T = 1$ sec in Fig. 5) is due to the switching of the reference trajectory from a sinusoidal one to a constant value, a situation that cannot be generated by any smooth autonomous exosystem. The required torque is similar to the one needed if the arm was rigid (dashed line in Fig. 6).

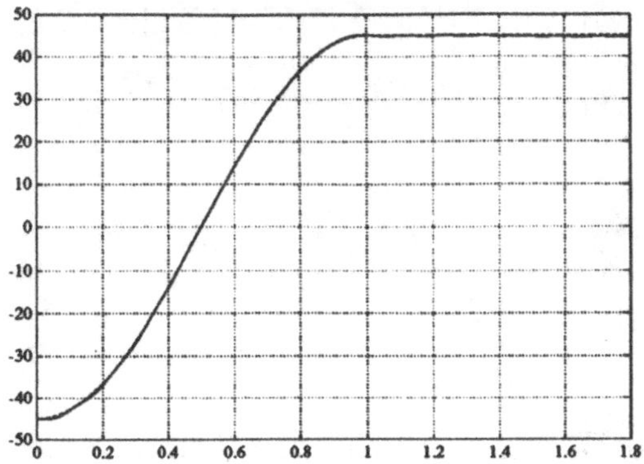


Fig. 1 – Tracking with inversion control ($\lambda = 0.3$)

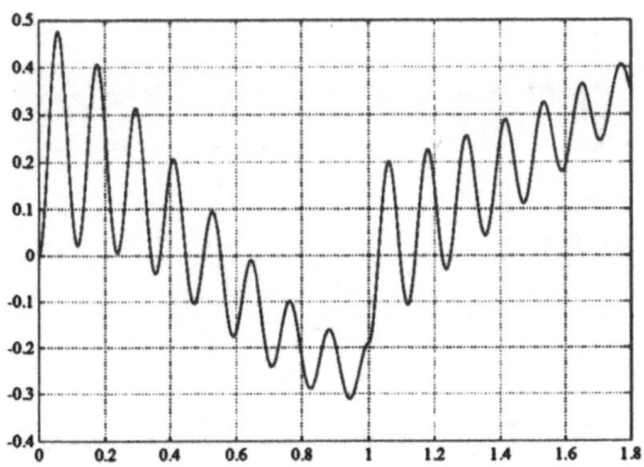


Fig. 2 – Trajectory error with inversion control ($\lambda = 0.3$)

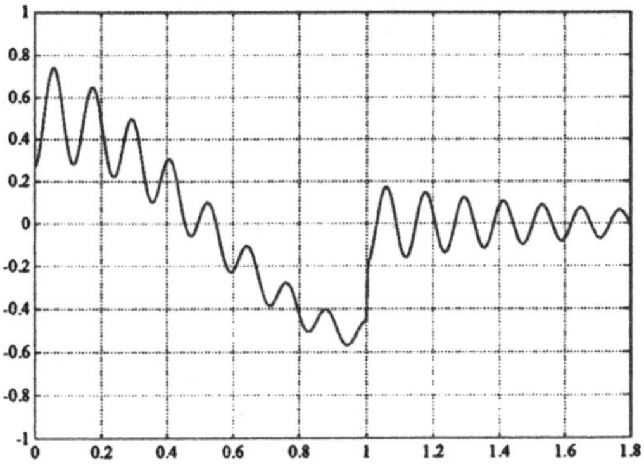


Fig. 3 – Input torque with inversion control ($\lambda = 0.3$)

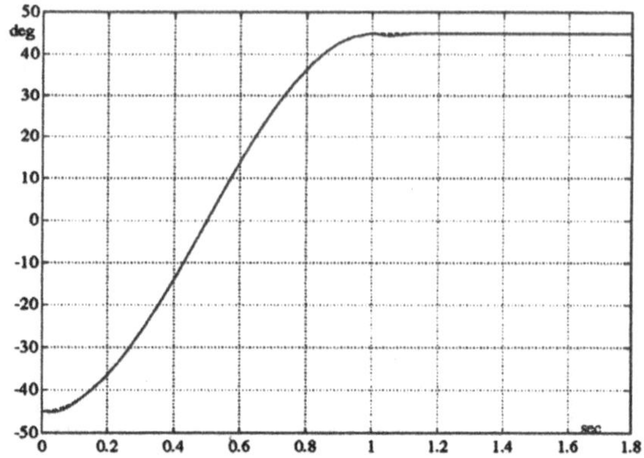


Fig. 4 – Tracking with nonlinear regulation

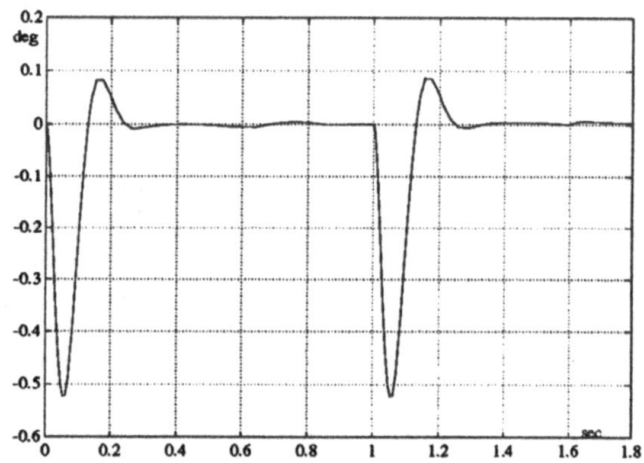


Fig. 5 – Trajectory error with nonlinear regulation

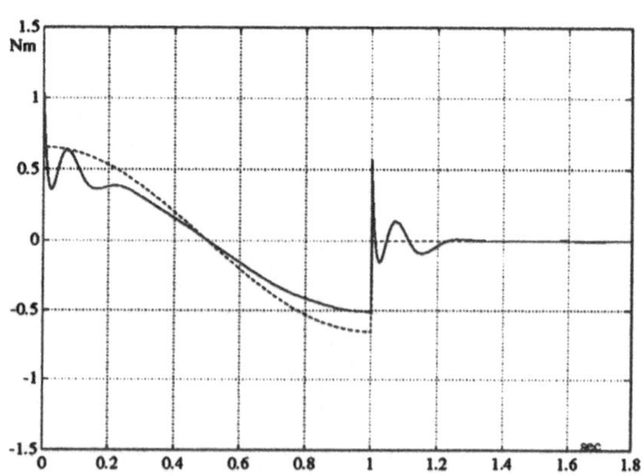


Fig. 6 – Input torque with nonlinear regulation

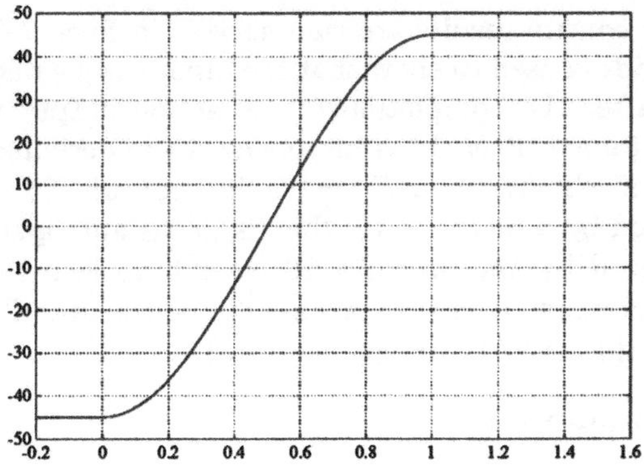


Fig. 7 – Tracking with learning control

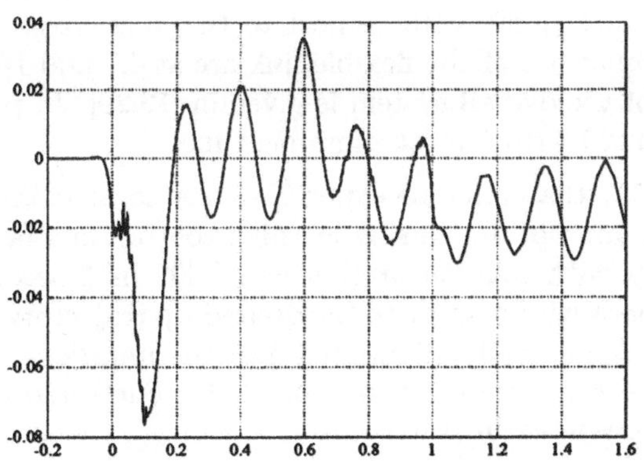


Fig. 8 – Trajectory error with learning control

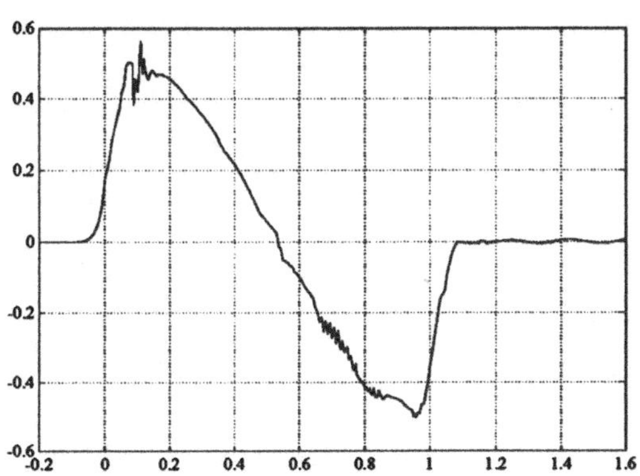


Fig. 9 – Input torque with learning control

Finally, learning control results are summarized in Figs. 7–9, where a time shift of 0.2 seconds is used to show that the input torque begins before the start and ends after the completion of the actual output trajectory. These results are obtained after 50 trials, using an underlying PD joint feedback with $F_P = 45$ Nm/rad and $F_D = 90$ Nm·sec/rad. The maximum error is almost reduced by a factor of 10, thanks to the anticipative torque action. This is realized by the use of a 10 msec time lead in the filter $\alpha(s)$. The very limited residual oscillations are due to the overall frequency cut-off at 13 Hz introduced by both $\alpha(s)$ and $\beta(s)$.

### 4.2 Experimental validation

The above control strategies are currently being implemented on the experimental testbed available in the Robotics Laboratory at DIS. The robot is a two-link planar manipulator with a very flexible forearm, driven by direct drive DC motors. The links are 0.3 and 0.7 meters long, with inertias of 0.447 and 0.303 kg·m$^2$ with respect to the joint axes. The first two clamped eigenfrequencies of the flexible link are at 4.7 and 14.4 Hz. A detailed description of the overall system is given in [16,29]. In particular, control laws are executed with 5 msec sampling time.

Figures 10 and 11, that refer to experiments performed keeping the first (rigid) link fixed, are obtained using learning control on a fast quintic polynomial trajectory with a rest-to-rest slew of 90° in 1 second. The typical lag and overshoot with respect to the desired tip trajectory of a pure PD joint feedback are practically eliminated over the iterations. After 27 trials, the maximum error is reduced to about 0.3°. Further results with learning control are presented in [13].

As a preliminary step towards end-effector regulation for the full two-link flexible arm, nonlinear regulation of desired *joint* trajectories has been implemented first [29].

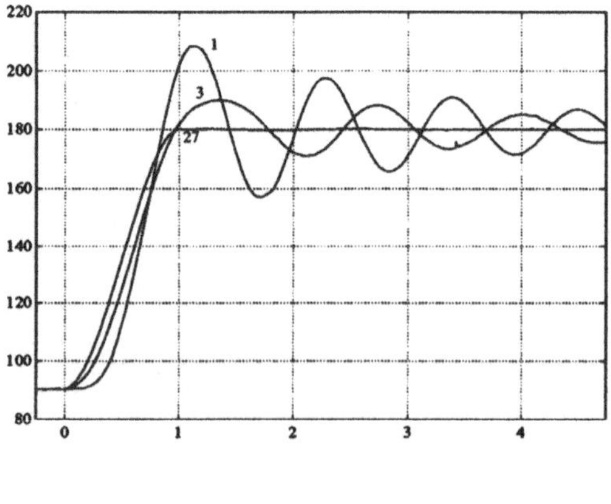


Fig. 10 – Tip tracking with learning control (*one-link experiment*)

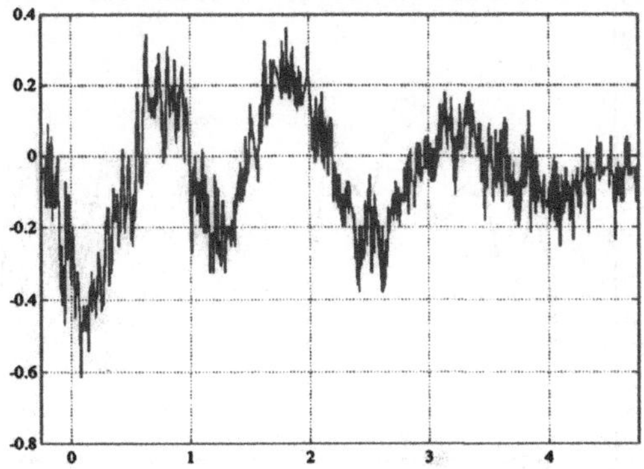


Fig. 11 – Tip error with learning control (*one-link experiment*)

Figures 12–14 illustrate the performance on a very demanding bang-bang acceleration trajectory; in particular, the second link undergoes a 4 seconds motion of 360°, with reversal at the midpoint. The stabilizing PD gains are $F_P = \text{diag}\{11.5, 6\}$ and $F_D = \text{diag}\{2, 0.8\}$. As expected, in contrast to the good accuracy obtained at the joint level (Fig. 12), the end-effector behavior is not satisfactory, with a maximum angular deflection of 7° (Fig. 13). Note that the deformation profile is similar to the one computed by simulation (dashed line) with an identified model of the robot arm. The torque oscillations in Fig. 14 compensate the effects of link deflections so to obtain a quasi-rigid joint motion. Only joint position and velocity measures are used in these experiments.

We point out that the same control hardware is ready to be used for end-effector regulation. In fact, once the joint trajectories associated to the desired tip motion are computed from the regulator equations (viz. $\theta_d = y_d - \delta_d$), these can be provided as reference for the joint controllers together with the proper feedforward.

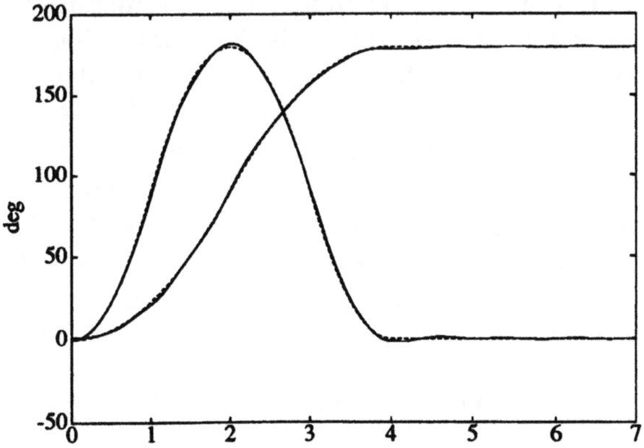


Fig. 12 – Joint trajectory regulation (*two-link experiment*)

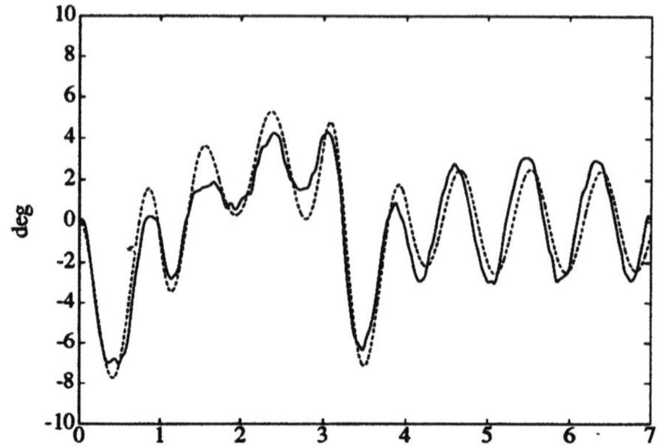


Fig. 13 – Tip angular error with joint regulation (*two-link experiment*)

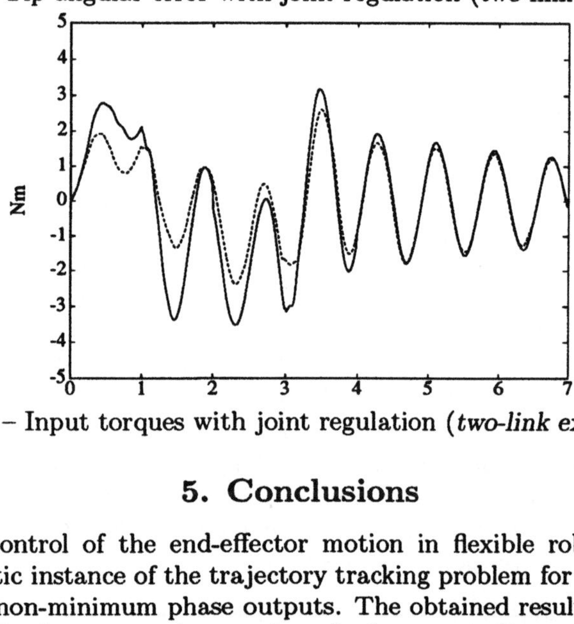


Fig. 14 – Input torques with joint regulation (*two-link experiment*)

## 5. Conclusions

Accurate control of the end-effector motion in flexible robot arms is a paradigmatic instance of the trajectory tracking problem for nonlinear systems with non-minimum phase outputs. The obtained results support the confidence in the proposed control methods as appealing solutions to this problem, in particular those based on nonlinear regulation and iterative learning theory. On-going research is devoted to the realization of robust (practical) tracking controllers for flexible robots. We are currently investigating the design of observers for the flexible deformation velocity. Within the same framework, we will evaluate the feasibility of end-effector trajectory regulation using only tip measurements and nonlinear dynamic output feedback. On-line identification of dynamic parameters (e.g. a tip payload) is also being studied, for inclusion in an adaptive regulation scheme.

## Acknowledgements

This paper is based on works supported by the *Ministero dell'Università e della Ricerca Scientifica e Tecnologica*, under 40% funds, and by the

*Consiglio Nazionale delle Ricerche*, contract no. 91.01946.PF67 (*Progetto Finalizzato Robotica*). We would like to thank other researchers who did participate to parts of the work reported here, and in particular L. Lanari, P. Lucibello, S. Panzieri, M. Poloni, and B. Siciliano.

## REFERENCES

[1] P.K. Nguyen, R. Ravindran, R. Carr, D.M. Gossain, and K.H. Doetsch, *Structural Flexibility of the Shuttle Remote Manipulator System Mechanical Arm*, SPAR Aerospace Ltd. Rep., Tech. Info. of AIAA, 1982.

[2] T.J. Tarn, A.K. Bejczy, A. Isidori, and Y. Chen, "Nonlinear feedback in robot arm control," *Proc. 23rd IEEE Conf. on Decision and Control*, Las Vegas, NV, pp. 736–751, 1984.

[3] R.H. Cannon, Jr. and E. Schmitz, "Initial experiments on the end-point control of a flexible one-link robot," *Int. J. of Robotics Research*, vol. 3, no. 3, pp. 62–75, 1984.

[4] C. Byrnes and A. Isidori, "Local stabilization of critically minimum phase nonlinear systems," *Systems and Control Lett.*, vol. 11, no. 1, pp. 9–17, 1988.

[5] A. De Luca, P. Lucibello, and G. Ulivi, "Inversion techniques for trajectory control of flexible robot arms," *J. of Robotic Systems*, vol. 6, no. 4, pp. 325–344, 1989.

[6] A. De Luca and B. Siciliano, "Trajectory control of a non-linear one-link flexible arm," *Int. J. of Control*, vol. 50, no. 5, pp. 1699–1715, 1989.

[7] S.K. Madhavan and S.N. Singh, "Inverse trajectory control and zero dynamics sensitivity of an elastic manipulator," *Proc. 1991 American Control Conf.*, Boston, MA, pp. 1879–1884, 1991.

[8] A. De Luca and B. Siciliano, "Inversion-based nonlinear control of robot arms with flexible links," submitted to *AIAA J. of Guidance, Control, and Dynamics*, 1991.

[9] A. De Luca, L. Lanari, and G. Ulivi, "Output regulation of a flexible robot arm," in *Analysis and Optimization of Systems*, A. Bensoussan and J.L. Lions (Eds.), Lecture Notes in Control and Information Sciences, vol. 144, pp. 833–842, Springer-Verlag, Berlin, 1990.

[10] A. De Luca, L. Lanari, and G. Ulivi, "End-effector trajectory tracking in flexible arms: Comparison of approaches based on regulation theory," in *Advanced Robot Control*, C. Canudas de Wit (Ed.), Lecture Notes in Control and Information Sciences, vol. 162, pp. 190–206, Springer-Verlag, Berlin, 1991.

[11] F. Pfeiffer, "A feedforward decoupling concept for the control of elastic robots," *J. of Robotic Systems*, vol. 6, no. 4, pp. 407–416, 1989.

[12] M. Poloni and G. Ulivi, "Iterative trajectory tracking for flexible arms with approximate models," *Proc. 5th Int. Conf. on Advanced Robotics*, Pisa, I, pp. 108–113, 1991.

[13] M. Poloni and G. Ulivi, "Iterative learning control of a one-link flexible manipulator," *Prep. 3rd IFAC Symp. on Robot Control*, Vienna, A, pp. 273–278, 1991.

[14] W.J. Book, "Recursive Lagrangian dynamics of flexible manipulator arms," *Int. J. of Robotics Research*, vol. 3, no. 3, pp. 87–101, 1984.
[15] P.B. Usoro, R. Nadira, and S.S. Mahil, "A finite-element Lagrangian approach to modeling lightweight flexible manipulators," *ASME J. of Dyn. Syst., Meas., and Control*, vol. 108, pp. 198–205, 1986.
[16] P. Lucibello and G. Ulivi, "Design and realization of a two link direct drive robot with a very flexible forearm," Rap. 16.89, Dipartimento di Informatica e Sistemistica, Università di Roma "La Sapienza", 1989.
[17] P. Lucibello and G. Ulivi, "An optical angular transducer for flexible robots arms," *IFAC Symp. on Low Cost Automation*, pp. 113–116, Milano, I, 1989.
[18] A. De Luca, P. Lucibello, and F. Nicolò, "Automatic symbolic modeling and nonlinear control of robots with flexible links," *Proc. IEE Int. Work. on Robot Control: Theory and Applications*, Oxford, UK, pp. 62–70, 1988.
[19] X. Ding, T.J. Tarn, and A.K. Bejczy, "A novel approach to the modelling and control of flexible robot arms," *Proc. 27th IEEE Conf. on Decision and Control*, Austin, TX, pp. 52–57, 1988.
[20] L. Meirovitch, *Analytical Methods in Vibrations*, Macmillan, 1967.
[21] A. De Luca and B. Siciliano, "Closed-form dynamic model of planar multi-link lightweight robots," *IEEE Trans. on Systems, Man, and Cybernetics*, vol. 21, no. 4, Jul. 1991.
[22] H. Goldstein, *Classical Mechanics*, Addison-Wesley, Reading, 1980.
[23] A. Isidori, *Nonlinear Control Systems*, 2nd Edition, Springer Verlag, Berlin, 1989.
[24] A. De Luca, "Zero dynamics in robotic systems," in *Nonlinear Synthesis*, C.I. Byrnes and A. Kurszanski (Eds.), Progress in Systems and Control Series, Birkhäuser, Boston, 1991.
[25] A. Isidori and C. Byrnes, "Output regulation of nonlinear systems," *IEEE Trans. on Automatic Control*, vol. 35, no. 2, pp. 131–140, 1990.
[26] E. Schmitz, "Modeling and control of a planar manipulator with an elastic forearm," *Proc. 1989 IEEE Int. Conf. on Robotics and Automation* Scottsdale, AZ, pp. 894–899, 1988.
[27] S. Arimoto, T. Naniwa, and H. Suzuki, "Selective learning with a forgetting factor for robotic motion control," *1991 IEEE Int. Conf. on Robotics and Automation*, Sacramento, CA, pp. 728–733, 1991.
[28] A. De Luca, G. Paesano, and G. Ulivi, "A frequency-domain approach to learning control: Implementation for a robot manipulator," *IEEE Trans. on Industrial Electronics*, to appear.
[29] A. De Luca, L. Lanari, P. Lucibello, S. Panzieri, and G. Ulivi, "Control experiments on a two-link robot with a flexible forearm," *Proc. 29th IEEE Conf. on Decision and Control*, Honolulu, HI, pp. 520–527, 1990.

Alessandro De Luca, Fernando Nicolò, Giovanni Ulivi
Dipartimento di Informatica e Sistemistica (DIS)
Università degli Studi di Roma "La Sapienza"
Via Eudossiana 18, 00184 Roma, Italy

# Variable Structure, Bilinear and Intelligent Control with Power System Application

R.R. MOHLER, Y. WANG, R.R. ZAKRZEWSKI, and V. RAJKUMAR

*Dedication. On this 20th anniversary of our first U.S.-Italy Seminar on Variable-Structure Synthesis in Sorrento, it is an opportunity to express my gratitude to Professor Antonio Ruberti for this enlightening scientific and cultural collaboration.*

**Abstract.** A brief background on variable-structure and bilinear systems relates this research to the first U.S.-Italy seminar on this topic. Sliding-mode control of a flexible ac transmission system (FACTS) is analyzed and shown to be quite effective for a simplified single-machine infinite-but model (SMIB). For cases where system complexities may cause robustness problems, a bilinear self-tuning control and a neural-network based control are introduced.

## 1. Introduction

An excellent tutorial on variable-structure control which emphasizes sliding modes is given in [1]. It is shown that such sliding-mode control can be designed so that system performance is largely independent of plant dynamics. About 30 years ago, Emelyanov and his associates investigated variable-structure systems in some detail and derived elaborate switching policies for second-order processes in particular. This work was finally published in book form in 1970 [2]. One interesting application of a sliding-mode (with chattering) to state-space (power) constrained optimal control for nuclear-powered rocket propulsion was independently developed about the same time [3,4].

Meanwhile, bilinear systems (BLS) research grew out of the Los Alamos project and was sponsored by the National Science Foundation. As a special form of variable-structure system (e.g., linear system with controlled damping), appropriately designed BLS were shown to exhibit superior performance and improved controllability over their linear subclass [5-7].

As a result of a cooperative research program with Ruberti at the University of Rome (La Sapienza), a number of publications evolved on variable structure systems and BLS. This created a substantial base of theory and multidisciplinary applications, some of which were reported in the proceedings of three U.S.-Italy seminars [9-11]. But a variety of outstanding contributions, too numerous to list here, were stimulated by this cooperation and made by investigators throughout the world. A recent tutorial of this work emphasizes practice [12].

The present paper demonstrates the role of variable-structure control and bilinear-based self-tuning control to a highly nonlinear electric power system.

## 2. FACTS

Flexible a.c. transmission systems (FACTS) has the potential to provide a substantially increased electric power level at a stable and reliable state without a big increase in the number of transmission lines. Today's electric power networks represent the largest system ever engineered by man. One network may involve thousands of generators of different types and sizes interconnected with numerous load centers, all of which may be thousands of miles apart. FACTS use parametric control along with conventional additive control already available by automatic voltage regulators (AVR) and power system stabilizers (PSS). The parametric control, for example, may be accomplished in conjunction with energy dissipation in a braking resistor, energy storage in a line capacitor or superconducting toroid. Here a line series capacitor (SC) control is considered. Fixed SC has been installed on numerous lines throughout the world to improve static transmission capability. Associated subsynchronous resonances are notorious for such designs. Future designs, however, include thyristor controlled capacitors which may be manipulated for good dynamic performance along with improved transmission quality and level.

Traditionally, power system control has been designed by classical linear methodology and more recently by linear adaptive control methods [13,14].

Even if the power system is assumed linear, the additional FACTS devices in reality would make it a BLS. Of course, for large transient stability studies, the process is more highly nonlinear. Here, a simplified representation by means of a SMIB model is studied in order to form a basis for more complex multimachine analyses and simulations.

## 3. Variable Structure Control

As shown in [15] and noted above, line-reactance control in general leads to a nonlinear system model with parametric control. A generalization of the power system yields the following model which is linear in control but nonlinear in state $x(t)$:

$$\dot{x} = A(x) + B(x)u \ . \tag{1}$$

Here $x(t)$, $A(\bullet)$, $B(\bullet)$ are appropriate vector functions, and $u(\bullet)$ is a scalar control.

For the maximum principle, the optimal control with control magnitude constraint (e.g., capacitance) is bang-bang with the possibility of singular arcs over portions of the trajectory. Similarly, state constraints (e.g., maximum allowable swing angle) can lead to similar arcs. Such processes may be approximated by variable-structure controllers with sliding-mode arcs.

For convenience, a linear sliding-mode arc is assumed with

$$S(x) = c^T x \ , \tag{2}$$

where $c^T$ is an appropriate row vector. Then, the sliding-mode control from $S(x) = 0$ becomes

$$u = -[c^T B(x)]^{-1} c^T A(x) \ .$$

Furthermore, from (1) and (2) it can be shown with a quadratic Liapunov function that the sliding-mode control with exponential approach may be generated by

$$u = -[c^T B(x)]^{-1} [c^T A(x) + esgnS + kS] \ , \tag{3}$$

where $\epsilon$, $k$ are appropriate constants to minimize chatter and provide good response.

Typical responses without control and with this SC variable structure control (VSC) are given in Figure 1. Table 1 summarizes the corresponding gains in electric power transfer capabilities. The SC VSC allowed nearly twice the maximum power with simple AVR or without control. These responses assume a 3-phase fault is applied at $t = 0$, fault line cleared at $t = 70$ ms, and successfully reclosed at $t = 350$ ms. A control delay time of 0.03 sec (in applying the control admittance) had little effect on performance but did require a bit more control action.

While SC control of the multimachine system has just begun, preliminary results for the BR-VSC of the WNAPS (Western North American Power System) representation are encouraging.

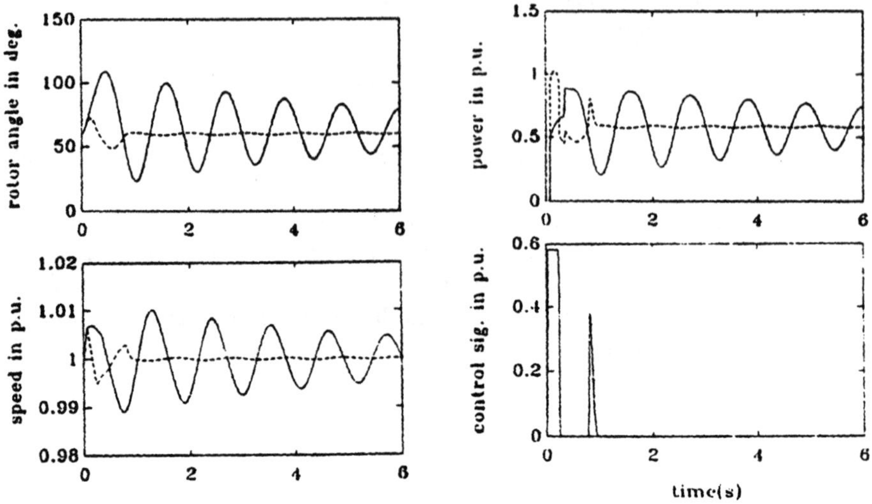


**Figure 1. System response with series capacitor variable structure control (___ no control, --- ideal VSC)**

**Table 1. Power Transfer Capabilities**

| Cases | Without Any Control | With Simple AVR* Kv = -10 $T_c$ = 0.1 s | Series Capacitor VSC $T_c$ = 0.0 s | Series Capacitor VSC $T_c$ = 0.03 s |
|---|---|---|---|---|
| $P_{max}$ | 0.6551 | 0.6703 | 1.2064 | 1.2036 |

*Kv and $T_c$ are gain and time constant of the automatic voltage regulator, respectively.

## 4. Time-Series Model-Reference Adaptive Control

Nonlinear dynamics of power systems such as the WNAPS and the nature of proposed FACTS control devices suggest the need for nonlinear adaptive control. The above VSC is one approach to such adaptation. But it may have some robustness problems with respect to assumed model and measurement assumptions which are under investigation.

Nonlinear time-series, model-reference adaptive control currently is being studied by means of a plant-identification algorithm in conjunction with an optimal model-reference adaptive controller. Low-order autoregressive moving-average (ARMA) models including polynomial and bilinear (in state measurements and control) structures have shown some degree of success for high-order SMIB systems and more complex representations are planned. Comparisons are made here with more conventional linear model reference adaptive controllers (MRAC).

With real power as the measured output, one obtains by Euler's approximation a nonlinear time-series structure with time-varying (updated) coefficients to represent system (1). Then a 4th-order approximation bilinear model has the following time-series form:

$$y(k) = \sum_{i=1}^{4} a(i)y(k-1) + \sum_{i+1}^{4} b(i)y(k-i)u(k-i) + \sum_{i=1}^{4} c(i)u(k-1) \tag{4}$$

where $a(i)$, $b(i)$, $c(i)$ are appropriate (possibly time-dependent) parameters, $y(k)$ = plant output, and $u(k)$ is the additional series capacitor reactance added/removed. Equation (4) normally can be put in standard bilinear discrete state-space form. While the additive control term in (4) does not follow directly from (1), it is added for generalization of the results.

Equation (4) can be expressed in compact form as

$$y(k+1) = \phi^T \theta \,. \tag{5}$$

Since they appear linearly, the parameters of the model in (4) can be estimated using recursive least squares (RLS) estimation with

$$\begin{aligned} \theta_{k+1} &= \theta_k + L_k\left[y_{k+1} - \phi_k^T \theta_k\right] \\ L_{k+1} &= \frac{P_k \phi_{k+1}}{1 + \phi_{k+1}^T P_k \phi_{k+1}} \\ P_{k+1} &= P_k - \frac{P_k \phi_{k+1} \phi_{k+1}^T P_k}{1 + \phi_{k+1}^T P_k \phi_{k+1}} \end{aligned} \tag{6}$$

where $P_k$ is a covariance matrix. Here $y_{k+1}$ refers to actual complex nonlinear system measured power at $k+1$th instant and the subsequent subscript sample-time notation designates estimates based on this information.

To check out the validity of the BARMA model (4), an 11-th order nonlinear model of a power system equipped with a series capacitor, and an IEEE type-1 excitation system, was subjected to a random, binary switching of the capacitor, and the proposed model was identified on-line from the measurements of power and capacitance.

To compare the performance of the bilinear model vis-a-vis an equivalent linear ARMA model, the system was identified as both the BARMA model of Eq. (4) and as a linear model of the form:

$$(7) \qquad y(k) = \sum_{i=1}^{n} a(i)y(k-1) + \sum_{i=1}^{n} c(i)u(k-1) .$$

It was observed that the bilinear model shows insignificant growth of the residual loss function [16] compared with the linear model. The comparisons are significantly less pronounced when the loss function is computed during the identification of lower order linear and bilinear models. However, the bilinear model is always more accurate. This would be expected since the linear model is a special case of the bilinear model.

The bilinear time-series model (4) now is used to design a bilinear self-tuning controller (BSTC) for the power system. For this purpose, a one-step-ahead cost function can be defined for (4) as:

$$(8) \qquad J(k) = 1/2\ [y(k) - y^*(k)]^2 + \rho/2\ [u(k-1)]^2$$

where $y^*(k)$ is the desired output at the next sampling instant, a constant, in the case of regulation, and $\rho$ is a weight imposed to keep the control "smooth."

The minimization of this function with respect to $u(k$-1) with $dJ(k)/du(k$-1) = 0 yields the self-tuning control

$$(9) \qquad \begin{aligned} u(k-1) = -\ [&a(1)y(k-1) + a(2)y(k-2) + a(3)y(k-3) \\ &+ a(4)y(k-4) + b(2)y(k-2)u(k-2) \\ &+ b(3)y(k-3)u(k-3) + b(4)y(k-4)u(k-4) \\ &+ c(2)u(k-2) + c(3)u(k-3) + c(4)u(k-4) \\ &- y^*(k)]/[\gamma + b(0)] \end{aligned}$$

where

$$b(0) = c(1) + b(1)\,y(k-1)\,, \quad \gamma = \rho/b(0)\,.$$

The parameters $a(i)$, $b(i)$, and $c(i)$ are estimated recursively by the RLS identifier (6). Of course, measured real power $y_k$ from the complex 11th-order model replaces $y(k)$ in (9).

Figure 2 demonstrates the accuracy of the BARMA model compared with the inaccurate ARMA model. Only real power measurement was used to formulate control action. The response for the linear case suggests that the controller attempts to linearize the closed loop system unsuccessfully with the allowed control action. For critical test purposes, field voltage limits were selected so that the system without SC control was stable for a one-cycle short circuit at the generator terminals but was unstable for a two-cycle fault.

The nonlinear self-tuning controller, NSTC, is based on a 4th-order linear, polynomial (generalized bilinear with added terms squared and cubic in power). The added nonlinear power terms in the reference model make the NLSTC system somewhat less oscillatory than that with the BLSTC.

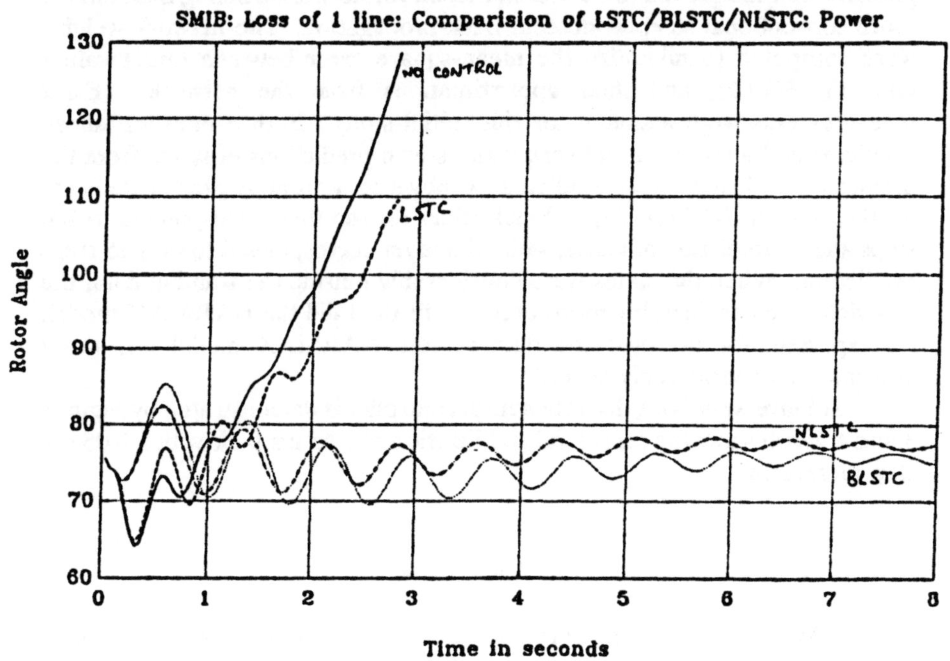


Figure 2. STC comparisons

## 5. Intelligent Control

The previously discussed nonlinear adaptive controls may be considered in the case of intelligent processes. Adaptation and learning keynote their operation. For example, the BLSTC discretely updates its model approximation and the corresponding control policy based on new information. But in this section we consider the use of "neural" networks (NN) to "learn" system dynamics which is followed by an appropriate control adaptation [17].

The idea here is to approximate arbitrarily close any finite dimensional continuous function by a multilayered feedforward network [18]. In particular, the SMIB model is studied with SC control. It was found that such a power system may be modeled by a NN with one hidden layer, and the model compares favorably with a nonlinear ARMA (NARMAX) representation. The latter also requires appropriate windowing of the data set for most accurate results.

The network used in this study had $\omega(k)$, $\delta(k)$, $X(k)$ as inputs and $\omega(k+1)$, $\delta(k+1)$ as outputs. Here, $\omega$ is rotor speed, $\delta$ rotor swing angle, and $X$ is admittance. It has one hidden layer in 10 units. A variant of conjugate gradient method proved to be more efficient for network training than did the more conventional steepest-descent back-propagation. The network weights were computed to minimize the mean-square error between target values $\omega(k+1)$, $\delta(k+1)$, and their approximations from the network. Exact measurements were assumed for this initial study. Both a one-step ahead predictor and a recurrent network with its own predictions of states from the previous time instant were studied and shown to accurately predict the state for 0.05 sec and 0.25 sec steps. Final values for the first corresponded to 100 steps ahead prediction of states, while the latter corresponded to only 25 steps prediction. When the states were considerably outside the training data, the NN yielded a considerably more accurate fit than did the NARMAX model. As expected, for fits over the data set the NARMAX model was more accurate and considerably faster.

An inverse-model, discrete neural controller is demonstrated by Figures 3 and 4 which show the successful stabilization of a 65 ms fault with a 0.05 sec step increment.

## 6. Conclusion

FACTS control for troublesome faults by means of SC manipulation is studied here for a SMIB system. The VSC performs extremely well under ideal conditions with rotor swing angle fed back to the controller. The NN-based intelligent control generates a similar response and generally is quite

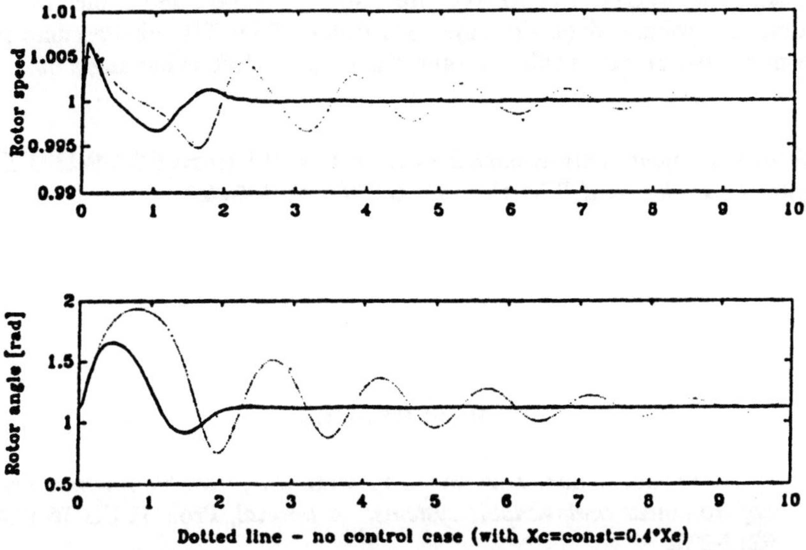


Figure 3. Rotor response with neural controller, 0.05 sec step

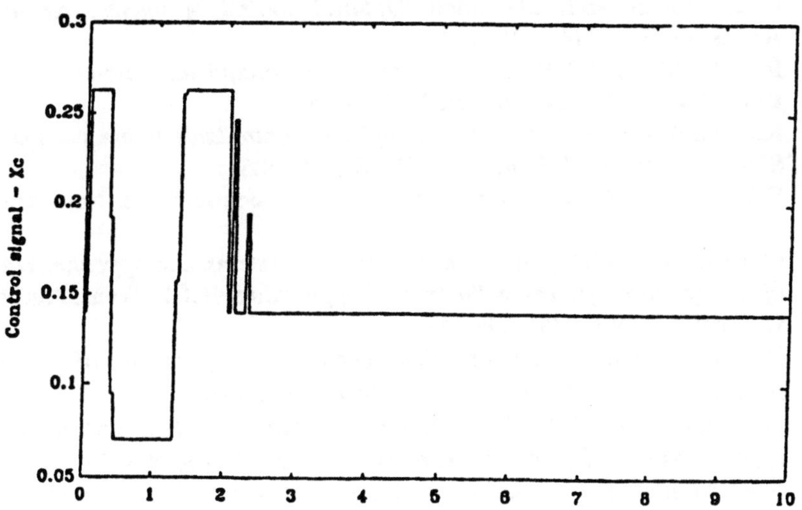


Figure 4. Neural control signal, 0.05 sec step

robust. The BLSTC and NLSTC are also very robust, and if the states are within the training data, the latter are faster. The STC also assume noisy power measurements while the others assume perfect swing angle data.

*Acknowledgement: This research is supported by NSF Grant ECS89-1373, EPRI Contract RR4006, and BPA Contract DEBI79-90B008423.*

## REFERENCES

[1] R.A. DeCarlo, S.H. Zak, and G.P. Matthews, *Variable structure control of nonlinear multivariable systems: A tutorial*, Proc. IEEE 76 (1988), 212-232.

[2] S.V. Emelyanov, *Theory of variable structure systems*, Science Press, Moscow, 1970.

[3] R.R. Mohler, *Optimal control of nuclear reactor processes*, Los Alamos Sci. Lab. Rept. LA-3257-MS, Los Alamos, NM, 1965.

[4] R.R. Mohler and C.N. Shen, *Optimal control of nuclear reactors*, Academic Press, New York, 1970.

[5] R.R. Mohler and R.E. Rink, *Control with a multiplicative mode*, J. Basic Engr. (ASME Trans.) 91 (1969), 201-206.

[6] R.E. Rink and R.R. Mohler, *Completely controllable bilinear systems*, SIAM J. Control & Optimiz. 6 (1968), 477-486.

[7] R.R. Mohler, *Bilinear control processes*, Academic Press, New York, 1973.

[8] C. Bruni, G. DiPillo, and G. Kouh, *Bilinear systems: An appealing class of nearly linear systems in theory and application*, IEEE Trans. Autom. Control AC 19 (1974), 334-348.

[9] R.R. Mohler and A. Ruberti, eds., *Theory and applications of variable structure systems*, Academic Press, New York, 1972.

[10] A. Ruberti and R.R. Mohler, eds., *Variable structure systems with application to biology and economics*, Springer-Verlag, New York, 1975.

[11] R.R. Mohler and A. Ruberti, eds., *Recent developments on variable structure systems, biology and economics*, Springer-Verlag, New York, 1978.

[12] R.R. Mohler, *Nonlinear systems: Vol. II Application to bilinear control*, Prentice-Hall, Englewood Cliffs, NJ, 1991.

[13] D.A. Pierre, *Perspectives on adaptive control of power systems*, IEEE Trans. PS 2 (1987), 387-396.

[14] D.J. Trudnowski, D.A. Pierre, J.R. Smith, and R. Adys, *Coordination of multiple adaptive PSS units using a decentralized control scheme*, IEEE Summer Power Meeting, San Diego, 1991 (to appear in Trans. PS).

[15] Y. Wang, R.R. Mohler, R. Spée, and W. Mittelstadt, *Variable structure controllers for power system transient stability*, IEEE Summer Power Meeting, San Diego, 1991 (to appear in Trans. PS).

[16] G.C. Goodwin and K.S. Sin, *Adaptive filtering, prediction and control*, Prentice-Hall, Englewood Cliffs, NJ, 1984.

[17] K.S. Narendra and K. Parthasarathy, *Identification and control of dynamical systems using neural networks*, IEEE Trans. Neural Networks NN 1 (1990), 4-27.

[18] K. Hornik, M. Stinchcombe, and H. White, *Universal approximations of an unknown mapping and its derivatives using multilayer feedforward networks*, Neural Networks 3 (1990), 551-60.

# On Modeling and Adaptive Control of Underwater Robots

A.K. RAMADORAI, T. J. TARN

*This paper is dedicated to Professor Antonio Ruberti.*

**Abstract**

**In this paper issues relating to modeling of underwater vehicles and manipulators are discussed. The dynamic modeling procedure enables inclusion of effects of buoyancy and viscous friction on the robot. A regressor is identified and it relates robot torques to a vector of unknown dynamic parameters of the arm and properties of the underwater environment like density and viscocity. An adaptive control scheme for task level servo to achieve free trajectory tracking is discussed.**

## 1. Introduction

Deployment of underwater Robotic manipulators and Remotely Operated Vehicles for scientific and commerical purposes has become popular. For recent studies in development of underwater robots one can see [7,13,14,15]. Some of the applications are underwater oil exploration, automated underwater cable-laying and marine life exploration. The most demonstrative and captivating endeavours in underwater robotics were the attempts to salvage the black box the Air-India jet that crashed off England coast and salvage of the wreckage of the Titanic. These applications have proven beyond doubt the technical prowess in the area of tele-supervised control of underwater robots. For completely autonomous operation of robots underwater, it is mandatory to use model-based dynamic control schemes. This not only calls for accurate modeling of the manipulator itself, but also that of the underwater environment. The latter is difficult to model in a real-world setting and hence the modeling errors that creep in will degrade performance. The variations in temperature, pressure and flow field variations bring about changes in buoyancy, viscocity and currents respectively. Other elements such as tether pull, drag and lift may also be contributory factors. To overcome these problems adaptive control laws must be formulated.

Much work has appeared in literature in study and development of adaptive control of robot manipulators. Slotine and Li [2], present an adaptive control algorithm which consists of a proportional and derivative feedback part and a full dynamics feedforward part. It is computationally simple and does not require joint acceleration or inversion of the estimated Inertia matrix. Craig et al., [1, 3] derive an adaptive version of the computed torque method for the control of robot manipulators. This method requires measurement of joint acceleration and inversion of the estimated Inertia matrix. Spong et al., [4] realize an alternative formulation of the law derived in [3], where the assumption of the boundedness of the inverse of the estimated inertia matrix in dropped.

The computed torque like methods of adaptive control schemes are essentially exact linearizing and decoupling schemes. While the results in [2, 3, 4] are derived in joint space, they may very well be derived in task space. Exact linearizing and decoupling schemes in task space have been developed and experimentally validated [8, 9]. They have shown "robustness" of the control law in simulation and experiments with respect to robot parameter modeling errors.

Task driven schemes for robot manipulators has distinct advantages. They are more general in theoretical development. They are more intuitive in nature and the task planner interacts more directly with the controller and are a step towards intelligent control. Being based on Differential Geometry, they fit naturally into planning schemes which essentially involve characterization of curves in space.

Application of adaptive dynamic control schemes in underwater robotics was discussed in [6]. Here adaptive extensions of sliding control for precise control of underwater vehicles is presented. Experimental results validate the claims in the paper and show that the laws compensate for unanticipated changes in buoyancy, current disturbances and tether pull.

In this paper the possibility of using adaptive control laws for control of underwater vehicles and manipulators is explored. Issues in modeling of underwater vehicles and manipulators are addressed. The modeling strategy allows explicit computation of a regressor that multiplies the vector of unknown quantities to give torques. The vector of unknown quantities includes the robot parameters as well as parameters of water like density and viscocity. They are then used for formulation of parameter update laws and adaptive control laws based on already existing results for a task driven control scheme. Section 2 discusses modeling issues of immersed underwater objects and manipulators. Section 3 contains task level adaptive control laws. Conclusions follow in section 4.

## 2. Dynamic Modeling of Underwater Robots

In this section a dynamic model for an object immersed underwater is developed based on rigid body spatial notation. Objects immersed in water experience effects of buoyancy and underwater flow yields. The effects of buoyancy is discussed here and a method to model it is explored.

*Archimedes principle* states [11] that: "*A body wholly or partly immersed in a fluid is buoyed up with a force equal to the weight of the fluid displaced by the body.*" The fluid exerts pressure on every part of the body's surface in contact with the fluid. The pressure is greater on the parts immersed more deeply. The resultant of all the forces is an upward force called buoyant force. The pressure on each part of the surface certainly does not depend on the material the body is made of. This pressure will result in an upward force acting through the center of buoyancy in the body which is the center of gravity of the liquid displaced. It is very likely that this center of buoyancy is different from the center of gravity of the immersed body. The center of gravity of the body depends on the mass distribution of the body, while the center of buoyancy depends on the density distribution of the fluid. However, if the object and the fluid have homogeneous mass and density distribution respectively the two centers may coincide.

When objects like robot arms are immersed, the links have non-uniform mass distribution, even if the overall shape of the link is symmetric. It is also assumed that the water in which the object is immersed is of a constant density. Therefore the two centers of gravity do not coincide.

The buoyant force is equal to the weight of fluid displaced. The weight of fluid displaced is the product of density and volume. The following notations are introduced (see Figure 1) for the dynamic modeling.

$C_w =$ World reference frame

$C_0 =$ Object attached reference frame at center of gravity

$C_b =$ Frame fixed at center of buoyancy

$R_0 =$ Rotation matrix between $C_w$ and $C_0$

$I_3 =$ $(3 \times 3)$ identity matrix

$m =$ mass of object

$I_0 =$ $(3 \times 3)$ inertia matrix about $C_0$

$I_b = R_0 I_0 R_0^t$, a $(3 \times 3)$ inertia matrix

$V_a =$ volume of body

$\varphi =$ density of water

$m_a = \varphi V_a =$ mass of water displaced

$I_a =$ Inertia matrix induced due to water displacement

$v =$ linear velocity of $C_0$ with respect to $C_w$

$w=$ Angular velocity of $C_0$ with respect to $C_w$.

$f=$ Applied force

$m=$ Appled moment

Here $C_0$ and $C_b$ have their corresponding axes parallel. The force due to buoyancy on the object can be viewed as an added mass and inertia matrix to the object. Now the dynamic equation for the immersed object are derived.

Let $\hat{v} = (w^t, v^t)^t \epsilon \mathbb{R}^6$ ; $\hat{\mathrm{f}} = (\mathrm{f}^t, \mathrm{m}^t)^t \epsilon \mathbb{R}^6$

$$\hat{I}_b = \begin{pmatrix} 0 & mI_3 \\ I_0 & 0 \end{pmatrix} + \begin{pmatrix} m_a\overline{BP}\times & m_aI_3 \\ I_a + \overline{PB} \times m_a\overline{BP}\times & \overline{PB} \times m_a \end{pmatrix}$$

$${}_w\hat{R}_b = \begin{bmatrix} R_0 & 0 \\ \overline{OP} \times R_0 & R_0 \end{bmatrix}; \quad {}_b\hat{R}_w = \begin{bmatrix} R_0^t & 0 \\ R_0\overline{PO}\times & R_0^t \end{bmatrix};$$

$$\hat{I} = {}_w\hat{R}_b \, \hat{I}_b \, {}_b\hat{R}_w$$

Here '$\times$' indicates vector cross-product operation. The spatial cross-product operator $\hat{X}$ is defined as

$$\begin{bmatrix} a \\ b \end{bmatrix} \hat{X} = \begin{bmatrix} a\times & 0 \\ b\times & a\times \end{bmatrix}.$$

The dynamic equation now becomes [10]:

$$\hat{I}\,\dot{\hat{v}} + \hat{v}\,\hat{X}\,\hat{I}\,\hat{v} = \hat{f} \tag{1}$$

If the volume $V_a$ of the object is known, then from the shape of the object and the constant density $\varphi$ of the liquid, $\overline{BP}$ can be found out. Also $m_a$ and $I_a$ are linear in $\varphi$. Therefore if $r\epsilon\mathbb{R}^6$ is chosen as the generalized coordinates of the object such that $\hat{v} = \Omega(r)\dot{r}$, then the dynamic equations can be written as

$$M(r)\ddot{r} + C(r,\dot{r}) = \hat{f}.$$

Here $M(r)$ and $C(r,\dot{r})$ contains the added mass and inertia terms due to the effect of buoyancy. Viscous friction term may be added so that:

$$M(r)\ddot{r} + C(r,\dot{r}) + F_v\dot{r} = \hat{f}. \tag{2}$$

This can be viewed also as a model for an underwater moving vehicle. $\hat{f}$ can be generated by a set of redundant or non-redundant actuators (thrusters).

In the model of the underwater object, certain parameters like load may be unknown or not known accurately. They have to be estimated. Let $\theta\epsilon\mathbb{R}^r$ be the parameters that are unknown. It was shown that the parameters might include water density and viscocity. Then density of water changes with temperature, pressure and depth. Therefore at any point of time it is difficult to determine it. Larger the volume of the object, the larger the modeling error due to density variations. Thus density variations can cause unanticipated changes in buoyancy. Also the viscous friction terms may change underwater. In general a regressor $R(r, \dot{r}, \ddot{r})$ can be identified. The regressor multiples a unique vector of unknown parameters of the arm as well as that of the environment to give torques. [12] gives a detailed study of identifying such a matrix.

$$M(r)\ddot{r} + C(r, \dot{r}) + F_v\dot{r} = \hat{f} = R(r, \dot{r}, \ddot{r})\,\theta. \tag{3}$$

A more refined model can be used to model fluid drag, lift and other such terms [5].

The same method of modeling can be extended to model robot manipulators with multiple links under water. Robot manipultors can be thought of as an articulated mass and the effect of buoyant force can be modeled for each link separately if its volume and shape is known accurately.

The aspects of modeling are exactly the same as in standard Newton-Euler recursion based model except for the added mass and inertia terms.

The Newton Euler recursion equations are given as follows [5]:

For $j = n$ to 1 do

$$\begin{aligned} m_j \cdot \left(\frac{d^jV_j}{dt} +^j w_j \times^j V_j\right) +^j M_j \cdot \left(\frac{d^jV_j}{dt} + \frac{d^jw_j}{dt} \times^j r_j\right) \\ +^j w_j \times \left[{}^jM_j \cdot ({}^jV_j +^j W_j \times^j r_j)\right] \\ =^j F_{j-1.j} +^j F_{j+1,\,j} +^j f_j - m_j^j g_e +^j F_{ej} \end{aligned} \tag{4}$$

$$\begin{aligned} &({}^jI_j +^j J_j) \cdot \frac{d^jw_j}{dt} +^j w_j \times \left[({}^jI_j +^j J_j)\ {}^j w_j\right] +^j r_j \times \left[{}^jM_j \cdot \left(\frac{d^jV_j}{dt} + \frac{d^jw_j}{dt} \times^j r_j\right)\right] \\ &+^jV_j \times \left[{}^jM_j \cdot ({}^jV_j +^j w_j \times^j r_j)\right] + \left\{\left[{}^jM_j \cdot ({}^jV_j +^j w_j \times^j r_j)\right] \times^j r_j\right\} \times^j w_j \\ &=^j N_{j-1.j} +^j N_{j+1.j} -^j g_j \times^j F_{j-1.j} + ({}^jI_j -^j g_j) \times^j F_{j+1.j} + ({}^jb_j -^j g_j) \times^j f_j +^j N_{ej} \end{aligned} \tag{5}$$

where

$m_j$: Mass of link $j$

$I_j$: Tensor of inertia around the mass centre of link $j$

$V_j$: Translating velocity of the mass centre of link $j$

$r_j$: Directional vector toward the added mass centre from the mass centre of link $j$

$M_j$: Added mass tensor of link $j$

$J_j$: Added tensor of inertia around the added mass centre of link $j$

$w_j$: Angular velocity of link $j$

$g_j$: Directional vector toward the mass centre from joint $j$

$l_j$: Directional vector toward joint $(j+1)$ from joint $j$

$b_j$: Directional vector toward the centre of buoyancy for link $j$

$f_j$: Buoyancy of link $j$

$F_{j-1.j}$: Action force through the joint toward link $j$ from link $(j-1)$

$F_{j+1.j}$: Action force through the joint toward link $j$ from link $(j+1)$

$N_{j-1.j}$: Moment of action toward link $j$ from link $(j-1)$

$N_{j+1.j}$: Moment of action toward link $j$ from link $(j+1)$

$F_{ej}$: Other external forces exerted on link $j$

$N_{ej}$: Other external moments around the mass centre acting on link $j$

$g_e$: Vector of gravitational acceleration.

In equations (4) and (5) the left superscript '$j$' indicates quantities expressed in frame fixed in link $j$.

It can be clearly seen that the corresponding regessor for the arm can be easily identified. Then the robot manipulator equation will look like

$$D(q)\ddot{q} + N(q,\dot{q}) + F_v\dot{q} = \tau = R(q,\dot{q},\ddot{q})\,\theta. \tag{6}$$

$\tau$, $q$, $\dot{q}$, $\ddot{q}$ $\in \mathbb{R}^n$ are joint torque, position, velocity and acceleration respectively. Here the terms $D(q)$ and $N(q,\dot{q})$ are modified inertia matrix and centrifugal and Coriolis force terms respectively which include added mass and inertia terms due to underwater effects. The model that contributes to the viscous friction is still primitive at this stage. In general $F_v$ is a diagonal matrix, with unequal diagonal entries. Thus every link is assigned a value depending upon the shape of the link. This model of viscous friction is very rudimentary. $R(q, \dot{q}, \ddot{q}) \in \mathbb{R}^{n\times r}$ is the regressor that is a function of the joint position, velocity and acceleration. It multiplies the vector of unknown parameters $\theta \in \mathbb{R}^r$ of the robot and underwater environment to yield joint torque.

Thus in the process of modeling, we have identified the unknown parameters of robot as well as the underwater environment such as density and viscocity and an explicit way to construct the regressor is outlined. This can be used to form adaptive control laws and parameter update laws based on already existing results.

## 3. Adaptive Control Laws

Many adaptive control schemes for free trajectory tracking of robot manipulators with joint servo have been discussed [1, 2, 3, 4]. All these schemes involve the manipulator regressor. In the case of underwater robot manipulators the manipulator regressor relates the robot joint torques to both unknown dynamic parameters of the arm as well as that of the underwater environment. The identification of such a regressor was discussed in section 2.

Many robot tasks are specified in task space. Hence task servo schemes are preferred. Besides they have numerous qualitative and quantitative benefits. Some of them were stated in section 1. In this section, an adaptive control scheme with task space servo based on results by Spong et al., [4] is proposed.

Let the dynamic model of a non-redundant manipulator be given by:

$$D(q)\ddot{q} + N(q,\dot{q}) + F_v\dot{q} = \tau = R(q,\dot{q},\ddot{q})\,\theta \tag{7}$$

$$Y = h(q); \qquad q,\ \tau,\ y\ \in \mathbb{R}^n.$$

Let $J_h(q) \in \mathbb{R}^{(n\times n)}$ be the corresponding output Jacobain. $\theta \in \mathbb{R}^r$ is a vector of unknown dynamic parameters of the robot and that of the environment. The problem of tracking a trajectory $\ddot{Y}^d(t)$, $\dot{Y}^d(t)$, $Y^d(t)$; $\quad t_0 \le t \le t_f$ can be achieved by control laws of the form:

$$\begin{aligned} &D(q)J_h^{-1}(q)\left[\ddot{Y}^d(t) + K_{vt}\left(\dot{Y}^d(t) - \dot{Y}\right) + K_{pt}\left(Y^d(t) - Y\right) - \dot{J}_h(q)\dot{q}\right] \\ &\qquad + N(q,\dot{q}) + F_v\dot{q} \end{aligned} \tag{8}$$

or

$$D(q)\left[\ddot{q}^d(t) + K_{v_j}\left(\dot{q}^d(t) - \ddot{q}\right) + K_{p_j}\left(q^d(t) - q\right)\right] + N(q,\dot{q}) + F_v\dot{q}. \tag{9}$$

(8) is a task servo scheme and (9) is a joint servo scheme. In (9) $\ddot{q}^d(t), \dot{q}^d(t)$, $q^d(t)$; $\quad t_0 \le t \le t_f$ are computed using inverse kinematics, inverse output Jacobian and the time derivative of the output Jacobian. Both are exact linearizing and decoupling schemes which yield asymptotically stable error dynamics [9]. However, due to modeling errors, linear and decoupled error dynamics cannot be achieved. This is clearly seen from the performance plots in [9], where the error profiles are nonlinear. If a regressor is identified, it can be used to formulate parameter update laws and adaptive control laws.

Spong et al., [4] derive adaptive control laws of the form:

$$\begin{aligned} \tau = &D_0(q)\left[\ddot{q}^d(t) + K_{v_j}\left(\dot{q}^d(t) - \dot{q}\right) + K_{p_j}\left(q^d(t) - q\right) + \delta q\right] \\ &+ N_0(q,\dot{q}) + F_{v_0}\dot{q} \end{aligned} \tag{10}$$

$D_0, N_0, F_{v_0}$ are apriori nominal estimates of $D, N$ and $F_v$ repsectively with $\theta = \theta_0$. $\delta q$ is an additional outer loop that compensates for deviations $\Delta D, \Delta N, \Delta F_v$ where $\Delta(\cdot) = (\cdot)_0 - (\cdot)$. On the same lines a task space control law can be formulated as:

$$\tau = D_0(q) J_h^{-1}(q) \left[ \ddot{Y}^d(t) + K_{vt} \left( \dot{Y}^d(t) - \dot{Y} \right) + K_{pt} \left( Y^d(t) - Y \right) - \dot{J}_h(q)\dot{q} + \delta y \right]$$

$$+ N_0(q, \dot{q}) + F_{v_0} \dot{q}.$$

(11)

Assume $J_h$ can be estimated accurately and is nonsingular. Combining (11) and (7) we get (denote $e_t = Y^d - Y$):

(12) $D_0 J_h^{-1} \left[ \ddot{e}_t + K_{vt}\dot{e}_t + K_{pt} e_t + \delta y \right] = -\Delta D \ddot{q} - \Delta N - \Delta F_v \dot{q} = -R \Delta \theta;$

So we have

$$\ddot{e}_t + K_{vt}\dot{e}_t + K_{pt}e_t + \delta y = -J_h D_0^{-1} R \Delta\theta = \Phi_0 \Delta\theta; \quad -J_h D_0^{-1} R = \Phi_0.$$

Set $\delta y = \Phi_0 \Delta\hat{\theta}$ to obtain

(13) $$\ddot{e}_t + K_{vt}\dot{e}_t + K_{pt}e_t = \Phi_0 \Delta\tilde{\theta}; \qquad \Delta\tilde{\theta} = \left( \Delta\theta - \Delta\hat{\theta} \right)$$

Rewriting (13) in state space form we get:

(14) $$\dot{x} = Ax + B\Phi_0 \Delta\tilde{\theta}$$

where

$$x = \begin{pmatrix} e_t \\ \dot{e}_t \end{pmatrix}; \qquad B = \begin{bmatrix} 0 \\ I \end{bmatrix};$$

$$A = \begin{bmatrix} 0 & I \\ -K_{pt} & -K_{vt} \end{bmatrix}.$$

Choose a Lyapunov candidate:

(15) $$V = x^T P x + \Delta\tilde{\theta}^t \Gamma \Delta\tilde{\theta}; \qquad \Gamma = \Gamma^t > 0$$

and $P$ is a unique symmetric positive definite solution to $A^t P + PA + Q = 0$, for a given symmetric positive definite $Q$. An update law

(16) $$\dot{\Delta\hat{\theta}} = -\Gamma^{-1} \Phi_0^t B^t P x$$

will guarantee $\dot{V} \leq 0$ and hence as in [4], $x \to 0$ when $t \to \infty$, if all signals remain bounded.

Therefore asymptotic error convergence in task space is achieved. Note that here we need non-singularity of the Jacobian $J_h$ and also the joint acceleration information for this adaptive control law.

## 4. Conclusion

In this paper, Adaptive Control, with task space servo, of underwater robot manipulators is discussed. The effect of buoyancy is modeled to get a dynamic equation that is similar in structure and properties to a free space robot manipulator. The dynamic model incorporates the effects of added mass, added inertia and viscous friction. The underwater robot model directly involves underwater properties like density and viscocity. A procedure to explicitly identify a regressor that relates joint torques to an unknown parameter vector is outlined.

The method is general and can be used to model underwater vehicles also. Accurate modeling of the underwater environment requires a lot of effort. Issues such as flow fields caused by the motion of the arm itself are complex and have to be studied. The modeling procedure dictates the regressor and the regressor may be used for adaptive control laws. Such adaptive control laws are imperative when performing autonomous (no human supervisory control) tasks underwater. The model presented in the paper is rather rudimentary and can only compensate for changes in density and viscocity of the water. The shapes of the links might also give rise to drag and lift, which have to be modeled. Thus there are a lot of modeling issues and related control laws which have to be investigated to enable autonomous robot tasks underwater.

Recently neural networks have gained wide acceptance in adaptive control. The use of neural networks for identification and control of nonlinear dynamic systems is proposed in [16]. A hierarchical intelligent control structure for multiple autonomous undersea vehicles feature in [17]. These studies suggest that use of neural networks for control of underwater robots will lead to good control strategies and is worthwhile investigating.

## REFERENCES

[1] J.J. Craig, *Adaptive Control of Mechanical Manipulators*, Addison-Wesley Publishing Company, 1988.

[2] J.J.E. Slotine and W. Li., *On the Adaptive Control of Robot Manipulators*, The Intl. Journal of Robotics Research, Vol. 6, No. 3, Fall 1987, pp. 49–59.

[3] J.J. Craig, P. Hsu and S.S. Sastry, *Adaptive Control of mechanical manipulators*, Proc. of 1986 IEEE Robotics and Automation Conference, pp. 190–195.

[4] M.W. Spong and R. Ortega, *On Adaptive Inverse Dynamics Control of Rigid Robots*, IEEE Trans of Robotics and Automation, Vol. 35, No. 1, January 1990, pp. 92–95.

[5] K. Ioi and K. Itoh, *Modelling and simulation of an underwater manipulator*, Trans. of Advanced Robotics, Vol. 4, No. 4, 1990, pp, 303–317.

[6] D.R. Yoerger and J.J. E Slotine, *Adaptive sliding control of an experimental underwater vehicle*, Proc. of 1991 IEEE Intl. Conf. on Robotics and Automation, Sacramento, CA, 1991, pp. 2746–2751.

[7] J. Xinseng and F.X. Sheng, *Research and development of underwater robot in China*, Proc. of Intl. Workshop on Advanced Robot, Aug. 1991, Beijing, pp. 1.18–1.35.

[8] T.J. Tarn, A.K. Bejczy, A. Isidori, Y.L. Chen, *Nonlinear feedback in robot arm control*, Proc of 23rd IEEE Conference on Decision and Control, Las Vegas, 1984 , pp. 736–751.

[9] T.J. Tarn, A.K. Bejczy, G.T. Marth and A.K. Ramadorai, *Six degree-of-freedom task space control for the PUMA 560 manipulator: An experimential study*, Preprints (Part 1) of IFAC/IFIP/IMACS Symposium on Robot control, SYROCO 1991, Vienna, Austria, pp. 177–182.

[10] R. Featherstone, *Robot dynamics algorithms*, Kluwer Academic Publications, 1987.

[11] D. Halliday and Resnick, *Fundamentals of Physics*, 3rd. edition, New York Wiley, 1988.

[12] P. Khosla and T. Kanade, *Parameter Identification of robot dyanmics*, Proc. of IEEE Conf. on Decision and Control, Dec. 1985, pp. 1754–1760.

[13] G.T. Russel, R.M. Dunbar , *Intelligent Control and Communication Systems for Underwater Vehicles*, Proc. of Mobile Robots for Subsea Environments, International Advanced Robotics Program, October 23–26, 1990, pp. 1–14.

[14] D. Maddalena, W. Prendin, A. Terribile, *Supervisory Controlled Telemanipulation System: Test Results and Future Evolution*, Proc. of Mobile Robots for Subsea Environments, International Advanced Robotics Program, October 23–26, 1990, pp. 97–108.

[15] G. Bartolini, M. Brenco, G. Cannata, G. Casalino, G. Garibotto, G. Sardini, *Advanced Simulation and Control Techniques of Underwater Robots*, Proc. of Mobile Robots for Subsea Environments, International Advanced Robotics Program, October 23–26, 1990, pp. 163–170.

[16] K.S. Narendra, *Adaptive Control Using Neural Networks*, Neural Networks for Control, Edited by W.T. Miller, R.S. Sutton, P.J. Werbos, MIT Press, 1990, pp. 115–142.

[17] M. Herman, J.S. Albus, T.H. Hong, *Intelligent Control for Multiple Autonomous Undersea Vehicles*, Neural Networks for Control, Edited by W.T. Miller, R.S. Sutton, P.J. Werbos, MIT Press, 1990, pp. 427–510.

**Acknowledgement**

This work was supported in part by the National Science Foundation under Grants IRI-9106317 and DMC-861593.

**A.K. Ramadorai and T.J. Tarn**
**Washington University**
**Campus Box 1040**
**St. Louis, Missouri 63130**
**U.S.A**

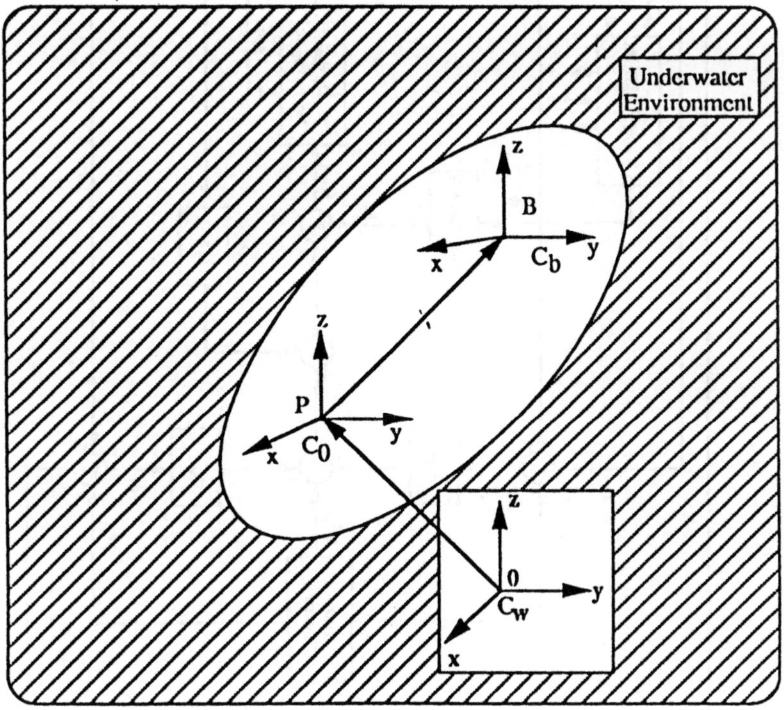


**Figure 1. Object Immersed Underwater.**

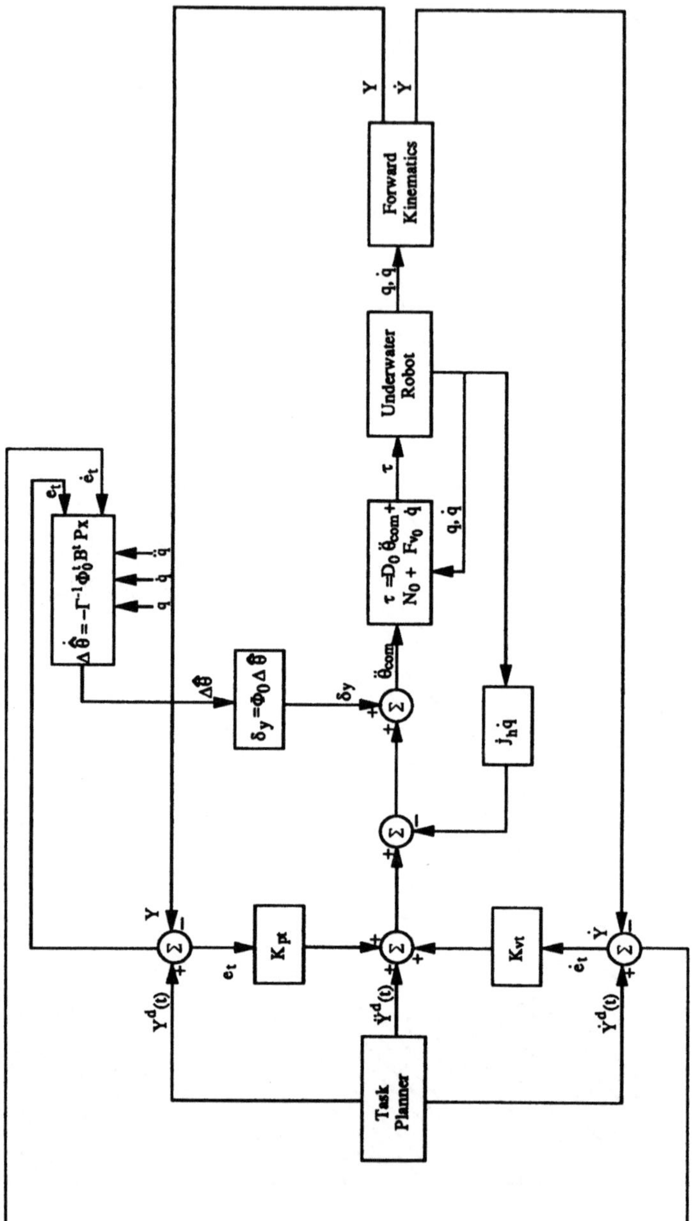


**Figure 2. Adaptive Task Servo Scheme for Underwater Robot.**

# A Generalization of the Kalman-Yakubovic Lemma

A. V. BALAKRISHNAN

Abstract

We present the following infinite-dimensional (Hilbert space) generalization of the "positive real" Lemma of Kalman-Yakubovic [1]. Let $\mathcal{H}$ denote a function space which is a Hilbert space under an inner product which is real on real functions. Let $A$ be the infinitesimal generator of an analytic semigroup $S(t)$, $t \geq 0$ of class $\mathcal{H}[-\Psi, \Psi]$, $0 < \Psi < \frac{\pi}{2}$, with compact resolvent, mapping real elements into real elements. Let $C$, $B$ be real elements. Suppose $(A, B)$ is controllable, and

$$Re\,[C, R(i\omega, A)\,B] \geq 0, \qquad \omega \text{ real}$$

where $R(\lambda, A)$ is the resolvent of $A$. Then we can define a new inner product $[\ ,\ ]_E$ on $\mathcal{H}$ in which $A$ is dissipative and further:

$$[C, R(\lambda, A)\,B] = [B, R(\lambda, A)\,B]_E\,, \qquad Re\,\lambda \geq 0.$$

The techniques of proof differ substantially from the finite dimensional version [1,2]. The application is to positive real compensators in the stabilization of flexible structures.

1. R.E. Kalman. "Lyapunov Functions for the Problem of Lur'e in Automatic Control," *Proc. Nat. Acad. Sci. U.S.A.*, Vol. 49 (1963), pp. 201-205.

2. B. D. O. Anderson. "A System Theory Criterion For Positive Real Matrices," J. SIAM Control, Vol. 5, No. 2 (1967), pp. 171-182.

# PSICO Today

SERGIO BITTANTI

*A stroll in the literature on periodic systems identification, control and optimization over more than 20 years*

*In honour of Professor Antonio Ruberti*

It is a common experience that celebrations constitute a break in the usual commitments of daily life, during which the mind is led to indulge over the developments of the past. This especially applies to me now, at the beginning of these few lines written for the 65th birthday of Professor Antonio Ruberti. In the far 1970, when I was working at the *laurea* thesis in Electrical Engineering at the Politecnico di Milano, the name of Professor Ruberti was already associated with the forward line of research activity in systems and control, an area I discovered thanks to the classes given at the Politecnico by Professor Emanuele Biondi and Professor Guido Guardabassi. To my young mind of those days, *automaticaworld* appeared as a spring of genuine science in a paraphernalia of engineering techniques. The thesis subject was evolving around the stability and structural properties of linear systems with periodically varying coefficients. This was my first contact with the realm of PSICO (Periodic Sistems Identification, Control and Optimization), a contact which had to prolong in the decades to come.

In those times, the attention on this subject stemmed from the observation that the performance of some chemical plants, usually operated in steady-state conditions, improved under periodic operation. PSICO theory, however, is not limited to periodic control of time invariant plants, but encompasses the analysis and control of systems which are intrinsically periodic; in principle, it also includes periodic control of (possibly nonperiodically) time-varying plants and nonlinear systems.

Periodic systems can be seen as an intermediate class between time- invariant and time-varying systems with any type of variability. However, not so much credit should be placed in the *naive* idea that the corresponding theory is just a not-so- difficult generalization of the one relative to time-invariant systems. As an example of the difficulties which can be encountered, one can mention the well known Kalman proposition of 1969, according to which, if a periodic system is completely controllable, all states can be driven to zero in a period at most. Although a number of authors tried to prove this conjecture, and some even published a 'proof' of it, a simple counterexample was subsequently worked out showing that the conjecture is not valid in general. Subtleties of this types are not uncommon, and frequently distributed along the PSICO road.

The aim of this paper is to supply a (selected) list of papers on the subject organized in topics of major interest. As such, the paper should be seen as an easy-to-use reference tool. Yet, due to space limitations, we cannot claim any completeness in quotation. In particular, the application oriented papers have been omitted.

## 2.

The interest in PSICO theory need hardly be emphasized: In nature, many phenomena are periodic, from tides to planet movements, from bird migrations to sunspot numbers. Periodic manmade signals are also common, especially in telecommunication networks (sinusoidal carriers in modulation systems, periodic scanning, and so on.) It is also worthwhile noticing that periodicity arises in multivariable time-invariant signals and systems whenever the variables are sampled at different (but commensurable) rates (*multirate systems.*)

In systems and control, the early days of periodic systems analysis and control can be traced back to the seventies, when a criterion to test whether a periodic controller can lead to better performances than the optimal steady-state one was worked out (Π *test.*) In those days, *periodic optimal control* was a main subject of investigation, and the very Π *test* can be seen as a product of the necessary optimality conditions. Ever since, the theory has developed along various streams. The *structural properties* (reachability, controllability, stabilizability, and their dual ones) have been deeply studied, and both modal and Grammian criteria are now available. This opened the road to the studies on the *canonical decomposition* of periodic systems, as well as the analysis of *periodic Lyapunov* and *periodic Riccati equations.* In the last years these tools have been most useful in the analysis of cyclostationary signals and in the realization problem context. Furthermore, thanks to periodic control ideas, major advances have been possible both in *robust control* and in *pole assignment* techniques. For instance, it has been shown that pole assignment can be achieved by periodic output feedback in many instances when this is impossible by means of time-invariant controllers.

Notwithstanding the noticeable degree of maturity reached in the field, PSICO is live and thrilling more than ever; many interesting problems are still open, both in the analysis of periodic systems and periodic control of time-invariant or time varying systems; they can be indicated to young (and less young) researchers as an exiting source of active thinking for the years to come. So, *arrivederci* in one or two decades, to discuss again, all together, the new developments of the field.

*ACKNOWLEDGEMENTS: This paper has been supported by MURST - National Research Project* Model identification, system control, signal processing, *CNR -*

*Centro di Teoria dei Sistemi, and CNR - Special Project* Algorithms and architectures for robust and adaptive identification and control.

## REFERENCES

### Structural properties and canonical decomposition

Bittanti S., *Deterministic and stochastic linear periodic systems*, in Time Series and Linear Systems, S. Bittanti ed. Springer-Verlag, Lecture Notes in Control and Information Sciences, Vol. 86, 1986, 141-182.

Bittanti S. and P. Bolzern, *Can the Kalman canonical decomposition be performed for a discrete-time linear periodic system?*, 1st Latin American Conference on Automatica, Campina Grande, Brasil, 1984, 449-453.

Bittanti S. and P. Bolzern, *Canonical decomposition of discrete-time linear systems*, 23th Conference on Decision and Control, Las Vegas, 1984, 1984, 1737,1738.

Bittanti S. and P. Bolzern, *Four equivalent notions of stabilizability of periodic linear systems*, 3rd American Control Conference, San Diego, 1984, 1321-1323.

Bittanti S. and P. Bolzern, *Reachability and controllability of discrete-time linear systems*, IEEE Transactions on Automatic Control, 30, 1985, 399-491.

Bittanti S. and P. Bolzern, *Stabilizability and detectability of linear periodic systems,* Systems and Control Letters, 6, 1985,141-145, plus *Addendum*, Systems and Control Letters, 7, 1985, 73.

Bittanti S. and P. Bolzern, *On the structure theory of discrete-time linear systems*, International Journal of Systems Sciences, 17, 1986, 33-47.

Bittanti S., P. Colaneri and G. De Nicolao, *Discrete- time periodic systems: a note on the reachability and controllability interval legth*, Systems and Control Letters, 8, 1986, 75-78.

Bittanti S., P. Colaneri and G. Guardabassi, *H and K-controllability of linear periodic systems*, 22nd Conference on Decision and Control, S. Antonio, 1983, 1376-1379.

Bittanti S., G. Guardabassi, C. Maffezzoni, and L. Silverman, *Periodic systems: controllability and the matrix Riccati equation*, SIAM Journal on Control and Optimization, 16, 1978, 37-40.

Bolzern P., *Criteria for reachability, controllability and stabilizability of discrete-time linear periodic systems*, 5th Polish-English Seminar on Real-Time Process Control, Radzjejovice, 1986, 69-83.

Brunovsky R.W., *Controllability and linear closed loop controls in linear periodic systems*, Journal of Differential Equations, 6, 296-313, 1969.

D'Alessandro P., A. Isidori and A. Ruberti, *A new approach to the theory of canonical decomposition of linear dynamical systems*, SIAM J. Control, 11, 1, 1973, 143-158.

Grasselli O.M., *A canonical decomposition of linear periodic disrete-time systems*, International Journal of Control, 40, 201-214, 1984.

Grasselli O.M. and S. Longhi, *Sottospazi invarianti controllati e sottospazi di controllabilita' per sister lineari periodici a tempo discrete*, Int. Rep. 2/85, University di Ancona, 1985.

Kalman R.E., *Theory of regulators for linear plants*, in R.E. Kalman, P.L. Falb and M.A. Arbib: Topics in Mathematical System Theory, McGraw Hill, 1969.

## Periodic Riccati Equation and Periodic Lyapunov Equation

Bekir E. and R.S. Bucy, *Periodic equilibria for matrix Riccati equations*, Stochastics, 2, 1976, 1-104.

Bittanti S., *The Riccati equation in Control, Systems, and Signals*, Lecture Notes of the IEEE/IFAC/SIAM Workshop on The Riccati Equation in Control, Systems, and Signals, Como; Pitagora Editrice, 1989.

Bittanti S., P. Bolzern and P. Colaneri, *The extended periodic Lyapunov lemma*, Automatica, 5, 1985, 603-605.

Bittanti S., P. Colaneri and G. De Nicolao, *The periodic Riccati equation: existence of a periodic positive semidefinite solution*, 26th Conference on Decision and Control, Los Angeles, 1987, 293-294.

Bittanti S., P. Colaneri and G. De Nicolao, *Two techniques for the solution of the discrete-time periodic Riccati equation,* in Linear Algebra in Signals, Systems and Control, B.N. Datta, C.R. Johnson, M.A. Kaashoek, R.J. Plemmons, E.D. Sontag eds., SIAM Publications, 315-331, 1988.

Bittanti S., P. Colaneri and G. De Nicolao, *The difference periodic Riccati equation for the periodic prediction problem*, IEEE Transactions on Automatic Control, $AC-33, 1988, 706-712$.

Bittanti S., P. Colaneri and G. De Nicolao, *A note on the maximal solution of the periodic Riccati equation*, IEEE Transactions on Automatic Control, 34, 12, 1989, 1316-1319.

Bittanti S., P. Colaneri and G. De Nicolao, *An algebraic Riccati equation for the discrete-time periodic prediction problem*, Systems and Control Letters, 14, 141-182, 1990.

Bittanti S., P. Colaneri and G. De Nicolao, *The periodic Riccati equation*, in The Riccati Equation, S. Bittanti, A.J. Laub, J.C. Willems Eds., Springer-Verlag, 1991.

Bittanti S., P. Colaneri and G. Guardabassi, *Analysis of the periodic Lyapunov and Riccati eqaution via canonical decomposition*, SIAM Journal on Control and Optimization, 24, 6, 1986, 1138-1149.

Bittanti S., P. Colaneri and G. Guardabassi, *Periodic solutions of periodic Riccati equations*, IEEE Transactions on Automatic Control, 29, 7, 1984, 665-667.

Bolzern P. and P. Colaneri, *Existence and uniqueness conditions for periodic solutions of the discrete-time periodic Lyapunov equation*, 25th Conference on Decision and Control, Athens, 1986, 1439-1443.

Bolzern P. and P. Colaneri, *The periodic Lyapunov equation*, SIAM Journal on Matrix Analysis and Applications, 9, 4, 1988, 499-512.

Bucy R.S., *Structural stability for the Riccati equation*, SIAM Journal on Control, 13, 4, 1975, 749- 761.

*Da* Prato G., *Periodic solutions of infinite dimensional Riccati equations*, Rendiconti Accademia Nazionale dei Lincei, 1984.

de Souza C.E., *Riccati differential equation in optimal filtering of periodic nonstabilizable systems*, International Journal of Control, 46, 4, 1987, 1235-1250.

de Souza C.E., *Existence conditions and properties for the maximal periodic solution of periodic Riccati difference equations*, International Journal of Control, 50, 1989, 731-742.

de Souza C.E., *Periodic strong solution for the optimal filtering problem of linear discrete-time periodic systems*, IEEE Transactions on Automatic Control, 36, 3, 1991, 333-338.

de Souza C.E. and G.C. Goodwin, *Periodic Solutions of matrix Riccati equation in optimal filtering of nonstabilizable periodic systems*, IFAC World Congress, Munich, 1990, 9.249-9.254.

Emre E. and G. Knowles, *A Newton-like approximation algoritmh for the steady-state solution of the Riccati equation for time-varying*, Proceedings of the International Symposium MTNS- 85, Stockholm, 1985.

Hernandez V. and L. Jodar, *Boundary problems and periodic Riccati equations*, IEEE Transactions on Automatic Control, 11, 1985, 1131-1133.

Hewer G.A., *Periodicity, detectability and the matrix Riccati equation*, SIAM Journal of Control, 13, 6, 1235-1251, 1975.

Kano H. and T. Nishimura, *Periodic solutions of matrix Riccati equations with detectability and stabilizability*, International Juornal of Control, 29, 1979, 471-487.

Kano H. and T. Nishimura, *Positive-definite periodic solutions of matrix Riccati equations in continuous time*, International Symposium MTNS-87.

Nishimura *T* and H. Kano, *Periodic strong solutions of periodically time-varying matrix Riccati equations*, International Journal of Control, 49, 193-205, 1989.

Shayman M.A., *Varieties of invariant subspaces and the algebraic Riccati equation*, Harvard University, Thesis, 1980.

Shayman M.A., *On the phase portrait of the matrix Riccati equation arising from the periodic control problem*, SIAM Journal on Control and Optimization, 23, 1985, 717-751.

Speyer J.L., *Periodic Riccati differential equation and Periodic regulator*, 26th Conference on Decision and Control, Los Angeles, 1987, 288-292.

Speyer J. L., *The Riccati differential equation and the accessory problem in the calculus of variation*, in The Riccati equation in Control, Systems, and Signals, S. Bittanti ed., Pitagora Editrice, 1989, 2-8.

## Identification, Signal Processing and Time Series Analysis

Bittanti S., *Periodic predictors for cyclostationary processes*, in Modelling, Robustness and Sensitivity Reduction in Control Systems, R. Curtain Ed., Springer-Verlag, 1987, 239-249.

Bittanti S., P. Bolzern, G. De Nicolao, L. Piroddi and D. Purassanta, *A minimum prediction error algorithm for estimation of periodic ARMA models*, European Control Conference, Grenoble, 1991, 1200-1203.

Bittanti S. and G. De Nicolao, *Periodic ARMA models: Optimal prediction and minimum phase condition.* 11th IFAC World Congress, Tallin, 1990, 96-111.

Bittanti S. and G. Guardabassi, *Optimal cycling for parameter identification*, 4th IFAC Symposium on Identification and System Parameter Estimation, Tblisi, 3, 1976, 590-603.

Gardner W.A., *Characterization of cyclostationary random processes*, IEEE Transactions on Information Theory, 21, 1, 1975, 4-14.

Gardner W. A:, *Exploitation of spectral redundancy in cyclostationary signals*, IEEE Signal Processing Magazine, 8, 2, 1991, 14-37.

Gardner W.A., *The spectral correlation theory of cyclostationary time series*, Signal Processing, 11, 1986, 13-36.

Miamee A.G. and H. Salehi, *On the prediction of periodically correlated stochastic processes*, in Multivariate Analysis, V.P.R. Krishnaiah ed., North Holland, 1980, 167-179.

Pagano M., *Periodic and multiple autoregression*, Annals of Statistics, 6, 1978, 1310-1317.

Vecchia A.V., *Maximum likelihood estimation for periodic autoregressive moving average models*, Technometrics, 27, 4, 1085, 375-384.

## Optimal Control

Artstein Z. and A. Leizarowitz, *Tracking periodic signals with the overtaking criterion*, IEEE Transactions on Automatic Control, 30, 11, 1985, 1123-1126.

Bittanti S., G. Fronza, G. Guardabassi and C. Maffezzoni, *A maximum priciple for periodic optimization*, Ricerche di Automatica, 3, 2, 170-179, 1972.

Bittanti S., G. Fronza and G. Guardabassi, *Periodic optimization of linear systems under control power constraints*, Automatica, 9, 1973, 269-271.

Bittanti S., G. Fronza and G. Guardabassi, *Periodic optimization of linear systems under control power constraint*, Automatica, 9, 269-271, 1973.

Bittanti S. and G. Guardabassi, *Optimal cyclostationary control: a parameter optimization frequency-domain approach*, IFAC World Congress, Kyoto, 1981, VI-191 - VI-196.

Bittanti S. and G. Guardabassi, *Optimal cyclostationary control: the LQG approach*, 20th IEEE Conference on Decision and Control, San Diego, 1981, 166-167.

Bittanti S. and G. Guardabassi, *Optimal periodic control and periodic systems analysis: a survey*, 25th Conference on Decision and Control, Athens, 1986, 1417-1423.

Bittanti S., A. Locatelli and C. Maffezzoni, *Second variation methods in periodic optimization*, Journal of Optimization Theory and Applications, 14, 1, 1974, 31-49.

Bittanti S. and C. Maffezzoni, *Structural properties of a Hamiltonian system in a periodic optimization problem*, Automation and Remote Control, 36, 6, Part 1, 1975, 877-884 ( translated from Avtomatika i Telemekhanika, 6, 1975, 5-13).

Chuang C.H., J.L. Speyer and J.V. Breakwell, *An asymptotic expansion about the chattering solution of an optimal periodic control problem*, 26th Conference on Decision and Control, Athens, 1987.

Chuang C.H., J.L. Speyer and J.V. Breakwell, *An asymptotic expansion for an optimal relaxation oscillator*, SIAM Journal on Control and Optimization, 26, 3, 1988.

Colonius F., *Optimality for periodic control of functional differential systems.* Journal of Mathematical Analysis and Applications, 1985.

Colonius F., *A global maximum principle for periodic control of functional differential equations*, 25th Conference on Decision and Control, Athens, 1986, 1424-1427.

Dorato P. and H.K. Knudsen, *Periodic optimization with applications to solar energy control*, Automatica, 15, 1979, 673-676.

Dorato P. and A.H. Levis, *Optimal linear regulators: the discrete time case.* IEEE Transactions on Automatic Control, 6, 613-620, 1971.

Evans R.T., *Optimal periodic control theory*, F.J. Seiler Research Laboratory, Air Force Systems Command, Report SRL$-TR-80-0024,1980$.

Gilbert E.G., *Optimal periodic control: A solution set theory od necessary and sufficient conditions*, IFAC World Congress, Helsinki, 1978.

Gilbert E.G., *Optimal periodic control: A general theory of necessary conditions.* SIAM Journal on Control and Optimization, 15, 717-746, 1977.

Horn F.J.M. and R.C. Lin, *Periodic Processes: A variational approach*, Industrial Engineering Chemical Processes Design Devices, 5, 1, 21-30, 1967.

Maffezzoni C., *Hamilton-Jacobi theory for periodic control problems*, Journal of Optimization Theory and Applications, 14, 21-29, 1974.

Matsubara M., Y. Nishimura and T. Takahashi, *Optimal periodic control of lumped parameter systems*, Journal of Optimization Theory and Applications, 13, 1, 1974

Markus L., *Optimal control of limit cycles or what control theory can do to cure a heart attack or to cause one.* Symposium on Ordinary Differential Equations, Minneapolis, W.A. Harris, Y. Sibuya eds., Springer-Verlag, 1973.

Marzollo A. (ed.), *Periodic optimization*, Springer- Verlag, 1972.

Noldus E., *A survey of optimal periodic control of continuous systems.* Journal A, 16, 11-16, 1975.

Speyer J.L. and R.T. Evans, *A second variational theory of optimal periodic processes*, IEEE Transactions on Automatic Control, 29, 138-148, 1984.

Yakubovich V.A., *Linear quadratic optimization problems and frequency theorem for periodic systems*, Siberian Mathematics Journal, 27, 1986, 181-200 (Russian version in Sibirsk. Mat. Zh., 27, 4, 1986.)

Yakubovich V.A., *Dichotomy and absolute stability of nonlinear systems with periodically nonstationary linear part*, Systems and Control Letters, 11, 1988, 221-228.

Yakubovich V.A. and V. M. Starzhinskii, *Linear differential equations with periodic coefficients*, J. Wiley, 1975 (russian edition, 1972).

## Stabilization, regulation, pole assignment and all that

Chammas A.B. and C.T. Leondes, *On the design of linear time invariant systems by periodic output feedback, Part I and Part II,* International Journal of Control, 27, 6, 1978, 885-903.

Chammas A.B. and C.T. Leondes, *Pole assignment by piecewise constant output feedback*, International Journal of Control, 44, 1986, 1661-1673.

Colaneri P., *Zero-error regulation of discrete- time linear periodic systems*, Systems and Control Letters, 15, 1990, 161- 167.

Colaneri P., *Output stabilization via pole placement of discrete-time linear periodic systems*, IEEE Transactions on Automatic Control, $AC-36, 6, 1991, 739-742$.

Davis J. H., *Stability conditions derived from spectral theory: discrete-time systems with periodic feedback*, SIAM Journal of Control, 10, 1972, 1-13.

Evans M.E., *Pole assignment with periodic output feedback: computer assisted study of a 2-dimensional linear system*, International Journal of Systems Science, 19, 12, 1988, 2661- 2672.

Feintuch A., P. Khargonekar, A. Tannenbaum, *On the sensistivity minimization problem for linear time-varying periodic systems*, SIAM journal on Control and Optimization, 1986.

Francis B.A. and T.T. Georgiu, *Stability theory for linear time invariant plants with periodic digital controllers*, IEEE Transactions on Automatic Control, $AC-33, 9, 1988, 820-832$.

Grasselli O.M. and F. Lampariello, *Dead-beat control of linear periodic discrete-time systems*, International Journal of Control, 33, 6, 1091-1106.

Grasselli O.M., *Dead-beat observers of reduced order for linear periodic discrete-time systems with inaccessible inputs*, International Journal of Control, 40, 4, 731-745, 1984.

Grasselli O.M. and S. Longhi, *Disturbance localization with dead-beat control for linear periodic discrete-time systems*, International Journal of Control, 44, 1986.

Grasselli O.M. and S. Longhi, *Linear function dead-beat observers with disturbance localization for linear periodic discrete-time systems*, International Journal of Control, 45, 1986.

Hamed M., Al- Rahmani and G.F. Franklin, *Linear periodic Systems: eigenvalue assignment using discrete periodic feedback*, IEEE Transactions on Automatic Control, 34, 1, 1989, 99-103.

Hernandez V. and A. Urbano, *Pole-assignment problem for discrete-time linear periodic systems*, International Journal of Control, 46, 2, 1987, 687-697.

Hernandez V. and A. Urbano, *Pole-placenment problem for discrete-time linear periodic systems*, International Journal of Control, 50, 1, 1989, 361-371.

Kabamba P.T., *Monodromy eigenvalue assignment in linear periodic systems*, IEEE Transactions on Automatic Control, $AC-31, 1986, 950-952$.

Kabamba P.T., *Control of linear systems using generalized sampled data hold functions*, IEEE Transactions on Automatic Control, $AC-24, 1987, 772-783$.

Kaczorek T., *Pole placement for linear discrete-time systems by periodic output-feedbacks*, Systems and Control Letters, 6, 1985.

Khargonekar P.P., K. Poola and A. Tannenbaum, *Robust control of time-inveraint plants using periodic compensation*, IEEE Transactions on Automatic Control, $AC-30, 11, 1088-1096, 1985$.

Kern G., *Linear closed-loop control in linear periodic systems with application to spin-stabilized bodies*, International Journal of Control, 31, 5, 1981, 905-916.

Willems J.L., V. Kucera and P. Brunovski, *On the assignment of invariant factors by time-varying strategies*, Systems and Control Letters, 5, 1984, 75-80.

## Periodic versus time-invariant and time varying controllers

Bailey J.E. and F.J.M. Horn, *Comparison between two sufficient conditions for improvement of an optimal steady state process by periodic operation*, Journal of Optimization Theory and Applications, 7, 5, 1071.

Bernstein D.S. and E.G. Gilbert, *Optimal periodic control: the Π test revisited*, IEEE Transactions on Automatic Control, $AC-25, 673-684, 1980$.

Bittanti S., G. Fronza and G. Guardabassi, *Periodic control: a frequency domain approach*, IEEE Transactions on Automatic Control, 18, 33-38, 1973.

Bittanti S., G. Fronza and G. Guardabassi, *Optimal steady state versus periodic operation in discrete systems*, Journal of Optimization Theory and Applications, 18, 4, 1976.

Chapellat H., M. Dahleh and S. P. Battacharrya, *Structure and optimality of multivariable periodic controllers*, University of California at Santa Barbara, Int. Report UCSB-ME-91-6, 1991.

Chapellat H. and M. Dahleh, *Analysis of time- varying control strategies for optimal disturbance rejection and robustness*, University of California at Santa Barbara, Int. Report UCSB-ME-91-9, 1991.

Colonius F., *The high frequency Pi-Criterion for retarded systems.* IEEE Transactions on Automatic Control, 11, 1045-1048, 1985.

Guardabassi G., *Optimal steady-state versus periodic control*, Ricerche di Automatica, 2, 240-252, 1971.

Guardabassi G., *The optimal periodic control problem*, Journal A, 17, 75-83, 1976.

Horn F.J.M. and R.C. Lin, *Periodic processes: a variational approach*, Industrial and Engineering Chemistry Process Design and Development, 6, 1967, 21-30.

Khargonekar P.P., K. Poola and A. Tannenbaum, *Robust control of time-invaraint plants using periodic compensation*, IEEE Transactions on Automatic Control, 30, 11, 1985, 1088-1096.

Watanabe N., Y. Nishimura and M. Matsubara, *Singular control test for optimal periodic control problems*, IEEE Transactions on Automatic Control, 1976, 609,610.

## Multirate Digital Control

Albertos P., V. Hernandez, J. Tornero and P. Morant, *Block multirate input output model for continuous time linear periodic systems.* IEEE Confernce on Decision and Control, 1988.

Araki M. and T. Hagiwara, *Pole assignment by multirate sampled-data output feedback,* International Journal of Control, 44, 1986, 1661-1673.

Al Rahmani H.M. and G.F. Franklin, *A new optimal multirate control of linear periodic systems*, IEEE Transactions on Automatic Control, 35, 1990, 406-415.

Berg M.C., N. Amit and J.D. Powell, *Multirate digital control system design*, IEEE Transactions on Automatic Control, 33, 1988, 1139-1150.

Colaneri P. and G. De Nicolao, *Optimal stochastic control of multirate sampled-data systems*, European Control Conference, Grenoble, 1991, 2519-2523.

Colaneri P., R. Scattolini and N. Schiavoni, *The LQG problem for multirate sampled-data systems*, 28th Conference on Decision and Control, Tampa, 1989, 469-474.

Colaneri P., R. Scattolini and N. Schiavoni, *LQG optimal control of multirate sampled-data systems*, IEEE Transactions on Automatic Control (to appear).

Colaneri P., R. Scattolini and N. Schiavoni, *Stabilization of multirate sampled data systems*, Automatica 26, 1990, 377-380.

Glasson D. P., *A new technique for multirate digital control design and sample rate selection*, Journal of Guidance, 5, 4, 1982, 379-382.

Hagiwara T. and M. Araki, *Design of stable state feedback controller based on multirate sampling of the plant output*, IEEE Transactions on Automatic Control, 33, 1988, 812- 819.

Lennartson B., *On the design of stochastic control systems with multirate sampling*, Chalmers University of Technology, School of Electrical and Computer Engineering, Tech. Rep. 161, 1986.

Jury E. J., and F.J. Mullin, *The analysis of sampled data control systems with a periodically time-varying sampling rate*, IRE Transactions on Automatic Control, 24, 1959, 15-21.

Meyer R.A. and C.S. Burrus, *A unified analysis of multirate and periodically time-varying digital filters*, IEEE Transactions on Circuits and Systems, 22, 1975, 162-167.

Ritchey V.S. and G.F. Franklin, *A stability criterion for asynchronous multirate linear systems*, IEEE Transactions on Automatic Control, 34, 1989, 529-535.

Serrano E.J. and P.J. Ramadge, *Sampled disturbace decoupling with stability using multirate control*, IEEE Transactions on Automatic Control, 36, 1991, 1061-1064.

Sergio Bittanti
*Politecnico di Milano*
*Dipartimento di Elettronica*
*Piazza Leonardo da Vinci 32*
*20133 Milano (Italy)*
*FAX ++39.2.23993587*

# Pulse Driven Dynamical Systems

R. W. Brockett

Dedicated to Professor Antonio Ruberti on the occasion of his $65^{th}$ birthday

**Abstract**

Implicit in much of the ongoing work in neural computing is the contention under certain circumstances it is useful to think about computer programs and/or computing hardware in terms of analog processing rather than insisting on a digital point of view. The arguments that have been advanced thus far are largely inconclusive, in part because there is no readily applicable formal basis for such studies. In this note we suggest a topological method for organizing the input/output analysis of dynamical systems in such a way as to facilitate comparisons of this type.

## 1 Introduction and Examples

Subtleties related to the interplay between "discrete" objects and "continuous" objects, lies at the heart of most difficulties encountered in applying digital computing to understanding or control of physical phenomena. In addition, an important, yet poorly understood, aspect of biological computation involves the interaction between "discrete" pulses (spikes) and the more slowly changing potentials and ionic concentrations found within the cell body [1]. Of course an interplay between the discrete and the continuous is central to an important branch of mathematics that has emerged in response to some difficult questions in geometry, i.e. algebraic topology. While there is a large literature on the latter there appears to be very little in the way of foundational work attempting to relate models for continuous computing with models for discrete computing. In this paper we suggest some ideas that appear to be useful in this way.

Let $C^k_m[0,\infty)$ denote the set of functions with $k$ continuous derivatives mapping $[0,\infty)$ to $\mathbf{R}^m$. Given $F \subset \mathbf{R}^m$ we let $A^k_m(F)$ denote the subset of $C^k_m[0,\infty)$ consisting of those functions that together with their first $k$

*This work was supported in part by the National Science Foundation under Engineering Research Center Program, NSF D CDR-8803012 and by the US Army Research Office under grant DAAL03-86-K-0171, and DARPA grant AFOSR-89-0506.

derivatives have a limit as $t$ goes to infinity with the limiting value of the function being in $F$. (Of course, the higher derivatives must approach zero.) An input/output system of the form

$$\dot{x}(t) = f(x(t), u(t)) \quad ; \quad x(0) = x_0 \quad ; \quad x(t) \in \mathbf{R}^n$$

will be said to have property $P_1$ if for each $u \in A_m^0(F)$ there exists exactly one solution to the differential equation with that solution approaching an equilibrium point as $t$ goes to infinity. If, in addition, it happens that for each $\alpha \in F$ there are either a finite or a countably infinite number of solutions of $f(x, \alpha) = 0$, say $x_1(\alpha), x_2(\alpha), \ldots$ then we say that the system has property $P_2$. Finally, if for each $x_i$ and each $\alpha \in F$ it follows that the Jacobian of $f(\cdot, \alpha)$ when evaluated at $(x_i(\alpha), \alpha)$ has eigenvalues whose real parts are not zero, we will say that the system has property $P_3$. Our goal is to investigate the way in which $\lim_{t\to\infty} x(t)$ depends on the past values of $u$ for systems having these properties.

Suppose that $f(\cdot,\cdot)$ is such that $\|f(x,u)\| \leq m(u) + k(u)\|x\|$ so that there can be no finite escape times and suppose that $\|f(x,u) - f(\bar{x},u)\| \leq k(u)\|x-\bar{x}\|$ so that there exists a unique solution. In this case the solution of the differential equation depends smoothly on the right-hand side provided that $f\colon \mathbf{R}^n \times \mathbf{R}^m \to \mathbf{R}^n$ is continuous.

Systems having these properties can be thought of as dividing $A_m^0(F)$ into a disjoint union $A_m^0(F) = \cup A_i$ with $A_i$ being the subset of $A_m^k(F)$ such that for $u \in A_m^0(F)$ we have

$$\lim_{t\to\infty} x(t) \to x_i(\lim_{t\to\infty} u(t))$$

We note, for future reference, that if the Jacobian $\partial f/\partial x$, evaluated at $x_i$, has eigenvalues with negative real parts and if $u \in A_m^0(F)$ drives the system to $x_i(u(\infty))$ then there exists $\epsilon > 0$ such that for $\sup_t |u(t) - \hat{u}(t)| < \epsilon$, $\hat{u}$ drives $x(t)$ to $x_i(\hat{u}(\infty))$.

**Example 1:** Let $F$ be an arbitrary subset of $\mathbf{R}^m$ and let $A$ be an $n$ by $n$ matrix with all its eigenvalues in the left half-plane. If $u \in A_m^0(F)$ then as $t$ goes to infinity all solutions of

$$\dot{x} = Ax + bu$$

approach a constant, $A^{-1}bu(\infty)$. Hence $x(\infty)$ depends only on $u(\infty)$ and not on any other aspect of the path defined by $u$.

**Example 2:** Let $F = [-a, a]$ with $|a| < 1/2$ and let $x$ satisfy the equation

$$\dot{x}(t) = -\sin(x(t)) + u(t) \quad ; \quad x(0) = 0$$

This equation serves to divide the space $A_1^0(F)$ into a countable union of sets $A_1^0(F) = \cup A_n$ according to the statement: $u \in A_n$ implies that $x(t)$ approaches $\pi n + \sin^{-1}(u(\infty))$. The following qualitative statements can be made about $A_n$. If $u(t)$ is never less than $-a$ and if it consists of a series of positive-going, narrow spikes whose areas are close to $2\pi$, and if the spikes are spaced far enough apart, then the formula

$$x(t) - x(\tau) = \int_{\tau}^{t} \sin(x(\sigma))d\sigma + \int_{\tau}^{t} u(\sigma)\, d\sigma$$

makes it clear that a particular spike will increase $x(t)$ by about $2\pi$. Thus $x(t)$ will always be near an integral multiple of $2\pi$. The equilibrium point corresponding to $-a \leq u(t) < a$ will be stable and, if the period between pulses is sufficiently long, $x(t)$ will relax enough between pulses to ensure that the effect of the next pulse will be to put $x(t)$ in the correct domain of attraction. Thus a $u$ with $n$ spikes will belong to $A_{2n}$. Such inputs will not drive $x(t)$ to $A_n$ for $n$ odd; these equilibria are unstable.

Similar statements apply to

$$\dot{x}(t) = u(t)\sin x(t) + a \quad ; \quad x(0) = 0$$

with $|a| < 1/2$ provided that $F$ is taken to be $F = \{x||x| > 1\}$. Again we can get a rough approximation to the $A_i$. The idea is that if $|\dot{u}(t)|$ is not too large and if $|u(t)|$ is greater than $a$ except for small periods of time then, for the most part, $x$ and $u$ will be related by $\sin x(t) \approx a/u(t)$. The solution $x$ will then count the number of times that the value of $u$ passes through zero. (See [4] for more details.)

These examples illustrate the fact that although it may be a difficult matter to characterize the sets $A_i$ concretely it is sometimes possible to find reasonably large open subsets within the $A_i$.

The next example exploits the idea of associating an "angle" with the input.

**Example 3:** Consider

$$\begin{aligned}
\dot{x}_1 &= x_2 \\
\dot{x}_2 &= -\alpha x_2 - \beta x_1 + u \\
\dot{x}_3 &= ((u - \alpha\dot{x} - \beta x)x - \dot{x}^2)/(\dot{x}^2 + x^2)
\end{aligned}$$

Where we limit $u$ to an annulus in phase space of the form $a \leq \dot{u}^2(t) + u^2(t) \leq 1$ with $a$ near one. In this case $x_3$ is the winding number of the curve $(x, \dot{x})$ relative to $(x_0, \dot{x}_0) = (0, 0)$. If $\beta$ is large and if the system has adequate damping this will be the same as the winding number of $(u, \dot{u})$ relative to $(0, 0)$.

**Example 4:** Say $(u, v) \in C^2[0, \infty)$ and consider

$$\begin{aligned}
\dot{x}_1(t) &= u_1(t) \\
\dot{x}_2(t) &= u_2(t) \\
\dot{x}_3(t) &= (x_1(t)u_2(t) - x_2(t)u(t))(x_1^2(t) + x_2^2(t))^{-1}
\end{aligned}$$

assuming that $x_1^2(t) + x_2^2(t) > 0$ for all time. In this case $x_3(t)/2\pi$ is the winding number with respect to the point $(0,0)$ of the curve $x_1(t) = \int_0^t u_1(\sigma)d\tau + x_1(0), x_2(t) = \int_0^t u_2(\tau d\tau + x_2(0)$. Notice that in this case the equilibrium solutions are not asymptotically stable. In a similar way we can consider

$$\dot{x} = (x^2 + y^2)u$$
$$\dot{y} = (x^2 + y^2)v$$
$$\dot{z} = xv - yu$$

If we ask that $(x(0), y(0))$ be nonzero then $x(t), y(t)$ will be nonzero for all time and $z$ is a winding number associated with a curve whose tangent is $(\dot{x}/(x^2 + y^2), \dot{y}/(x^2 + y^2))$.

As example 4 shows, computing discrete invariants of the input path based on closed, but not exact, one-forms, such as the differential of $\tan^{-1} \dot{u}/u$, has a major defect in that the equilibrium points of the differential equation will not be asymptotically stable.

# 2 Differential Inclusions

In defining the fundamental group of a connected topological space one considers the space of (continuous) closed curves that start and end at a particular point, say $x_0$. These curves are represented in parameterized form with a real parameter running from zero to one. This gives them an orientation and allows them to be concatenated using the rule $u_1 \circ u_2 = u_3$ with $u_3(t) = u_1(2t)$ for $0 \leq t \leq 1/2$ and $u_3(t) = u_2(2t-1)$ for $1/2 < t \leq 1$.

We consider a modification of this idea making a construction that associates a semigroup with certain set of equivalence classes of inputs in such a way as to facilitate the analysis of systems driven by pulse trains. Let $u \in C_1^1[0, \infty)$ and let $M$ be a connected open subset of $\mathbf{R}^2$. Suppose that for each $t \in [0, \infty)$ the point $(u(t), \dot{u}(t)) \in M$. If $u(t)$ and $\dot{u}(t)$ approach a limit as $t$ goes to infinity then the "lift" of $u$ to $(u, \dot{u})$ defines a parameterized curve in $\mathbf{R}^2$ which will be closed if $(u(0), \dot{u}(0)) = (u(\infty), \dot{u}(\infty))$. Not all paths in a given $M$ will be lifts of functions $u: [0, \infty) \rightarrow (-\infty, \infty)$. For example the tangent to the curve in $M$ that comes from lifting a $u$ in $C_1^1[0, \infty)$ must make an angle with the positive $u$-axis that is between $-\pi/2, \pi/2$ if $\dot{u}(t) > 0$ if $\dot{u}(t) > 0$ and an angle with the $u$-axis that is between $\pi/2$ and $3\pi/2$ and if $\dot{u}(t) < 0$. (If $\dot{u}(t) = 0$ then there is no condition.)

If $\gamma_1$ and $\gamma_2$ are two closed curves in $M$ we might wish to think of them as being homotopic if there exists a one parameter family of functions $u_\epsilon$ defined for $1 \leq \epsilon \leq 2$ and depending continuously on $\epsilon$ such that the lifts of functions $u_\epsilon: [0, \infty) \rightarrow (\infty, \infty)$ are in $M$ with $u_1$ lifting to $\gamma_1$ and $u_2$ lifting to $\gamma_2$. It is clear that this is a more demanding condition that asking that $\gamma_1$ and $\gamma_2$ be homotopic in the ordinary sense. Later on we will make a precise definition of this type.

The condition $(u(t), \dot{u}(t)) \in M$ can be thought of as imposing some $u(t)$-dependent constraint on $\dot{u}(t)$. For example, $M$ might be defined by a set of inequalities $\dot{u}(t) < \phi_i(u)(t) i = 1, 2, \ldots, r$. In this sense the requirement that $(u(t), \dot{u}(t)) \in M$ is a generalization of the idea of a differential equation. In the literature one sees such constraints referred to as **differential inclusions**.

The ideas just sketched can be thought of as a first order version of a more general theory in which the inclusion involves higher derivatives as well. We now argue that both mathematical and empirical evidence supports the idea that second order differential inclusions are the natural ones for use in this context. We consider the empirical evidence first. One of the goals of this study is to model and learn from biological phenomena. In particular we would like to understand the merits of a signaling strategy that uses spike trains for longer distance communication and uses levels for local computation. We want our treatment of signal constraints to be general enough to handle pulses in a flexible way. There are two reasons why the spike formation process is not adequately modeled by a first order differential inclusion. The first involves the need to force the spike to decay rapidly from its peak value. The second concerns the need to force the presence of a refractory period between spikes. Thus, although for $a$ slightly smaller than one, the inclusion

$$M = \{(u, \dot{u}) | a < (u(t) - 1)^2 + \dot{u}^2(t) < 1\}$$

insures that $u$ will take the form of a series of raised sine wave fragments interspersed with intervals on which $u$ is nearly constant, 2 or 0, e.g.

$$u(t) \approx \begin{cases} 1 - \cos t & ; \quad 0 \leq t \leq \pi \\ 2 & ; \quad \pi \leq t \leq \pi + a \\ 1 - \cos(t + \pi - a) & ; \quad \pi + a \leq t \leq\!< \pi + a + b \\ \ldots\ldots & ; \quad \ldots\ldots \end{cases}$$

it is not possible to use first order inclusions to force $u(t)$ to return promptly from a value near 2 to a value near zero, nor is it possible to use first order inclusion to force $u(t)$ to stay near zero for some "rest period" before firing again. In this respect first order inclusions are inadequate. A simple modification that meets the basic requirements is to define a subset of $(u(t), \dot{u}(t), \ddot{u}(t)) \in \mathbf{R}^3$, by

$$M = \{(u, \dot{u}, \ddot{u}) | \quad a < (u(t) - 1)^2 + \dot{u}^2(t) < 1: \quad \ddot{u}(t) < f(u)\}$$

By taking $f$ to be, for example, $f(u) = -u^2(u - 1)/4$ we force $u$ to be a sequence of pulses separated by rest period during which $u(t)$ is near zero. Because $f(2) \approx -1$ and $f(0) \approx 0$ this enforces a quick decay from $u(t) \approx 2$ and enforces a certain refractory period after each pulse.

# 3 Semigroups and Differential Inclusions

We now define a modified version of the fundamental group of a space. To do so it is necessary to assume from now on that $F$ is the disjoint union of a set of simply connected components $F = \cup F_i$. We will say that $u \in A_m^k(F)$ defines a **quasi-closed curve** if $u(0)$ and $u(\infty)$ belong to the same connected component of $F$. The motivation for this definition is that if $u$ is quasi-closed then there exists a closed curve whose lift agrees with that of $u$ in $M - F_i$ provided that $u(0)$ and $u(\infty)$ belong to $F_i$.

**Definition:** Given $M \subset (\mathbf{R}^m)^k$ we let $B_m^k(M, F_i)$ denote the subset of $A_m^k(F)$ consisting of quasi-closed curves whose lifts belong to $M$. We say that two elements of $B_m^k(M, F_i)$ are **$l$-homotopic** if there is a homotopy $u_\epsilon$ such that for all $t$ and $1 \le \epsilon \le 2$ $(u_\epsilon(t), u_\epsilon^{(1)}(t), \ldots, u_\epsilon^{(k)}(t)) \in M, u_\epsilon(0) \in F_i$ and $u_\epsilon(\infty) \in F_i$. We let $[u]_M^{F_i}$ denote the equivalence class of functions in $B(M, F_i)$ that contain $u$ with $u_1 \approx u_2$ if $u_1$ and $u_2$ are $l$-homotopic.

Define a multiplication on the space of equivalence classes of curves by selecting representative curves $u_1$ and $u_2$ such that $u_1(t)$ is constant for $t$ larger than some finite time, say $t_1$. (This is possible because $M$ is open.) Then we let

$$u_1 \circ u_2 = u_3$$

with

$$\begin{aligned} u_3(t) &= u_1(t) \quad ; \quad 0 \le t \le t_1 \\ u_3(t) &= u_2(t - t_1) \quad ; \quad t_1 \le t \le t_1 + t_2 \end{aligned}$$

It can be shown that this gives rise to a multiplication on the equivalence classes.

Let $M \subset (\mathbf{R}^m)^k$ be an open connected set. Define $\hat{\pi}(M, F_i)$ to be the set of equivalence classes of quasi-closed curves $u \in A_m^k(F)$ starting in $F_i$, ending in $F_i$ and having lifts $(u(t), u^{(1)}(t), \ldots, u^{(k)}(t)) \in M$. We give $\hat{\pi}(M, F_i)$ a semigroup structure defining the product of two equivalence classes as above. Call $\hat{\pi}(M, F_i)$ the **$l$-semigroup** of $B_m^k(M, F_i)$.

# 4 Finite Automata and Digital Systems

In automata theory one deals with the free monoid generated by a finite alphabet $\{a_1, a_2, \ldots, a_n\}$. This is the set of all finite strings, including the empty string, thought of as a semigroup with multiplication being concatenation.

**Definition:** If $\dot{x}(t) = f(x(t), u(t))$ has property $P_3$ and if $M$ is an open subset of $(\mathbf{R}^m)^k$ such that all inputs that lift to quasi-closed $l$-homotopic paths lead to the same steady state we will say that the system is **digital**. In this case we can think of the system as defining a map of $\hat{\pi}(M, F_i)$ to the steady state of $x_i$.

**Theorem:** If $u \in A_1^2(0,a)$ and if $M$ is

$$M = \{(u,\dot{u},\ddot{u}) | 1 - a^2 < (u-1)^2 + \dot{u}^2 < 1 : \ddot{u} < u^2(u-1)/4\}$$

then for $a$ sufficiency small the system

$$\dot{x}(t) = -2a \sin x(t) + u(t) \quad ; \quad x(0) = 0$$

Defines a homomorphism from $\hat{\pi}_1(M,(0,a))$ to $\mathbf{Z}^+$ given by

$$x(\infty) = (\text{winding number of}\, u) 2\pi + \sin^{-1}(u(\infty))$$

# References


[1] Eric Kendel and James H. Schwartz, *Principles of Neural Science*, Elsevier, New York, 1985.

[2] A. F. Murray et al., "Pulsed Silicon Neural Networks-Following the Biological Leader", in *VLSI Design of Neural Networks*, U. Ramacher, U. Ruckert, Eds., Kluwer Academic Publisher, Boston, 1991.

[3] J. R. Munkres, *Topology*, Prentice Hall, Engelwood Cliffs, N.J., 1975

[4] R.W. Brockett, *Smooth Dynamical Systems Which Realize Arithmetical and Logical Operations*, in *Lecture Notes in Control and Information Sciences. Three Decades of Mathematical Systems Theory.* (H. Nijmeijer and J. M. Schumacher, eds.) Springer-Verlag, Berlin, 1989, pp. 19-30.

[5] R. W. Brockett, "On the Asymptotic Properties of Solutions of Differential Equations with Multiple Equilibria," *Journal of Differential Equations*, Vol. 18 (1982) pp. 249-262.

# Polynomial MMSE Deconvolution and its Duality with LQGR

L. CHISCI and E. MOSCA

**Abstract.** The problem of linear *minimum mean-square error* (MMSE) multichannel deconvolution of sampled signals from noisy observations is approached via matrix polynomial equations. The general solution is given in terms of a left spectral factorization and a pair of bilateral Diophantine equations. The first Diophantine equation is obtained by imposing optimality of the deconvolution filter whereas the second ensures stability of the filter, should the signal model be unstable. The proposed solution encompasses classical Wiener as well as stationary Kalman filtering, prediction and fixed-lag smoothing. Its duality with the polynomial equations for LQG regulation (LQGR) is discussed.

## 1. Introduction

Deconvolution consists of estimating the input of a linear system given an observed output corrupted by noise. This problem has a variety of applications in many fields of engineering including seismology, digital communication, radio astronomy, underwater acoustics and control.

This paper addresses the problem of linear *minimum mean-square error (MMSE)*, also referred to as *minimum variance*, multichannel deconvolution in a stochastic stationary framework providing a polynomial-equation procedure for the design of the optimal deconvolution filter. It is assumed throughout the paper that all the signals are stationary and all the models - the system model, the noise model and the input signal model - are linear time-invariant and known. Even if these assumptions are seldom fulfilled in applications ( blind and adaptive deconvolution are actually more realistic ways to look at the input estimation problem), the off-line filter design procedure proposed in this paper is nevertheless relevant both as a basic reference for real deconvolution problems and as a starting point for possible adaptive implementations.

A polynomial approach to MMSE deconvolution has been recently reported for the *single-channel* (scalar) case [1,2]. This paper extends these results to the *multichannel* case. The multichannel extension accounts for possible multiple inputs to be estimated and/or multiple outputs to be observed, and therefore broadens the range of applications of the deconvolution problem. Further, as pointed out in [3, Sect. 11.2], multichannel deconvolution may provide improvements with respect to single-channel deconvolution whenever multiple sensors are available and the respective measurements are cross-correlated. In such a case, instead of averaging the available measurements into a single measurement, one can fully exploit the available information by solving a multichannel problem.

This paper provides, at the best of the authors' knowledge, the first complete polynomial solution to the MMSE multichannel deconvolution problem. The optimal deconvolution filter is obtained by first solving a matrix spectral factorization problem, and next finding the minimum degree solution with respect to a "dummy" polynomial matrix of a pair of bilateral Diophantine equations. The first Diophantine equation is obtained by imposing optimality of the deconvolution filter, whereas the second ensures stability of the filter, should the signal model be unstable. MMSE multichannel deconvolution can be alternatively solved in a state-space form via a Riccati-equation procedure [3] without the need of the stationarity assumption, or, in the stationary case, with the polynomial-equation procedure given in [4]. However, to cast the deconvolution problem in a state-space form, a suitable state augmentation is required so as to express the unknown input signal as a linear combination of state components. In contrast with the polynomial approach of this paper, the state augmentation technique may considerably increase the design computational complexity, expecially for high order system, signal and noise models.

The results of this paper cover Wiener (output) as well as stationary Kalman (state) filtering, prediction and smoothing problems. A polynomial solution to the discrete-time multichannel Wiener problem, based on two Diophantine equations, has been presented in [5] both for the filtering and the prediction cases. In [6] both the discrete-time and continuous-time Wiener filters are considered but the solution given therein uses a single Diophantine equation, being restricted to stable signal models. The polynomial approach to Kalman filtering was addressed for continuous-time systems in [7]. This reference provides a polynomial procedure for computing the stationary Kalman gain which is valid in the special case of a completely observable system. Another relevant contribution is [5] where, though the general discrete-time case is addressed, the Kalman gain is given again under the above observability assumption. Furthermore, [5] shows how to compute the transfer-matrix of the Kalman filter and the $k$-step predictor by polynomial equations. A direct polynomial solution for the Kalman gain which no longer requires the above observability assumption is given in [4], where the connection with the classical Riccati equation approach is also discussed.

The outline of the paper is as follows. In Sect. 2, after the introduction of the notation and terminology used throughout the paper, the problem formulation is stated. In Sect. 3, the optimal MMSE deconvolution filter is derived by a polynomial equation approach. In Sect. 4, the main results on the solvability of the problem are stated. Next, Sect. 5 briefly explores the duality with *linear quadratic Gaussian regulation (LQGR)*. Finally, Sect. 6 ends the paper.

## 2. Notation and problem formulation

*A. Notation and terminology*

The reader is referred to [8] for the basic facts on the theory of polynomial equations and its use in the context of linear quadratic optimization problems. The notation and terminology used here, which adheres as much as possible to that of [8], is summarized hereafter.

- For a matrix $M$, $M'$ denotes its transpose, and $tr\{M\}$ its trace;
- $E[\ ]$ denotes expectation;
- $\delta_{ij} = \begin{cases} 0, & \text{for } i \neq j \\ 1, & \text{for } i = j \end{cases}$ denotes the Kronecker delta function;
- $a \vee b$ denotes the maximum between $a$ and $b$;
- the indeterminate $d$ denotes the *unit delay operator*, i.e. $ds(t) \triangleq s(t-1)$;
- $s^t \triangleq \{s(j) : j \leq t\}$ denotes the set of all $s$-observations up to time $t$.
- $\Psi_s(d) \triangleq \sum_{j=-\infty}^{\infty} E[s(t+j)s'(t)]\, d^j$ denotes the power spectrum of the stationary sequence $\{s(t)\}$;
- Given a sequence matrix $S(d) = \sum_{j=-\infty}^{\infty} S_j d^j$:
  - $S^*(d) \triangleq S'(d^{-1})$ denotes its conjugate;
  - $S^{-*}(d) \triangleq (S^*(d))^{-1}$;
  - $\prec S(d) \succ \triangleq S_0$ denotes extraction of the zero-power coefficient from $S(d)$;
  - $[S(d)]_+ \triangleq \sum_{j=0}^{\infty} S_j d^j$ denotes extraction of the *causal* part from $S(d)$;
  - $tr\{\prec S(d)S^*(d) \succ\}$ is the squared $H_2$-norm of $S(d)$;
  - *ord* $S(d)$ denotes the minimum $d$-power present in $S(d)$;
  - $S(d)$ is causal (anticausal) if *ord* $S(d) \geq (\leq) 0$;

- $S(d)$ is strictly causal (anticausal) if *ord* $S(d) > (<)0$.

- Given a polynomial matrix $P(d) = \sum_{j=0}^{\partial P} P_j d^j$:
  - $\partial P$ denotes its degree;
  - $P(d)$ is *(strictly) Hurwitz* if it is square and *det* $P(d)$ has no zeros (on and) inside the unit circle;
  - $P(d)$ is *regular* if it is square and its leading matrix coefficient $P_{\partial P}$ is non-singular;
  - $P(d)$ is *causal* if it is square and its zero-power matrix coefficient $P_0$ is non-singular.

- Given a rational (transfer-) matrix $R(d)$:
  - polynomial matrices $N_R(d)$ and $D_R(d)$, respectively $D_L(d)$ and $N_L(d)$, such that $R(d) = N_R(d)D_R^{-1}(d)$, respectively $R(d) = D_L^{-1}(d)N_L^{-1}(d)$, denote a *right*, respectively *left*, *matrix-fraction-description (mfd)* of $R(d)$;
  - $R(d)$ is causal if the denominator of any irreducible left or right mfd of $R(d)$ is a causal polynomial matrix;
  - $R(d)$ is (externally, BIBO) stable if the denominator of any irreducible left or right mfd of $R(d)$ is a strictly Hurwitz polynomial matrix.

*B. Problem formulation*

A linear time-invariant discrete-time stochastic system is considered

$$\begin{cases} x(t) = \mathcal{H}_{x|w,u}(d) \begin{bmatrix} w(t) \\ u(t) \end{bmatrix} = \mathcal{H}_{x|w}(d)\, w(t) + \mathcal{H}_{x|u}(d)\, u(t) \\ y(t) = \mathcal{H}_{y|x,v}(d) \begin{bmatrix} x(t) \\ v(t) \end{bmatrix} = \mathcal{H}_{y|x}(d)\, x(t) + \mathcal{H}_{y|v}(d)\, v(t) \end{cases} \tag{1}$$

where: $x(t)$ is the unknown signal to be estimated, $y(t)$ is the observed output, $u(t)$ is a known signal, $\{w(t)\}$ and $\{v(t)\}$ are w.s. stationary random sequences with zero-mean and covariances

$$E\left\{ \begin{bmatrix} w(t) \\ v(t) \end{bmatrix} \begin{bmatrix} w'(\tau) & v'(\tau) \end{bmatrix} \right\} = \begin{bmatrix} \Psi_w & 0 \\ 0 & \Psi_v \end{bmatrix} \delta_{t\tau} . \tag{2}$$

Here $x(t)$, $y(t)$, $u(t)$ , $w(t)$ and $v(t)$ denote *multichannel (vector)* signals of dimensions $n_x$, $n_y$, $n_u$, $n_w$ and, respectively, $n_v$, while $\mathcal{H}_{y|x}(d)$, $\mathcal{H}_{y|v}(d)$, $\mathcal{H}_{x|w}(d)$, $\mathcal{H}_{x|u}(d)$ denote rational transfer-matrices of compatible dimensions.

In addition, the following preliminary assumptions are made

**A.1)** $\Psi_w > 0$ and $\Psi_v > 0$;

**A.2)** $\{y(t)\}$ is a full-rank process, viz. $E[y(t)y'(t)] > 0$;

**A.3)** Any unstable root of the characteristic polynomial associated with a minimal state-space representation of $\mathcal{H}_{y|w,v,u}$ is also an unstable root of the characteristic polynomial associated with a minimal state-space representation of $\mathcal{H}_{y|u}$ with the same multiplicity.

Assumption A.3) implies that there exists a stabilizing dynamic output feedback $u(t) = -\mathcal{K}(d)y(t)$. From now on, it will be understood that such a stabilizing component is used, whenever $\mathcal{H}_{x|w}$ and $\mathcal{H}_{y|wv}$ are unstable, in order to make the relevant signals stationary. Nevertheless, in the sequel $u(t)$ will be set equal to zero without loss of generality in that it will be assumed that the effect of $u(t)$ on $x(t)$ is properly taken into account in a separate way. Stationary Kalman filtering is a relevant example of the above setup. In fact, in such a case

$$\begin{cases} x(t+1) &= Fx(t) + Gw(t) + G_u u(t) \\ y(t) &= Hx(t) + v(t) \end{cases}$$

Here, if $(F, G_u)$ and $(F, H)$ are stabilizable and, respectively, detectable pairs, a transfer-matrix $\mathcal{K}(d)$ can be chosen in such a way that $u(t) = -\mathcal{K}(d)y(t)$ stabilizes the system.

The MMSE deconvolution problem consists of finding, whenever it exists, a linear time-invariant stable estimator of $x(t)$ from $y^{t-k}$, viz.

$$\hat{x}(t|t-k) = \mathcal{H}^{(k)}_{\hat{x}|y}(d)\, y(t-k)\,, \qquad \mathcal{H}^{(k)}_{\hat{x}|y} \text{ stable} \tag{3}$$

which makes *finite* and *minimum* the mean-square error

$$J \triangleq tr\{E[\tilde{x}(t|t-k)\,\tilde{x}'(t|t-k)]\}\ . \tag{4}$$

where $\tilde{x}(t|t-k) \triangleq x(t) - \hat{x}(t|t-k)$ denotes the estimation error. According to whether $k < 0$, $k = 0$ or $k > 0$ we have a (fixed-lag) *smoothed*, *filtered* or *predictive* deconvolution problem, respectively.

## 3. Polynomial solution

We now proceed to the derivation of the polynomial equations which are needed to solve the above-stated problem. By the sake of simplicity, we shall drop the argument $d$ unless required to avoid possible confusion.

### *A. System matrix-fraction-descriptions*

Let us introduce the following mfd's

$$\begin{array}{rcl} [\mathcal{H}_{y|x} \;\; \mathcal{H}_{y|v}] & = & D^{-1}[B \;\; C_v] \\ \mathcal{H}_{x|w} & = & A^{-1}C_w \end{array} \tag{5}$$

where $A, B, C_v, C_w$ and $D$ are polynomial matrices, in $d$, of compatible dimensions. Then the system (1), with $u(t) \equiv 0$, can be rewritten as

$$\left\{ \begin{array}{rcl} D\,y(t) & = & B\,x(t) + \zeta(t) \\ A\,x(t) & = & \xi(t) \end{array} \right. \tag{6}$$

where $\{\xi(t)\}$ and $\{\zeta(t)\}$ are stationary sequences defined by

$$\begin{array}{rcl} \xi(t) & = & C_w\, w(t) \\ \zeta(t) & = & C_v\, v(t) \end{array} \tag{7}$$

with power spectra given by

$$\begin{array}{rcl} \Psi_\xi & = & C_w \Psi_w C_w^* \\ \Psi_\zeta & = & C_v \Psi_v C_v^* \,. \end{array} \tag{8}$$

*B. Cost-function expansion*

Using the above definitions, the estimation error $\tilde{x}(t|t-k)$ becomes

$$\tilde{x}(t|t-k) = \mathcal{H}_{\tilde{x}|\xi}\,\xi(t) + \mathcal{H}_{\tilde{x}|\zeta}\,\zeta(t) \tag{9}$$

where $\mathcal{H}_{\tilde{x}|\xi}$ and $\mathcal{H}_{\tilde{x}|\zeta}$ are the following transfer-matrices

$$\begin{array}{rcl} \mathcal{H}_{\tilde{x}|\xi} & \triangleq & (I - d^k \mathcal{H}^{(k)}_{\hat{x}|y} D^{-1} B) A^{-1} \\ \mathcal{H}_{\tilde{x}|\zeta} & \triangleq & -d^k\, \mathcal{H}^{(k)}_{\hat{x}|y} D^{-1} \end{array} \tag{10}$$

Thus, the quadratic cost $J$ in (4) can be conveniently expressed as:

$$J = tr\left\{ \prec \Psi_{\tilde{x}}(d) \succ \right\} = tr\left\{ \prec \mathcal{H}_{\tilde{x}|\xi} \Psi_\xi \mathcal{H}^*_{\tilde{x}|\xi} + \mathcal{H}_{\tilde{x}|\zeta} \Psi_\zeta \mathcal{H}^*_{\tilde{x}|\zeta} \succ \right\} \tag{11}$$

where we have used the fact that $w(t)$ and $v(t)$, and, therefore, $\xi(t)$ and $\zeta(t)$ are mutually uncorrelated. Define a left-coprime mfd $A_1^{-1}B_1$ of $BA^{-1}$, viz.

$$A_1^{-1}B_1 = BA^{-1} \tag{12}$$

Then, substituting (10) into (11) and taking into account (12), one gets

$$\begin{array}{rcl} J & = & tr\Big\{ \prec (A^{-1}\mathcal{H}^{(k)}_{\hat{x}|y} D^{-1} A_1^{-1})\; (A_1 \Psi_\zeta A_1^* + B_1 \Psi_\xi B_1^*)\; (A^{-1}\mathcal{H}^{(k)}_{\hat{x}|y} D^{-1} \\ & & A_1^{-1})^* \; - \; (A^{-1}\mathcal{H}^{(k)}_{\hat{x}|y} D^{-1} A_1^{-1})\; (d^{-k} A^{-1} \Psi_\xi B_1^*)^* \; - \\ & & (d^{-k} A^{-1} \Psi_\xi B_1^*)\; (A^{-1}\mathcal{H}^{(k)}_{\hat{x}|y} D^{-1} A_1^{-1})^* \; + \; A^{-1}\Psi_\xi A^{-*} \succ \Big\} \end{array} \tag{13}$$

Let $E(d)$ be a Hurwitz polynomial matrix solving the following *left spectral factorization* problem

$$\boxed{EE^* \;=\; A_1 C_v \Psi_v C_v^* A_1^* \;+\; B_1 C_w \Psi_w C_w^* B_1^*} \tag{14}$$

Recall from [8] that such a matrix $E$ exists if and only if

$$rank\,[A_1 C_v \quad B_1 C_w] \;=\; n_y$$

Then substituting (14) in (13) and completing the squares, one finds

$$J \;=\; J_1 + J_2 \tag{15}$$

with

$$J_1 \;\triangleq\; tr\{\prec LL^* \succ\} \tag{16}$$

$$J_2 \;\triangleq\; tr\{\prec A^{-1}\Psi_\xi (I - B_1^* E^{-*} E^{-1} B_1 \Psi_\xi) A^{-*} \succ\} \tag{17}$$

$$L \;\triangleq\; \mathcal{H}^{(k)}_{\hat{x}|y} D^{-1} A_1^{-1} E - d^{-k} A^{-1} C_w \Psi_w C_w^* B_1^* E^{-*} \tag{18}$$

*C. Optimization*

Since $J_2$ in (17) does not depend on the choice of $\mathcal{H}^{(k)}_{\hat{x}|y}$, the problem amounts to finding a causal stable transfer-matrix $\mathcal{H}^{(k)}_{\hat{x}|y}$ minimizing $J_1$. Note that $J_1$ equals the square of the $H_2$-norm of the $n_x \times n_y$ sequence matrix $L(d)$ in (18). In turn (18) has two additive terms. The first term $L_1 \triangleq \mathcal{H}^{(k)}_{\hat{x}|y} D^{-1} A_1^{-1} E$ is a causal sequence matrix, whereas the second one $L_2 \triangleq d^{-k} A^{-1} C_w \Psi_w C_w^* B_1^* E^{-*}$ is a two-sided sequence matrix, viz. the cascade of the anticausal mfd $(E^{-1} B_1 C_w)^*$ with the causal mfd $A^{-1} C_w \Psi_w$ plus the $k$-steps shift $d^{-k}$. Then, in order to find the optimal causal filter $\mathcal{H}^{(k)}_{\hat{x}|y}$, we try to additively decompose $L(d)$ in terms of a causal sequence matrix $L_+(d)$ plus a strictly anticausal sequence matrix $L_-(d)$ of possibly finite $H_2$-norm, viz.,

$$L(d) \;=\; L_+(d) \;+\; L_-(d) \tag{19}$$

Since, from causality of $L_+$ and strict anticausality of $L_-$, $tr\{\prec L_-(d) L_+^*(d) \succ\} = tr\{\prec L_+(d) L_-^*(d) \succ\} = 0$ it follows that

$$J_1 \;=\; tr\{\prec L_+(d) L_+^*(d) \succ\} \;+\; tr\{\prec L_-(d) L_-^*(d) \succ\} \tag{20}$$

where the second term is bounded provided that $L_-^*(d)$ is analytic in $|d| < 1$.

In order to carry out the causal-anticausal decomposition (19), we proceed as follows. First, it is convenient to re-express $L(d)$ solely in terms of the indeterminate $d$. To this end, let $p$ be any positive integer such that

$$p \;\geq\; \begin{cases} max\,\{\partial E, \partial(B_1 C_w) + k\}\,, & k > 0 \\ max\,\{\partial E, \partial(B_1 C_w)\}\,, & k \leq 0 \end{cases} \tag{21}$$

and define

$$\overline{P}(d) \triangleq d^p P^*(d) \tag{22}$$

Consequently,

$$L_2 = A^{-1} C_w \Psi_w \overline{d^k B_1 C_w} \, \overline{E}^{-1} . \tag{23}$$

where, by the choice of $p$ in (21), $\overline{E}$ and $\overline{d^k B_1 C_w}$ are both polynomial matrices in $d$. Next consider a solution, provided it exists, of the following *bilateral* Diophantine equation [8]

$$Y\overline{E} + AZ = C_w \Psi_w \overline{d^k B_1 C_w} \tag{24}$$

with $\partial Z < \partial \overline{E}$. Using (24) in (23), we get for $L$ the desired decomposition (19) with

$$L_+ = \mathcal{H}^{(k)}_{\hat{x}|y} D^{-1} A_1^{-1} E - A^{-1} Y \tag{25}$$

$$L_- = - Z\overline{E}^{-1} \tag{26}$$

While causality of $L_+$ follows at once from causality of $L_1$ and $A^{-1}$, strict anticausality of $L_-$ is proved by showing that $L_-^*$ is strictly causal. In fact, since $E$ is Hurwitz and $\partial Z < p = \partial \overline{E}$, $L_-^* = E^{-1} d^p Z^*$ is a strictly causal mfd. In addition, provided that $E$ is strictly Hurwitz, $L_-^*$ is analytic in $|d| < 1$, i.e., $L_-$ has finite $H_2$-norm. Thanks to the above decomposition one finds

$$J_1 = J_3 + J_4 \tag{27}$$

where

$$J_3 \triangleq tr \left\{ \prec Z\overline{E}^{-1} \overline{E}^{-*} Z^* \succ \right\}$$

does not depend on $\mathcal{H}^{(k)}_{\hat{x}|y}$, and

$$J_4 \triangleq tr \left\{ \prec L_+ L_+^* \succ \right\} \tag{28}$$

Hence, minimizing $J$ amounts to choosing $\mathcal{H}^{(k)}_{\hat{x}|y}$ so as to minimize (28). Thus, setting $L_+ = 0$, from (25) one obtains the optimal transfer-matrix

$$\mathcal{H}^{(k)}_{\hat{x}|y} = A^{-1} Y E^{-1} A_1 D \tag{29}$$

By simple algebraic manipulations, the corresponding MMSE turns out to be

$$J_{min} = J_2 + J_3 = tr \left\{ \prec A^{-1} \left( C_w \Psi_w C_w^* - YY^* \right) A^{-*} \succ \right\} \tag{30}$$

Provided that (24) has a unique solution $(Y, Z)$ with $\partial Z < \partial \overline{E}$, (29) uniquely specifies the optimal MMSE deconvolution filter. However this is not the case, in general, whenever $A$ is non-Hurwitz. According to what done in [8] for the dual case of LQGR, a second Diophantine equation, to be satisfied along with (24), must be deduced from stability requirements. In fact, it remains to be ensured that $J_{min}$ in (30) is finite, i.e. that the estimation error $\tilde{x}(t|t-k)$ is mean-square bounded, and that the deconvolution filter $\mathcal{H}^{(k)}_{\hat{x}|y}$ is stable.

Unlike the previous derivation of the Diophantine equation (24) based on optimality arguments, to derive the second Diophantine equation we shall proceed in a separate way for the prediction ($k > 0$) and the filtering/smoothing ($k \leq 0$) cases. The reason for this distinction is that, while in the prediction case mean-square boundedness of the estimation error can be reduced to the stability of a suitable feedback system, this does not apply to the filtering/smoothing case. Hereafter, we shall first deal with stability of the MMSE predictor using standard results on stability of feedback systems in a way that is completely analogous to the one used for the dual LQGR. This will enlighten the duality between MMSE prediction and LQGR. Subsequently, we shall address stability of the MMSE filter/smoother using more general arguments which could of course be used also in the prediction case.

*D. Stability: prediction case*

In order to derive the second Diophantine equation in the case $k > 0$, we can proceed as follows. Let us introduce the prediction error of $z(t) \triangleq D\, y(t)$ given $y^{t-k}$ as

$$\tilde{z}(t|t-k) \triangleq z(t) - B\, \hat{x}(t|t-k) \tag{31}$$

Taking into account (31) and (3), one finds

$$\mathcal{H}^{(k)}_{\hat{x}|y} = \mathcal{H}^{(k)}_{\hat{x}|\tilde{z}}(I + d^k B \mathcal{H}^{(k)}_{\hat{x}|\tilde{z}})^{-1} D \tag{32}$$

where the transfer-matrix $\mathcal{H}^{(k)}_{\hat{x}|\tilde{z}}$ is implicitly defined by

$$\hat{x}(t|t-k) = d^k \mathcal{H}^{(k)}_{\hat{x}|\tilde{z}}\, \tilde{z}(t|t-k) \tag{33}$$

Thus, the problem amounts to finding $\mathcal{H}^{(k)}_{\hat{x}|\tilde{z}}$ so as to minimize (28). A convenient parametrization for $\mathcal{H}^{(k)}_{\hat{x}|\tilde{z}}$ is as follows

$$\mathcal{H}^{(k)}_{\hat{x}|\tilde{z}} = A^{-1} N_2 M_2^{-1} \tag{34}$$

where $N_2$ and $M_2$ are rational matrices. With the above setup, the estimation error $\tilde{x}(t|t-k)$ can be regarded as being generated by the following feedback system:

$$\begin{cases} \tilde{x}(t|t-k) = A^{-1}\left(\xi(t) - d^k N_2 M_2^{-1}\, \tilde{z}(t|t-k)\right) \\ \tilde{z}(t|t-k) = B\, \tilde{x}(t|t-k) + \zeta(t) \end{cases} \tag{35}$$

with plant $\mathcal{P} = BA^{-1}$ and compensator $\mathcal{K} = d^k N_2 M_2^{-1}$. Note that *ord* $\mathcal{P}(d)\mathcal{K}(d) \geq k > 0$ implies *well-posedness* of the feedback system (35). Provided that $\mathcal{P}$ and $\mathcal{K}$ are free of unstable hidden modes, mean-square boundedness of $\tilde{x}(t|t-k)$ is, therefore, equivalent to external stability of the above feedback system. This, in turn, is fulfilled if and only if $N_2$ and $M_2$ are stable rational matrices and satisfy the Diophantine equation

$$A_1 M_2 \; + \; d^k B_1 N_2 \; = \; I \tag{36}$$

Using (29) and (34) in (32), and taking into account (36), one finds $N_2 = YE^{-1}$. Next, by (36)

$$\begin{aligned} M_2 &= A_1^{-1}\left(I - d^k B_1 Y E^{-1}\right) = A_1^{-1}\left(E\overline{E} - d^k B_1 Y \overline{E}\right)\overline{E}^{-1} E^{-1} \\ &= \left(C_v \Psi_v \overline{A_1 C_v} + d^k BZ\right)\overline{E}^{-1} E^{-1} \end{aligned} \tag{37}$$

where we have used (24), (14) and (12). In order to make $\overline{A_1 C_v}$ a polynomial matrix in $d$, the integer $p$ must also satisfy, in addition to (21), the inequality

$$p \geq \partial(A_1 C_v) \tag{38}$$

For mean-square boundedness, $M_2$ must be stable from which, by (37), $(C_v \Psi_v \overline{A_1 C_v} + d^k BZ)\overline{E}^{-1}$ must be stable. Thus, the anti-Hurwitz polynomial matrix $\overline{E}$ has to be a *right-divisor* of $C_v \Psi_v \overline{A_1 C_v} + d^k BZ$, i.e. the following Diophantine equation must also be satisfied

$$X\overline{E} \; - \; d^k BZ \; = \; C_v \Psi_v \overline{A_1 C_v} \tag{39}$$

for some polynomial matrix $X$. Premultiplying (24) by $d^k B_1$ and (39) by $A_1$, summing the two resulting equations and taking into account (12) and (14), the following *unilateral* Diophantine equation [8] is also derived

$$A_1 X \; + \; d^k B_1 Y \; = \; E \tag{40}$$

Equation (39) or, equivalently, (40) is a necessary requirement for mean-square boundedness. In turn, mean-square boundedness is fulfilled if $E$ is strictly Hurwitz. Note that mean-square boundedness implies stability of $\mathcal{H}_{\tilde{x}|\zeta}$ which, in turn, implies, by (10), stability of the deconvolution filter $\mathcal{H}^{(k)}_{\hat{x}|y}$.

*E. Stability: filtering/smoothing case*

Assuming $k \leq 0$ the feedback system under consideration is, in general, not well-posed. Thus, in such a case the above treatment no longer applies. Hence we proceed as follows. For $k = -\ell \leq 0$ the estimation error of $x(t)$ given $y^{t+\ell}$, is more conveniently rewritten as

$$\tilde{x}(t|t+\ell) = \widetilde{\mathcal{H}}_{\tilde{x}|\xi}\, \xi(t+\ell) \; + \; \widetilde{\mathcal{H}}_{\tilde{x}|\zeta}\, \zeta(t+\ell) \tag{41}$$

where $\widetilde{\mathcal{H}}_{\tilde{x}|\xi}$ and $\widetilde{\mathcal{H}}_{\tilde{x}|\zeta}$ are the following causal transfer-matrices.

$$\begin{aligned}\widetilde{\mathcal{H}}_{\tilde{x}|\xi} &\triangleq A^{-1}(d^{\ell}I - YE^{-1}B_1)\\ \widetilde{\mathcal{H}}_{\tilde{x}|\zeta} &\triangleq -A^{-1}YE^{-1}A_1\end{aligned} \tag{42}$$

Accordingly

$$\tilde{z}(t|t+\ell) \triangleq B\,\tilde{x}(t|t+\ell) + \zeta(t) = [\mathcal{H}_{\tilde{z}|\zeta} \;\; \mathcal{H}_{\tilde{z}|\xi}]\begin{bmatrix}\zeta(t+\ell)\\ \xi(t+\ell)\end{bmatrix} \tag{43}$$

with

$$\begin{aligned}[\mathcal{H}_{\tilde{z}|\zeta} \;\; \mathcal{H}_{\tilde{z}|\xi}] &= [d^{\ell}I - BA^{-1}YE^{-1}A_1 \quad BA^{-1}(d^{\ell}I - YE^{-1}B_1)]\\ &= A_1^{-1}(d^{\ell}I - B_1YE^{-1})\,[A_1 \;\; B_1]\\ &= A_1^{-1}(d^{\ell}E\overline{E} - B_1Y\overline{E})(E\overline{E})^{-1}\,[A_1 \;\; B_1]\\ &= (d^{\ell}C_v\Psi_v\overline{A_1C_v} + BZ)(E\overline{E})^{-1}\,[A_1 \;\; B_1]\end{aligned} \tag{44}$$

where we have used (41), (42), (43), (12), (14) and (24). Note that mean-square boundedness of $\tilde{x}(t|t+\ell)$ implies mean-square boundedness of $\tilde{z}(t|t+\ell)$ in (43) which, in turn, implies stability of the transfer-matrix in (44). Next, $A_1$ and $B_1$ being left-coprime, $E\overline{E}$ and $[A_1 \; B_1]$ must also be left-coprime. Hence, the anti-Hurwitz factor $\overline{E}$ can only be cancelled out by the polynomial matrix $(d^{\ell}C_v\Psi_v\overline{A_1C_v}+BZ)$, i.e., the following Diophantine equation must hold

$$X\overline{E} - BZ = d^{\ell}C_v\Psi_v\overline{A_1C_v} \tag{45}$$

for some polynomial matrix $X$. Proceeding in the same way as to derive (40) from (39), from (45) one also gets

$$A_1X + B_1Y = d^{\ell}E \tag{46}$$

Equation (45) or, equivalently, (46) is the required necessary condition for mean-square boundedness in the filtering/smoothing case. A sufficient condition, whose proof is given in [9], follows hereafter.

**Lemma 1.** *Provided that*
**i)** *The* greatest common right divisors (GCRD) *of $A$ and $B$ are strictly Hurwitz,*
**ii)** *$E$ is strictly Hurwitz,*
*the estimation error $\tilde{x}(t|t+\ell)$ is mean-square bounded.*

## 4. Main Results

For reference comodity, we rewrite the relevant Diophantine equations in a unified form which holds for both the prediction and smoothing/filtering cases:

$$Y\overline{E} + AZ = C_w\Psi_w \overline{d^k B_1 C_w} \quad (47)$$
$$X\overline{E} - d^{k\vee 0}BZ = d^{(-k)\vee 0}C_v\Psi_v\overline{A_1 C_v} \quad (48)$$
$$A_1 X + d^{k\vee 0}B_1 Y = d^{(-k)\vee 0}E \quad (49)$$

where barred polynomial matrices are defined according to (22) and

$$p \triangleq \max\{\partial E, \partial(A_1 C_v), \partial(B_1 C_w) + (k \vee 0)\}$$

The following theorem sums up the above and some additional results proved in [9].

**Theorem 1.** *Consider the MMSE deconvolution problem* (1) – (4). *Then,*

1. *The problem is solvable if and only if the GCRD's of A and B are stable.*
2. *Provided that the solvability condition is fulfilled, the optimal deconvolution filter is given by* (29) *where the polynomial matrix Y is computed by first solving the spectral factorization problem* (14) *w.r.t. E, and next finding* $(X, Y, Z)$ *as the* minimum degree *solution w.r.t. Z of the coupled Diophantine equations* (47) *and* (48), *or* (47) *and* (49), *or, if* $B_1$ *has full column rank,* (48) *and* (49).
3. *If* $\det(\overline{E})$ *and* $\det(A)$ *are coprime then Y, along with Z, can be obtained by the minimum degree solution w.r.t. Z of* (47).
4. *Mean-square boundedness and filter stability are guaranteed provided that* $A_1 C_v$ *and* $B_1 C_w$ *have no left common factors with zeros on* $|d| = 1$.

## 5. Duality with LQGR

It is well known that LQ regulation and Kalman filtering are dual problems. In fact, the optimal LQ feedback gain is a function of the solution to the control Riccati equation while the Kalman filter gain is a function of the solution to the filter Riccati equation.

This duality concept can also be recognized by comparing the two sets of polynomial equations solving the LQ regulation problem [10] and, respectively, the Kalman filtering problem [4]. It is not immediate to relate the LQG (dynamic) regulation problem to any specific estimation problem. Comparing the polynomial equations of this paper (14), (24) and (39) with the corresponding ones in [8], respectively, (6.2.7), (6.2.16*b*) and (6.2.16*a*), it is seen that MMSE deconvolution and LQGR are dual problems.

## 6. Conclusions

A polynomial solution to the MMSE multichannel deconvolution problem has been presented. Several authors have recently considered polynomial solutions to MMSE estimation problems. The present work provides, at the best of the authors' knowledge, the most general solution. In fact, it extends the results of [1,2] to the multivariable case, and generalizes the contributions to Wiener and Kalman filtering given in [5,6] and, respectively, [4,5,7].

In this paper, the solution is given in terms of two coupled Diophantine equations rather than a single one as in [1,6,7]. The need of using two equations has a conceptual interest, in that it corresponds to optimizing and stabilizing the deconvolution filter, as well as practical relevance even if the coprimeness condition of Lemma 5 is fulfilled. In fact, it can be conjectured that the presence of almost common factors in (47) can yield - similarly to what reported in [11] for the dual problem of LQG dynamic compensation - a higher numerical sensitivity if (47) only is used instead of both (47) and (48), or (47) and (49).

## REFERENCES

[1 ] Ahlen A. and M. Sternad (1989). Optimal deconvolution based on polynomial methods. *IEEE Trans. Acoust., Speech, Signal Processing*, **37**, 217-226.

[2 ] Chisci L. and E. Mosca (1991). A general polynomial solution to the MMSE deconvolution problem. *IEEE Trans. Signal Processing*, **39**, 962-965.

[3 ] Mendel J. (1983). *Optimal seismic deconvolution*, Academic Press, New York.

[4 ] Chisci L. and E. Mosca (1991). Polynomial equations for the MMSE state estimation. *IEEE Trans. Aut. Control*, to appear.

[5 ] Grimble M.J. (1985). Polynomial systems approach to optimal linear filtering and prediction. *Int. J. of Control*, **41**, 1545-1564.

[6 ] Roberts A.P. and N.M. Newmann (1988). A polynomial approach to Wiener filtering. *Int. J. of Control*, **47**, 681-696.

[7 ] Kucera V. (1981). New results in state estimation and regulation. *Automatica*, **17**, 745-748.

[8 ] Kucera V. (1979). *Discrete Linear Control*, J. Wiley, Chichester.

[9 ] Chisci L and E. Mosca (1991). Polynomial approach to MMSE multichannel deconvolution. *Tech. Rep. DSI 17/91*, DSI, Universitá di Firenze, Firenze, Italy.

[10 ] Casavola A., M.J. Grimble, E. Mosca and P. Nistri (1991). Continuous-time LQ regulator design by polynomial equations. *Automatica*, **27**, 555-558.

[11 ] Peterka V. (1984). Predictor-based self-tuning control. *Automatica*, **20**, 39-50.

The authors are with:
Dipartimento di Sistemi e Informatica
Universitá di Firenze
Via di Santa Marta 3
50139 Firenze, Italy.

# Topics in 2D Systems Theory

E.FORNASINI and G.MARCHESINI

## 1. Introduction

Since the early 1970s two dimensional (2D) models have been attracting the attention of the scientific community working in the area of dynamical systems theory. Several different motivations underlie this research field. The inspiring idea originated in the framework of two-dimensional data processing (using parallel computer structures) from the necessity of modelling dynamical processes parametrized by two independent variables (e.g. time and one space coordinate). Beyond that, very nice system theoretical reasons exist for dealing with this kind of models, whose dynamics evolves on partially ordered sets and where the extension of "classical" systems notions does not appear as a practicable road.

The conceptual and formal difficulties envisaged in the theory of 2D systems are well known and descend from the fact that most definitions and operations, which are natural in 1D theory, have to be totally revised or abandoned when one deals with signals having support in $\mathbf{Z} \times \mathbf{Z}$.

Our aim in this contribution is to discuss some research perspectives in the areas of modelling, realization and control of 2D systems where a significant set of results is already available in the literature.

## 2. Modelling

Dynamical 2D systems have been introduced as a natural tool for representing the processing of discrete signals which are two-dimensional, in the sense that they are functions of two independent variables. Since the support of 2D signals is the discrete plane $\mathbf{Z} \times \mathbf{Z}$, many different orderings can be conceived for representing cause-effect relations. These lead to mathematical models that exhibit characteristics which are not always comparable.

In the early works [1-4] the quarter plane causal structure was a priori assumed in the model, as a consequence of the choice of the product ordering in $\mathbf{Z} \times \mathbf{Z}$. In some recent works [5-7] the partial ordering definition and the induced causal structure are not a priori given, and rely on the analysis of the system trajectories that reveals the underlying causality relations. In some cases [8], inspired by the process used for discretizing partial differential equations, signal constraints have to be assumed on a closed contour in $\mathbf{Z} \times \mathbf{Z}$. Thus the solution essentially relies on the evolution of a 1D system and the two-dimensional causality does not play an essential role in the computation of the response.

Models which exhibit a quarter plane causality have been initially investigated from an input/output point of view in the framework of 2D filters theory [9], where two-dimensional filters are represented by proper rational functions in two indeterminates of the following type

$$W(z_1, z_2) = \frac{\sum_{i+j\geq 1} n_{ij} z_1^i z_2^j}{1 + \sum_{i+j\geq 1} d_{ij} z_1^i z_2^j} \tag{1}$$

The idea of associating 2D state space models with two-dimensional filters originated very naturally. However, since the beginning it appeared that the "canonical" technique based on the Nerode equivalence leads to an infinite dimensional state space [3,4] So, following sometimes heuristic procedures, several finite dimensional models have been introduced, where two notions of state play different roles:

1. *local states* $x(h, k)$ belong to a finite dimensional vector space. They enter in the state updating equation and determine the value of the output;

2. *global states* $\mathcal{X}_h = \{x(i + h, -i), i \in \mathbf{Z}\}$ provide the initial conditions on a separation set of $\mathbf{Z} \times \mathbf{Z}$. These belong to an infinite dimensional vector space, which provides an extension of the space of Nerode equivalence classes.

The most common state model with quarter plane causality is represented by the following equations [10]:

$$\begin{aligned} \mathbf{x}(h+1, k+1) =& A_1\mathbf{x}(h, k+1) + A_2\mathbf{x}(h+1, k) \\ &+ B_1\mathbf{u}(h, k+1) + B_2\mathbf{u}(h+1, k) \\ \mathbf{y}(h, k) =& C\mathbf{x}(h, k) \end{aligned} \tag{2}$$

where $\mathbf{x}(h, k) \in \mathbf{R}^n, \mathbf{u}(h, k) \in \mathbf{R}^m, \mathbf{y}(h, k) \in \mathbf{R}^p$ are the values of the local state, the input and the output at $(h, k) \in \mathbf{Z} \times \mathbf{Z}$. Since the local state at $(h + 1, k + 1)$ is computed by solving a first order difference equation, the system (2), denoted by $\Sigma_1 = (A_1, A_2, B_1, B_2, C)$, is named a first order system.

This model has been extensively studied in its general form (2) and under some constraints on the system matrices. The most popular particularized version of (2) is the Roesser's model [1,11], where the local state space $X$ is assumed as the direct sum of two vector spaces $X^h$ and $X^v$, and the matrices of the model are constrained to have the following partitioned form

$$A_1 = \begin{bmatrix} A_{11}^{(1)} & A_{12}^{(1)} \\ 0 & 0 \end{bmatrix}, A_2 = \begin{bmatrix} 0 & 0 \\ A_{21}^{(2)} & A_{22}^{(2)} \end{bmatrix}, B_1 = \begin{bmatrix} B_1^{(1)} \\ 0 \end{bmatrix}, B_2 = \begin{bmatrix} 0 \\ B_2^{(2)} \end{bmatrix} \tag{3}$$

Second order models are less frequently used: the typical structure of their equation is given by [3,4]:

$$\begin{aligned} \mathbf{x}(h+1,k+1) &= A_1\mathbf{x}(h,k+1) + A_2\mathbf{x}(h+1,k) + A_0\mathbf{x}(h,k) + B\mathbf{u}(h,k) \\ \mathbf{y}(h,k) &= C\mathbf{x}(h,k) \end{aligned} \tag{4}$$

In particular, the Attasi's model [2] has the structure (4) with $A_1$ and $A_2$ commutative matrices and $A_1A_2 = -A_0$. It realizes separable filters only and constitutes an interesting second order model, as the underlying theory is very close to the 1D theory.

Recently the "behaviour" approach has been extended to 2D systems. Following this theory, a 2D system is defined by a family $\mathcal{B}$ of "admissible functions" (behaviour), defined over the discrete plane. These functions are characterized by the property of belonging to the kernel of a polynomial matrix $M(z_1, z_2)$ in two variables

$$\mathcal{B} = \{\mathbf{w} = \sum_{i,j \in \mathbf{Z}} w_{ij} z_1^i z_2^j | M\mathbf{w} = 0\} \tag{5}$$

Associated with the external description provided by the behaviour $\mathcal{B}$ different "internal" representations can be given by introducing the so called *latent* variables models. State variables constitute a particular type of latent variables, that hold the memory of the system with respect to a notion of "past" introduced on $\mathbf{Z} \times \mathbf{Z}$. When a state description is possible, i.e. when the notions of past, present and future are allowed by the structure of $\mathcal{B}$, the behaviour is called *markovian*. Since there is'nt any "natural" direction for the evolution in $\mathbf{Z} \times \mathbf{Z}$, the markovian property appears more general than the familiar quarter plane causality and has been exploited in the analysis of non-causal 2D dynamics [5-7].

## 3. Realization

In the theory of 1D systems minimal realizations of a given transfer function are reachable and observable. They are algebraically equivalent and can be canonically computed using the Nerode equivalence. This very simple picture does not fit with the 2D situation. We are not very far from the truth by saying that the minimal realization problem is the bottleneck of the entire 2D theory and that appropriate tools for investigating this problem have not yet been set up. In particular, the reachability and observability analysis, relative to local and global states, is not useful for constructing minimal realizations. Actually, local reachability and observability do not necessarily imply minimality [12]. On the other side, globally reachable and observable realizations [13] are minimal, but the converse is not true. It should also be noticed that in some cases (see Example below) globally reachable and observable realizations do not exist [13].

*Example 1.*[14] The transfer function $W(z_1, z_2) = z_1^2 + z_2^2$ does not admit globally reachable and observable realizations, independently of the field of the entries of the system matrices.

Several approaches to the minimal realization problem have been undertaken. One of these consists in associating an Hankel matrix with the impulse response and using a suitable version of the Ho's algorithm. It was originally conceived for scalar systems, but it can be extended to the multivariable case without any conceptual difficulty. Given a formal power series $s = \sum_{ij} s_{ij} z_1^i z_2^j$ in the commuting variables $z_1$ and $z_2$, the Hankel matrix $\mathcal{H}(s)$ is an infinite matrix with entries indexed on the commutative semigroup of monic monomials in two indeterminates. The element indexed by $(z_1^i z_2^j, z_1^h z_2^k)$ is given by the coefficient $s_{i+h,j+k}$. It has been proved [15] that, except for the case of series expansion of separable transfer functions, the rank of $\mathcal{H}(s)$ is infinite, even when $s$ is rational. So any direct implementation of the Ho's algorithm to obtain minimal realizations of a rational transfer function becomes quite problematic.

A way to overcome this difficulty [12] descends from the observation that the Hankel matrix associated with a noncommutative rational series has finite rank. This allows to use linear algorithms to construct a minimal matrix representation of the series. More precisely, given the alphabet $\{\xi_1, \xi_2\}$ and the noncommutative monoid $\{\xi_1, \xi_2\}^*$, consider a strictly proper rational series

$$\sigma = \sum_{w \in \{\xi_1, \xi_2\}^* \setminus \{\emptyset\}} (\sigma, w) w$$

with noncommutative indeterminates $\xi_1$ e $\xi_2$, and introduce the matrix $\mathcal{H}(\sigma)$ indexed in $\{\xi_1, \xi_2\}^*$, where the $(w_1, w_2)$ entry is the coefficient $(\sigma, w_1 w_2)$. Ho's algorithm enables us to represent $\sigma$ in the form

$$\sigma(\xi_1, \xi_2) = C(I - A_1\xi_1 - a_2\xi_2)^{-1}(B_1\xi_1 + B_2\xi_2) \tag{6}$$

and to obtain a representation (6) whith $A_1$ and $A_2$ having minimal dimension $\delta(\sigma)$. If the series expansion of the transfer function $W(z_1, z_2)$ is the commuative image of $\sigma$, the system $\Sigma_1 = (A_1, A_2, B_1, B_2, C)$ provides a realization of $W(z_1, z_2)$.

**Thus a method to obtain the minimal realizations of $W(z_1, z_2)$ consists in finding the set of noncommutative rational series $\sigma(\xi_1, \xi_2)$ having $W(z_1, z_2)$ as commutative image and $\delta(\sigma)$ minimal, and then in constructing their representations (6). The algorithm for obtaining the set of such noncommutative series is intrinsically nonlinear and evidentiates that**

**i) the dimension of minimal realizations of $W(z_1, z_2)$ depends on the ground field of the elements of the system matrices;**

**ii) two minimal realizations are not algebraically equivalent if they are computed applying Ho's algorithm to different noncommutative series having $W(z_1, z_2)$ as commutative image.**

*Example 2.* **Consider the following transfer function**

$$W(z_1, z_2) = \frac{2z_1z_2}{1 + z_1^2 + z_2^2}$$

**It admits a minimal realization of type (2)**

$$A_1 = \begin{bmatrix} -i & 0 \\ 0 & i \end{bmatrix}, A_2 = \begin{bmatrix} 0 & i \\ i & 0 \end{bmatrix}, B_1 = \begin{bmatrix} 0 \\ 0 \end{bmatrix}, B_1 = \begin{bmatrix} i \\ -i \end{bmatrix}, B_1 = \begin{bmatrix} 1 & 1 \end{bmatrix}$$

**on the complex field. However it can be shown that there are no realizations of type (2) on the real field which have dimension two.**

*Example 3.* **The following systems**

$$A_1 = \begin{bmatrix} 1 & 1 \\ 0 & 1 \end{bmatrix}, A_2 = \begin{bmatrix} 1 & 0 \\ 0 & 1 \end{bmatrix}, B_1 = \begin{bmatrix} 0 \\ 0 \end{bmatrix}, B_2 = \begin{bmatrix} 0 \\ 1 \end{bmatrix}, C = \begin{bmatrix} 1 & 0 \end{bmatrix}$$

**and**

$$A_1 = \begin{bmatrix} 1 & 0 \\ 0 & 1 \end{bmatrix}, A_2 = \begin{bmatrix} 1 & 1 \\ 0 & 1 \end{bmatrix}, B_1 = \begin{bmatrix} 0 \\ 1 \end{bmatrix}, B_2 = \begin{bmatrix} 0 \\ 0 \end{bmatrix}, C = \begin{bmatrix} 1 & 0 \end{bmatrix}$$

**are minimal realizations of the same transfer function, but are not algebraically equivalent.**

**The use of noncommutative power series sheds also some light on a property of 2D hidden modes. Given an irreducible 2D transfer function $n(z_1, z_2)/d(z_1, z_2)$ and a state space realization $\Sigma = (A_1, A_2, B_1, B_2, C, D)$ of $W(z_1, z_2)$, the hidden modes are the common factors of $C\text{adj}(I - A_1z_1 - A_2z_2)(B_1z_1 + B_2z_2)$ and $\det(I - A_1z_1 - A_2z_2)$.**

**Since in the 1D case hidden modes are associated with unreachable and/or unobservable subsystems, whose existence prevents a realization from being minimal, a naive extension of 1D theory would suggest that minimal 2D realizations are free of hidden modes. Actually cancellations of 2D polynomials between $C\text{adj}(I - A_1z_1 - A_2z_2)(B_1z_1 + B_2z_2)$ and the characteristic polynomial $\det(I - A_1z_1 - A_2z_2)$ are always connected with**

the existence of plane curves where at least one of the PBH controllability and reconstructibility matrices is not full rank. This fact however is not in contradiction with the minimality of the realization and consequently the above intuition is false. This is illustrated by Example 4, which is based on the following multistep procedure:

i) express $W(z_1, z_2)$ as the product $W_2W_1$ of two irreducible transfer functions that exhibit the cancellation of a nonconstant polynomial $c(z_1, z_2)$.

ii) construct two minimal realizations $\Sigma_1$ and $\Sigma_2$ of $W_1(z_1, z_2)$ and $W_2(z_1, z_2)$, respectively

iii) perform the series connection of $\Sigma_1$ and $\Sigma_2$. This provides a realization of $W(z_1, z_2)$, whose characteristic polynomial includes the factor $c(z_1, z_2)$.

The noncommutative power series $\sigma_{21}$ that corresponds to the series connection of $\Sigma_1$ and $\Sigma_2$, is the product of the noncommutative power series $\sigma_2$ and $\sigma_1$ associated with $\Sigma_2$ and $\Sigma_1$ respectively. The cancellation of a commutative polynomial $c(z_1, z_2)$ in $W_2W_1$ needs not imply that cancellations arise in the product $\sigma_2\sigma_1 = \sigma_{21}$, when $\sigma_1$ and $\sigma_2$ are expressed via finite sums, products and inverses of noncommutative polynomials. This suggests that the series connection of $\Sigma_1$ and $\Sigma_2$ may be a minimal matrix representation of $\sigma_{21}$, irrespective of cancellations in $W_2W_1$.

*Example 4.* Consider the transfer function $W(z_1, z_2) = (z_1 + z_2)^3 + z_2^2 + z_2$ and its factorization into

$$W(z_1, z_2) = W_1(z_1, z_2)W_2(z_1, z_2) = \left[\frac{(z_1 + z_2)^3 + z_2^2 + z_2}{1 + z_2}\right][1 + z_2]$$

Starting from a minimal realization $\Sigma_1$ of $W_1(z_1, z_2)$, given by

$$\bar{A}_1 = \begin{bmatrix} 0 & -1 & 0 \\ 0 & 0 & -1 \\ 0 & 0 & 0 \end{bmatrix}, \bar{A}_2 = \begin{bmatrix} 0 & -1 & 0 \\ 0 & 0 & -1 \\ 0 & 0 & -1 \end{bmatrix}, \bar{B}_1 = \begin{bmatrix} 0 \\ 0 \\ 1 \end{bmatrix}, \bar{B}_2 = \begin{bmatrix} 1 \\ 0 \\ 1 \end{bmatrix},$$
$$\bar{C} = [1 \quad 0 \quad 0],$$

and a minimal realization $\Sigma_2$ of $W_2(z_1, z_2)$, given by $\hat{A}_1 = [0], \hat{A}_2 = [0], \hat{B}_1 = [0], \hat{B}_2 = [1], \hat{C} = [1], \hat{D} = [1]$, we compute the series connection of $\Sigma_1$ and $\Sigma_2$

$$A_1 = \begin{bmatrix} \hat{A}_1 & 0 \\ B_1\hat{C} & \bar{A}_1 \end{bmatrix}, A_2 = \begin{bmatrix} \hat{A}_2 & 0 \\ B_2\hat{C} & \bar{A}_2 \end{bmatrix}, B_1 = \begin{bmatrix} \hat{B}_1 \\ B_1\hat{D} \end{bmatrix}, B_2 = \begin{bmatrix} \hat{B}_2 \\ B_2\hat{D} \end{bmatrix},$$
$$C = [0 \quad \bar{C}]$$

The system above constitutes a minimal realization of $W$. The proof of this fact is rather long [16], and involves a detailed analysis of the matrix pairs

$A_1, A_2$ in $\mathbf{C}^{3\times 3}$ that satisfy the finite memory condition $\det(I - A_1 z_1 - A_2 z_2) = 1$. Note that $\Sigma_1$ and $\Sigma_2$ provide minimal matrix representations of $\sigma_1 = (\xi_1+\xi_2)^2(1+\xi_2)^{-1}(\xi_1+\xi_2)+\xi_2$ and $\sigma_2 = 1+\xi_2$ respectively, and $\Sigma$ provides a minimal MR of $\sigma = (1+\xi_2)[(\xi_1+\xi_2)^2(1+\xi_2)^{-1}(\xi_1+\xi_2)+\xi_2]$. No cancellation arises in the above expression because of the noncommutativity of the factors $(1+\xi_2)$ and $(\xi_1+\xi_2)$.

Every strictly proper rational function can be realized by systems (2) and (4). However, if we assume that the system matrices satisfy structural constraints, only subclasses of the class of rational transfer functions can be realized. In summary we have the following:

- the transfer function of an Attasi's model is a separable rational function; conversely, the whole class of separable proper rational functions can be realized using Attasi's models. It can be shown that, given a separable transfer function $W(z_1, z_2)$, the rank of its Hankel matrix $\mathcal{H}(W)$ provides the minimal dimension of the realizations having the Attasi's structure. Also, minimal realizations are algebraically equivalent and can be computed using linear algorithms based on the Ho's procedure.
- consider model (4) and assume that the system matrices satisfy the constraints

$$A_0 = 0,\ A_1A_2 = A_2A_1. \tag{7}$$

In this case the class of transfer functions which can be realized coincides with the set of rational functions having series expansions of the following form [17]

$$W(z_1, z_2) = \Big[n_0(z_1, z_2) + \sum_j \frac{n_j(z_1, z_2)}{(1 - a_{1j}z_1 - a_{2j}z_2)^{\nu_j}}\Big] z_1 z_2 \tag{8}$$

with $\deg n_j < \nu_j$ and $n_0 \in \mathbf{R}[z_1, z_2]$. The minimal realization algorithm is based on linear procedures.

- transfer functions having series expansion (8) do not exhaust the set of transfer functions whose denominators factorize as the product of first order polynomials. The realization of a transfer function in this set is always accomplishable by systems (2) and (4), with $A_1, A_2$ (and $A_0$) simultaneously triangularizable [18]. Algorithms for constructing minimal realizations in this class are not yet known.

## 4. Control

Comparing with the 1D situation, we have that, as a consequence of the partial ordering structure, the class of 2D feedback schemes which can be implemented is much wider. Actually we can conceive control procedures where the values of the state and/or the output at $(h, k)$ influence the input values at points which are not causally related to $(h, k)$. In these cases the resulting closed loop system looses the quarter plane causality, which could be undesirable in the context of the synthesis problem we are dealing with.

A further general remark is in order about the static or dynamic nature of the feedback schemes. For the 2D systems the solution of most control problems requires to introduce causal dynamic compensators that realize state-input or output-input recursive relationships of the following form

$$\mathbf{u}(h,k) = \sum H_{ij}\mathbf{u}(h-i,k-j) + \sum K_{ij}\mathbf{x}(h-i,k-j)$$

However, as we shall see, in the case of optimal control the input at $(h, h)$ is generated by a static noncausal control law that involves an infinite number of local states. Obviously this gives some inconveniences from the realization point of view, so that it might be preferable to resort to suboptimal causal dynamic compensators.

### 4.1 Feedback stabilization

By definition, a 2D system $\Sigma = (A_1, A_2, B_1, B_2, C)$ is internally stable if, for any global state $\mathcal{X}_0 = \{\mathbf{x}(i,-i),\ i \in \mathbf{Z}\}$ with $\sup_i ||\mathbf{x}(i,-i)|| < \infty$, the free evolution of the local states satisfies

$$\lim_{h+k\to+\infty} \mathbf{x}(h,k) = 0. \tag{10}$$

Accordingly, a stabilizing output feedback compensator is a non strictly proper $p$ inputs, $m$ outputs 2D system $\bar{\Sigma} = (\bar{A}_1, \bar{A}_2, \bar{B}_1, \bar{B}_2, \bar{C}, \bar{D})$ that makes internally stable the closed loop system resulting from the feedback interconnection of $\Sigma$ and $\bar{\Sigma}$.

Both stability and stabilizability properties can be checked by analyzing the intersection of suitable polynomial varieties with the closed unit bidisk

$$\mathcal{P}_1 = \{(z_1, z_2) : |z_1| \le 1, |z_2| \le 1\}.$$

The following results have been obtained in [19,20]:

i) $\Sigma$ is internally stable if an only if the characteristic polynomial

$$\det(I - A_1 z_1 - A_2 z_2) \tag{11}$$

associated with the pair $(A_1, A_2)$ is devoid of zeros in $\mathcal{P}_1$.

ii) $\Sigma$ admits a stabilizing output feedback compensator if and only if

$$[\mathcal{V}(\mathcal{R}) \cup \mathcal{V}(\mathcal{O})] \cap \mathcal{P}_1 = \emptyset \tag{12}$$

Here $\mathcal{V}(\mathcal{R})$ and $\mathcal{V}(\mathcal{O})$ denote the complex varieties of the ideals generated by the maximal order minors of the following polynomial matrices in two variables:

$$\mathcal{R} = [\, I - A_1 z_1 - A_2 z_2 \quad B_1 z_1 + B_2 z_2 \,] \tag{13}$$

and

$$\mathcal{O} = \begin{bmatrix} I - A_1z_1 - A_2z_2 \\ C \end{bmatrix} \tag{14}$$

The meaning of condition (12) can be fully understood by examining the analogous condition in the 1D framework. It is well known that a discrete time 1D system $(A, B, C)$ is stabilizable via dynamic output feedback if and only if

$$[\mathcal{V}(\mathcal{R}_1) \cup \mathcal{V}(\mathcal{O}_1)] \cap \mathcal{D}_1 = \emptyset,$$

where $\mathcal{D}_1$ denotes the closed unit disk $\{z : |z| \leq 1\}$, and $\mathcal{R}_1$ and $\mathcal{O}_1$ are the PBH controllability and reconstructibility matrices

$$\mathcal{R}_1 = [\, I - Az \quad Bz \,], \qquad \mathcal{O}_1 = \begin{bmatrix} I - Az \\ C \end{bmatrix}.$$

If $N_R(z)D_R^{-1}(z)$ is a right coprime MFD of the transfer matrix of $(A, B, C)$, so that $\det D_R$ divides $\det(I - Az)$, then $[\mathcal{V}(\mathcal{R}_1) \cup \mathcal{V}(\mathcal{O}_1)]$ is the zero set of $\det(I - Az)/\det D_R$, i.e.

$$\mathcal{V}(\mathcal{R}_1) \cup \mathcal{V}(\mathcal{O}_1) = \mathcal{V}\left(\frac{\det(I - Az)}{\det D_R}\right) \tag{15}$$

This implies that the points of $[\mathcal{V}(\mathcal{R}_1) \cup \mathcal{V}(\mathcal{O}_1)]$ are naturally associated with the hidden modes of the system and 1D stabilizability reduces to have all hidden modes converging to zero.

The situation is quite different in the 2D case. Given any right factor coprime MFD $N_R(z_1, z_2)D_R^{-1}(z_1, z_2)$ of the transfer matrix $W(z_1, z_2)$ of $\Sigma$, the ideal of the maximal order minors of $[\, N_R \quad D_R \,]$ does not depend on the particular right coprime MFD and coincides with the analogous ideal of any left coprime MFD of $W(z_1, z_2)$ (see [20]). This ideal will be unambiguously denoted by $\mathcal{I}(W)$ and the points of the corresponding finite, possibly nonempty, variety $\mathcal{V}(W)$ are called "rank singularities" of $W(z_1, z_2)$.

The variety $\mathcal{V}(W)$ turns out to be of great importance in investigating the relationship between the set of points $[\mathcal{V}(\mathcal{R}) \cup \mathcal{V}(\mathcal{O})]$ and the variety of the characteristic polynomial $\mathcal{V}(\det(I - A_1z_1 - A_2z_2))$. For the 2D case this relationship is more complex than (15), essentially because $\mathcal{V}(W)$ is an invariant subset of $[\mathcal{V}(\mathcal{R}) \cup \mathcal{V}(\mathcal{O})]$ with respect to the specific realization taken into account.

In [21] the following result has been obtained:

**Theorem 1** *Let $W(z_1, z_2)$ be the transfer matrix of a 2D system $\Sigma = (A_1, A_2, B_1, B_2, C)$ and $N_R(z_1, z_2)D_R^{-1}(z_1, z_2)$ be a right coprime MFD of $W$. Then the variety of the maximal order minors of* (13) *and* (14) *is given by,*

$$\mathcal{V}(\mathcal{R}) \cup \mathcal{V}(\mathcal{O}) = \mathcal{V}(W) \cup \mathcal{V}\left(\frac{\det(I - A_1z_1 - A_2z_2)}{\det D_R(z_1, z_2)}\right) \tag{16}$$

Eqn. (16) shows that the set of critical points for 2D stabilizability includes both the hidden modes variety, i.e. the algebraic curve associated with the polynomial $h(z_1, z_2) := \det(I - A_1 z_1 - A_2 z_2)/\det D_R(z_1, z_2)$, and the variety of rank singularities $\mathcal{V}(W)$.

The difference between the structure of (15) and (16) has important consequences on the stabilizability of the state space realizations of input/output maps given by transfer matrices. Clearly, minimal realizations in the 1D case are stabilizable, since no hidden modes are left in the system and therefore the polynomial matrices $\mathcal{R}_1$ and $\mathcal{O}_1$ are full rank everywhere. On the other hand, as a consequence of Theorem 1, the 2D stabilizability condition (12) requires that both $\mathcal{V}(h)$ and $\mathcal{V}(W)$ do not intersect $\mathcal{P}_1$. For a given transfer matrix $W(z_1, z_2)$ it is always possible to compute a realization which fulfils the requirement that $\mathcal{V}(h) \cap \mathcal{P}_1 = \emptyset$; nevertheless, if $\mathcal{V}(W) \cap \mathcal{P}_1 \neq \emptyset$, there are no stabilizable realizations of $W$, since the set of the rank singularities $\mathcal{V}(W)$ is independent of the realization.

Theorem 1 contains a first result about the constraints which have to be fulfilled by the closed loop polynomial variety, but it does not specify to what extent this variety can be modified using output feedback. The key result on the assignability of the closed loop polynomial variety is given by the following theorem.

**Theorem 2** [21] *Let $\Sigma = (A_1, A_2, B_1, B_2, C)$ be a realization of a strictly proper transfer matrix $W(z_1, z_2)$. For any output feedback compensator $\tilde{\Sigma}$, the variety of the closed loop polynomial $\bar{\Delta}(z_1, z_2)$ satisfies the inclusion*

$$\mathcal{V}(\bar{\Delta}) \supseteq \mathcal{V}(h) \cup \mathcal{V}(W)$$

*Viceversa, given any algebraic curve $\mathcal{C}$ that includes $\mathcal{V}(h) \cup \mathcal{V}(W)$ and excludes the origin, there exists a compensator $\tilde{\Sigma}$ such that $\mathcal{V}(\bar{\Delta}) = \mathcal{C}$.*

At this point two problems naturally arise:

i) given a polynomial $c(z_1, z_2)$ in $\mathbf{R}[z_1, z_2]$, decide about the assignability of the variety $\mathcal{C} := \mathcal{V}(c)$

ii) if $\mathcal{V}(c)$ is assignable, find algorithms for realizing a compensator $\tilde{\Sigma}$

The solution of the first problem consists in verifying

$$(0,0) \notin \mathcal{C} \tag{17}$$

$$\mathcal{V}(h) \subseteq \mathcal{C} \tag{18}$$

$$\mathcal{V}(W) \subseteq \mathcal{C} \tag{19}$$

Checking (17) is trivial. Moreover, once the polynomial $h(z_1, z_2)$ has been computed, we can easily verify (18) using a linear algorithm to see if $h$

divides $c^{\deg h}$. The condition (19) can be checked by first computing a set of generators of $\mathcal{I}(W)$ and successively exploiting them for constructing a pair of commuting matrices $M_1$ and $M_2$, with the property

$$p(z_1, z_2) \in \mathcal{I}(W) \quad \Leftrightarrow \quad p(M_1, M_2) = 0. \tag{20}$$

Thus $\mathcal{V}(W) \subseteq \mathcal{C}$ if and only if $c(M_1, M_2)$ is a nilpotent matrix. For the construction of $M_1$ and $M_2$ the reader is referred to [22].

It remains to show how to compute the polynomial $h$ and a set of generators for $\mathcal{I}(W)$ starting from the system matrices $A_1, A_2, B_1, B_2, C$.

For this, let

$$[C\text{adj}(I - A_1 z_1 - A_2 z_2)(B_1 z_1 + B_2 z_2)]\,[I_m \det(I - A_1 z_1 - A_2 z_2)]^{-1} = \bar{N}\bar{D}^{-1}$$

be a MFD of the transfer matrix $W(z_1, z_2)$. The generators set can be obtained by evaluating the maximal order minors $m_1, m_2, \ldots m_t$ in $[\,\bar{N}^T \quad \bar{D}^T\,]$ and then by eliminating their g.c.d. $d(z_1, z_2)$. Thus $h$ is given by

$$h = \frac{\det(I - A_1 z_1 - A_2 z_2)}{\det D_R} = \frac{\det(I - A_1 z_1 - A_2 z_2) d(z_1, z_2)}{\det \bar{D}}$$

As far as the second problem is concerned, suppose that a variety $\mathcal{C} = \mathcal{V}(c)$ that fulfils conditions (17)-(19) has been given, and suppose we want to synthesize a compensator $\Sigma$ that produces a closed loop polynomial $\Delta$ whose variety is $\mathcal{C}$. The procedure can be summarized as follows:

- Evaluate a right coprime MFD $N_R D_R^{-1}$ of $W$. This can be performed by using the primitive factorization algorithm [11] or any other algorithm that does not require primitive factorizations [23].
- Compute the maximal order minors $m_1, m_2, \ldots m_t$.
- Compute an integer $r$ and a Gröbner basis $g_1, g_2, \ldots g_w$ such that $c^r = \sum_i m_i g_i$. A technique for performing this step has been presented in [22].
- Solve the Bézout equation $c^r I_m = X D_R + Y N_R$.
- Use the realization algorithm given in [21] for computing a coprime realization of $X^{-1} Y$.

## 4.2 Optimal control

The compensator synthesis procedure illustrated in section 4.1 enables us to obtain any preassigned variety of the closed loop polynomial, provided the constraints specified by Theorem 2 are satisfied. However the pole placement design essentially affects the asymptotic behaviour and exhibits a poor control of the short term system response. In this section we shall tackle the problem from a different point of view and outline a state feedback synthesis procedure based on the minimization of a quadratic cost functional $J$. The system which constitutes the end result of the optimal design is not merely internally stable, but satisfies additional requirements on the state and input evolutions that are summarized by $J$.

Assume that the initial global state $\mathcal{X}_0 = \{\mathbf{x}(i, -i), i \in \mathbf{Z}\}$ is an $\ell_2$ sequence and consider the cost functional

$$J = \sum_{h+k \geq 0} \mathbf{x}^T(h,k)Q\mathbf{x}(h,k) + \mathbf{u}^T(h,k)R\mathbf{u}(h,k) \tag{21}$$

with $Q \geq 0$ and $R > 0$. Theorems 3 and 4 below provide a complete solution to the following optimal control problems:

1. given $\mathcal{X}_0$, derive conditions for the existence and the uniqueness of an input function $\mathbf{u}(\cdot,\cdot)$ that minimizes $J$;
2. whenever these conditions are satisfied, explicitly compute the optimal input function and the corresponding value of $J$.

The following theorem [24] shows that the existence and the uniqueness of a stabilizing optimal control reduce to rank conditions on polynomial matrices in two variables.

**Theorem 3** *For any $\mathcal{X}_0 \in \ell_2$, there exists an $\ell_2$- solution of the optimal control problem. (i.e. an input $\mathbf{u}(\cdot,\cdot)$ in $\ell_2$ such that the corresponding state evolution $\mathbf{x}(\cdot,\cdot)$ is in $\ell_2$ and the value of $J$ is minimized) if and only if the polynomial matrix* (13) *has full rank on the set*

$$\mathcal{M} = \{(z_1, z_2) \in \mathbf{C} \times \mathbf{C} : |z_1| = |z_2| \leq 1\}$$

*and the polynomial matrix*

$$\begin{bmatrix} I - A_1 z_1 - A_2 z_2 \\ Q \end{bmatrix} \tag{22}$$

*has full rank on the unit torus*

$$\mathcal{T}_1 = \{(z_1, z_2) \in \mathbf{C} \times \mathbf{C} : |z_1| = |z_2| = 1\}$$

When the conditions of Theorem 3 are fulfilled, the optimal control law is obtained via an algebraic Riccati equation whose coefficients are polynomial matrices in one variable.

**Theorem 4** *Assume that* (13) *is full rank on $\mathcal{M}$ and* (22) *is full rank on $\mathcal{T}_1$. Then the following algebraic Riccati equation (ARE-z)*

$$P(z) = Q + (A_1^T + A_2^T z^{-1})P(z)(A_1 + A_2 z) - (A_1^T + A_2^T z^{-1})P(z)(B_1 + B_2 z) \cdot$$
$$\cdot \left[R + (B_1^T + B_2^T z^{-1})P(z)(B_1 + B_2 z)\right]^{-1} (B_1^T + B_2^T z^{-1})P(z)(A_1 + A_2 z)$$

*in the unknown matrix $P(z)$ admits a unique solution in an open annulus that includes the unit circle $\gamma_1$, with the following properties:*

1. $P(e^{j\omega}) = P^*(e^{j\omega}) \geq 0, \quad \forall \omega \in [0, 2\pi]$
2. *the matrix*

$$K(z) := -\left[R + (B_1^T + B_2^T z^{-1})P(z)(B_1 + B_2 z)\right]^{-1} \cdot \\ \cdot (B_1^T + B_2^T z^{-1})P(z)(A_1 + A_2 z)$$

*is analytic in an open annulus that includes $\gamma_1$.*

*The coefficients of its Laurent series expansion*

$$K(z) = \sum_{i=-\infty}^{+\infty} K_i z^i \tag{23}$$

*provide a stabilizing feedback law $u(h,k) = \sum_{i=-\infty}^{+\infty} K_i x(h+i, k-i)$ and the minimum value of $J$ is given by*

$$J_{\min} = \frac{1}{2\pi} \int_0^{2\pi} \hat{\mathcal{X}}_0^*(\omega) P(e^{j\omega}) \hat{\mathcal{X}}_0(\omega)\, d\omega \tag{24}$$

*where $\hat{\mathcal{X}}_0(\omega)$ denotes the Fourier transform of the $\ell_2$-sequence $\mathcal{X}_0$.*

The proofs of theorems 3 and 4 are quite long and the interested reader is referred to [24]. We shall give here two examples. The first one shows how the solvability conditions based on the rank of (13) and (22) reflect into the analytic structure of $P(e^{j\omega})$. The second one gives an idea of some difficulties involved in the computation of the feedback matrices $K_i$, even in dimension 1.

*Example 5.* Assume in (2) $m = n = 1, A_1 = B_1 = B_2 = 1, A_2 = -1$ and in (21) $R = Q = 1$. In this case the solution of (ARE-z) can be obtained in closed form as

$$P(z) = \frac{-1 \mp \sqrt{5 + 2z + 2z^{-1}}}{-2(2 + z + z^{-1})}$$

Letting $z = e^{j\omega}$, the first solution is negative and the second is given by

$$P(e^{j\omega}) = \frac{1}{-1 + \sqrt{5 + 4\cos\omega}}$$

Since we are looking for nonnegative solutions, we consider only the second one, which is positive for $\omega \in [0, 2\pi]$, except at $\omega = \pi$, where $P(e^{j\omega})$ diverges. Actually, this is not surprising because (13) is not full rank at

$(1/2, -1/2) \in \mathcal{M}$. Hence there exist initial global states in $\ell_2$ that do not admit a stabilizing optimal feedback law.

*Example 6* Let's change the sign of $A_2$ in the previous example. In this case the unique positive definite solution of (ARE-z) along $\gamma_1$ is given by

$$P(e^{j\omega}) = \frac{2}{\sqrt{1+16(1+\cos\omega)^2} - (3+4\cos\omega)}$$

and the corresponding feedback matrix is

$$K(e^{j\omega}) = \frac{4(1+\cos\omega)}{1+\sqrt{1+16(1+\cos\omega)^2}}$$

Since $P(e^{j\omega})$ attains its minimum value at $\omega = \pi$, we have

$$J_{\min}(\mathcal{X}_0) = \frac{1}{2\pi}\int_0^{2\pi} P(e^{j\omega})||\hat{\mathcal{X}}_0(\omega)||^2\, d\omega \geq ||\mathcal{X}_0||^2 P(e^{j\pi})$$

and $J_{\min}(\mathcal{X}_0)$ can be made arbitrarily close to the lower bound, if we consider initial global states whose spectral content is concentrated in a narrow neighbourhood of $e^{j\pi}$.
The computation of the values of $K_h$ depends on the evaluation of the following integrals

$$K_h = \frac{1}{2\pi}\int_0^{2\pi} \frac{4(1+\cos\omega)\cos(\omega h)}{1+\sqrt{1+(1+\cos\omega)^2}}\, d\omega, \qquad h = 0, 1, \ldots$$

When infinitely many $K_h$s are different from zero, the optimal feedback law cannot be implemented by a finite dimensional device and the resulting closed loop system is an half plane causal 2D system, whose updating equation requires in principle to cope with an infinite dimensional state vector. To overcome storage and computation problems, it seems natural to investigate whether, in case (2) satisfies the rank condition of theorem 3, the stabilizing feedback matrix could be constrained to have all elements in the bilateral polynomials ring $\mathbb{R}[z, z^{-1}]$. An obvious advantage of this control law is that $\mathbf{u}(h, k)$ would only depend on a finite number of local states, which makes the closed loop system weakly causal [25].

The question above can be positively answered. Actually, the bilateral polynomial matrix

$$K_N(z) = \sum_{i=-N}^{N} K_i z^i$$

obtained by truncation of the Laurent series (23) gives a stabilizing state feedback, provided that $N$ is large enough. Even more, when $N$ diverges

**and $\ell_2$ initial states are considered, the corresponding cost functional $J_N$ asymptotically converges to the minimum value $J_{\min}$.**

## REFERENCES


1. R.P.Roesser, *A discrete state-space model for linear image processing*, IEEE Trans. Aut. Contr., vol. AC-20, pp.1-10, 1975.
2. S.Attasi, *Systèmes linéaires homognes à deux indices* , Rap. Laboria, 31, 1973.
3. E.Fornasini and G.Marchesini, *Algebraic realization theory of two-dimensional filters*, in Variable Structure Systems (A.Ruberti, R.Mohler eds.), L.N. in Econ. and Math. Sys., vol. 111, Springer, 1974.
4. E.Fornasini and G.Marchesini, *State space relization theory of two-dimensional filters*, IEEE Trans. Aut. Contr., vol. AC-21, pp. 484-92, 1976.
5. P.Rocha and J.C.Willems, *State for 2D systems* Lin. Alg. Appl., Special Issue on Linear Systems, vol.122, 1989.
6. P.Rocha, *Controllability of 2D systems*, IEEE Trans. Aut. Contr, vol AC-36, pp.413-423, 1991.
7. E.Fornasini, P.Rocha and S.Zampieri, *State realization of 2D finite dimensional autonomous systems*, Proc. of ECC'91, pp.715-19, Grenoble, 1991.
8. B.Lévy, M.Adams and A.Willsky *Solution and linear estimation of 2D nearest neighbour models*, IEEE Proc., vol78, n.4, pp627-41, Apr. 1990.
9. J.L.Shanks, S.Treitel and J.H.Justice, *Stability and synthesis of two dimensional recursive filters* , IEEE Trans. Audio El., vol AU-20, pp.115-28, 1972.
10. E.Fornasini and G.Marchesini, *Doubly indexed dynamical systems: state space models and structural properties* , Math. Sys. Th., vol.12, pp.59-72, 1978.
11. S.Kung, B.Lévy and T.Kailath, *New results in 2-D systems theory. Part I and Part II* , Proc. IEEE, vol 65, n.6, 1977.
12. E. Fornasini, *On the relevance of noncommutative power series in spatial filters realization* , IEEE Trans. Circ. Sys., vol. CAS-25, pp. 290-99, 1979.
13. E.Fornasini and G.Marchesini, *Global properties and duality in 2D systems* , Sys. Contr. Letters, vol.2, pp.30-38, 1982.
14. E.Fornasini and G.Marchesini, Structure and properties of 2D systems in "Multidimensional Systems", S.Tzafestas ed., M.Dekker publ., pp.33-88, 1986.
15. M.Fliess, *Matrices de Hankel*, J. Math. Pures Appl., vol. 53, pp.197-224, 1974.

16. E.Fornasini and S.Zampieri, *A note on the state space realization of 2D FIR transfer functions*, Systems and Control Letters, vol.16, n.2, pp.17-22, 1991.
17. M.Bisiacco, E.Fornasini and G.Marchesini *2D systems feedback compensation: an approach based on commutative linear transformations*, Lin. Alg. Appl., vol. 121, pp. 135-50, 1989.
18. M.Bisiacco, E.Fornasini and G.Marchesini, *Some connections between algebraic properties of pairs of matrices and 2D realizations*, in Analysis and Optimization of Systems (A.Bensoussan, J.L.Lions eds.), Springer L.N. Contr. Inf. Sci., vol 63, part 2, pp. 117-29, 1984.
19. E.Fornasini and G.Marchesini, *Stability analysis of 2D systems*, IEEE Trans. Circ. Sys., vol. CAS-27, pp.1210-17, 1980.
20. M. Bisiacco, *On the state reconstruction of 2D systems*, Systems and Control Letters, vol. 5, pp. 347-353, 1985.
21. M. Bisiacco, E. Fornasini and G. Marchesini, *Dynamic regulation of 2D systems: a state space approach*, Lin. Alg. Appl., special issue on Linear Systems, vol. 122-24, pp. 195-218, 1989.
22. M. Bisiacco, E.Fornasini and G.Marchesini *2D Systems feedback compensation: an approach based on commutative linear transformations*, Lin. Alg. Appl., vol. 121, pp.135-50, 1989.
23. Y.S.Lai and C.T.Chen, *Coprime fraction computation of 2D rational matrices*, IEEE Trans. on Aut. Contr., vol. AC-32, pp.333-36, 1987.
24. M. Bisiacco and E. Fornasini, *Optimal control of 2D systems*, SIAM Jour. Contr. and Optimiz., vol.28, pp.582-601, June 1990.
25. J. P. Guiver and N.K. Bose, *Causal and weakly causal 2D filters with applications in stabilization*, in "Multidimensional Systems Theory", pp. 52-100, Reidel Publ. Co., 1985.

Ettore Fornasini and Giovanni Marchesini
Dipartimento di Elettronica e Informatica
Università degli Studi di Padova
via Gradenigo 6/A, 35131 Padova, Italy

# Complexity and Feedback

A. LEPSCHY and U. VIARO

## 1. Introduction

In many scientific fields an increasing attention is being devoted to problems connected with complexity, as testified, for example, by the recent book by Nicolis and Prigogine [1].

Complexity is not an easy notion to define; to this end different points of view may be adopted. Perhaps, only *structured* sets pertain to complexity whereas unstructured sets (like the set of all the molecules of a gas in a given volume) do not. In fact, not only the single elements of the system are of interest but also, and above all, their relationships, so that the famous statement "the whole is more than the sum of its parts" applies very well. By using the terminology of Bertalanffy [2], distinction is to be made between *summative* characteristics and *constitutive* characteristics: complexity arises when the latter appear.

To the notion of complexity is related the one of *emergence* in that, on passing from a given level of complexity to a higher level, features and behaviours "emerge" that were not present at the lower levels even if the system elements are always of the same kind.

As happens in every research field, one may follow either a centrifugal direction, by considering systems of increasing complexity and supplying the tools for their study, or a centripetal direction, by tracing out the lowest levels for which it is possible to speak of complexity. This direction has been followed by the present authors [3–6] who have come to believe that the first level of complexity may be found in feedback systems.

The opinion might seem objectionable. In its support the following considerations will be developed in order to show how feedback can be regarded, by using the efficacious expression of Waddington [7], a powerful "tool for thought" which may account for complex behaviours in a simple way and proves useful not only in the physical and chemical sciences and their related technological applications but also in mathematics, biology, and the sciences of man and society.

In this sense, feedback models can be regarded as cybernetic models according to the suggestion of Dennis Gabor [8] who, next to Wiener classical definition, interpreted cybernetics as "the science of goal-directed, purposeful activities". Thus, according to him, cybernetic models are automata; however, one must not think that any automatic device can be considered as a cybernetic model. This is not the case, for instance, with rigidly programmed devices such as Vaucanson's duck, Jacquard's loom, or even modern computers which, "most unfortunately, are often called electronic brains" (this is Gabor's opinion and the choice of this example is indicative).

According to his definition of cybernetics, Gabor thinks that a cybernetic model must be capable of following a goal without completely determining its behaviour. Obviously, however, not all models with these characteristics are equivalent and it becomes natural to establish a hierarchy of cybernetic devices considered as possible models of natural systems.

At the lower level, we find very simple devices capable of achieving a simple and precise goal within a limited range of the relevant variables. The example presented by Gabor is the tracking servomechanism capable of attaining the aimed result even if the movement of the target is not known a priori and varies according to random characteristics, making forecasts difficult. Furthermore, the result can be obtained even if environmental conditions change and if signals are corrupted by disturbances and noises.

At higher levels, we find automata capable of achieving more complex goals up to the case of devices whose end is to "survive" through a continual learning process, even in the presence of very unstable environmental conditions.

In the mentioned paper, Gabor points out the possibility to conceive higher levels, as von Neumann had already done by considering devices capable of reproducing and evolving. However, Gabor limited himself to the consideration of devices which, at that time, were already constructed or could have been easily built.

Gabor paper is outdated, since it was written almost 40 years ago. However, the idea of including feedback structures among cybernetic models (even if at an elementary level) is certainly still valid.

## 2. Emergence of new features in simple feedback systems

As already said, an index of complexity is the emergence of new features. In fact, the systems represented by graphs with feedback loops exhibit properties that are not found in systems formed from the same elements but represented by acyclic graphs. This may be observed even in the simple case of systems comprising linear, time-invariant, finite-dimensional, strictly proper, dynamic elements only. By connecting such elements in cascade and/or in parallel, each input-output transfer function exhibits poles that may only belong to the set of the poles of all system elements. If, on the contrary, feedback loops are present, the input-output transfer functions also include poles different from those of the connected elements. In particular, if integrators and linear, static elements are connected in cascade and/or parallel, the relevant step responses will be polynomials in the time variable, whereas, if loops are present, the relevant step responses will also include transcendental functions.

The emergence of new features, which is typical of the signal flow graphs with feedback loops, arises in the flow charts containing *while loops* as well.

For example, if an arithmetical unit capable of performing additions, subtractions and multiplications is available, divisions may also be performed by adopting an algorithm whose flow chart includes loops. Similarly, if the arithmetical unit can perform all four operations, then square roots, cubic roots etc. can also be performed by using an algorithm whose flow chart includes loops, and so on.

A general approach to the problem has been adopted by Dana Scott [9] in a paper where problems of the mathematical theory of computation are studied within the framework of lattice theory. In this paper, among other things, Scott analyses in formal terms the equivalence between a while loop and an infinite flow diagram.

The first time that the possibility of substituting a feedback loop for an infinite process has been conceived, may be traced in a well known and often quoted series of notes, presented by William Thomson (Lord Kelvin) [10–12] at the Royal Society of London in 1876, about the mechanical solution of the second order linear differential equation:

$$\frac{d}{dx}\left(\frac{1}{p(x)}\cdot\frac{dy}{dx}\right) = -y(x) \quad , \quad y(0) = 0 \quad , \quad \left.\frac{dy}{dx}\right|_{x=0} = C\,p(0)$$

by computing the elements

$$y_i(x) = \int_0^x p(z)\left\{C - \int_0^z y_{i-1}(t)\,dt\right\} dz \quad , \quad i = 2, 3, \ldots$$

of a sequence which converges, under suitable assumptions, to the solution of the equation.

In his paper, Lord Kelvin writes: "Now let my brother's [J. Thomson] integrator be applied to find

$$C - \int_0^z y_1(t)\,dt$$

and let its result feed ... a second machine, which shall find the integral of the product of its result into $p(z)\,dz$. The second machine will give out continuously the value of $y_2$. Use again the same procedure with $y_2$ instead of $y_1$, and then $y_3$, and so on ... So far I had gone and was satisfied, feeling

I had done what I wished to do for many years. But then came a pleasing surprise. Compel agreement between the function fed into the double machine and that given out by it." Here Lord Kelvin shows how this equality can be obtained by means of a mechanical connection between the output of the second integrator and the input of the first one. Then he says: "Thus I was led to a conclusion which was quite unexpected; and it seems to me very remarkable that the general differential equation of the second order with variable coefficients may be ... continuously, and in a single process solved by a machine."

## 3. Historical development of technological feedback systems

The word "feedback" has been coined around 1920 in the frame of wireless engineering. In this field, the application of feedback led to the design of electronic amplifiers and oscillators with peculiar characteristics and the so-called classical theory of feedback systems was formulated thanks to the contributions of Nyquist [13], Black [14] and Bode [15].

During World War II, this theory was first applied to the design of servomechanisms.

Since then, feedback theory has mainly been exploited in control applications. It must be observed, however, that regulating and control systems based on feedback had been built much earlier.

According to Otto Mayr [16], the first instance in which the principle of feedback was applied to an artificial device is the water clock built by Ktesibios in the third century B.C.; from this prototype originate some interesting medieval Islamic clocks too.

Instances of feedback control may also be recognised in devices due to Philon of Byzantium (second half of the third century B.C.) and to Heron of Alexandria (presumably, first century A.D.).

The present authors [17] have shown that the medieval verge-and-foliot clocks too can be given an explanatory model based on feedback.

Anyway, a role that is particularly important in the history of feedback in technology is played by the temperature regulators built by Cornelis Drebbel (1572–1633). It is not by chance that Drebbel has been credited with the invention of the thermometer: a temperature regulator must indeed include a temperature transducer in the feedback path.

Drebbel's thermostat precedes by half a century the particular velocity regulating mechanism of the pendulum clock realized by Ch. Huygens in 1657 but already conceived by Galilei.

By following the history of the technological applications of feedback, mention should be made of Papin's pressure-cooker (1681) and of Watt's centrifugal governor (1788) that uses the well-known speed regulating flyball mechanism already present in windmills.

Watt's age coincides, at least in Great Britain, with the beginning of the industrial revolution. As this revolution progresses, feedback devices for control purposes become more and more popular and sophisticated. It must be observed, however, that the diffusion of these techniques did not correspond to a complete awareness of its very nature. Even Maxwell's celebrated paper [18], which is considered as one of the first scientific contributions to the theory of control and is characterized by a neat approach leading to valid results, does not seem to refer to the closed-loop structure of the considered governors.

Such awareness, as already mentioned, is acquired in the field of electronics only in the third and fourth decades of our century.

## 4. The role of feedback in explanatory models

As is known, the interest of feedback is not limited to its use for the design of certain "artificial" systems in technological applications, because it also supplies a suggestive model for interpreting many natural phenomena or even phenomena, such as those studied in economics, which, though not being traditionally considered as "natural", in their spontaneous development cannot be considered as the result of a conscious use of feedback (even if this technique may suggest, as has indeed happened, control interventions intended to modify the spontaneous course of these phenomena in a desired sense).

About these models it is useful to make some preliminary considerations. It is usual to classify mathematical models (or, more generally, non-physical models including those conceptual models where the formalisation phase is at its beginning stage and the quantification phase has yet to be approached) into classes at whose extremes we may place *explanatory* and *predictive* models.

The first ones correspond to the formalization of a hypothesis and their interest consists much more in the elegance with which this hypothesis is translated into formal relations than in their ability to account, in a reasonably reliable way, for the relevant phenomena, as they actually take place in the real world. A predictive model, instead, may be considered valid if it is effectively usable not only for summarising what is already known but also to provide reliable forecasts.

It often happens, in such a context, that the main advantage of an explanatory model consists in its *simplicity*, while almost always the condition for a predictive model to be suited to its purpose is that it is adequately complex and articulated. Referring to explanatory models, it may be useful to remember what was noted for the case of the tortoises of Walter Gray Walter and other more or less famous cybernetic animals towards the end of the 1940's and at the beginning of the 1950's. The creators of these machines, exhibiting a soomorphic behaviour, did not mean to suggest that the mechanism used by them was that actually present "in nature" but they thought it interesting to show how an apparently complex behaviour could be obtained by combining few devices whose operation mechanism is very simple.

In the following, we shall limit ourselves to the use of explanatory models that are deliberately elementary. From these, obviously, the reader must not expect a detailed, realistic and thorough description of the phenomena, as this would imply a far greater model complexity. Nevertheless, the examples we shall present have, we feel, their own intrinsic interest because they allow us to verify how feedback systems, in spite of the simplicity of their components, may exhibit various kinds of behaviour, different from those of the subsystems which influence each other; in such a way, even if these are very simple, linking them into feedback loops may give rise to many interesting overall behaviours.

## 5. Feedback models in the sciences of nature

Speaking about explanatory feedback models in the field of physics entails some difficulties since it is often arbitrary to distinguish between the situations in which feedback is inherent and those in which feedback arises from our interpretation.

Let us refer, for example, to Newton's second law of motion, usually expressed in the form:

$$F = mA$$

which may be phrased as "the force $F$ applied to a body having mass $m$ makes it accelerate with an acceleration $A$ proportional to the force itself". In this case, we recognize a forward path that leads from $F$ to $A$ and do not see any feedback path from $A$ to $F$. Yet, according to d'Alembert's principle, this law of dynamics can be presented as a law of statics expressing the balance of two opposing forces:

$$F - mA = 0$$

which corresponds to the feedback diagram of Fig. 1. Here, the forward path, accounts for the fact that a difference, if any, between the action $F$ and the reaction $mA$ would give rise to a change of $A$ such as to annihilate the above difference: in this case, the forward path is formed by an infinite gain element.

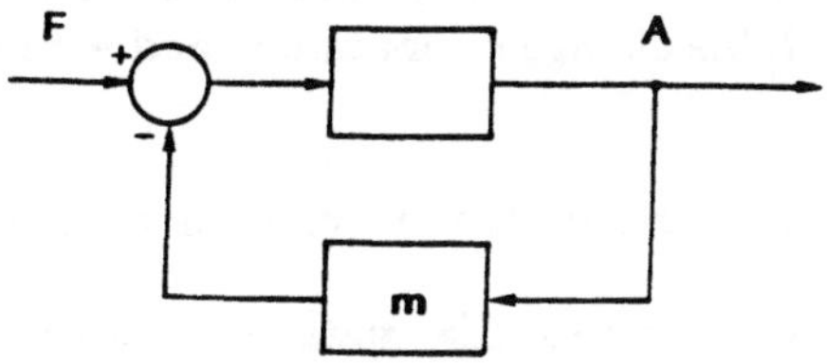


Fig. 1

Feedback diagrams of the same kind can be adopted every time Newton's third law of motion applies (*Principia*, 58:13): *To every action there is always opposed a reaction; or the mutual actions of two bodies upon each other are always equal and directed to contrary parts.*

The principle of action and reaction has its counterparts in other fields of physics and chemistry.

A particular case, very familiar to people involved in electromagnetics, is the Lenz (1804–1865) rule usually stated in the following terms: *A current brought into action by an induced electromotive force always produces effects which oppose the inducing action.*

An almost identical statement corresponds to Le Chatelier's principle in physical chemistry: *In a system at equilibrium, the change of a parameter causes a transformation which, if it took place independently, would produce a change of the considered parameter in the opposite direction.*

All these principles express that, in case of disturbance, the system develops forces which counteract the disturbance and restore a state of equilibrium; they are derivations from the principle of minimum effect.

More suggestive and interesting are, perhaps, the feedback models used in biology, in particular those accounting for homeostasis.

Population dynamics and ecology are other fields in which feedback models may be used for explanatory purposes and, sometimes, for predictive purposes as well.

By now public opinion seems to be well aware of the feedback relations among the relevant variables. We like to mention, in this regard, some excerpts from an essay by M. Ceruti and C. Testa [19], which did not appear in a scientific magazine but in a political journal: "... what the sciences of the living say about the dynamics of the biosphere and of each ecosystem denies a clear distinction between organisms and environment. Organisms contribute to the history and character of the environment to the same extent as the environment contributes to the history and character of organisms. Emphasis should thus be given to the ideas of mutual evolution, construction and production of the organisms and the environment. The only way to maintain certain equilibria between organisms and environments is that of maintaining the richness of interactions through the continual, unavoidable change of organisms and environment".

## 6. Feedback models in the sciences of man and society

Concerning a hard science like mathematical economics, we shall only mention the importance of feedback in economy models as those of Samuelson and Hicks [20,21] and of Phillips [22,23].

A greater interest deserve the feedback models in the so-called soft sciences in which a formal and quantitative approach is not usual. In this field we shall mention first the work of Lanchester [24] concerning the evolution of military conflicts; in the simplest version of such a model, we assume that each one suffers losses proportional to the size of the opposing army; therefore, denoting by $y_1$ and $y_2$ the sizes of the two armies, we have:

$$dy_1/dt = -k_2\, y_2$$

$$dy_2/dt = -k_1\, y_1$$

where the coefficients $k_1$ and $k_2$ are positive numbers as high as the corresponding armament is lethal. The related block diagram exhibits a feedback loop.

Another interesting model has been suggested by Richardson [25] in order to represent the armaments race in quantitative terms. In the two-power case, if the armament levels are indicated by $y_1$ and $y_2$, the simplest form of the model is the following:

$$dy_1/dt = a_1\, y_2 - b_1\, y_1 + c_1$$

$$dy_2/dt = a_2\, y_1 - b_2\, y_2 + c_2$$

where the coefficients $a_i$ are called *defense* coefficients, the coefficients $b_i$ are called *fatigue-and-expense* coefficients and the coefficients $c_i$ are called *attitude* coefficients and are positive in the case of an antagonistic attitude and negative in the case of a cooperative attitude ("positive" and "negative" obviously refer to the algebraic and not to the ethical meaning of the terms). In this case, the block diagram exhibits three interconnected feedback loops.

Feedback loops are also present in the diagrams representing the results of the studies of Maruyama [26,27] on the Danish national character and on the algebra of interpersonal relations. In particular, he considers attitudes which favour other attitudes in the same people in a direct or, more often, mediated way; therefore, closed loops are present which induce a mutual strengthening of these attitudes. The attitudes so strengthened merge to form the typical "character" of a more or less numerous group of people and, in this particular case, that of an entire nation.

Concerning the effects of feedback in the political and historical phenomena, we may perhaps be allowed to quote an academic diversion, to which one of the present authors has contributed [28], aiming at offering an interpretation in terms of system and control theory of some aspects of Arnold Toynbee's ideas on the history of civilisations. It has also been noted [29] that a simplified form of this feedback model is susceptible of valid interpretations even outside this context.

If the indulgence asked for self-citation still holds, we may also recall a complex model about electoral behaviour [30] where feedback plays an important role.

## 7. Feedback models and philosophical problems

Let us now consider some philosophical aspects related to the use of feedback models.

In the case of biological systems, it has been noted that such models allow a *mechanistic* interpretation (where the effects depend only on *efficient causes*, to use a traditional term of Aristotelian-Scholastic derivation) of phenomena in which there is, at least apparently, a *finalistic* behaviour (i.e., a goal-oriented operation, where the effects appear to be influenced by *final causes*); this is typical of the feedback models for homeostasis, tropism and other goal-seeking behaviours.

Feedback can, however, offer suggestions on how to formalize problems that are the immediate object of a philosophical reflection. This will be illustrated with reference to the dialectics of *thesis*, *antithesis* and *synthesis* of Hegelian idealism. According to the vulgate version of it, dialectics consists in posing an "abstract and limited" concept (thesis), suppressing this concept as something finite, passing to its opposite (antithesis), and finally to the synthesis of the preceding determinations, which retains what in each one is valid and, at the same time, surpasses them both. Hegel actually writes about the intellectual, the dialectical and the speculative (or rational positive) moment, but the dialectics is not linked to the second only, but rather to the whole movement and, owing to the identity of rational and real, is the key not only of the scientific progress but also of the "becoming" of reality. The latter moves and "becomes" dialectically so that triads of thesis, antithesis and synthesis are identified everywhere: the antithesis consists in the negation, being "other" with respect to the thesis, and the synthesis is the reconstruction of a unity and the becoming true of the first and the second.

This is not the place to say whether or how far this presentation corresponds to Hegel's thought. However things may be, the dialectics of thesis/antithesis/synthesis constitutes a commonplace in the diffusion of many trends of modern thought, particularly those of the more or less orthodox forms of idealism and marxism.

This "popular" version can be presented in graphic terms as in Fig. 2a, where the process that leads from the thesis $T$ to the antithesis $A$ is represented by block $S_1$ and the one leading from the thesis and the antithesis to the synthesis $S$ is represented by block $S_2$.

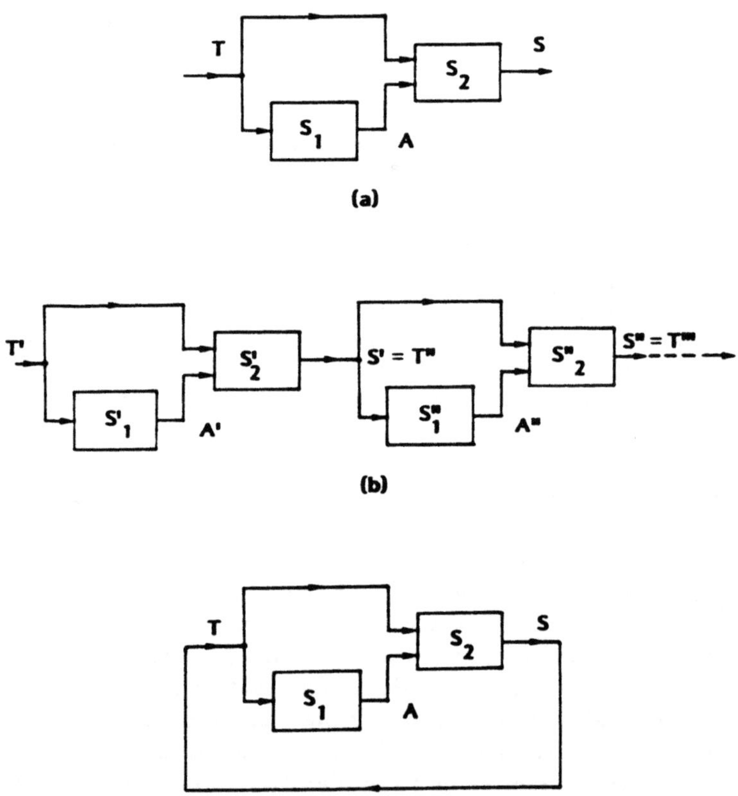


Fig. 2

Dialectical progress, however, is something continuous, where the synthesis arrived at in the first phase is reproposed as a new thesis, which generates a new antithesis and combines with it, hence giving place to another synthesis, and so on indefinitely.

This is roughly outlined in Fig. 2b, where the chain should be indefinitely prolonged.

This is precisely the situation encountered in the already mentioned mechanical computer of W. Thomson [10–12] and in the indefinite graphs studied by D. Scott [9] that are equivalent to a feedback loop. The same procedure can be applied in the present case, thus arriving at the diagram of Fig. 2c: the synthesis $S$ obtained according to the diagram of Fig. 2a is reported as a thesis $T$ at the input of the same diagram.

In this way, feedback interprets the dialectical process satisfactorily, also accounting for that feature (not always easy to illustrate "in words") according to which the passage from thesis and antithesis to synthesis, the posing of synthesis as a new thesis etc. are to be considered as moments *not of a temporal but of a logical* sequence.

At the same time, this allows us to understand the compatibility of such point of view with the not too few examples in which, in the historical development, one passes, through temporally distinct and successive phases, from a position initially adopted to another position which exhibits contrasting and opposing characteristics, until one finally tends towards a more balanced, in a sense, intermediate, position. In fact, such a behaviour characterizes many feedback systems where the evolution of the output exhibits damped oscillations: a feature which is quite familiar to those who deal with feedback control systems.

This way of reasoning is appropriate to the situation where $T$, $A$ and $S$ correspond to physical variables that take on quantifiable values, expressed for instance by real numbers. The same considerations hold in economics. Such a situation may also be accepted in the context of Hegel's conceptions and is typical of Engel's *Dialektik der Natur.* One may therefore associate with the blocks of Fig. 2c functional relationships which link inputs to outputs.

If we adopt the assumption, certainly restrictive but often acceptable in the context of a more or less rough approximation, that such relationships are linear and time-invariant, the blocks in question can be characterised by means of transfer functions. In Fig. 3a the transference that links the thesis to the antithesis is indicated by $-F_1$ (the minus sign clearly shows the opposition between the two), and the relationship that links thesis and antithesis to synthesis is represented (necessarily, given the assumptions) by a linear combination: $S = F_{2T}T + F_{2A}A$ ; finally, the transference representing the process by which the synthesis is posed as a new thesis is indicated by $F_3$.

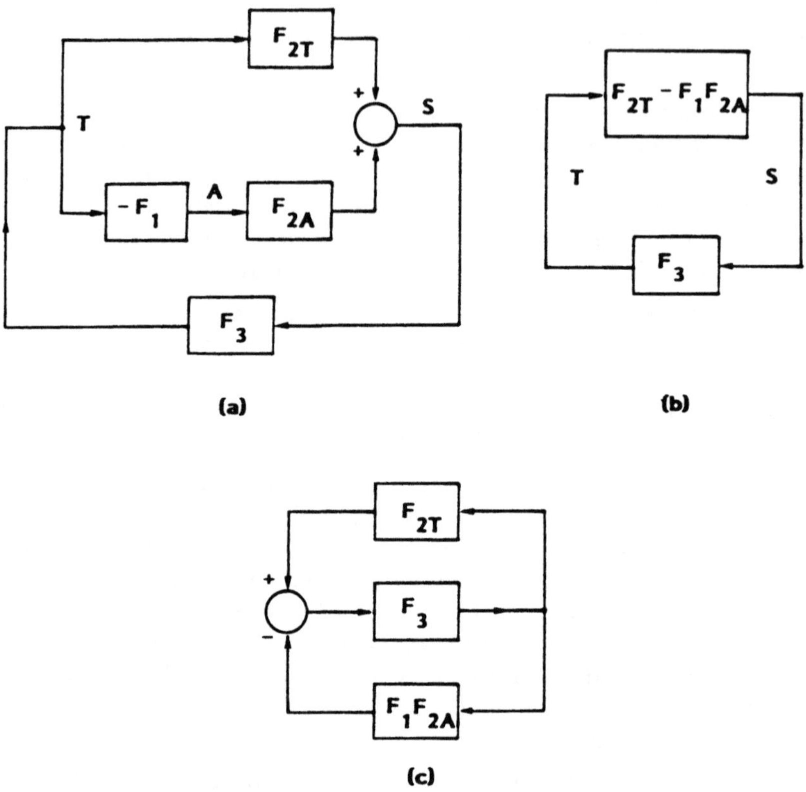


Fig. 3

Fig. 3b reabsorbs the link between thesis and synthesis in a single block of transference $F_{2T} - F_1F_{2A}$. Alternatively, the diagram of Fig. 3c reorganises the blocks of Fig. 3a in a different way, by putting block $F_3$ in a central position; block $F_2$ is the feedback path of a positive feedback loop, whereas the feedback path of a negative feedback loop is formed by a block of transference $F_1F_{2A}$. The above considerations allow us to understand how rather simple transfer functions $F_1$, $F_{2T}$, $F_{2A}$ and $F_3$ render the model suitable to simulate a very large number of different phenomena which may therefore easily be interpreted in terms of Hegelian dialectics.

Feedback may profitably be used to state in a more satisfactory way other problems of interest in philosophy. Let us consider, for instance, a classic dispute regarding the methodology of science. According to the naive inductivism, induction leads from the observed facts (O.F.) to the theory

(T), and deduction allows us to make forecasts (F) and derive explanations (E), which may be represented with the open-loop diagram of Fig. 4a.

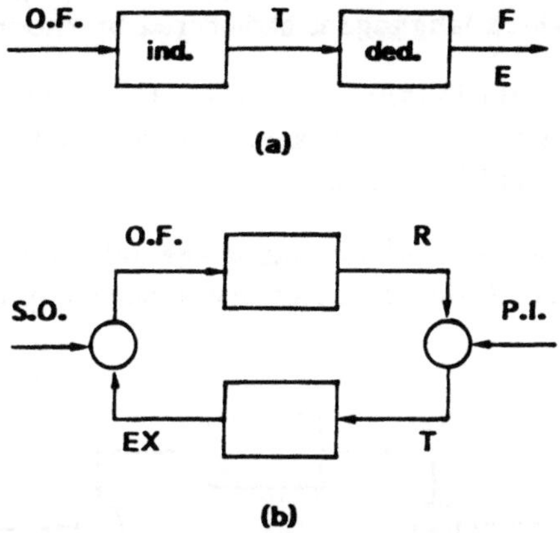


Fig. 4

It has been objected, however, that observations depend to some extent on the theory, observative propositions presuppose it, and experiments are guided by it. Therefore, one may reasonably resort to the feedback diagram of Fig. 4b in which two distinct inputs are present, i.e., the spontaneous observations (S.O.) and the preliminary ideas on the related phenomena (P.I.) that are often unconscious. The observed facts (O.F.) not only come from the spontaneous observations but also from the experiments (EX) performed according to the theory (T); the last, in turn, is enriched, enlarged, modified or, if necessary, replaced by a different theory because of reflections (R) on the observed facts. The diagram of Fig. 4b seems to be interesting since it also accounts both for aspects that are more directly related to the methodology of science and for aspects that are connected with the psychology of discovery.

Obviously, a very simple diagram may only explain the general features of the phenomena under consideration (in other words, it may be regarded as a simple explanatory model); a more detailed description requires more complex diagrams whose kernel, however, is the previous simple model.

These considerations also apply to the following example, based on a simplified version of the sketch presented in [31] concerning the hypotheses on the relationship between language and thought. With a rough simplification, four hypotheses may be considered:

- the so-called strong version of the Sapir-Worf hypothesis (or, rather, a somewhat arbitrary extension of it) according to which language itself gives shape to ideas (different *Weltanschauungen* correspond to different languages);

- the strong version of the Piaget hypothesis (cognitive hypothesis) according to which language is structured by thought and reflects it;
- the genetic-evolutionary hypothesis of Vygotsky for which thought and language develop simultaneously to a continual bidirectional dynamics of mutual influences;
- the Boas hypothesis which claims the substantial independence and the absence of a tight and necessary relationship between thought and language.

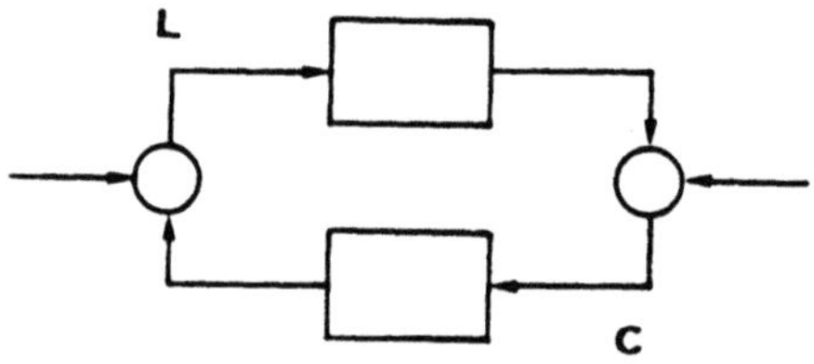


Fig. 5

If we adopt an eclectic approach, we may suggest the feedback diagram of Fig. 5, where two inputs are present which, according to the ideas of Franz Boas, correspond, respectively, to the component of language L irreducible to the influences of culture C and to the component of thought irreducible to the influences of language. The language, however, contributes to culture through a feedback path according to the ideas of Edward Sapir and Benjamin Worf, in the same way as culture contributes to language through the complementary feedback path according to the ideas of Jean Piaget. Finally, the feedback configuration of the overall diagram turns out to be consistent with the ideas of Lev Semonovich Vygotsky.

## REFERENCES

[1] G. Nicolis and I. Prigogine, *Exploring Complexity. An Introduction*, R. Piper Gmb-H & Co., München, 1987.

[2] L. von Bertalanffy, *General Systems Theory*, G. Brasiller, New York, 1968.

[3] A. Lepschy, *La retroazione: una tecnica e uno "strumento per pensare"*, Scienza e Cultura, 6 (1983), 159–214.

[4] A. Lepschy, *Mathematical models and feedback in the study of complex systems*, in E. Belardinelli (ed.), Imola Conference on University and Research, Martello, Milano, 1990, 117–124.

[5] R. Calimani and A. Lepschy, *Feedback*, Garzanti, Milano, 1990.

[6] A. Lepschy, G.A. Mian, and U. Viaro, *La retroazione: una tecnica e un modo di interpretare fenomeni di varie scienze*, Nuova Secondaria, **8** (1991), 21–32, 49–54, 59–64.

[7] C.H. Waddington, *Tools for Thought*, Cape, London, 1977.

[8] D. Gabor, *Models in cybernetics*, in I Modelli nella Tecnica, Accademia Nazionale dei Lincei, Roma, 1956, vol. 2, 236–274.

[9] D. Scott, *The lattice of flow diagrams*, in E. Engeler (ed.), Symposium on Semantics of Algorithmic Languages, Springer, Berlin, 1971, 311–372.

[10] W. Thomson, *On an instrument for calculating the integral of the product of two given functions*, Proc. Royal Society, **24** (1876), 266–268.

[11] W. Thomson, *Mechanical integration of the linear differential equation of the second order with variable coefficients*, Proc. Royal Society, **24** (1876), 269–271.

[12] W. Thomson, *Mechanical integration of general linear differential equation of any order with variable coefficients*, Proc. Royal Society, **24** (1876), 271–273.

[13] H. Nyquist, *Regeneration theory*, Bell Syst. Tech. J., **11** (1932), 120–147.

[14] H.S. Black, *Stabilized feedback amplifiers*, Bell Syst. Tech. J., **13** (1934), 1–13.

[15] H.W. Bode, *Relations between attenuation and phase in feedback amplifier design*, Bell Syst. Tech. J., **19** (1940), 421–454.

[16] O. Mayr, *The Origins of Feedback Control*, The M.I.T. Press, Cambridge (Mass.), 1970.

[17] A. Lepschy, G.A. Mian, and U. Viaro, *Feedback control in ancient water and mechanical clocks*, IEEE Trans. Educ.; to appear.

[18] J.S. Maxwell, *On governors*, Proc. Royal Society, **16** (1867/68), 270–283.

[19] M. Ceruti and C. Testa, *Gli otto peccati mortali della cultura verde*, MicroMega, **3/91** (1991), 7–15.

[20] P. Samuelson, *Interaction between the multiplier analysis and the principle of acceleration*, Review of Econ. Statistics, **21** (1939), 75–79.

[21] J.R. Hicks, *A Contribution to the Theory of the Trade Cycle*, Clarendon Press, London, 1950.

[22] A.W. Phillips, *Stabilisation policy in a closed economy*, The Economic Jour., **64** (1954), 290–323.

[23] A.W. Phillips, *Stabilisation policy and time-forms of lagged responses*, The Economic Jour., **67** (1957), 265–277.

[24] F.W. Lanchester, *Mathematics in warfare*, in The World of Mathematics, Simon and Schuster, New York, 1956, vol. 4, 2138–2157.

[25] L.F. Richardson, *Arms and Insecurity*, Stevens, London, 1960.

[26] M. Maruyama, *The multilateral mutual causal relationships among the modes of communication, sociometric pattern and the intellectual orientation in the Danish culture*, Phylon, **22** (1961), 41–58.

[27] M. Maruyama, *Algebra of interpersonal interaction*, Methodos, **13** (1962), 25–36, 49–50.

[28] A. Lepschy and S. Milo, *Historical event dynamics and "A Study of History" by Arnold Toynbee*, Scientia, **70** (1976), 35–50.

[29] C. Cobelli, A. Lepschy, and S. Milo, *Models of societal and historical phenomena*, IV Polish - Italian Meeting on Mathematical System Theory in Management and Economics, Internal Report n° 71, Istituto di Elettrotecnica e di Elettronica, Università di Trieste, Trieste, 1978.

[30] C. Cobelli, G. Langella, and A. Lepschy, *A system approach to the modelling of voting behaviour*, in J. Gutenbaum and M. Niezgòdka (eds.), Applications of System Theory to Economics, Management and Technology, Panstwowe Wydawnictwo Naukowe, Warszawa-Lodz, 1980, 189–199.

[31] E. De Felice, *Le Parole d'Oggi*, Mondadori, Milano, 1984.

*A. Lepschy, U. Viaro*
*Department of Electronics and Informatics*
*University of Padova*
*via Gradenigo 6/A*
*35131 Padova, Italy*

# Point Estimation, Stochastic Approximation, and Robust Kalman Filtering

SANJOY K. MITTER and IRVIN C. SCHICK

*Dedicated to Antonio Ruberti on the occasion of his sixty-fifth birthday.*

**Abstract.** Significantly non-normal noise, and particularly the presence of outliers, severely degrades the performance of the Kalman Filter, resulting in poor state estimates, non-white residuals, and invalid inference. An approach to robustifying the Kalman Filter based on minimax theory is described. The relationship between the minimax robust estimator of location formulated by Huber, its recursive versions based on the stochastic approximation procedure of Robbins and Monro, and an approximate conditional mean filter derived via asymptotic expansion, is shown. Consistency and asymptotic normality results are given for the stochastic approximation recursion in the case of multivariate time-varying stochastic linear dynamic systems with no process noise. A first-order approximation is given for the conditional prior distribution of the state in the presence of $\varepsilon$-contaminated normal observation noise and normal process noise. This distribution is then used to derive a first-order approximation of the conditional mean estimator for the case where both observation and process noise are present.

## 1. Introduction

Kalman Filtering has found an exceptionally broad range of applications, not only for estimating the state of a dynamic system in the presence of process and observation noise, but also for simultaneously estimating model parameters, choosing among several competing models, and detecting abrupt changes in the states, the parameters, or the form of the model. It is a remarkably versatile estimator, originally derived via orthogonal projections as a generalization of the Wiener filter to non-stationary processes, then shown to be optimal in a variety of settings: as the weighted least-squares solution to a regression problem, without regard to distributional assumptions; as the Bayes estimator assuming Gaussian noise, without regard to the cost functional; and as the solution to various game theoretic problems.

Nevertheless, the Kalman Filter breaks down catastrophically in the presence of heavy-tailed noise, i.e. outliers. Even rare occurrences of unusually large observations severely degrade its performance, resulting in poor state estimates, non-white residuals, and invalid inference. A robust version of the Kalman Filter would have to satisfy two objectives: be as nearly optimal as possible when there are no outliers (under "nominal" conditions); and be resistant to outliers when they do occur (track the underlying trajectory without being unduly affected by spurious observations).

Below, the notation $\mathbf{L}(\underline{x})$ denotes the probability law of the random vector $\underline{x}$, $\mathbf{N}(\underline{\mu}, \Sigma)$ denotes a multivariate normal distribution with mean $\underline{\mu}$ and covariance $\Sigma$, and $\mathbf{N}(\underline{x}; \underline{\mu}, \Sigma)$ is the corresponding probability density function.

Consider the model

$$\underline{z}_n = H_n \, \underline{\theta}_n + D_n \, \underline{v}_n, \tag{1.1}$$

where

$$\underline{\theta}_{n+1} = F_n \, \underline{\theta}_n + \underline{w}_n, \tag{1.2}$$

$n = 0, 1, \cdots$ denotes discrete time; $\underline{\theta}_n \in \mathbf{R}^q$ is the system state, with a random initial value distributed as $\mathbf{L}(\underline{\theta}_0) = \mathbf{N}(\overline{\underline{\theta}}_0, \Sigma_0)$; $\underline{z}_n \in \mathbf{R}^p$ is the observation (measurement); $\underline{w}_n \in \mathbf{R}^q$ is the process (plant) noise distributed as $\mathbf{L}(\underline{w}_n) = \mathbf{N}(\underline{0}, Q_n)$; $\underline{v}_n \in \mathbf{R}^p$ is the observation (measurement) noise distributed as $\mathbf{L}(\underline{v}_n) = F$, with $E[\underline{v}_n] = \underline{0}$ and $E[\underline{v}_n \underline{v}_n^{\mathrm{T}}] = R$; $\{F_n\}$, $\{H_n\}$, $\{D_n\}$, $\{Q_n\}$, $\Sigma_0$ and $R$ are known matrices or sequences of matrices with appropriate dimensions; $\overline{\underline{\theta}}_0 \in \mathbf{R}^q$ is a known vector; and finally $\underline{\theta}_0$, $\underline{w}_n$, and $\underline{v}_n$ are mutually independent for all $n$.

The Kalman Filter is the estimator $\hat{\underline{\theta}}_n$ of the state $\underline{\theta}_n$ given the observations $\{ \underline{z}_1, \cdots, \underline{z}_n \}$, and obeys the well-known recursion

$$\hat{\underline{\theta}}_{n+1} = F_n \, \hat{\underline{\theta}}_n + K_{n+1} \, \underline{\gamma}_{n+1}, \tag{1.3}$$

where

$$\underline{\gamma}_{n+1} = \underline{z}_{n+1} - H_{n+1} \, F_n \, \hat{\underline{\theta}}_n \tag{1.4}$$

is the innovation at time $n+1$ and

$$\Gamma_{n+1} = H_{n+1} \, M_{n+1} \, H_{n+1}^{\mathrm{T}} + D_{n+1} \, R \, D_{n+1}^{\mathrm{T}} \tag{1.5}$$

is its covariance,

$$K_{n+1} = M_{n+1} H_{n+1}^{\mathrm{T}} \Gamma_{n+1}^{-1} \tag{1.6}$$

is the gain,

$$M_{n+1} = F_n \Sigma_n F_n^{\mathrm{T}} + Q_n \tag{1.7}$$

is the *a priori* estimation error covariance at time $n+1$ (i.e. before updating by the observation $\underline{z}_{n+1}$), and

$$\Sigma_{n+1} = ( I - K_{n+1} H_{n+1} ) M_{n+1} \tag{1.8}$$

is the *a posteriori* estimation error covariance at time $n+1$ (i.e. after updating). The inital condition is

$$\hat{\underline{\theta}}_0 = \overline{\underline{\theta}}_0. \tag{1.9}$$

As is clear from (1.3)-(1.4), the estimate is a linear function of the observation, a characteristic that is optimal only in the case of normally distributed noise (Goel and DeGroot [4]). Similarly, (1.6)-(1.8) show that the gain and covariance are independent of the data, a property related once again to the assumption of normality. Finally, in the Gaussian case $F = \mathbf{N}( \underline{0}, R )$, the residual (innovation) sequence $\{ \underline{\gamma}_1 , \cdots , \underline{\gamma}_n \}$ is white and is distributed as $\mathbf{L}(\underline{\gamma}_i) = \mathbf{N}( \underline{0}, \Gamma_i )$. When $F$ is not normal, on the other hand, the state estimation error can grow without bound (since the estimate is a linear function of the observation noise), the residual sequence becomes colored, and residuals become non-normal. Thus, not only is the estimate poor, but furthermore invalid inference would result from utilizing the residual sequence in the case of significant excursions from normality.

Past efforts to mitigate the effects of outliers on the Kalman Filter range from *ad hoc* practices such as simply discarding observations for which residuals are "too large," to more formal approaches based on non-parametric statistics, Bayesian methods, or minimax theory. The purpose of this paper is to review robust recursive estimation in the context of Huber's theory of minimax robust estimation. The relationship between robust point estimation, recursive robust estimation by means of stochastic approximation, and approximate conditional mean estimation based on asymptotic expansion, is described. This provides a rigorous basis for sub-optimal filtering in the presence of non-Gaussian noise.

## 2. Robust Point Estimation

Let $( \mathbf{R}, \mathbf{B}, \lambda )$ be a measure space, where $\mathbf{R}$ is the real line, $\mathbf{B}$ the Borel $\sigma$-algebra, and $\lambda$ the Lebesgue measure. Let $F$ be a zero-mean probability measure on $( \mathbf{R}, \mathbf{B} )$ such that $F$ is absolutely continuous with respect to $\lambda$ and admits the density $f(x) := dF(x) / dx$ a.s. in accordance with the Radon-Nikodym theorem.

For some positive integer $n$, let $\{ v_1, \cdots, v_n \}$ be a sample of independent random variates taking values in $\mathbf{R}$, with common distribution $F$. Let $\theta \in \boldsymbol{\Theta} \subseteq \mathbf{R}$ be a location parameter, and define the observations $z_i$ by

$$z_i := \theta + v_i, \qquad (2.1)$$

for $i = 1, \cdots, n$. Let $\mathbf{R}^n$ be the product of $n$ copies of $\mathbf{R}$, and let $T_n : \mathbf{R}^n \to \boldsymbol{\Theta}$ be an estimator for the parameter $\theta$.

A broad class of such estimators are solutions $T_n( z_1, \cdots, z_n )$ to maximization problems of the form

$$\max_{T \in \boldsymbol{\Theta}} \sum_{i=1}^{n} \rho( z_i - T ), \qquad (2.2)$$

for some suitably chosen real-valued function $\rho$. For instance, if $\rho(x) := \log f(x)$, then the solution of (2.2) is the maximum likelihood estimate; if $\rho(x) := - \| x \|^2$, it is the least squares estimate; if $\rho(x) := - | x |$, it is the minimum modulus estimate, i.e. the median.

Robust estimation answers the need raised by the common situation where the distribution function $F$ is not precisely known. A class of solutions to such problems is based on *minimax theory*: the distribution $F$ is assumed to be a member of some set of distributions, and the best estimator is sought for the least favorable member of that set, in terms of some given measure of performance. While this approach is pessimistic, since the true distribution may well not be the least favorable one, it has the advantage of providing an optimum lower bound on performance. Minimax theory has been used as a conservative approach to hypothesis testing and decision problems in the presence of statistical indeterminacy; the first to formulate a minimax theory of robust estimation was apparently Huber [5]-[9].

A suitable measure of performance for the robust estimation of a location parameter is the *asymptotic variance*. This choice has several advantages. First, as is usually the case, asymptotic analytical results are considerably easier to obtain than small sample results. Furthermore, under certain conditions, the estimator can be shown to be asymptotically normal, which has the added benefit of making possible hypothesis testing and the construction of confidence intervals. Second, the sample variance is strongly dependent on the tails of the distribution; indeed, for any estimator whose value is always contained within the convex hull of the observations, the supremum of its actual variance is infinite. Thus, the asymptotic variance is a better performance measure than the sample variance. Third, the asymptotic variance is related to the Fisher Information through the Cramér-Rao inequality, and the Fisher Information lends itself well to algebraic manipulation.

The procedure, then, is as follows. It is postulated that the unknown distribution function $F$ is a member of a certain set $\mathbf{P}$ of distributions on

( **R**, **B** ). The least favorable distribution is that member of **P** leading to the largest asymptotic variance, or, equivalently (provided that the Cramér-Rao lower bound is achieved), the one minimizing the Fisher Information. Since the maximum likelihood estimator is known to achieve the Cramér-Rao lower bound, demonstrating that the least favorable distribution and the maximum likelihood estimator associated with it are a saddle point yields a minimax robust estimator. (For a theorem that provides regularity conditions under which a distribution-estimator pair is a saddle-point solution, see Verdú and Poor [25].)

The existence of a least favorable distribution has been investigated by several researchers; indeed, one of the primary tasks of minimax theory is deriving sufficient conditions for the existence of such distributions. In general, proofs of existence involve some topological restrictions that are problematical since in many cases the sets of probability distributions of interest are not tight, so that their closures are not compact in the weak topology. To circumvent this difficulty, Huber proposes to endow the set **P** with the "vague" topology, defined as the weakest topology such that maps $P \rightarrow \int \psi \, dP$ are continuous for all continuous functions $\psi$ with compact support. Let $I(P)$ denote the Fisher Information for the distribution $P$, and suppose that every $P \in \mathbf{P}$ admits a density in accordance with the Radon-Nikodym theorem. In this framework, the existence and uniqueness of the least favorable distribution in **P** are established by the following theorem due to Huber:

**Theorem 2.1** *If* **P** *is vaguely compact and convex, then there is a* $P_0 \in \mathbf{P}$ *minimizing* $I(P)$. *If, furthermore,* $0 < I(P_0) < \infty$ *and the support of the corresponding density* $f_0$ *is convex, then* $P_0$ *is unique.*

**Proof** See Huber [5:86-90], [6:81-85], [9:79-81]. QED

Let $\rho$ be a continuous, convex, real-valued function of a real variable, whose derivative $\psi$ exists a.e. and takes both negative and positive values. An alternative way of stating (2.2), provided that $\Theta$ is an open set, is

$$\sum_{i=1}^{n} \psi(z_i - T) = 0 \tag{2.3}$$

at $T = T_n(z_1, \cdots, z_n)$, where $\psi(z - T) := \alpha \, \partial\rho(z - T) / \partial T$ a.e., and $\alpha$ is an arbitrary constant. Choosing $\alpha = -1$ for aesthetic reasons, it follows that for the case of the maximum likelihood estimator associated with the least favorable density,

$$\psi(z - T) = -\frac{\partial}{\partial T} \log f_0(z - T) \tag{2.4}$$

$$= \frac{f_0'(z - T)}{f_0(z - T)} \tag{2.5}$$

a.s., provided that the derivatives exist. Let

$$\xi(T) := \int \psi(z-T)\, dF(z) \tag{2.6}$$

denote the expectation of $\psi$ with shift $T$, provided that it exists. Note the relationship between (2.3) and (2.6). The following lemma, due to Huber, establishes the existence of the expectation in (2.6), and the fact that it crosses zero:

**Lemma 2.2** *If there is a $T^*$ such that $\xi(T^*) < \infty$ exists, then $\xi(T)$ exists for all $T$ (though it is not necessarily finite), is monotone decreasing with $T$, and takes both positive and negative values.*

**Proof** A proof of existence is suggested in Huber [9:48]; for the rest of the proof, see Huber [5], [6:64-65]. QED

Given the conditions of Lemma 2.2, the following theorem, also due to Huber, establishes the consistency and asymptotic normality of the estimator $T_n : \mathbf{R}^n \to \Theta$ defined above:

**Theorem 2.3** *If $\xi(T)$ exists and there is a $T^*$ such that $0 < \xi(T)$ for $T < T^*$ and $\xi(T) < 0$ for $T^* < T$, and if*

$$\int |\psi(z-T)|\, dF(z) < \infty, \tag{2.7}$$

*then $T_n(z_1, \cdots, z_n) \to T^*$ as $n \to \infty$ almost surely and in probability (i.e. $T_n$ is consistent).*

*If, moreover, $\xi(T^*) = 0$, $\xi(T)$ is continuous, differentiable and strictly monotone in a neighborhood of $T^*$, and if*

$$0 < \int \psi^2(z-T)\, dF(z) < \infty \tag{2.8}$$

*is continuous in a neighborhood of $T^*$, then*

$$\mathbf{L}(\sqrt{n}\,(T_n - T^*)) \to \mathbf{N}\left[0, \frac{\int \psi^2(z-T^*)\, dF(z)}{(\xi'(T^*))^2}\right] \tag{2.9}$$

*as $n \to \infty$ (i.e. $T_n$ is asymptotically normal).*

**Proof** See Huber [5], [6:66-72]; also [9:45-50]. QED

Finally, the relationship of the results of Theorem 2.3 to the true distribution and location parameter is established by the following corollary:

**Corollary 2.4** *If the conditions of Theorem 2.3 are satisfied, and if the true underlying distribution is $F = P_0$, then*

$$\mathbf{L}(\sqrt{n}\,(T_n - \theta)) \to \mathbf{N}\left[0, \frac{1}{I(P_0)}\right] \tag{2.10}$$

*as* $n \to \infty$ *(i.e.* $T_n$ *is asymptotically efficient).*

**Proof** See Huber [5:72-73]; also Schick [22:34-36]. QED

A convenient model of indeterminacy, proposed by Huber [5], is the $\varepsilon$-*contaminated normal neighborhood*

$$\mathbf{P}_\varepsilon := \{ (1-\varepsilon) \, \mathrm{N}(0,1) + \varepsilon H : \; H \in \mathbf{S} \}, \tag{2.11}$$

where S is the set of all probability distributions symmetric with respect to the origin, and $0 \le \varepsilon < 1$ is the known fraction of "contamination." Note that the presence of outliers in a nominally normal sample can be modeled here by a distribution $H$ with tails heavier than normal. The least favorable distribution in this neighborhood is given by the following theorem:

**Theorem 2.5** *For the set* $\mathbf{P}_\varepsilon$, *the distribution minimizing the Fisher Information is given by*

$$f_\varepsilon(x) := \begin{cases} (1-\varepsilon)\,\mathrm{N}(k;0,1)\, e^{kx+k^2} & x < -k \\ (1-\varepsilon)\,\mathrm{N}(x;0,1) & -k \le x \le k \\ (1-\varepsilon)\,\mathrm{N}(k;0,1)\, e^{-kx+k^2} & k < x \end{cases} \tag{2.12}$$

*where* $k$ *is related to the fraction of contamination* $\varepsilon$ *by*

$$2\left[ \frac{\mathrm{N}(k;0,1)}{k} - \int_{-\infty}^{-k} \mathrm{N}(x;0,1)\,dx \right] = \frac{\varepsilon}{1-\varepsilon}. \tag{2.13}$$

**Proof** Outlines of a proof can be found in Huber [6:87-89], [9:84-85]. QED

It follows from (2.5) and (2.12) that

$$\psi_\varepsilon(x) = \begin{cases} -k & x < -k \\ x & -k \le x \le k \\ k & k < x \end{cases} \tag{2.14}$$

a.s. Thus, the transformation $\psi_\varepsilon(x)$ leaves its argument unaffected if it is within some predefined range, and truncates it if it goes beyond that range. It is easy to see by integrating (2.14) that the corresponding $\rho_\varepsilon$ is quadratic in the center and linear in the tails, so that the estimator defined by (2.3) and (2.14) represents in some sense a continuum between the sample mean and the sample median. As $\varepsilon \to 0$, (2.13) implies that $k \to \infty$, so that $\rho_\varepsilon(x) \propto x^2$ resulting in the sample mean (the least square estimate). As $\varepsilon \to 1$, on the other hand, $k \to 0$, and for small $k$, $\rho_\varepsilon(x) \propto |x|$ approximately, corresponding to the sample median (the minimum modulus estimate).

A drawback of this approach is that solving (2.3) involves "batch" processing with some kind of iterative procedure such as the Newton-Raphson method. In other words, it requires all the observations $\{ z_1, \cdots, z_n \}$ at once. Another drawback is that it assumes that the observations are identically distributed, i.e. that $\theta$ is constant. The next section describes a recursive method that updates the estimate every time an observation $z_i$ is received, and that allows for a linear time-variant location parameter.

## 3. Stochastic Approximation

It is possible to recursively maximize a stochastic function like (2.2), or find the root of a stochastic function like (2.3), by means of the *stochastic approximation* procedure based on the work of Robbins and Monro [21] and developed by many others. For general reviews of this methodology, see for instance Wasan [26:8-35], Nevel'son and Has'minskii [19:79-83, 88-94], or Kushner and Clark [10:19-47]. The use of stochastic approximation in the context of robust estimation was first proposed by Martin [11], Martin and Masreliez [12], Nevel'son [18], and Price and Vandelinde [20]. See also Englund, Holst, and Ruppert [2], who investigate the colored noise case.

For $\xi(T)$ defined in (2.6), suppose that there is a $T^*$ such that $0 < \xi(T)$ for $T < T^*$ and $\xi(T) < 0$ for $T^* < T$. Consider the recursion

$$T_{n+1} = T_n + a_n \psi( z_n - T_n ), \tag{3.1}$$

where $n = 1, 2, \cdots$, $\{a_n\}$ is a given real-valued sequence, and $T_1$ is an arbitrary (possibly random) starting point. There is a very considerable literature investigating conditions under which $T_n \to T^*$ as $n \to \infty$, as well as the asymptotic distribution of $T_n$. Note that since the value of $\psi( z_n - T_n )$ is random, it is necessary for the sequence $\{a_n\}$ to obey certain conditions in order to ensure convergence: it must tend towards zero at a rate sufficient for the error variance to vanish asymptotically, yet must not reach zero for $n < \infty$ since it must be able to compensate for any and all random perturbations due to the observations $\{z_n\}$. Indeed, there must at all times remain "an infinite amount of corrective effort" to converge to the correct limit, no matter where the estimate may have deviated (Young [28:34]).

A rather more general result than those in the literature is proven below, extending consistency and asymptotic normality results to the multivariate, time-varying case where the location parameter does not necessarily approach a limit.

For some integer $p$, let $( \mathbf{R}^p, \mathbf{B}, \lambda )$ be a measure space, where $\mathbf{R}$ is the real line, $\mathbf{B}$ the Borel $\sigma$-algebra, and $\lambda$ the Lebesgue measure. Let $F$ be a zero-mean probability measure on $( \mathbf{R}^p, \mathbf{B} )$ such that $F$ is absolutely continuous with respect to $\lambda$ and admits the density $f$ in accordance with the Radon-Nikodym theorem.

For some positive integer $n$, let $\{ \underline{v}_1, \cdots, \underline{v}_n \}$ be a sample of independent random variates taking values in $\mathbf{R}^p$, with common distribution $F$. Define the transformation

$$\underline{z}_n = H_n \underline{\theta}_n + D_n \underline{v}_n, \tag{3.2}$$

$n = 1, 2, \cdots$, where $\{H_n\}$ and $\{D_n\}$ are known sequences of matrices with $D_n \in \mathbf{R}^{p \times p}$ and $H_n \in \mathbf{R}^{p \times q}$, and $\underline{\theta}_n \in \mathbf{R}^q$ obeys the recursion

$$\underline{\theta}_{n+1} = F_n \underline{\theta}_n, \tag{3.3}$$

$n = 1, 2, \cdots$, where $\{F_n\}$ is a known sequence of matrices with $F_n \in \mathbf{R}^{q \times q}$, and $\underline{\theta}_0$ is an unknown (but finite) parameter.

Consider the recursion

$$\underline{T}_{n+1} = F_n \underline{T}_n + \left[ (D_{n+1}^{-1} H_{n+1})^{\mathrm{T}} (D_{n+1}^{-1} H_{n+1}) \right]^{-1} (D_{n+1}^{-1} H_{n+1})^{\mathrm{T}} A_n \underline{\psi} \left[ D_{n+1}^{-1} (\underline{z}_{n+1} - H_{n+1} F_n \underline{T}_n) \right] \tag{3.4}$$

(provided that all inverses exist), where $n = 1, 2, \cdots$, $\underline{T}_n \in \mathbf{R}^q$, $\{A_n\}$ is a given matrix sequence with $A_n \in \mathbf{R}^{q \times q}$, $\underline{T}_0$ is an arbitrary (possibly random) starting point, and $\underline{\psi}$ is related to the least favorable distribution by

$$\underline{\psi}(\underline{z} - \underline{T}) = -\underline{\nabla}_T \log f_0(\underline{z} - \underline{T}) \tag{3.5}$$

$$= -\frac{1}{f_0(\underline{z} - \underline{T})} \underline{\nabla}_T f_0(\underline{z} - \underline{T}) \tag{3.6}$$

a.s., within an arbitrary multiplicative constant. Furthermore, let

$$\underline{\xi}(\underline{T}) := E[\underline{\psi}(\underline{z} - \underline{T})] \tag{3.7}$$

as before. Let

$$\Sigma(\underline{T}) := E\left[ (\underline{\psi}(\underline{z} - \underline{T}) - \underline{\xi}(\underline{T}))(\underline{\psi}(\underline{z} - \underline{T}) - \underline{\xi}(\underline{T}))^{\mathrm{T}} \right], \tag{3.8}$$

and define

$$J(\underline{T}) := \left[ \frac{\partial}{\partial t_j} \xi_i(\underline{t}) \right]_{\underline{t} = \underline{T}} \tag{3.9}$$

to be the Jacobian of $\underline{\xi}(\underline{T})$, provided it exists.

Note that finding the least favorable distribution in the multivariate case is not trivial. The usual ordering of matrices (given $X, Y \in \mathbf{R}^{m \times m}$, $Y > X$ if and only if $Y - X > 0$, i.e. their difference is positive definite) is not a lattice ordering. Practically, this means that (in contrast to numbers on the real line) two non-equal matrices need not have an ordered relationship. Thus, finding the member of a class of distributions that minimizes the Fisher Information is not generally possible in the multivariate case. In the special case of *spherically symmetric* distributions, the multivariate extension is of course trivial: the least favorable distributions and influence-bounding functions are found coordinatewise, and everything else follows immediately.

The asymptotic behavior of the recursion $\underline{T}_n$ is established by the following theorem:

**Theorem 3.1** *Let $\underline{\xi}(\underline{T})$ exist for all $\underline{T}$, and for any $\delta > 0$ and all $q \times q$ matrices $M > 0$, let*

$$\sup_{\delta \leq \|\underline{T}\|} \underline{T}^{\mathrm{T}} M \, \underline{\xi}(\underline{T}) < 0. \tag{3.10}$$

*Assume there exists an $S_0 < \infty$ such that*

$$E\left[ \underline{\psi}(\underline{z} - \underline{T}) \, \underline{\psi}^{\mathrm{T}}(\underline{z} - \underline{T}) \right] \leq S_0 \tag{3.11}$$

*for all $\underline{T}$, and let $\{A_n\}$ be a sequence such that $A_n > 0$ for all $n$,*

$$\sum_{n=1}^{\infty} A_n = \infty, \tag{3.12}$$

*and*

$$\sum_{n=1}^{\infty} A_n^{\mathrm{T}} A_n < \infty. \tag{3.13}$$

*If there is an $\alpha < \infty$ such that for all $n$ and all $m$, with $0 \leq m \leq n$,*

$$\left[ \prod_{j=m}^{n} F_j \right]^{\mathrm{T}} \left[ \prod_{j=m}^{n} F_j \right] < \alpha I \tag{3.14}$$

*(where products are ordered by descending index), if there is a $\beta_1 > 0$ and a $\beta_2 < \infty$ such that*

$$\beta_1 I < D_n < \beta_2 I \tag{3.15}$$

*for all $n$, and finally if there is a $\gamma_1 > 0$ and a $\gamma_2 < \infty$ such that*

$$\gamma_1 I < H_n^{\mathrm{T}} H_n < \gamma_2 I \tag{3.16}$$

*for all $n$, then, given any $\underline{T}_0 < \infty$, $\underline{T}_n - \underline{\theta}_n \to 0$ as $n \to \infty$ a.s. (i.e. $\underline{T}_n$ is consistent).*

*If, moreover, $\underline{\xi}(0) = 0$, $\underline{\xi}(\underline{T})$ is continuous, differentiable and strictly monotone in a neighborhood of 0 with $\|J(0)\| < \infty$, if $\Sigma(0) > 0$, $\Sigma(\underline{T})$ is continuous and bounded in a neighborhood of 0, and finally if*

$$\limsup_{n \to \infty} n A_n < \infty, \tag{3.17}$$

*then*

$$\mathbf{L}(\Sigma_n^{-1/2} (\underline{T}_n - \underline{\theta}_n)) \to \mathbf{N}(0, I), \tag{3.18}$$

*where*

$$\Sigma_n = \left[ ( D_n^{-1} H_n )^{\mathrm{T}} ( D_n^{-1} H_n ) \right]^{-1} ( D_n^{-1} H_n )^{\mathrm{T}}$$
$$\left[ I + A_{n-1} J(0) \right] ( D_n^{-1} H_n F_{n-1} ) \Sigma_{n-1}$$
$$( D_n^{-1} H_n F_{n-1} )^{\mathrm{T}} \left[ I + A_{n-1} J(0) \right]^{\mathrm{T}}$$
$$( D_n^{-1} H_n ) \left[ ( D_n^{-1} H_n )^{\mathrm{T}} ( D_n^{-1} H_n ) \right]^{-1}$$
$$+ \left[ ( D_n^{-1} H_n )^{\mathrm{T}} ( D_n^{-1} H_n ) \right]^{-1} A_{n-1} \Sigma(0) A_{n-1}^{\mathrm{T}}$$
$$\left[ ( D_n^{-1} H_n )^{\mathrm{T}} ( D_n^{-1} H_n ) \right]^{-1} \tag{3.19}$$

*with*

$$\Sigma_0 = 0 \tag{3.20}$$

*(i.e. $\underline{T}_n$ is asymptotically normal).*

**Proof** For the sake of legibility, the case $H_n = D_n = I$ for all $n$ is treated below. The extension to the general case is straight-forward.

The proof of consistency is a generalization of Blum [1]. Defining

$$Y_n := E\left[ ( \underline{T}_{n+1} - \underline{\theta}_{n+1} )^{\mathrm{T}} ( \underline{T}_{n+1} - \underline{\theta}_{n+1} ) - ( \underline{T}_n - \underline{\theta}_n )^{\mathrm{T}} ( \underline{T}_n - \underline{\theta}_n ) \,\middle|\, \underline{T}_1, \cdots, \underline{T}_n \right], \tag{3.21}$$

it can be shown that the sequence

$$\left\{ ( \underline{T}_n - \underline{\theta}_n )^{\mathrm{T}} ( \underline{T}_n - \underline{\theta}_n ) - \sum_{j=1}^{n-1} Y_j \right\} \tag{3.22}$$

is a martingale. Establishing first that the expectation of the absolute value of (3.22) is bounded for all $n$, it follows by virtue of a martingale convergence theorem that the sequence (3.22) converges almost surely. It is then shown that each of the two terms in (3.22) does so as well, by using monotonicity and boundedness to prove that

$$\lim_{n \to \infty} \left[ ( \underline{T}_{n+1} - \underline{\theta}_{n+1} )^{\mathrm{T}} ( \underline{T}_{n+1} - \underline{\theta}_{n+1} ) - ( \underline{T}_n - \underline{\theta}_n )^{\mathrm{T}} F_n^{\mathrm{T}} F_n ( \underline{T}_n - \underline{\theta}_n ) + 2 ( F_n \underline{T}_n - \underline{\theta}_{n+1} )^{\mathrm{T}} A_n \underline{\xi}( F_n \underline{T}_n - \underline{\theta}_{n+1} ) \right] = 0 \tag{3.23}$$

w.p.1. Some manipulation and the Chebychev inequality then imply that there exists a subsequence $\{n_m\}$ such that

$$\lim_{m\to\infty} \left( F_{n_m} \underline{T}_{n_m} - \underline{\theta}_{n_m+1} \right)^{\mathrm{T}} \left[ \prod_{k=n_m+1}^{\infty} F_k \right]^{\mathrm{T}} \left[ \prod_{k=n_m+1}^{\infty} F_k \right] A_{n_m} \underline{\xi}( F_{n_m} \underline{T}_{n_m} - \underline{\theta}_{n_m+1} ) = 0 \tag{3.24}$$

w.p.1, whence it follows that

$$\lim_{m\to\infty} ( \underline{T}_{n_m} - \underline{\theta}_{n_m} ) = 0 \tag{3.25}$$

w.p.1. Substitution into (3.23) then generalizes (3.25) to all $n$, proving consistency.

The proof of asymptotic normality is a generalization of Fabian [3], and is based upon the convergence of the characteristic function. From the continuity and differentiability of $\underline{\xi}(\underline{T})$ in a neighborhood of 0 by hypothesis,

$$\underline{\xi}(\underline{T}) = J(0)\, \underline{T} + O( \| \underline{T} \|^2 ) \tag{3.26}$$

for small enough $\| \underline{T} \|$. By virtue of consistency, (3.4) may thus be rewritten as

$$\underline{T}_{n+1} - \underline{\theta}_{n+1} = \left[ I + A_n J(0) + A_n O_p( \| F_n \underline{T}_n - \underline{\theta}_{n+1} \| ) \right] F_n ( \underline{T}_n - \underline{\theta}_n ) + A_n \left[ \underline{\psi}( \underline{z}_{n+1} - F_n \underline{T}_n ) - \underline{\xi}( F_n \underline{T}_n - \underline{\theta}_{n+1} ) \right] \tag{3.27}$$

w.p.1 for large enough $n$. It is first established that, defining

$$A( n, \delta_2, \underline{T}, \underline{\theta} ) := \left\{ \underline{z} : \| \underline{\psi}(\underline{z} - \underline{T}) - \underline{\xi}(\underline{T} - \underline{\theta}) \|^2 \geq \delta_2 n \right\} \tag{3.28}$$

for some $\delta_2 > 0$,

$$\lim_{n\to\infty} \int_{A(n, \delta_2, F_n \underline{T}_n, \underline{\theta}_{n+1})} \| \underline{\psi}( \underline{z} - F_n \underline{T}_n ) - \underline{\xi}( F_n \underline{T}_n - \underline{\theta}_{n+1} ) \|^2 \, dP_n(\underline{z}) = 0 \tag{3.29}$$

w.p.1, which is analogous to Lindeberg's condition for asymptotic normality.

Taylor approximations are then constructed for the characteristic functions of the terms in the recursion (3.27), and it is shown that the characteristic function of $\underline{T}_n - \underline{\theta}_n$ approaches that of a normal distribution as $n \to \infty$. This is aided by constructing the recursion

$$\zeta_{n+1}(\underline{s}) = \zeta_n \left[ F_n^{\mathrm{T}} \left( I + A_n\, J(0) + o_p(n^{-1}) \right)^{\mathrm{T}} \underline{s} \right] \left[ 1 - \frac{1}{2}\, \underline{s}^{\mathrm{T}} A_n\, \Sigma(0)\, A_n^{\mathrm{T}}\, \underline{s} \right], \tag{3.30}$$

subject to the initial condition

$$\zeta_0(\underline{s}) = e^{i\, \underline{s}^{\mathrm{T}} (\underline{T}_0 - \underline{\theta}_0)}, \tag{3.31}$$

and showing that $\zeta_n$ is asymptotically equivalent to the characteristic function of $\underline{T}_n - \underline{\theta}_n$. Finally, it is shown that the limiting variance is given by (3.19) by constructing a recursion that yields the variance in the exponent of the limiting characteristic function, and proving that it is asymptotically equivalent to the asymptotic variance of $\underline{T}_n - \underline{\theta}_n$. (For a more detailed proof, see Schick [22:92-106].)

QED

It is clear that one can do no better recursively than in batch mode; in other words, it is not possible to do better by considering the observations one at a time than by considering them all at once. Thus, the asymptotic variance of the recursive estimator is no smaller than that of the Huber estimator of Section 2, but it can be shown that the two are asymptotically equivalent for the right choice of gains $\{A_n\}$. (See for instance Schick [22:67].) Note in passing that if the true distribution is the least favorable one, then this choice of gain sequence results in an asymptotically efficient estimator.

This section shows the relationship between robust point estimation and robust recursive estimation. However, the estimator of Theorem 3.1 corresponds to a linear dynamic model with no process noise. In other words, it is an estimator of a location parameter that varies in a deterministic and known manner. While there may be instances that require such models, the absence of process noise makes this a special case of limited application. Not only is process noise often physically present, but it is also a useful abstraction that compensates for small and unsystematic modeling errors. The following section addresses the case where process noise is present.

## 4. Conditional Mean Estimation

As before, let

$$\underline{z}_n = H_n\, \underline{\theta}_n + D_n\, \underline{v}_n, \tag{4.1}$$

but let the location parameter now be random and obey the recursion

$$\underline{\theta}_{n+1} = F_n \, \underline{\theta}_n + \underline{w}_n , \tag{4.2}$$

$n = 1, 2, \cdots$, with all parameters and distributions as specified in Section 1. Again, let $\{ \underline{v}_1, \cdots, \underline{v}_n \}$ be a sample of independent random variates taking values in $\mathbf{R}^p$, with common distribution $F \in \mathbf{P}_\varepsilon$ (a multivariate version of (2.11)) having positive and bounded variance $R$.

In this case, asymptotic variance (or alternatively the Fisher Information) is not a meaningful measure of performance. The *conditional mean* estimator, on the other hand, is well known to have several desirable properties, such as unbiasedness and minimum error variance. The first derivation of a robust approximate conditional mean estimator in the present context is due to Masreliez and Martin [15]-[16], and is based on Masreliez [13]-[14]; some generalizations are provided by West [27].

A key assumption made by these and other authors is that at each $n$, the conditional probability distribution of the state $\underline{\theta}_n$ given past observations $\{ \underline{z}_0, \cdots, \underline{z}_{n-1} \}$ is normal. This assumption allows some algebraic manipulations that yield an elegant stochastic approximation-like estimator. However, while it has been shown in simulation studies to be a good approximation of the true conditional density, it is only strictly correct for finite $n$ in the special case where $F$ is normal (see Spall and Wall [24]), which is clearly of no interest here.

In this section, a first-order approximation of the conditional distribution prior to updating, $p( \underline{\theta}_n \mid \underline{z}_0, \cdots, \underline{z}_{n-1} )$, is derived for the case where $F$ is known. (The extension of this result to the least favorable distribution remains an open problem at this writing.) Although conditional normality is never exactly satisfied in the presence of non-normal noise, it is shown that the zeroeth-order term in a Taylor series representation of the distribution is indeed normal. The small parameter around which the Taylor series is constructed involves $\varepsilon$, the fraction of contamination. This approximation is then used, in an extension of Masreliez's theorem, to derive a first-order approximation of a robust conditional mean estimator.

Note first that the Kalman Filter recursion is exponentially asymptotically stable under certain conditions. This property ensures that the effects of past outliers are attenuated rapidly enough as new observations become available. The stability of the Kalman Filter recursions has been studied by several researchers; the following theorem is due to Moore and Anderson:

**Theorem 4.1** *Let the matrix sequences* $\{F_n\}$, $\{H_n\}$, $\{Q_n\}$, *and* $\{D_n\}$ *be bounded above, and let* $\{D_n\}$ *also be bounded below. Let there exist positive integers* $t$ *and* $s$ *and positive real numbers* $\alpha$ *and* $\beta$ *such that for all* $n$,

$$\sum_{i=n}^{n+t} \left[ \prod_{j=n}^{i-1} F_j \right]^{\mathrm{T}} H_i^{\mathrm{T}} ( D_i \, R \, D_i^{\mathrm{T}} )^{-1} H_i \left[ \prod_{j=n}^{i-1} F_j \right] > \alpha I \tag{4.3}$$

*(i.e. the system is completely observable) and*

$$\sum_{i=n-s}^{n} \left[ \prod_{j=i+1}^{n} F_j \right] Q_i \left[ \prod_{j=i+1}^{n} F_j \right]^{\mathrm{T}} > \beta I \tag{4.4}$$

*(i.e. the system is completely controllable).*

*Then, given any* $\underline{\tilde{\theta}}_0 < \infty$ *, and defining the closed-loop recursion*

$$\underline{\tilde{\theta}}_{n+1} = ( I - K_{n+1} H_{n+1} ) F_n \, \underline{\tilde{\theta}}_n , \tag{4.5}$$

*(where* $K_n$ *is the Kalman gain defined in equation (1.6)), there exist* $\lambda > 0$ *and* $0 < \delta < 1$ *such that*

$$|| \, \underline{\tilde{\theta}}_n \, || < \lambda \, \delta^n , \tag{4.6}$$

*(i.e. the filter is exponentially asymptotically stable).*

**Proof** See Moore and Anderson [17]. QED

This result is used in the following, slightly different form.

**Corollary 4.2** *Let the conditions of Theorem 4.1 be satisfied, and let a* $0 < \phi < \infty$ *exist such that for all* $n$*,*

$$\left\| \prod_{j=1}^{n} F_j \right\| < \phi \tag{4.7}$$

*(i.e. the system is uniformly stable). For* $i = 1, 2$*, let*

$$\underline{\theta}_{n+1}^i = F_n \, \underline{\theta}_n^i + K_{n+1}^i ( \underline{z}_{n+1} - H_{n+1} F_n \, \underline{\theta}_n^i ) \tag{4.8}$$

$$K_n^i = M_n^i H_n^{\mathrm{T}} ( H_n M_n^i H_n^{\mathrm{T}} + D_n \, R \, D_n^{\mathrm{T}} )^{-1} \tag{4.9}$$

$$M_{n+1}^i = F_n P_n^i F_n^{\mathrm{T}} + Q_n \tag{4.10}$$

$$P_n^i = ( I - K_n^i H_n ) M_n^i \tag{4.11}$$

*be two Kalman Filters with respective initial state means* $\underline{\theta}_0^i$ *and covariances* $M_0^i$*,* $i = 1, 2$*. Then, there is a* $0 < \delta < 1$ *such that for any finite* $\underline{\theta}$*,*

$$\mathrm{N}( \underline{\theta} ; \underline{\theta}_n^1 , M_n^1 ) = \mathrm{N}( \underline{\theta} ; \underline{\theta}_n^2 , M_n^2 ) + O( \delta^n ). \tag{4.12}$$

**Proof** Since $\mathrm{N}( \underline{x} ; \underline{\mu}, \Sigma )$ is everywhere continuously differentiable with respect to $\underline{\mu}$ and $\Sigma$ except at $\Sigma = 0$, and moreover since it can be shown that $M_n^i$ is bounded away from 0 for all $n$, it is possible to write a first-order Taylor series expansion of $\mathrm{N}( \underline{\theta} ; \underline{\theta}_n^1 , M_n^1 )$ around $\underline{\theta}_n^2$ and $M_n^2$. But Theorem 4.1 implies that the respective differences between $M_n^i$ and between $\underline{\theta}_n^i$ for $i = 1, 2$ are each $O(\delta^n)$ or less. The result follows immediately. (For a detailed proof, see Schick [22:122-124].) QED

A first-order approximation of the conditional probability distribution $p(\underline{\theta}_n \mid \underline{z}_0, \cdots, \underline{z}_{n-1})$ of the state $\underline{\theta}_n$ given past observations $\{\underline{z}_0, \cdots, \underline{z}_{n-1}\}$ is given by the following theorem:

**Theorem 4.3** *Let the conditions of Theorem 4.1 and Corollary 4.2 be satisfied for the system given by equations (4.1)-(4.2), and let $\delta$ be a real number for which (4.6) holds. Let $\omega$ be the smallest integer such that*

$$\delta^{\omega} \leq \varepsilon. \tag{4.13}$$

*If*

$$\omega\,\varepsilon < 1 \tag{4.14}$$

*and if the distribution $H$ has bounded moments, then*

$$\begin{aligned} p(\underline{\theta}_n \mid \underline{z}_0, \cdots, \underline{z}_{n-1}) &= (1-\varepsilon)^n \kappa_n \kappa_n^0 \mathrm{N}(\underline{\theta}_n; \underline{\theta}_n^0, M_n^0) \\ &\quad + \varepsilon(1-\varepsilon)^{n-1} \kappa_n \sum_{i=1}^{n} \kappa_n^i \mathrm{N}(\underline{\theta}_n; \underline{\theta}_n^i, M_n^i) \\ &\quad \int \mathrm{N}(\underline{z}_{i-1} - \underline{\xi}; H_{i-1}\underline{v}_n^i + H_{i-1}V_n^i(\underline{\theta}_n - \underline{\theta}_n^i), \\ &\quad H_{i-1}W_n^i H_{i-1}^{\mathrm{T}} - H_{i-1}V_n^i M_n^i V_n^{i\,\mathrm{T}} H_{i-1}^{\mathrm{T}}) \\ &\quad h(\underline{\xi})\, d\underline{\xi} \\ &\quad + O_p(n^2\varepsilon^2) \end{aligned} \tag{4.15}$$

*for all $n \leq \omega$, and*

$$\begin{aligned} p(\underline{\theta}_n \mid \underline{z}_0, \cdots, \underline{z}_{n-1}) &= (1-\varepsilon)^{\omega} \kappa_n \kappa_n^0 \mathrm{N}(\underline{\theta}_n; \underline{\theta}_n^0, M_n^0) \\ &\quad + \varepsilon(1-\varepsilon)^{\omega-1} \kappa_n \sum_{i=n-\omega+1}^{n} \kappa_n^i \mathrm{N}(\underline{\theta}_n; \underline{\theta}_n^i, M_n^i) \\ &\quad \int \mathrm{N}(\underline{z}_{i-1} - \underline{\xi}; H_{i-1}\underline{v}_n^i + H_{i-1}V_n^i(\underline{\theta}_n - \underline{\theta}_n^i), \\ &\quad H_{i-1}W_n^i H_{i-1}^{\mathrm{T}} - H_{i-1}V_n^i M_n^i V_n^{i\,\mathrm{T}} H_{i-1}^{\mathrm{T}}) \\ &\quad h(\underline{\xi})\, d\underline{\xi} \end{aligned}$$

$$+ \; O_p(\omega^2 \varepsilon^2) \tag{4.16}$$

*for all* $n \geq \omega$, *where, for* $i = 1, 2, \cdots$ *and* $n > i$,

$$\underline{\theta}_n^i = F_{n-1} \underline{\theta}_{n-1}^i + F_{n-1} M_{n-1}^i \, H_{n-1}^{\mathrm{T}} \Gamma_{n-1}^{i \; -1} \left( \underline{z}_{n-1} - H_{n-1} \underline{\theta}_{n-1}^i \right) \tag{4.17}$$

$$M_n^i = F_{n-1} P_{n-1}^i \, F_{n-1}^{\mathrm{T}} + Q_{n-1} \tag{4.18}$$

$$P_n^i = M_n^i - M_n^i H_n^{\mathrm{T}} \Gamma_n^{i \; -1} H_n M_n^i \tag{4.19}$$

$$\Gamma_n^i = H_n M_n^i H_n^{\mathrm{T}} + D_n D_n^{\mathrm{T}} \tag{4.20}$$

$$V_n^i = V_{n-1}^i \, P_{n-1}^i \, F_{n-1}^{\mathrm{T}} \, M_n^{i \; -1} \tag{4.21}$$

$$\underline{\nu}_n^i = \underline{\nu}_{n-1}^i + V_{n-1}^i \, M_{n-1}^i \, H_{n-1}^{\mathrm{T}} \, \Gamma_{n-1}^{i \; -1} \left( \underline{z}_{n-1} - H_{n-1} \underline{\theta}_{n-1}^i \right) \tag{4.22}$$

$$W_n^i = W_{n-1}^i - V_{n-1}^i \, M_{n-1}^i \, H_{n-1}^{\mathrm{T}} \, \Gamma_{n-1}^{i \; -1} H_{n-1} M_{n-1}^i \, V_{n-1}^{i \;\; \mathrm{T}} \tag{4.23}$$

$$\kappa_n^i = \kappa_{n-1}^i \, \mathrm{N}( \underline{z}_{n-1}; \, H_{n-1} \underline{\theta}_{n-1}^i, \, \Gamma_{n-1}^i ) \tag{4.24}$$

*subject to the initial conditions*

$$\underline{\theta}_i^i = F_{i-1} \underline{\theta}_{i-1}^0 \tag{4.25}$$

$$M_i^i = F_{i-1} M_{i-1}^0 \, F_{i-1}^{\mathrm{T}} + Q_{i-1} \tag{4.26}$$

$$V_i^i = M_{i-1}^0 \, F_{i-1}^{\mathrm{T}} \, M_i^{i \; -1} \tag{4.27}$$

$$\underline{\nu}_i^i = \underline{\theta}_{i-1}^0 \tag{4.28}$$

$$W_i^i = M_{i-1}^0 \tag{4.29}$$

$$\kappa_i^i = \kappa_{i-1}^0 \tag{4.30}$$

*for* $i > 0$, *and*

$$\underline{\theta}_0^0 = \overline{\underline{\theta}}_0 \tag{4.31}$$

$$M_0^0 = M_0 \tag{4.32}$$

$$\kappa_0^0 = 1. \tag{4.33}$$

*The normalization constant satisfies*

$$\kappa_n^{-1} = ( 1 - \varepsilon )^n \, \kappa_n^0 + \varepsilon ( 1 - \varepsilon )^{n-1} \sum_{i=1}^{n} \kappa_n^i \int \mathrm{N}( \underline{z}_{i-1} - \underline{\xi}; \, H_{i-1} \underline{\nu}_n^i, \, H_{i-1} W_n^i H_{i-1}^{\mathrm{T}} ) \, h(\underline{\xi}) \, d\underline{\xi} \tag{4.34}$$

*for all* $n \leq \omega$, *and*

$$\kappa_n^{-1} = (1-\varepsilon)^{\omega}\,\kappa_n^0 + \varepsilon(1-\varepsilon)^{\omega-1} \sum_{i=n-\omega+1}^{n} \kappa_n^i \int N(\underline{z}_{i-1}-\underline{\xi};\, H_{i-1}\underline{v}_n^i,\; H_{i-1} W_n^i H_{i-1}^{\mathrm{T}})\, h(\underline{\xi})\, d\underline{\xi} \qquad (4.35)$$

*for all* $n \geq \omega$.

**Proof** Equation (4.15) is first established by induction. There remains to show that (4.16) holds for $n > \omega$, i.e. that the number of terms in (4.15) does not increase without bound as $n \to \infty$. Corollary 4.2 and the Chernoff bound are used to demonstrate that terms in (4.15) for $j \leq n - \omega$ are "absorbed" into the zeroeth-order term with an exponentially vanishing error term. Finally, a combinatorial argument establishes that the order of the error term is $\omega^2\varepsilon^2$. (For a detailed proof, see Schick [22:130-144]; in addition, an abbreviated proof is given in Schick and Mitter [23].) QED

It is interesting to note that Equations (4.17)-(4.20) are a bank of Kalman Filters, each starting at a different time $i = 0, 1, 2, \cdots$: the cases $i > 0$ correspond to Kalman Filters skipping the $i$th observation, while the case $i = 0$ is based on all observations. Equations (4.21)-(4.23) are a bank of optimal fixed-point smoothers, each estimating the state at a different time $i = 0, 1, 2, \cdots$, based on all preceeding and subsequent observations. Thus, each term in the summations on the right-hand sides of (4.15)-(4.16) is a Kalman Filter that skips one observation, coupled with an optimal smoother that estimates the state at the time the observation is skipped.

Loosely defining a random variable distributed as $H$ as an "outlier," the first term in (4.15)-(4.16) corresponds to the event that "there has been no outlier among the first $n$ observations," and each term in the summation to the event "there has been exactly one outlier among the first $n$ observations, at time $i-1$." Higher-order terms correspond to the occurrence of two or more outliers, and are absorbed into the error term.

Evidently, as $n \to \infty$, the probability of the event that only a finite number of outliers occur vanishes for any $\varepsilon > 0$. That the density can nevertheless be approximated by the first-order expression in (4.16) is due to the exponential asymptotic stability of the Kalman Filter: $\omega$ represents a "window size" beyond which the effects of older observations have sufficiently attenuated.

The approximate conditional prior probability distribution given by Theorem 4.3 is now used in an extension of a theorem due to Masreliez, resulting in a first-order approximation of the conditional mean estimator.

Let $h$ denote the Radon-Nikodym derivative of the contaminating distribution $H$ with respect to the Lebesgue measure, provided it exists. Let

$$\underline{T}_n := E\,[\,\underline{\theta}_n \mid \underline{z}_0, \cdots, \underline{z}_n\,] \tag{4.36}$$

and

$$\Sigma_n := E\,[\,(\underline{\theta}_n - \underline{T}_n)(\underline{\theta}_n - \underline{T}_n)^{\mathrm{T}} \mid \underline{z}_0, \cdots, \underline{z}_n\,] \tag{4.37}$$

respectively denote the *a posteriori* conditional mean and conditional variance of $\underline{\theta}_n$. In addition, let the score function (the additive inverse of the gradient of the logarithm) for the conditional probability of $\underline{z}_n$ given that no outliers occurred during the first $n-1$ observations be denoted by

$$\underline{\psi}_n^0(\underline{z}_n) := -\underline{\nabla}_{\underline{z}_n} \log p(\,\underline{z}_n \mid \underline{z}_0, \cdots, \underline{z}_{n-1}, \\ \eta_0 = 0, \cdots, \eta_{n-1} = 0\,). \tag{4.38}$$

Similarly, for $i = 1, 2, \cdots$ and all $n \geq i$, let

$$\underline{\psi}_n^i(\underline{z}_{i-1}) := -\underline{\nabla}_{\underline{z}_{i-1}} \log p(\,\underline{z}_{i-1} \mid \underline{z}_0, \cdots, \underline{z}_{i-2}, \underline{z}_i, \cdots, \underline{z}_n, \\ \eta_0 = 0, \cdots, \eta_{i-1} = 1, \cdots, \eta_n = 0\,) \tag{4.39}$$

denote the score function for the conditional probability of $\underline{z}_{i-1}$ given that no outliers occurred among the remaining $n-2$ observations. Finally, for $i = 0, 1, 2, \cdots$ and all $n \geq i$, let

$$\Psi_n^i(\,\underline{z}\,) := \underline{\nabla}_{\underline{z}}\, \underline{\psi}_n^{i\,\mathrm{T}}(\,\underline{z}\,) \tag{4.40}$$

denote the additive inverse of the Hessian of the logarithm of the conditional probability, i.e. the Jacobian of $\underline{\psi}_n^i$.

A first-order approximation of the conditional mean estimator $\underline{T}_n$ of the state $\underline{\theta}_n$ given past and present observations $\{\,\underline{z}_0, \cdots, \underline{z}_n\,\}$ is given by the following theorem:

**Theorem 4.4** *Let the conditions of Theorem 4.1, Corollary 4.2, and Theorem 4.3 be satisfied for the system given by equations (4.1)-(4.2). If h exists and is bounded and differentiable a.e., then*

$$\underline{T}_n = (1-\varepsilon)^n\, \kappa_{n+1}\, \pi_n^0\, \underline{T}_n^0 + \varepsilon\,(1-\varepsilon)^{n-1}\, \kappa_{n+1} \sum_{i=1}^{n} \pi_n^i\, \underline{T}_n^i \\ + O_p(\,n^2\varepsilon^2\,) \tag{4.41}$$

*for all* $n \leq \omega$, *and*

$$\underline{T}_n = (1-\varepsilon)^{\omega}\, \kappa_{n+1}\, \pi_n^0\, \underline{T}_n^0 + \varepsilon\,(1-\varepsilon)^{\omega-1}\, \kappa_{n+1} \sum_{i=n-\omega+1}^{n} \pi_n^i\, \underline{T}_n^i \\ + O_p(\,\omega^2\varepsilon^2\,) \tag{4.42}$$

*for all* $n \geq \omega$, *where*

$$\underline{T}_n^0 = \underline{\theta}_n^0 + M_n^0 H_n^{\mathrm{T}} \underline{\psi}_n^0(\underline{z}_n - H_n \underline{\theta}_n^0) \tag{4.43}$$

$$\underline{T}_n^i = \underline{\theta}_n^i + M_n^i H_n^{\mathrm{T}} \Gamma_n^{i\;-1}(\underline{z}_n - H_n \underline{\theta}_n^i) + P_n^i V_n^{i\;\mathrm{T}} H_{i-1}^{\mathrm{T}} \underline{\psi}_n^i(\underline{z}_{i-1} - H_{i-1} \underline{v}_{n+1}^i) \tag{4.44}$$

$$\pi_n^0 = (1-\varepsilon)\,\kappa_{n+1}^0 + \varepsilon\,\kappa_n^0 \int \mathrm{N}(\underline{z}_n - \underline{\xi};\, H_n \underline{\theta}_n^0,\, H_n M_n^0 H_n^{\mathrm{T}})\, h(\underline{\xi})\, d\underline{\xi} \tag{4.45}$$

$$\pi_n^i = (1-\varepsilon)\,\kappa_{n+1}^i \int \mathrm{N}(\underline{z}_{i-1} - \underline{\xi};\, H_{i-1} \underline{v}_{n+1}^i,\, H_{i-1} W_{n+1}^i H_{i-1}^{\mathrm{T}})\, h(\underline{\xi})\, d\underline{\xi} \tag{4.46}$$

*and the score functions are given by*

$$\underline{\psi}_n^0(\underline{z}_n - H_n \underline{\theta}_n^0) = -\underline{\nabla}_z \log\Big[(1-\varepsilon)\,\mathrm{N}(\underline{z};\, H_n \underline{\theta}_n^0,\, \Gamma_n^0) + \varepsilon \int \mathrm{N}(\underline{z} - \underline{\xi};\, H_n \underline{\theta}_n^0,\, H_n M_n^0 H_n^{\mathrm{T}})\, h(\underline{\xi})\, d\underline{\xi}\Big]\Big|_{\underline{z} = \underline{z}_n} \tag{4.47}$$

$$\underline{\psi}_n^i(\underline{z}_{i-1} - H_{i-1} \underline{v}_{n+1}^i) = -\underline{\nabla}_z \log \int \mathrm{N}(\underline{z} - \underline{\xi};\, H_{i-1} \underline{v}_{n+1}^i,\, H_{i-1} W_{n+1}^i H_{i-1}^{\mathrm{T}})\, h(\underline{\xi})\, d\underline{\xi}\,\Big|_{\underline{z} = \underline{z}_{i-1}} \tag{4.48}$$

*with* $\underline{\theta}_n^i$, $M_n^i$, $P_n^i$, $\Gamma_n^i$, $V_n^i$, $\underline{v}_n^i$, $W_n^i$, $\kappa_n^i$, *and* $\kappa_n$ *as defined in equations (4.17)-(4.24) and (4.34)-(4.35), subject to the initial conditions (4.25)-(4.33). Furthermore,*

$$\Sigma_n = (1-\varepsilon)^n\, \kappa_{n+1}\, \pi_n^0\, \Sigma_n^0 + \varepsilon\,(1-\varepsilon)^{n-1}\, \kappa_{n+1} \sum_{i=1}^{n} \pi_n^i\, \Sigma_n^i + O_p(n^2 \varepsilon^2) \tag{4.49}$$

*for all* $n \leq \omega$, *and*

$$\Sigma_n = (1-\varepsilon)^{\omega}\, \kappa_{n+1}\, \pi_n^0\, \Sigma_n^0 + \varepsilon\,(1-\varepsilon)^{\omega-1}\, \kappa_{n+1} \sum_{i=n-\omega+1}^{n} \pi_n^i\, \Sigma_n^i + O_p(\omega^2 \varepsilon^2) \tag{4.50}$$

*for all* $n \geq \omega$, where

$$\Sigma_n^0 = M_n^0 - M_n^0 H_n^{\mathrm{T}}\, \Psi_n^0(\underline{z}_n - H_n \underline{\theta}_n^0)\, H_n M_n^0 + (\underline{T}_n - \underline{T}_n^0)(\underline{T}_n - \underline{T}_n^0)^{\mathrm{T}} \tag{4.51}$$

$$\Sigma_n^i = P_n^i - P_n^i V_n^{i\ \mathrm{T}} H_{i-1}^{\mathrm{T}} \Psi_n^i ( \underline{z}_{i-1} - H_{i-1} \underline{\nu}_{n+1}^i ) H_{i-1} V_n^i P_n^i + (\underline{T}_n - \underline{T}_n^i)(\underline{T}_n - \underline{T}_n^i)^{\mathrm{T}}, \qquad (4.52)$$

*and $\Psi_n^i$ is given by equation (4.43), subject to (4.47)-(4.48).*

**Proof** From Theorem 4.3, the conditional prior is the sum of an $O_p(1)$ normal distribution and $\omega$ $O_p(\varepsilon)$ terms that each involve a normal distribution convolved with the contaminating distribution $h$. Furthermore, each term can be treated independently by virtue of linearity. Since the zeroeth-order term is normal, the corresponding term in the expression for the conditional mean can be found by direct application of the theorem due to Masreliez [14], and has the form of (4.43). The product of normal distributions in each $O_p(\varepsilon)$ term can be rewritten by grouping together the terms involving $\underline{\theta}_n$, after which Masreliez's theorem can once again be applied by changing the order of integration, yielding terms of the form (4.44). The terms in the expression for the conditional covariance are obtained analogously. (For a detailed proof, see Schick [22:147-157]; in addition, an abbreviated proof is given in Schick and Mitter [23].) QED

The estimator of Theorem 4.4 is a weighted sum of terms having the form of stochastic approximation equations. The robust filter of Masreliez and Martin [15]-[16] is approximately equivalent to the zeroeth-order term in (4.41)-(4.42), i.e. to $\underline{T}_n^0$, although the way in which they transform a one-step estimator into a recursion is *ad hoc* and violates their assumption of conditional normality at the next time step.

Note that the current observation $\underline{z}_n$ is processed by the influence-bounding function $\underline{\psi}_n^0$, i.e. $\underline{T}_n^0$ is robust against an outlier at time $n$. Similarly, each past observation $\underline{z}_{i-1}$ is processed by an influence-bounding function $\underline{\psi}_n^i$, i.e. $\underline{T}_n^i$ is robust against an outlier at time $i-1$. However, $\underline{T}_n^i$ is linear in $\underline{z}_n$. This is because while the $O_p(1)$ term corresponds to the event that there were no outliers among the most recent $\omega$ observations, so that the current observation could be one with probability $O(\varepsilon)$, the $O_p(\varepsilon)$ terms each correspond to the event that there was an outlier among the most recent $\omega$ observations, so that the probability that the current observation is an outlier is only $O_p(\varepsilon^2)$.

Both Theorem 4.3 and Theorem 4.4 are based on the assumption that outliers occur rarely relative to the dynamics of the filter. In the unlikely event that two outliers occur within less than $\omega$ time steps of each other, the estimate would be strongly affected. This implies that the estimator developed here is robust in the presence of rare and isolated outliers, but not when outliers occur in batches. Higher-order approximations for the conditional prior distribution and the conditional mean could be constructed to be robust against pairs, triplets, or higher numbers of outliers.

Unlike the Kalman Filter, the estimation error covariance in Theorem 4.4 is a function of the observations. Note, however, that the covariance is a function of a set of matrices $\{M_n^i\}$, $\{P_n^i\}$, $\{\Gamma_n^i\}$, $\{V_n^i\}$, and $\{W_n^i\}$, which are themselves independent of the observations. Thus, they can be pre-computed and stored, as is sometimes done with the Kalman Filter. This would drastically reduce the on-line computational burden. Moreover, the banks of parallel filters and smoothers are entirely independent of each other, so that this estimate appears to be well suited to parallel computation.

Note finally that, as can easily be verified, for $\varepsilon = 0$,

$$\underline{\psi}_n^0(\underline{z}_n - \underline{\theta}_n^0) = -\frac{\underline{\nabla}_{\underline{z}_n} \mathrm{N}(\underline{z}_n; H_n \underline{\theta}_n^0, \Gamma_n^0)}{\mathrm{N}(\underline{z}_n; H_n \underline{\theta}_n^0, \Gamma_n^0)} \tag{4.53}$$

$$= \Gamma_n^{0\ -1}(\underline{z}_n - H_n \underline{\theta}_n^0), \tag{4.54}$$

so that $\underline{T}_n$ reduces to the Kalman Filter when the observation noise is Gaussian.

## 5. Conclusion

This paper reviews Huber's minimax approach for the robust estimation of a location parameter, as well as its recursive extensions inspired by the stochastic approximation method of Robbins and Monro, and develops an approximate conditional mean estimator by constructing an asymptotic expansion for the conditional prior distribution around a small parameter involving the fraction of contamination in the observation noise.

It underscores the relationship between point estimation and filtering: both seek to obtain estimates of parameters based on observations contaminated by noise, but while the parameters to be estimated are fixed in the former case, they vary according to some (possibly stochastic) model in the latter. When the "location parameter" varies randomly, i.e. when process noise is present, the stochastic approximation technique cannot be used to obtain a consistent recursive estimator. Moreover, asymptotic performance measures make little sense in this case, and a conditional mean estimator is sought instead.

The derivation of the least favorable distribution in this context remains an open problem. The estimator presented here is therefore approximately Bayesian but not minimax. An approximation that has been suggested is to replace the convolution terms in Theorems 4.3 and 4.4 with Huber's least favorable distribution given in Theorem 2.5. Although this would result in a conservative estimator, its simplicity is quite appealing, and the results of simulation experiments have been favorable.

The approximate conditional mean estimator derived here is robust when outliers occur in isolation, but not when they occur in patches. Higher-order approximations to the conditional prior and conditional mean would result in estimators that are robust in the presence of patchy outliers, though at the expense of considerable additional complexity.

Other directions for future research include the application of time scaling to the problem of patchy outliers or other colored noise; the continuous-time case, for which the algebra promises to be more tractable; outliers in the process noise; fault detection and identification in the presence of outliers; and the asymptotic behavior of the approximate filters presented in this paper.

## REFERENCES

[1] Blum, J.R. (1954) "Multidimensional Stochastic Approximation Methods," *Ann. Math. Stat.*, 25, 4, 737-744.

[2] Englund, J.E., U. Holst, and D. Ruppert (1988) "Recursive M-estimators of Location and Scale for Dependent Sequences," *Scandinavian J. Statistics*, 15, 2, 147-159.

[3] Fabian, V. (1968) "On Asymptotic Normality in Stochastic Approximation," *Ann. Math. Stat.*, 39, 4, 1327-1332.

[4] Goel, P.K. and M.H. DeGroot (1980) "Only Normal Distributions Have Linear Posterior Expectations in Linear Regression," *J.A.S.A.*, 75, 372, 895-900.

[5] Huber, P.J. (1964) "Robust Estimation of a Location Parameter," *Ann. Math. Stat.*, 35, 1, 73-101.

[6] Huber, P.J. (1969) *Thèorie de l'Infèrence Statistique Robuste*, Presses de l'Université de Montréal (Montréal).

[7] Huber, P.J. (1972) "The 1972 Wald Lecture. Robust Statistics: a Review," *Ann. Math. Stat.*, 43, 4, 1041-1067.

[8] Huber, P.J. (1977) *Robust Statistical Procedures*, Society for Industrial and Applied Mathematics (Philadelphia, Pennsylvania).

[9] Huber, P.J. (1981) *Robust Statistics*, John Wiley (New York).

[10] Kushner, H.J. and D.S. Clark (1978) *Stochastic Approximation Methods for Constrained and Unconstrained Systems*, Springer-Verlag (Berlin and New York).

[11] Martin, R.D. (1972) "Robust Estimation of Signal Amplitude," *IEEE Trans. Information Theory*, IT-18, 5, 596-606.

[12] Martin, R.D. and C.J. Masreliez (1975) "Robust Estimation via Stochastic Approximation," *IEEE Trans. Information Theory*, IT-21, 3, 263-271.

[13] Masreliez, C.J. (1974) "Approximate Non-Gaussian Filtering with Linear State and Observation Relations," *Proc. Eighth Annual Princeton Conf. Information Sciences and Systems*, Dept. Electrical Engineering, Princeton University (Princeton, New Jersey), 398 (abstract only).

[14] Masreliez, C.J. (1975) "Approximate Non-Gaussian Filtering with Linear State and Observation Relations," *IEEE Trans. Automatic Control*, AC-20, 1, 107-110.

[15] Masreliez, C.J. and R.D. Martin (1974) "Robust Bayesian Estimation for the Linear Model and Robustizing the Kalman Filter," *Proc. Eighth Annual Princeton Conf. Information Sciences and Systems*, Dept. Electrical Engineering, Princeton University (Princeton, New Jersey), 488-492.

[16] Masreliez, C.J. and R.D. Martin (1977) "Robust Bayesian Estimation for the Linear Model and Robustifying the Kalman Filter," *IEEE Trans. Automatic Control*, AC-22, 3, 361-371.

[17] Moore, J.B. and B.D.O. Anderson (1980) "Coping with Singular Transition Matrices in Estimation and Control Stability Theory," *Int. J. Control*, 31, 3, 571-586.

[18] Nevel'son, M.B. (1975) "On the Properties of the Recursive Estimates for a Functional of an Unknown Distribution Function," in P. Révész (ed.), *Limit Theorems of Probability Theory* (Colloq. Limit Theorems of Probability and Statistics, Keszthely), North-Holland (Amsterdam and London), 227-251.

[19] Nevel'son, M.B. and R.Z. Has'minskii (1973) *Stochastic Approximation and Recursive Estimation*, American Mathematical Society (Providence, Rhode Island).

[20] Price, E.L. and V.D. Vandelinde (1979) "Robust Estimation Using the Robbins-Monro Stochastic Approximation Algorithm," *IEEE Trans. Information Theory*, IT-25, 6, 698-704.

[21] Robbins, H. and S. Monro (1951) "A Stochastic Approximation Method," *Ann. Math. Stat.*, 22, 400-407.

[22] Schick, I.C. (1989) "Robust Recursive Estimation of the State of a Discrete-Time Stochastic Linear Dynamic System in the Presence of Heavy-Tailed Observation Noise," Ph.D. thesis, Department of Mathematics, Massachusetts Institute of Technology (Cambridge, Massachusetts). Reprinted as Report LIDS-TH-1975, Laboratory for

Information and Decision Systems, Massachusetts Institute of Technology, May 1990.

[23] Schick, I.C. and S.K. Mitter (1991) "Robust Recursive Estimation in the Presence of Heavy-Tailed Observation Noise," submitted to *Ann. Stat.*

[24] Spall, J.C. and K.D. Wall (1984) "Asymptotic Distribution Theory for the Kalman Filter State Estimator," *Commun. Statist. Theor. Meth.*, 13, 16, 1981-2003.

[25] Verdú, S. and H.V. Poor (1984) "On Minimax Robustness: a General Approach and Applications," *IEEE Trans. Automatic Control*, AC-30, 2, 328-340.

[26] Wasan, M.T. (1969) *Stochastic Approximation*, Cambridge University Press (Cambridge, U.K.).

[27] West, M. (1981) "Robust Sequential Approximate Bayesian Estimation," *J. Royal Statistical Society*, B, 43, 2, 157-166.

[28] Young, P. (1984) *Recursive Estimation and Time-Series Analysis: an Introduction*, Springer-Verlag (Berlin and New York).

*Acknowledgments*: This research was supported in part by the U.S. Air Force Office of Scientific Research under contracts AFOSR-85-0227 and AFOSR-89-0276.

Sanjoy K. Mitter, Department of Electrical Engineering and Computer Science, and Laboratory for Information and Decision Systems, Massachusetts Institute of Technology, Cambridge, Massachusetts 02139.

Irvin C. Schick, Network Analysis Department, BBN Communications Division, Bolt Beranek and Newman, Inc., Cambridge, Massachusetts 02140.

# Canonical Representations of Nonlinear Discrete-time Systems

S. MONACO and D. NORMAND-CYROT

**Abstract** The paper deals with exponential representations for nonlinear discrete-time systems. It is shown how such representations provide a unified framework to study nonlinear discrete-time dynamics as well as sampled dynamics. Bilinear systems are shortly investigated. This paper has been written at the occasion of the 65th birthday of Professor Ruberti who, in the beginning of the 70s, promoted in Italy research activities in nonlinear control starting from bilinear systems.

## 1 Introduction

In spite of recent developments and an increasing interest, nonlinear control theory in discrete time is still less understood than in continuous time. There are many reasons for this such as the difficulty of setting local concepts, extensively studied in continuous-time case, and the lost of geometric properties even assuming an affine structure of the state equations with respect to the control.

Nonlinear control theory in discrete time has been initialized in [14] with the study of bilinear equations, which present the interest of involving matrix algebra. In fact, at first bilinear equations were considered, in both continuous-time and discrete-time domains as rather general with respect to their nonlinearity, but quite structurally simple "nearly linear systems." This last feeling was rapidly denied by the encountered difficulties and the fact that the theory, developed for analytic differential equations, is not facilitated when bilinear equations are considered. In the discrete-time case the situation is also more complicated, since studies, which are parallel to the continuous bilinear situation, involve state affine equations which are nonlinear with respect to $u$.

In the linear context, the above-mentioned difficulties obviously do not occur and, moreover, the linear structure is preserved under sampling. The result is that digital control problems, set on the basis of continuous-time models, are solved with respect to the sampled dynamics interpreted as a discrete model. This is quite difficult in a nonlinear context, where the structure is generally not preserved under sampling. Bilinear equations are transformed into state affine equations, linear analytic equations are transformed into general nonlinear difference equations.

A recently developed orientation of research concerns sampled nonlinear systems and the preservation of control properties under sampling [1, 10, 15]. The motivation is very practical, since most of the physical systems are modelled by differential equations while controlled by digital computers.

The object of this paper is to show the existence of exponential representations for discrete-time and sampled nonlinear dynamics. Such representations are called canonical, in the sense that they are identical in both cases except with reference to the involved vector fields. Another justification comes from the fact that these exponential forms are described in terms of vector fields and their Lie algebras, which have been shown to characterize controllability, invariance and control properties of the dynamics [4, 6, 7, 8, 9, 11]. According to these comments it is easily understood that very strong analogies and differences with the continuous-time situation can be emphasized and that the study of the preservation of control properties under sampling can be facilitated. A different approach to the one here proposed is pointed out in [2].

The paper is organized as follows. After the introduction (section 1) exponential forms are proposed in a discrete-time context (section 2) and a sampled context (section 3). In section 4 particular structures, as linear and bilinear equations, are treated. The properties of the canonical vector fields are used to fully characterize linear and bilinear dynamics. Several assumptions enabling to simplify the exponential form and the resulting dynamics are discussed, as well as the situation which is the most parallel to the continuous time case. Throughout the paper single input-single output systems are investigated, but all the results are easily extended to the multi input-multi output cases just requiring generalized notations.

## 2 Exponential Representation of Discrete-Time Dynamics

The object of this section is to associate an exponential representation to a nonlinear discrete-time dynamics. Such a representation exhibits a family of vector fields, in terms of which the analysis and the synthesis of control schemes are facilitated when dealing with discrete-time systems.

Consider a nonlinear discrete-time dynamics

$$x(k+1) = F(x(k), u(k)) \tag{1}$$

where $x \in M$ is a smooth $n$-dimensional manifold, $F : M \times \mathbb{R} \rightarrow M$, is an analytic function of both arguments. Assume that $(x_e, 0)$ is a fixed point of $F$ and denote by $F_0(.) := F(., 0)$ the drift term which is supposed to be invertible.

Denoting by $Id$ the identity function, the following basic result can be stated:

**Theorem 2.1** *[12] The discrete dynamics (1) can be represented as*

(2)
$$x(k+1) = e^{u(k)G^0(.,u(k))}(Id)\big|_{F_0(x(k))}$$

*where* $G^0(x,u) := M \to T_x M$ *is a smooth vector field parametrized by* $u$.

*Outline of the proof.* One has to show that the forced evolution around the drift can be formulated as

(3)
$$F(F_0^{-1}(x), u) = e^{uG^0(.,u)}(x)$$

First, compute the expansion of the left-hand side of (3) in powers of $u$ and then show that its formal logarithm results in an infinite series in $u$, the coefficients of which are vector fields.

For, let us introduce the following functions

(4)
$$\begin{aligned} G_1^0(x,u) &= \tfrac{\partial}{\partial\varepsilon}\big|_{\varepsilon=0} F(F^{-1}(x,u), u+\varepsilon) = \left(\tfrac{\partial}{\partial\varepsilon}\big|_{\varepsilon=0} F(., u+\varepsilon)\right)\big|_{F^{-1}(x,u)} \\ & G_1^0(x) := G_1^0(x,0) \end{aligned}$$

(5)
$$G_i^0(x) := \frac{\partial^{i-1}}{\partial u^{i-1}}\Big|_{u=0} G_1^0(x,u) \quad \forall i > 1$$

that is

(6)
$$G_1^0(x,u) = G_1^0(x) + \sum_{i\geq 1} \frac{u^i}{i!} G_{i+1}^0(x)$$

In the sequel, the Lie derivative associated to $G_i^0(x)$ will be denoted by $G_i^0$, and the identity operator by $I$.

In [12] it is shown that the Taylor expansion of the l.h. side of (3) takes the form

(7)
$$F(F_0^{-1}(x), u) = (I + \sum_{k\geq 1} \frac{u^k}{k!} \sum_{i_1+\cdots+i_m=k} c(i_1,\ldots,i_m) G_{i_1}^0 \circ \ldots \circ G_{i_m}^0)(x)$$

i.e. $F(F_0^{-1}(x), u)$ can be obtained by the application of a series of first order differential operators to the identity function. Since the coefficients $c(i_1,\ldots,i_m)$ verify the shuffle relations, from [13], by taking the formal logarithm expansion of the differential series into parentheses, we deduce that $uG^0(.,u)$ is a Lie element in the $G_i^0$'s; i.e. it admits a series representation in powers of $u$, the coefficients of which are Lie polynomials in the $G_i^0$'s of the form

(8)
$$uG^0(.,u) = \sum_{i\geq 1} \frac{u^i}{i!} B_i(G_i^0,\ldots,G_i^0)$$

$B_i$ is a specific homogeneous Lie polynomial of degree $i$ in the arguments and $G_k^0$, by convention, has degree $k$. More precisely, using formal notations, from

(9)
$$\log(1+X) = X - \frac{X^2}{2} + \frac{X^3}{3} - \frac{X^4}{4} + \ldots$$

applied to the right-hand side of (7) with

$$X = \sum_{k\geq 1} \frac{u^k}{k!} X_k = \sum_{k\geq 1} \frac{u^k}{k!} \sum_{i_1+\cdots+i_m=k} c(i_1,\ldots,i_m) G^0_{i_1} \circ \cdots \circ G^0_{i_m} \tag{10}$$

one obtains

$$B_r(G^0_1,\ldots,G^0_r) = r! \sum_{m=1}^{r} \frac{(-1)^{m+1}}{m} \sum_{i_1+\cdots+i_m=r} \frac{X_{i_1} \circ \cdots \circ X_{i_m}}{i_1!\ldots i_m!} \tag{11}$$

In (9) the powers of the formal indeterminate $X$ correspond to the usual composition of non-commuting differential operators. (11) completely specifies $B_r$ as a polynomial in the $G^0_i$'s, once the formal indeterminate $X_k$ is replaced by its expression, set in (10). As previously noted, from [13] we know that an adequate regrouping of the monomials appearing therein, give rise to a homogeneous Lie polynomial of degree $r$, so that from (8) $uG^0(.,u)$ a parameterized vector field results.

Unfortunately, a simple Lie decomposition of the general polynomial $B_i$ in (11) is not available, one obtains for the first terms

$$B_1 = G^0_1,\ B_2 = G^0_2,\ B_3 = G^0_3 + 1/2[G^0_1, G^0)2],\ B_4 = G^0_4 + [G^0_1, G^0_3] \tag{12}$$

From (2) an explicit representation for the state evolution, involving composition of exponentials, can be deduced.

For, arguing as in [12] (see also [7]), from $G^0_1(x,u)$ in (4) we define the following vector fields

$$\begin{aligned} G^k_1(x,u) &:= J(F^k_0)|_{F_0^{-k}(x)} G^0_1(F_0^{-k}(x),u));\ \forall k>0 \\ G^k_i(x) &:= J(F^k_0)|_{F_0^{-k}(x)} G^0_i(F_0^{-k}(x));\ i>1,\ \forall k>0 \end{aligned} \tag{13}$$

Defining in the same way $G^k(x,u)$ from $G^0(x,u)$ according to (8), it follows that

$$\begin{aligned} x(k) &= F(.,u(k-1))_\circ F_0^{-1} \circ F(.,u(k-2)) \circ F_0^{-2} \circ \cdots \\ &\quad \cdots \circ F_0^{k-1} \circ F(.,u(0)) \circ F_0^{-k} \circ F_0^{-k}(x_0) \\ &= e^{u(0)G^{k-1}(.,u(0))} \circ \cdots \circ e^{u(k-1)G^0(.,u(k-1))}(Id)|_{F^k_0(x_0)} \end{aligned} \tag{14}$$

The vector fields $G^k_i$ will be called the canonical vector fields associated to (1). We note that they have been introduced in [9] and [7] to characterize controlled invariant distributions and the controllability Lie algebra.

## 3 Exponential Representation under Sampling

Consider a linear analytic continuous-time system of the form:

$$\dot{x}(t) = f(x(t)) + u(t)g(x(t)) \tag{15}$$

where $x(t)$ belongs to $\mathbb{R}^n$, $f$ and $g : \mathbb{R}^n \to \mathbb{R}^n$ are analytic vector fields and the control is constant over time intervals of amplitude $\delta$ : $u(t) = u(k)$ for $k\delta \leq t < (k+1)\delta$, $k \geq 0$.

**Definition 3.1** *The nonlinear difference equation of the form:*

$$x(k+1) = F^\delta(x(k), u(k)) = F_0^\delta(x(k)) + \sum_{i\geq 1} \frac{u(k)^i}{i!} F_i^\delta(x(k)) \tag{16}$$

*is said to be sampled equivalent to (15) if and only if, at each sampling instant $t = k\delta$, the equality $x(k) = x(t)$ is verified whenever $x(0) = x(t = 0)$.*

Recurrent expressions for the functions $F_i^\delta$ can be obtained by means of integro-differential formulae based on the Poincaré expansion of the flow associated to (15) or by means of combinatoric relations as proposed in [12].

It has been shown that the following equality of formal series is verified

$$e^{-\delta f} e^{\delta f + \delta u g} = 1 + \sum_{i\geq 1} \frac{u^i}{i!} P_i^\delta(E_1^\delta, \ldots, E_i^\delta) \tag{17}$$

where the $P_i^\delta$ are homogeneous polynomials of degree $i$ in their arguments and the $E_i^\delta$'s are vector fields, which can be expressed as Lie elements in terms of the vector fields $f$ and $g$.

More precisely, denoting as in [3] by

$$\begin{aligned} E_1^\delta(x,u) : &= \frac{\partial}{\partial \varepsilon}\Big|_{\varepsilon=0} e^{-(\delta f + \delta u g)} e^{\delta f + \delta u g + \delta \varepsilon g}(x) \\ &= E_1^\delta(x) + \sum_{i\geq 1} \frac{u^i}{i!} E_{i+1}^\delta(x) \end{aligned} \tag{18}$$

where the vector fields $E_{i+1}^\delta(x)$ are defined as

$$E_{i+1}^\delta(x) := \frac{\partial^i}{\partial u^i}\Big|_{u=0} E_1^\delta(x,u) \tag{19}$$

the following formal combinatorial expressions have been exhibited in [12]:

$$\begin{aligned} E_1^\delta &= \frac{1 - e^{-\delta ad_f}}{\delta\, ad_f}(\delta g) \\ E_2^\delta &= \frac{1 - \delta ad_f - e^{\delta ad_f}}{\delta^2 ad_f^2} \omega \delta ad_g(\delta g) \\ &\ldots \end{aligned} \tag{20}$$

where "$\omega$" denotes the shuffle product [13], "*ad*" the usual Lie bracket operator and where the quotient must be interpreted as a formal cancellation. *Remark* It

is interesting to note that the vector fields (20) satisfy the following relations, which fully characterize their structure and play a crucial role when studying the inverse problem of the discretization. More precisely, denoting by $()'$ the derivative with respect to $\delta$, it can be proved that

$$(E_1^\delta)' = g \; - ad_f E_1^\delta \qquad (E_i^\delta)' = -(i-1)ad_g E_{i-1}^\delta - ad_f E_i^\delta \quad i \geq 2$$

*Remark* Denoting by $\mathcal{L}$ the Lie algebra generated by $f$ and $g$ and $\mathcal{L}_f$ the Lie ideal of $\mathcal{L}$ generated by $g$, it follows from (20) that

$$E_i^\delta \in \mathcal{L}_0^i \; i \geq 1$$

where $\mathcal{L}_0^1 = \mathcal{L}_0$ and for $i > 2$, $\mathcal{L}_0^i = [\mathcal{L}_0^1, \; \mathcal{L}_0^{i-1}]$, (decreasing sequence of ideals).

On the basis of (18)÷(20), a recurrent expression for the operators $P_i^\delta$ in (17) can be stated, enlightening their structure. Denoting by "+" the derivative with respect to $u$, one has:

**Proposition 3.1** *The $P_i^\delta(E_1^\delta, \ldots, E_i^\delta)$ are homogeneous polynomials of degree $i$ in the $E_j^\delta$ which can be obtained by means of the following recurrent formula:*

(21) $$P_i^\delta(E_1^\delta, \ldots, E_i^\delta) = E_1^\delta \circ P_{i-1}^\delta(E_1^\delta, \ldots, E_{i-1}^\delta) + (P_{i-1}^\delta)^+(E_1^\delta, \ldots, E_{i-1}^\delta)$$

*Proof.* Firstly, from (17) (18) we deduce:

(22) $$\frac{\partial}{\partial u} e^{\delta f + u\delta g} = \frac{\partial}{\partial \varepsilon}\Big|_{\varepsilon=0} e^{\delta f + (u+\varepsilon)\delta g} = e^{\delta f + u\delta g} E_1^\delta(\cdot, u)$$

that is:

(23) $$P_1^\delta(E_1^\delta) = E_1^\delta$$

Moreover, from (17), (22) and (23) we obtain:

$$\begin{aligned}
&\frac{\partial^2}{\partial u^2}\Big|_{u=0} e^{\delta f + u\delta g} = e^{\delta f} P_2^\delta(E_1^\delta(\cdot, u), \; E_2^\delta(\cdot, u)) \\
&= \frac{\partial}{\partial u}\Big|_{u=0}\left(\frac{\partial}{\partial u} e^{\delta f + u\delta g}\right) = \frac{\partial}{\partial u}\Big|_{u=0} e^{\delta f + u\delta g} E_1^\delta(\cdot, u) \\
&\quad \frac{\partial}{\partial u}\Big|_{u=0}(e^{\delta f + u\delta g}) P_1^\delta E_1^\delta + e^{\delta f} \frac{\partial}{\partial u}\Big|_{u=0} E_1^\delta(\cdot, u) \\
&= e^{\delta f} E_1^\delta P_1^\delta E_1^\delta + e^{\delta f} \frac{\partial}{\partial u}\Big|_{u=0} E_1^\delta(\cdot, u)
\end{aligned}$$

that is:

(24) $$P_1^\delta(E_1^\delta, E_2^\delta) = E_1^\delta P_1^\delta(E_1^\delta) + (P_1^\delta)^+ = E_1^\delta E_1^\delta + E_2^\delta$$

according to $E_2^\delta = (E_1^\delta)^+$ defined in (19). Iterating the reasoning, one recovers (21). ◁ *Remark* Note that, according to the rule of derivation of a product, $(P_k^\delta)^+$

can be deduced from $P_k^\delta$ by substituting $(E_{i_1+1}^\delta E_{i_2}^\delta \dots E_{i_m}^\delta + E_{i_i}^\delta E_{i_2+1}^\delta \dots E_{i_m}^\delta + \dots + E_{i_i}^\delta E_{i_2}^\delta \dots E_{i_m+1}^\delta)$ to the generic element $E_{i_i}^\delta E_{i_2}^\delta \dots E_{i_m}^\delta$.

It is important to note from (21) that the coefficients, appearing in the polynomial expressions of the $P_i^\delta$'s, satisfy the shuffle relations. Therefore, the exponential representation, stated in the following theorem, can be obtained with the same arguments as in section 2.

**Theorem 3.1** *The sampled dynamics (16) can be expressed as*

$$x(k+1) = e^{u(k)E^\delta(.,u(k))}(Id)\big|_{e^{\delta f}}(x(k)) \tag{25}$$

*where $uE^\delta(.,u)$is a smooth vector field parameterized by $u$.*

The proof is achieved arguing as in theorem 2.1 and substitute $P_i^\delta$ to $X_i$ into (11); (25) and (26) correspond to (2) and (8) with $uE^\delta(.,u)$ and $E_i^\delta$ at the place of $uG^0(.,u)$ and $G_i^0$. One gets

$$uE^\delta(.,u) = \sum_{i\geq 1} \frac{u^i}{i!} B_i(E_1^\delta, \dots, E_i^\delta)(.) \tag{26}$$

where $B_i(E_1^\delta, \dots, E_i^\delta)$ is a Lie polynomical of degree $i$ with respect to the indeterminates $E_1^\delta, \dots, E_i^\delta$;($(E_i^\delta)$ being of degree $i$). For the first terms, one obtains as previously

$$B_1^\delta = E_1^\delta, \quad B_2^\delta = E_2^\delta, \quad B_3^\delta = E_3^\delta + \frac{1}{2}[E_1^\delta, E_2^\delta], \quad B_4^\delta = E_4^\delta + [E_1^\delta, E_3^\delta], \dots$$

Proceeding further the analogies between the discrete-time and sampled dynamics, let us define as in (14) from $E_i^\delta$ the vector fields $(E_i^\delta)^k$, one has

$$(E_i^\delta)^k = e^{-\delta f} E_i^\delta e^{\delta f} = e^{k\,\delta ad_f} E_i^\delta, \qquad i \geq 1 \tag{27}$$

It is not difficult to verify that, from (25), the following explicity representation for the state evolution involving composition of exponentials can be deduced.

$$x(k) = e^{u(0)(E^\delta)^{k-1}(.,u(0))} \circ \dots \circ e^{u(k-1)E^\delta(.,u(k-1))}\big|_{e^{k\delta f}(x_0)} \tag{28}$$

with

$$(E^\delta)^k(.,u) = \sum_{i\geq 1} \frac{u^i}{i!} B_i\left((E_1^\delta)^k, \dots, (E_i^\delta)^k\right).$$

## 4 Some Structural Properties of the Canonical Vector Fields

Starting from the representations introduced in the previous sections, a study on the proprties of the canonical vector fields, when associated to linear and bilinear dynamics, is proposed hereafter. On these bases we formulate necessary and sufficient conditions for linear equivalence not yet enlightened in a discrete context. We recall that linear equivalence means the existence of a nonlinear coordinates transformation around an equilibrium point under which a given dynamics assumes a linear controllable representation. Once for all we note tha, even if not explicitely mentionnned, the results formulated have a local validity. In the second part of the section we study bilinear systems emphasizing the properties of the canonical vector fields which characterize discrete-time and sampled dynamics. The section ends with the study of a dynamics which corresepons to a simple exponential representation. Such a dynamics has a structure which is the most parallel to the linear analytic continuous-time one, in a sense that we will explain hereafter. This structure is also proposed to approximate a given nonlinear system. *A - Linear Systems*

Let us consider a single input linear system on $\mathbb{R}^n$ of the form

$$x(k+1) = F(x(k), u(k)) = Ax(k) + Bu(k) \tag{29}$$

where $A$ and $B$ are matrices of suitable dimensions.

Computing the vector fields defined in (5) and (13) it is easily verified that

$$G_1^0(x,u) = \frac{\partial}{\partial \epsilon}\Big|_{\epsilon=0} \left(AF^{-1}(x,u)\right) + Bu + B\epsilon = B = F_1^0(x) \tag{30}$$

Since $G_1^0\ (x,u)$ is constant and independent on $u$, it results that

$$G_i^0(x) = 0 \qquad \text{for} \qquad i \geq 2 \qquad\qquad \mathbf{A} \tag{31}$$

$$G_1^k(x) = A^k G_1^0(x) = A^k B = Cst. \quad \forall k \in Z \tag{32}$$

so that

$$\left[G_1^0,\ G_1^k\right](x) = 0 \qquad \forall k \in Z \qquad\qquad \mathbf{B} \tag{33}$$

From **A** one deduces the exponential represention of (29). That is

$$x(k) = e^{u(k)B}\,(Id)\,\big|_{Ax(k)} \tag{34}$$

and from **B** the evolution can be expressed as follows

$$\begin{aligned} x(k) &= e^{u(0)A^kB} \circ \ldots \circ e^{u(k-1)B}\,(Id)\,\big|_{A^k x_0} \\ &= e^{u(0)A^{k-1}B + \ldots + u(k-1)B}(Id)\big|_{A^{k-1}x_0} \end{aligned} \tag{35}$$

We have shown that conditions **A** and **B** are implied by the linearity, with a little more effort, a stronger and interesting result can be proved

**Theorem 4.1** *The exponential representation (2) describes a linear equivalent discrete-time dynamics if and only if*

*(i) conditions* **A** *and* **B** *hold,*

*(i) rank* $\left(G_1^0,\ G_1^1, \ldots, G_1^{n-1}\right)(x_e) = n.$

Let us now consider the sampling situation. For, setting in (15) $f(x) = Ax$ and $g(x) = B$, it can easily be verified from (18), (19) and (20) that

(36) $$E_1^\delta = \frac{e^{\delta A} - I}{A} B = \sum_{i \geq 1} \frac{\delta^i}{i!} A^{i-1} B = Cst.$$

$$E_i^\delta = 0, \qquad \forall i \geq 2$$

$$\left(E_1^\delta\right)^k = e^{k\delta A} E_1^\delta = Cst., \qquad \forall k \in Z$$

and thus

(37) $$\left[E_1^\delta,\ \left(E_1^\delta\right)^k\right] = 0, \qquad \forall k \in Z$$

(36) and (37) correspond to conditions **A** and **B** and characterize the linearity of the state equations which is presevered under sampling. In fact, recalling the conditions which ensure the linearity of a given continuous dynamics under coordinates transformation around an equilibrium point [16] and taking into account that coordinates transformation and sampling commute, it is not difficult to prove that:

**Corollary 4.2** *(i) and (ii) of theorem 4.1 hold on a sampled dynamics if and only if the continuous system is linear equivalent, i.e. the generic state representation is linear analytic and the conditions below hold*

*(i)'* $ad_g ad_f{}^k g = 0 \qquad \forall k \geq 1$

*(ii)'* $rank\left(g,\ ad_f g, \ldots, ad_f{}^{n-1} g\right)(x_e) = n$

*Remark* It must be noted that when sampling a linear analytic representation it can be proved that condition **A**, for $i = 2$, implies **A** and **B**. *B - Bilinear Systems*

Let us now consider a single input bilinear system with invertible drift term of the form

(38) $$x(k+1) = F\left(x(k), u(k)\right) = \left(A + Nu(k)\right) x(k) + Bu(k)$$

where $A, N$, and $B$ are matrices of suitable dimensions. One computes

$$G_1^0(x,u) = \frac{\partial}{\partial \epsilon}\Big|_{\epsilon=0} \left((A + N\epsilon + Nu)\left(F^{-1}(x,u)\right) + Bu + B\epsilon\right) = NF^{-1}(x,u) + B$$

with

$$F^{-1}(x,u) = A^{-1}x + \sum_{i\geq 1} (-1)^i u^i \left(A^{-1}N\right)^i A^{-1}(x - Bu)$$

Hence, one has

$$G_1^0(x) = NA^{-1}x + B \tag{39}$$

$$G_i^0(x) = (i-1)!(-1)^{i-1}\left(NA^{-1}\right)^{i-1}\left(NA^{-1}x + B\right) \tag{40}$$

$$\begin{aligned} G_i^k(x) &= A^k G_i^0\left(A^{-k}x\right) \\ &= (i-1)!(-1)^{i-1}A^k\left(NA^{-1}\right)^{i-1}\left(NA^{-1}A^{-k}x + B\right) \end{aligned} \tag{41}$$

It can easily be verified that:

$$\begin{aligned} G_i^k &= -(i-1)\,L_{G_1}^k\left(G_{i-1}^k\right) \\ &= -(i-1)\,J\left(G_{i-1}^k\right)G_1^k, \quad \forall k \in Z,\ \forall i \geq 2 \quad \mathbf{C} \end{aligned} \tag{42}$$

$$\left[G_i^0, G_j^0\right] = 0 \qquad \forall i,\ j \geq 1 \quad \mathbf{D} \tag{43}$$

so that

$$\left[G_i^k, G_j^k\right] = 0 \qquad \forall k \in Z \tag{44}$$

and

$$uG^k(.,u) = \sum_{i\geq 1} \frac{u^i}{i!} G_i^k \tag{45}$$

Hence, for a bilinear system we have

$$x(k) = e^{\sum_{i\geq 1}\frac{u(0)^i}{i!}G_i^{k-1}} \circ \ldots \circ e^{\sum_{i\geq 1}\frac{u(k-1)^i}{i!}G_i^0}(Id)\Big|_{A^k(x_0)} \tag{46}$$

Moreover, because of (44), the following decomposition holds:

$$e^{\sum_{i\geq 1}\frac{u^i}{i!}G_i^k} = e^{uG_1^k} \circ e^{\frac{u^2}{2}G_2^k} \circ \ldots \tag{47}$$

It can be proved that conditions **C** and **D** are sufficient to characterize a bilinear dynamics. More precisely

**Theorem 4.3** *The exponential represention (2) describes a bilinear discrete-time dynamics if and only if conditions* **C** and **D** hold.

*Remark* By assuming, without loss of generality, an homogeneous representation for a bilinear system, the relations simplify. One obtains from (40) and (41)

$$\begin{aligned} G_i^0(x) = G_i^0 x &= (i-1)!(-1)^{i-1}(NA^{-1})^i x,\ \forall i \geq 1 \\ G_i^k &= A^k G_i^0\, A^{-k} x,\ \forall k \in Z \end{aligned}$$

Let us now study the effects of sampling on a bilinear continuous-time system. For, referring to an homogeneous representation, let us set in (15), $f(x) = Ax$ and $g(x) = Nx$ (homogeneous case). After simple computations one obtains

$$\begin{aligned} E_1^\delta(x) &= \frac{e^{\delta ad_A}-1}{ad_A}(N)x \\ E_2^\delta(x) &= \frac{e^{\delta ad_A}-1-\delta ad_A}{ad_A^2}\omega\delta ad_N(N)x \\ &\dots \end{aligned}$$

and

$$(E_i^\delta)^k(x) = e^{\delta k ad_A}E_i^\delta(x) = (E_i^\delta)^k x$$

with $ad_A N = AN - NA$ the Lie product of matrices. *Remark* It is easily verified that condition **C** and **D** do not hold, this is reflected in the fact that the sampled version of a continuous bilinear dynamics does not preserve the structure. Let us rapidly illustrate this pathology expanding in powers of $\delta$ the vector fields $E_1^\delta$, $E_2^\delta$. One has

$$E_1^\delta = \delta N + \frac{\delta^2}{2}ad_A N + O(\delta^3) \qquad E_2^\delta = \frac{\delta^3}{3!}ad_N ad_A N + O(\delta^4)$$

and so

$$[E_1^\delta,\ E_2^\delta] = \frac{\delta^4}{3!}ad_N ad_N ad_A N + O(\delta^5)$$

with in general

$$ad_N ad_N ad_A\ N \neq 0.$$

As far as the preservation of the bilinear structure under sampling is preserved, according to (17) and (21) the following proposition can be proved.

**Corollary 4.4** *Conditions* **C** and **D** hold on a sampled dynamics if and only if the continuous dynamics is bilinear and

$$NA^iN = 0 \qquad \forall i \geq 0.$$

*Remark* When sampling a bilinear representation, condition **C** for $k = 0$, that is

$$E_i^\delta x = -(i-1)E_{i-1}^\delta E_1^\delta x \qquad \text{for i} \geq 2$$

imply **C** and **D**. For, condition **C** follows from the linearity of the drift term, moreover $[E_i^\delta, E_j^\delta] = 0,\ \forall i,j \geq 1$, because of the linearity of $E_i^\delta$.

*C - An Interesting Dynamics*

First of all, it must be emphasized that the affine structure with respect to $u$ of the function $F(x,u)$ in (1) does not much simplify the study with respect to the general situation, even if a recurrent characterization of the $G_i^0$'s can be given.

**Proposition 4.1** The exponential representation (2) describes a linear analytic dynamics if and only if

$$G_i^0 = -(i-1)L_{G_1^0}(G_{i-1}^0) = J(G_{i-1}^0)G_1^0 \quad \forall i \geq 2 \tag{48}$$

It is well known that, assuming a linear analytic structure in a discrete-time context, it does not simplify the study with respect to more general cases. In fact, it could be shown more precisely, but it can be understood from the analysis of the linear and the bilinear cases, that (48) does not imply any significant consequences on the behaviours of the system. It is interesting to note that (48) is nothing else but condition **C** for $k = 0$.

Starting from these considerations, let us briefly analyze an exponential representation of the form

$$x(k+1) = e^{u(k)G(.)}(Id)\big|_{F_0(x(k))} \tag{49}$$

The main feature of representation (49) is to make possible the extension of several results derived from the study of linear analytic continuous control systems by simply substituting $ad_f^k g$ by $G^k$ defined as

$$G^k(x,u) := J(F_0^k)\big|_{F_0^{-k}(x)} G^0(F_0^{-k}(x)) \tag{50}$$

Thus, the dynamics (50) can be interpreted as the discrete-time equivalent of a linear analytic continuous one.

In spite of this, it must be noted that the existence of such an exponential representation corresponds to particular properties on the function $F(x,u)$. In fact, it describes the exponential form of a discrete dynamics, which satisfies condition **A**.

Let us point out how condition **A** reflects on the state equation (1). Consider the Taylor expansion with respect to $u$ of (1), that is

$$F(x,u) = F_0(x) + \sum_{i\geq 1} \frac{u^i}{i!} F_i(x) \tag{51}$$

For the first terms, one computes the relations

$$\begin{aligned} F_2(x) &= J(F_1 \circ F_0^{-1})F_1(x) \\ F_3(x) &= J(F_2 \circ F_0^{-1})F_1(x) - 2J(J(F_1 \circ F_0^{-1})F_1 \circ F_0^{-1})F_1(x) \\ &\quad + J(F_1 \circ F_0^{-1})F_2(x) \end{aligned}$$

which imply

$$G_1(x) = F_1 \circ F_0^{-1}(x), \qquad G_2(x) = 0, \qquad G_3(x) = 0.$$

*Remark*It can easily be verified that (49) is the exponential representation associated to the dynamics

$$F(x,u) = F_0(x) + Bu$$

and in this case $G_1^0(x) = B = Cst$. As far as sampled systems are concerned, from (21) and corollary (29), it is not difficult to show that

**Proposition 4.1** *Linear analytic dynamics are described under sampling by exponential representations of the form (49) if and only if they are linearly equivalent.*

The previous arguments clearly specify the restrictions which characterize the representation (49) and explains the weak generality, when assuming it.

A different point of view consists in considering such an exponential representation as an approximation of discrete or sampled dynamics. We just note, referring to a sampled system, that the mentioned approximation does not correspond to the linearization of the continuous dynamics as proposition (29) might suggest. On the contrary, it must be considered as obtained from a nonlinear approximation, $f_a(x,u)$, of the dynamics $f(x) + ug(x)$.

## 5 Conclusion

It has been shown how exponential representations of discrete-time dynamics provide a unified approach for studying discrete-time and sampled systems. Such representations exhibit a family of vector fields defining a canonical structure.

The study of linear and bilinear dynamics and some elementary considerations starting from assumptions on the canonical structure enlighten the interest in developin gresearch activities in two directions: classification of discrete dynamics on the basis of the canonical structure, approximation schemes based on the exponential representation for analysis and synthesis of control problems.

## REFERENCES

[1] A. Arapostathis, B. Jakubczyk, H.G. Lee, Marcus and E.D. Contag, *The effect of sampling on linear equivalence and feedback linearization*, Systems and Cont. Letters, 13, 1989, 373–381.

[2] M. Fliess, *Should the theories for continuous-time and discrete-time linear and nonlinear systems really look alike?*, in "Nonlinear stems Control Design", A. Isidori Ed., Pergamon Press, Oxford, 1989.

[3] R. Goodman, *Lifting vector fields to nilpotent Lie groups*, J. Math. Pures et Appl., 57, 1978, 77–86.

[4] J.W. Grizzle, *Feedback linearization of discrete-time systems*, Systems and Cont. Letters, 9, 1987, 411–416.

[5] A. Isidori, *Nonlinear Control Systems: An Introduction*, (II ed.), Springer-Verlag, 1989.

[6] B. Jakubczyk*Feedback linearization of discrete-time systems*, Systems and Cont. Letters, 9, 1987, 411–416.

[7] B. Jakubczyk and E.D. Sontag, *Controllability of nonlinear discrete time systems; a Lie algebraic approach*, SIAM J. Cont. and Opt., 28, 1990.

[8] H.G. Lee, A. Arapostathis and S.I. Marcus, *On the linearization of discrete-time systems*, Int. J. Cont., 45, 1987, 1783–1785.

[9] S. Monaco and D. Normand-Cyrot, *Invariant distributions for discrete-time nonlinear systems*, Systems and Cont. Letters, 5, 1985, 191–196.

[10] S. Monaco and D. Normand-Cyrot, *Zero dynamics of sampled nonlinear systems*, Systems and Cont. Letters, 11, 1988, 229-234.

[11] S. Monaco and D. Normand-Cyrot, *Functional expansions for nonlinear discrete-time systems*, Math. Sys. Theo., 21, 1989, 235–254.

[12] S. Monaco and D. Normand-Cyrot, *A combinatorial approach to the nonlinear sampling problem*, in Lect. Notes in Cont. and Info. Sciences, A Bensoussan and J.L. Lions Eds., Vol. 144, Springer-Verlag, 1990, 788–797.

[13] R. Ree, *Lie elements and an algebra associated with shuffles*, Ann. of Maths., 68, 1958, 210–220.

[14] P. d'Alessandro, A. Isidori and A. Ruberti, *Realization and structure theory of bilinear dynamical systems*, SIAM J. Cont., 12, 1974, 517–537.

[15] E.D. Sontag, *An eigenvalue for sampled weak controllability of billinear systems*, Systems and Cont. Letters, 7, 1986, 313–316.

[16] H.J. Sussmann, *Lie brackets, real analycity and geometric control*, in "Differential Geometric Control Theory", R.W. Brond al. Eds., Birkhäuser, 1983, 1–116.

Work supported by the "Ministero della Ricerca Scientifica e Tecnologica" and by the "Centre National de la Recherche Scientifique".

Dipartimento di Informatica e Sistemistica, Università di Roma "La Sapienza"
Via Eudossiana 18, 00184 Rome, Italy

Laboratoire des Signaux et Systèmes, CNRS-ESE
Plateau de Moulon, 91190 Gif-sur-Yvette, France

# Stochastic Model Reduction by Aggregation

GIORGIO PICCI

*This paper is dedicated to Professor Antonio Ruberti, in occasion of his 65th birthday*

**Abstract**

**In this paper we continue the investigation of a particular model reduction methodology introduced in previous papers [1], [4]. The main ideas are vaguely inspired by Statistical Physics and some of the techniques are borrowed from Markovian representation of stationary processes (stochastic realization theory) as exposed e.g. in the surveys [6] and [7]. The model reduction technique applies to infinite dimensional conservative systems and is believed to be useful also as a theoretical tool for analysis of unmodelled dynamics e.g.in identification and adaptive control schemes . In this article we discuss an interesting class of aggregable system for which the physical interpretation of the abstract aggregaability conditions is particularly transparent.**

## 1 Problem formulation

For clarity of exposition the aggregation problem will be formulated in two dual prototype situations which concern observed autonomous systems (no inputs) or controlled systems without observation variables (i.e. without outputs). The general case of systems with both control inputs and observables can be treated by merging the solution techniques relative to the two cases above.

We are given a "large" ( this will actally mean *infinite dimensional* in the following) autonomous linear sytem $\Sigma_o$ evolving on a Hilbert space $\mathbf{H}$, which is observed through a finite number $m$, of observation channels

$$\Sigma_o : \quad \begin{cases} \dot{z}(t) = Fz(t) \\ y(t) = Hz(t) \end{cases} \tag{1}$$

where $z(t) \in \mathbf{H}$, $t \in \mathbf{R}$ is called the *microscopic state* of the system, and $y(t) \in \mathbf{R}^m$ is the (vector ) observable or *output* of the system at time $t$.

One would like to generate the output trajectories $t \rightarrow y(t)$ of $\Sigma_o$ by means of a *finite dimensional* system *with driving variables* of the type

$$\Sigma : \quad \begin{cases} \dot{x} = Ax + Kv \\ y = Cx \end{cases} \tag{2}$$

where $x(t) \in \mathbf{R}^n$, $K \in \mathbf{R}^{n\times p}$ and $v$ is a $p$–dimensional free (i.e. unconstrained) input variable ranging on some "memoryless" function space, i.e. a vector space of time functions closed with respect to concatenation, typically for example $L_p^2(\mathbf{R})$ [2].

In the following we shall feel free to call such a variable $v$ *white noise* and eventually interpret $\Sigma$ as a *Stochastic System*[1].

The study of the above representation problem, which we shall actually reformulate a bit more precisely in the next paragraph, besides being of intrinsic interest is also motivated by the desire of understanding (obviously, in a mathematically idealized context) the common engineering practice of describing complicated physical systems by *stochastic* low order models in which an additive white noise term is added to account for "modeling errors". This is for example a very common procedure in building state models for Kalman filtering. Using engineering terminology one might say that a "large" system is *aggregable* if the state vector $z$ can be decomposed into a "low frequency" and a "high frequency" (or parasytic) component, the high frequency part being, for an external observer, equivalently "lumpable" into an additive noise acting on the low frequency dynamics.

In order to avoid trivialities we shall assume that the microscopic system $\Sigma_o$ is *irreducible*, in the sense that one cannot generate the same family of output trajectories of $\Sigma_o$ by a "smaller" *autonomous* linear system (say with state space a proper subspace of $\mathbf{H}$).

The basic definition in this respect is the following.

**Definition 1** *If for all $t_o \in \mathbf{R}$ and for each initial condition $z(t_o) = z_o \in \mathbf{H}$ there are $x_o \in \mathbf{R}^n$ and a driving input trajectory $v$ for the system (2), such that the corresponding output trajectories $y_{z_o}$ and $y_{x_o,v}$ coincide in $t \geq t_o$ and, conversely, if every output trajectory of (2) coincides for $t \geq t_o$ with some output trajectory of (1), we shall say that the system $\Sigma_o$ is aggregable and $\Sigma$ is a* forward *aggregation of $\Sigma_o$.*

*A* backward *aggregation of $\Sigma_o$ is instead a finite dimensional system (2) for which the same happens for all $t \leq t_o$ for any initial time $t_o$.*

It descends from Definition 1 that the system $\Sigma_o$ in (1) is indistinguishable from a forward (resp. backward) aggregation for an observer having access to future (resp. past) trajectories of the output function $y$.

Dually, given a "large" control system $\Sigma_c$ evolving on a Hilbert space $\mathbf{H}$

$$\Sigma_c: \quad \dot{z}(t) = Fz(t) + Gu(t) \tag{3}$$

where the *microscopic state* of the system is $z(t) \in \mathbf{H}$, $t \in \mathbf{R}$ and the control input $t \to u(t)$, is a piecewise continuous function of time with values in $\mathbf{R}^p$.

[1] Actually, for aggregable systems the interpretation of $v$ as a sample path of a bona fide white noise process is possible, but the rigorous justification of this fact would require a long detour into Probability Theory on Hilbert Spaces and generalized stochastic processes which we would like to avoid in this expository paper.

In this case we ask under what conditions on $\Sigma_c$ there are *finite dimensional* subspaces $X$ of the Hilbert space $\mathbf{H}$ such that the "component" $x(t) := P^X z(t)$ of $z(t)$ relative to the subspace $X$, evolves in time according to a *finite dimensional* dynamic model of the type

$$\dot{x}(t) = Ax(t) + Bu(t) + Kv(t) \tag{4}$$

where $A, B, K$ are finite dimensional maps and $v$ is again a free (noise) variable ranging on some memoryless function space.

**Definition 2** *If for all $t_o \in \mathbf{R}$ and for each initial condition $z(t_o) = z_o \in \mathbf{H}$ there are $x_o \in \mathbf{X}$ and a driving input trajectory $v$ , such that the projection onto the subspace $X$, $x(t) := P^X z(t)$ of the solution $z(t)$ of (3) with initial condition $z_o$, coincides, for all $t \geq t_o$ and for all admissible control functions $u$, with the solution, $x_o(t)$, of the system (4) with initial state $x_o$, control function $u$ and driving input function $v$ and, converseley, if for all $t_o \in \mathbf{R}$ and $t \geq t_o$ each solution $x(t)$ of (4) can be interpreted as a projection onto a fixed subspace $X \subset \mathbf{H}$ of a solution of (3) driven by the same control $u$, we shall say that the linear control system $\Sigma_c$ is aggregable and (4) is a* forward *aggregtion of $\Sigma_c$.*

*If instead the same happens (for all $t_o$ and) for all $t \leq t_o$ then (4) is called a* backward *aggregation of $\Sigma_c$.*

In perfect duality to the autonomuos case, for a controller generating the future (resp. past) trajectories of the control function $u$, the system $\Sigma_c$ in (3) will be indistinguishable from a forward (resp. backward) aggregation. Indeed, for both systems the control variable acts in exactly the same way on the subspace $X$.

It is not difficult to show that irreducible linear systems can be aggregable only if they are *infinite dimensional* [4][2], moreover a natural "physical" condition for the existence of aggregate (i.e. thermodynamic) descriptions is that the microscopic dynamics should be *lossless* or *conservative* i.e. governed by a unitary, energy–preserving group. From the engineering point of view this condition might seem to reduce the applicability of aggregability to a rather special class of microscopic systems. Losslessness however can, at least abstractly, be met by imbedding the physical system into a *lossless dilation*, by a well known mathematical procedure [12].

# 2 A Canonical model of Aggregable System

A "canonical" model of linear conservative systems admitting aggregate description will be described in this section. It is a generalization of the so–called *Lamb model* which was proposed in 1900 as a mechanical model for explaining radiation. See [8],[9],[10].

[2]There is however a notion of approximate aggregability [3], where the "exact" equivalence discussed before is only required to hold in the limit

Consider a semi–infinite string tautly stretched at tension $\tau$, connected at the left-end point, situated at $x = 0$, to a lumped *conservative* mechanical load, for example composed of an arbitrary (but finite) number of point masses and linear springs connected together.

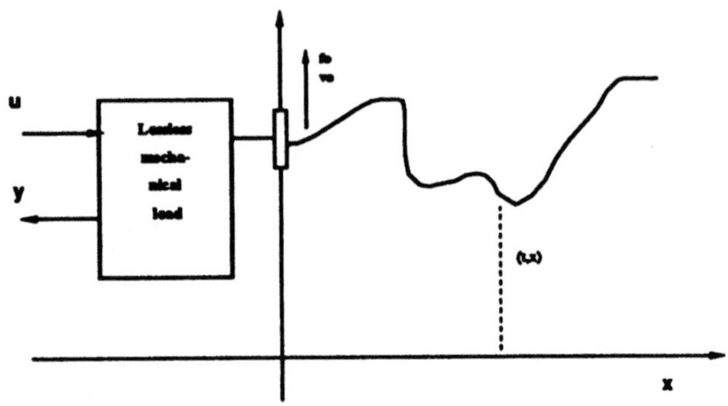


Figure 1: The semi–infinite string.

Let $\varphi(t, x)$ be the vertical deflection of the string at distance $x$ from the load and let

$$v(t,x) = \frac{\partial}{\partial t}\varphi(t,x), \quad f(t,x) = \tau\frac{\partial}{\partial x}\varphi(t,x) \tag{5}$$

be the vertical components of the velocity and of the tension of the string at $x$.

At $x = 0$ the pulling force $f_o(t) := f(t, 0)$ acts on the mechanical load iducing a motion along the vertical axis with velocity $v_o(t) := v(t, 0)$, related to the acting force by the mechanial impedence of the load. Using Laplace transforms we can write

$$\hat{v}_o(s) = Z_o(s)\hat{f}_o(s) \tag{6}$$

where $Z_o(s)$ is the mechanical impedence of the load.

The system can be described by the following state equations

$$\begin{bmatrix} \frac{\partial v}{\partial t} \\ \frac{\partial f}{\partial t} \end{bmatrix} = \begin{bmatrix} 0 & 1/\rho\frac{\partial}{\partial x} \\ \tau\frac{\partial}{\partial x} & 0 \end{bmatrix} \begin{bmatrix} v \\ f \end{bmatrix} \tag{7}$$

$$\dot{\mathbf{x}} = A\mathbf{x} + b_o f_o + bu \tag{8}$$

$$v_o = \mathbf{c}_o\mathbf{x} \tag{9}$$

where $\rho$ is the density of the string, $\mathbf{x} := \begin{bmatrix} q \\ p \end{bmatrix}$ is the state of the load ( of dimension $2n$ ) and we have introduced an external input force $u$ acting at an arbitrary location on the load. The last two equations (8) and (9), for

$u = 0$, can be thought as a realization of the mechanical impedence $Z_o(s)$ which will then be expressible as

$$Z_o(s) = c_o(sI - A)^{-1}b_o \tag{10}$$

Note that since $Z_o(s)$ is a *Lossless* impedence function, there will be no constant feedtrough term in the realization (10).

We assume that the *observed* output variable of the system is some linear combination of the state of the load $\mathbf{x}$ with perhaps a direct feedtrough term from the pulling force force of the string $f_o$ and the external input $u$

$$y = c\mathbf{x} + hf_o + ju \tag{11}$$

The string equation (7) is just the wave equation written in vector form. We assume suitable units have been chosen to insure $\tau/\rho = 1$ (so that the speed of propagation along the string is one). The evolution of the composite system (7),(8) can be seen as the evolution of a conservative Hamiltonian system $\dot{z} = Fz + gu$ with state (phase) vector

$$z := \begin{bmatrix} \mathbf{x} \\ v \\ f \end{bmatrix} \tag{12}$$

taking place in the phase space $\mathbf{H} := \mathbf{R}^{2n} \oplus L_2^2(\mathbf{R}_+)$. This space can be given a Hilbert space structure by introducing the *energy norm*

$$\| \begin{bmatrix} \mathbf{x} \\ v \\ f \end{bmatrix} \|^2 = 1/2\mathbf{x}^T\Omega\mathbf{x} + 1/2\int_0^{+\infty}(\rho v^2 + 1/\tau f^2)dx \tag{13}$$

where $\Omega$ is a symmetric nonnegative matrix representing the total energy (the Hamiltonian) of the load, a quadratic form in the state $\mathbf{x}$. By choosing $\mathbf{x}$ minimally we can always guarantee $\Omega > 0$.

The state evolution operator

$$F \;:=\; \begin{pmatrix} A & \begin{bmatrix} 0 & b_o\delta_o \\ 0 & 0 \end{bmatrix} \\ 0 & \begin{bmatrix} 0 & 1/\rho\frac{\partial}{\partial x} \\ \tau\frac{\partial}{\partial x} & 0 \end{bmatrix} \end{pmatrix} \tag{14}$$

where $\delta_o$ is the evaluation operator at $x = 0$, is skew-adjoint on its natural domain (of smooth functions satisfying the boundary conditions (9)) and generates a norm ( = energy ) -preserving group on $\mathbf{H}$.

Since the string subsystem obeys the wave equation we can express the displacement in the well known "scattering" form

$$\varphi(t, x) = a(t + x) + b(t - x) \qquad t \in \mathbf{R}, x \geq 0 \tag{15}$$

where the functions $a$ and $b$ are called the *incoming* and *outgoing* waves respectively. In the present setup it is actually only the derivatives $a'$ and

$b'$ which will enter the scattering representation of the state vector $(v, f)^T$ as determined by the initial data of (vertical) velocity and tension along the string at (say) time zero. In fact, putting $t = 0$ in (5), (15), we have

$$\begin{aligned} v_i(x) := v(0,x) &= a'(x) + b'(-x) \\ f_i(x) := f(0,x) &= a'(x) - b'(-x) \end{aligned} \tag{16}$$

Note that this system of equations determines only $a'(x)$ *for* $x \geq 0$ and $b'(x)$ *for* $x \leq 0$. Moreover, for arbitrary initial data $(v_i, f_i)^T$ in $L^2_2(\mathbf{R}_+)$ the restrictions $a'|_{x\geq 0}$ and $b'|_{x\leq 0}$ are essentially "free variables" in $L^2(\mathbf{R}_+)$ and $L^2(\mathbf{R}_-)$, respectively.

Now, we shall see that a reinterpretation of an idea of Scattering Theory, [11], leads to a mathematical descriptions of the boundary variables $v_o(t)$ and $f_o(t)$ by linear *finite dimensional* models driven by free $L^2$–input variables. These linear models could be called "stochastic" as they have the same structural properties of *stochastic models driven by white noise processes.*

The procedure starts with the identities

$$\begin{aligned} v_o(t) &= a'(t) + b'(t) \\ f_o(t) &= a'(t) - b'(t) \end{aligned} \tag{17}$$

and then uses the "steady–state" boundary condition at $x = 0$, relating $v_o(t)$ and $f_o(t)$ specified by (6). It is immediate to check that (17) are both interpretable as *output feedback* laws on the mechanical load system $(A, b_o, c_o)$ i.e. (17) can be written

$$f_o(t) = -v_o(t) + 2a'(t) \tag{18}$$

$$f_o(t) = v_o(t) - 2b'(t) \tag{19}$$

and it follows then that (18) and, respectively, (19), after substitution of (6) yield the following pair of representations of the "signal" $f_o$ [3]

$$\left\{ \begin{aligned} \dot{\mathbf{x}}(t) &= (A - b_o c_o)\mathbf{x}(t) + 2b_o a'(t) \\ f_o(t) &= -c_o\mathbf{x}(t) + 2a'(t) \end{aligned} \right. \tag{20}$$

and, respectively

$$\left\{ \begin{aligned} \dot{\mathbf{x}}(t) &= (A + b_o c_o)\mathbf{x}(t) - 2b_o b'(t) \\ f_o(t) &= c_o\mathbf{x}(t) - 2b'(t) \end{aligned} \right. \tag{21}$$

We shall now check that this is a bona fide *Forward–Backward pair of representations* of the "signal" $f_o$, in the spirit of stochastic modeling of stationary processes [7].

**Theorem 1** *Assume the realization (10) is minimal. Then the representations (20) and (21) are, respectively, asymptotically stable and antistable, in fact*

$$\Re e\lambda(A - b_o c_o) < 0, \qquad \sigma(A + b_o c_o) = -\sigma(A - b_o c_o) \tag{22}$$

[3] Of course we can obtain analogous representations also for the velocity $\mathbf{v}_o$

*the symbol $\sigma(A)$ denoting the spectrum of the matrix $A$.*

*Moreover the two representations are related by a "change of white noise input" formula of the type $\hat{b}' = K(s)\hat{a}'$ with $K(s)$ an inner function. In fact $K(s)$ is precisely the scattering function associated to the boundary condition (6), i.e.*

$$K(s) = \frac{Z_o(s) - 1}{Z_o(s) + 1} \tag{23}$$

**Proof.** It follows from (6),(18) and (19) that

$$\hat{f}_o(s) = (1 + Z_o(s))^{-1} 2\hat{a}'(s) \tag{24}$$
$$\hat{f}_o(s) = (1 - Z_o(s))^{-1} 2\hat{b}'(s) \tag{25}$$

and obviously (20) and (21) are minimal realizations of the two input–output relations above. Since $Z_o$ is *Positive Real* the same is true for $(1 + Z_o(s))^{-1}$ which must therefore have all poles strictly inside the left half plane. By minimality this implies that the first condition in (22) is in fact true.

Expressing the impedence of the load $Z_o$ as $Z_o = N_o/D_o$ we conclude from the argument above that all zeros of the polynomial $D_o(s) + N_o(s)$ must lie strictly inside the left half plane. Now it is well known that a lossless impedance must be the ratio of even and odd polynomials in $s$. From this it is not hard to see that the zeros of the denominator $D_o(s) - N_o(s)$, of the transfer function $(1 - Z_o(s))^{-1}$ are just the opposite of the zeros of $D_o(s) + N_o(s)$. This proves the second relation in (22).

The last statement of the theorem follows now easily from the above, after eliminating $f_o$ in (25). Q.E.D.

Note that, since the input function $a'$ is determined by the initial conditions of the string only on positive half lines $\{t \geq t_o\}$, the *Forward* representation (20) can only represent the time evolution of $f_o(t)$ on positive half lines $\{t \geq t_o\}$ of the time axis. The evolution of $f_o(t)$ *Backwards in time* is governed by a *different model*, namely (21), which does *not* correspond to the trivial change of direction of time transformation $t \to -t$ on (20). Hence we might say that the temporal evolution of the variable $f_o(t)$ is *irreversible in time*, a truly stochastic phenomenon which has no counterpart in finite dimensional deterministic dynamics.

The final step for the construction of aggregate models for the observed/controlled system is now almost trivial, being just a matter of combining together the "finite–dimensional subsystem" (8) and (11) with the models for the "noise process" $f_o$ which is acting as an additive random disturbing force both on the state and the output equations. Note that combining (8), (11) with, respectively, (20) and (21) we end up again with a forward–backward pair of finite dimensional linear models driven by the original control input $u$ and by a "white noise path-function" $a'$ (or $b'$). We shall omit the details of the computation.

**Acknowledgement.** It is a pleasure to thank Dr. Stefano Pinzoni for his sharp comments on an earlier version of this paper.

# References

[1] G. Picci, Stochastic aggregation, *proc.of the third Italian-German Symp.on the application of mathematics in industry, Siena June 1988*,V. Boffi and H. Neunzert eds. Teubner, Kluwer, Stuttgart, 1989.

[2] F.Fagnani and J.C. Willems, Controllabiity of $l^2$ systems, *Preprint*, University of Groningen,Aug. 1990.

[3] G. Picci and T.J. Taylor, Stochastic aggregation of linear Hamiltonian systems with microcanonical distribution, in *Realization and modeling in System Theory*, M.A. Kaashoek, J.H. van Schuppen, A.C.M. Ran eds., pp 513-520, Birkhauser Boston, 1990.

[4] G. Picci, Aggregation of linear systems in a completely deterministic framework, in *Three decades of Mathematical Systems Theory: A Collection of Surveys at the occasion of the Fiftieth Birthday of Jan C. Willems*, H. Neijmeijer, J.M. Schumacher eds. *Springer Lect. Notes in Control and Information Sciences*, **135**, pp 358-381, 1989.

[5] G. Picci, A Theory of Dynamic aggregation by stochastic models, *Proc.of the 11-th IFAC World Congress, Tallin, Estonia* **2**, pp 262–265 , 1990.

[6] A Lindquist and G. Picci, Realization Theory for multivariate stationary Gaussian processes *SIAM Journal on Control and Optimiz.*, **23**, pp1-50, 1985.

[7] A Lindquist and G. Picci, A geometric approach to modeling and estimation of linear stochastic systems *Journal of Math. Syst. Estimation and Control*, **1** ,1991.

[8] J.T.Lewis and L.C.Thomas, How to make a heat bath, in: *Functional integration and its applications*, A.M.Arthurs ed., Oxford, Clarendon 1975.

[9] H. Maasssen, On a class of quantum Langevin equations and the question of approach to equilibrium. *Thesis*, University of Groningen, 1982.

[10] J.T.Lewis and H. Maasssen, Hamiltonian models of classical and quantum stochastic processes, in *Quantum Probability and Applications to the Quantum theory of Irreversible Processes*, L. Accardi, A. Frigerio and V.Gorini eds. Springer Lect. Notes in Mathematics, **1055**, 1984.

[11] P.D.Lax and R.S.Phillips, *Scattering Theory*, Academic Press, New York, 1967.

[12] P.A.Fuhrmann, *Linear Operators and Systems in Hilbert Space*, Mc Graw Hill, Neww York, 1981.

Dipartimento Di Elettronica e Informatica,
Universitá di Padova
via Gradenigo 6/A
35131, Padova
Italy

# Feedback in a Behavioral Setting

JAN C. WILLEMS

## 1. Introduction

I consider it a privilege to contribute an article to this *Festschrift* at the occasion of the 65-th birthday of Professor Antonio Ruberti.

The late fifties and early sixties witnessed the birth of what is – even nowadays – called Modern Control Theory. Out of this grew the field of Mathematical System Theory. The impetus of this development came principally from researchers in North America. There was not much activity at that time in Europe. A notable exception was the *Istituto di Automatica* in Rome where under the direction of Prof. Ruberti a sizeable group was working in this area already in the mid-sixties. It is undoubtedly thanks to the momentum gained by this research effort that the field of System Theory has gained a strong foothold in Italian engineering departments.

I vividly recall attending a lecture by Professor Ruberti in the mid seventies. At that time he was already Rector of the University of Rome. In this lecture he explained his audience the philosophy followed in system theory as an approach to dynamical systems, in which the *interaction with the environment* is captured by the *input/output* structure, and in which the *memory* is coded in the *state*. Of course, he also emphasized the importance of the notion of *feedback*. These are also the themes of the present article.

Recent developments have led us to look at dynamical systems from a somewhat different vantage point. Central in this view is the *behavior* of a dynamical system. The behavior consists of all time trajectories compatible with the laws governing a dynamical system. Since in general a system will interact with its environment this behavior usually contains free input variables and bound output variables. However, which variables should be considered inputs may depend on the application. We view the external variables through which the system interacts with its environment as *manifest* variables. Models set up from first principle will invariably contain auxiliary (internal) variables. We call these *latent* variables. An example of latent variables are state variables. The state is a latent variable which captures the memory of a system. This framework with the behavior, behavioral equations, and manifest and latent variables as the central notions, is explained in detail in [1].

The purpose of this article is to cast the notion of feedback in this behavioral framework. We will view feedback simply as an interconnection in which the behavior after interconnection becomes the intersection of the behavior of the forward loop and that of the feedback loop. We will distinguish between *regular* and *singular* feedback. In regular feedback the forward loop and the feedback loop can be interconnected at any time instant, while with singular feedback interconnection requires a suitable preparation of the state. An example of singular feedback are controllers with differentiating action, for example PID control laws.

## 2. The Behavioral Approach

In the behavioral approach a dynamical system is viewed as simply any family of trajectories. Formally, a *dynamical system* $\Sigma$ is defined as a triple $\Sigma = (T, W, \mathfrak{B})$ witj $T$ a subset of $\mathbb{R}$ called the *time axis*, $W$ a set called the *signal space*, and $\mathfrak{B}$ a subset of $W^T$ (:= the set of all maps from $T$ to $W$), called the *behavior*. Thus $T$ denotes the collection of all time instances relevant to the dynamical model, $W$ denotes the space in which the time signals which the system produces take on their values, and $\mathfrak{B}$ denotes the collection of time signals which are compatible with the laws governing laws governing the dynamical system.

The ideas concerning interconnection and feedback put forward in this paper can be developped in a very general setting involving nonlinear, infinite dimensional, or discrete event systems. However, for the sake of concreteness, we will consider only *continuous time* ($T = \mathbb{R}$), *lumped* ($W = \mathbb{R}^q$), *linear, time invariant* systems, described by *ordinary differential equations.* $\Sigma = (\mathbb{R}, \mathbb{R}^q, \mathfrak{B})$ is said to be *linear* if $\mathfrak{B}$ is a linear subspace of $(\mathbb{R}^q)^{\mathbb{R}}$ (viewed as vector space with pointwise addition and multiplication) and *time invariant* if $\sigma^t\mathfrak{B} = \mathfrak{B}$ for all $t \in \mathbb{R}$ ($\sigma^t$ denotes the *t-shift* acting on $(\mathbb{R}^q)^{\mathbb{R}}$ and defined by $(\sigma^t w)(t') := w(t' + t)$).

## 3. Dynamical Systems

In this paper we will consider only systems represented by ordinary differential equations. However, we will introduce such representations as properties of the behavior. Let $\mathcal{L}^{loc}(\mathbb{R}, \mathbb{R}^q)$ denote the space of locally integrable functions from $\mathbb{R}$ to $\mathbb{R}^q$ equipped with the topology which ensures that $w_n \rightarrow 0$ iff $\int_{t_0}^{t_1} \| w_n(t) \| \, dt \rightarrow 0$ for all $t_o, t_1 \in \mathbb{R}$.

Let $\mathcal{S}$ be a closed shift invariant linear subspace of $\mathcal{L}^{loc}(\mathbb{R}, \mathbb{R}^q)$. Define $\mathcal{S}^\perp$ as the set of *annihilators* of $\mathcal{S}$. Thus $\mathcal{S}^\perp$ is a subset of the compact support

$(1 \times q)$ vector distributions on $\mathbf{R}$ with $(f \in \mathcal{S}^\perp) :\Leftrightarrow (f * \mathcal{S} = 0)$ where $*$ denotes convolution. Obviously $\mathcal{S}^\perp$ is also linear and shift invariant. Let $\mathcal{S}_0^\perp$ denote the subset of elements of $\mathcal{S}^\perp$ with support in the point $\{0\}$. We will call ***instantaneously specified*** if $(w \in \mathcal{S}^\perp) \Leftrightarrow (\mathcal{S}_0^\perp * \sigma^t w = 0$ for all $t \in \mathbf{R})$.

With $\mathcal{L}^q$ we will denote the family of dynamical systems $(\mathbf{R}, \mathbf{R}^q, \mathfrak{B})$ with $\mathfrak{B}$ a *linear shift invariant instantaneously specified* subspace of $\mathcal{L}^{loc}(\mathbf{R}, \mathbf{R}^q)$.

Let $\mathbf{R}[\xi]$ denote as usual the ring of polynomials with real coefficients in the indeterminate $\xi$. $\mathbf{R}^{n_1 \times n_2}[\xi]$ denotes the $(n_1 \times n_2)$ matrices with elements from $\mathbf{R}[\xi]$. $\mathbf{R}^{\bullet \times n}[\xi]$ denotes such matrices with $n$ columns and any (of course, finite) number of rows.

Let $R \in \mathbf{R}^{\bullet \times q}[\xi]$ and consider the following system of differential equations

$$(1) \qquad R\left(\frac{d}{dt}\right) w = 0$$

(1) defines the dynamical system $\Sigma = (\mathbf{R}, \mathbf{R}^q, \ker R\left(\frac{d}{dt}\right))$ with $R\left(\frac{d}{dt}\right)$ viewed as a map from $\mathcal{L}^{loc}(\mathbf{R}, \mathbf{R}^q)$ into the vector of distribution with the same number of rows as $R$. It is easy to see that $(\mathbf{R}, \mathbf{R}^q, \ker R\left(\frac{d}{dt}\right)) \in \mathcal{L}^q$. Conversely, it can be shown that $\Sigma \in \mathcal{L}^q$ iff there exists $R \in \mathbf{R}^{\bullet \times q}[\xi]$ such that $\Sigma = (\mathbf{R}, \mathbf{R}^q, \ker R\left(\frac{d}{dt}\right))$. Thus we will call (1) a *behavioral equation representation* of this $\Sigma \in \mathcal{L}^q$. In the language of [1], $\mathbf{R}^{\bullet \times q}[\xi]$ defines hence a *parametrization* of $\mathcal{L}^q$, meaning that there exists a surjective map $\pi : \mathbf{R}^{\bullet \times q}[\xi] \to \mathcal{L}^q$, in casu, $\pi$ is given by $\pi(R) := (\mathbf{R}, \mathbf{R}^q, \ker R\left(\frac{d}{dt}\right))$. However, this parametrization is not *trim* (meaning that $\pi$ is not injective).

The map $\pi$ induces in a natural way the equivalence relation $\underset{\pi}{\sim}$ on $\mathbf{R}^{\bullet \times q}[\xi]$ defined by $(R_1 \underset{\pi}{\sim} R_2) :\Leftrightarrow (\pi(R_1) = \pi(R_2))$. This equivalence relations leads to the notion of a $(\pi$-$)$*canonical form*, a *trim canonical form*, an *invariant*, and a *complete invariant*. These are defined formally in [1].

Next, consider $\mathbf{R}^{\bullet \times q}[\xi]$. Define the *column dimension map* $c : \mathbf{R}^{\bullet \times q}[\xi] \to \mathbf{Z}_+$ (:= the set of nonnegative integers) and the *rank map* $r : \mathbf{R}^{\bullet \times q}[\xi] : \{0, 1, \cdots, q\}$ in the obvious way. $\mathbf{R}^{\bullet \times q}[\xi]$ is said to be of *full row rank* $(frr)$ if $c(R) = r(R)$. Let $\mathbf{R}^{\bullet \times q}_{frr}[\xi]$ denote the set of elements of $\mathbf{R}^{\bullet \times q}[\xi]$ of full row rank. Consider next the *degree map* $d : \mathbf{R}^{\bullet \times q}[\xi] \to \mathbf{Z}_+$ as $d(R) :=$ the largest degree of all the polynomial elements of $R$. In addition to the degree, we will consider the *total degree map* $d_{tot} : \mathbf{R}^{\bullet \times q}[\xi] \to \mathbf{Z}_+$ defined by $d_{tot}(R) := \sum_{k=1}^{c(R)} d_k(R)$ with $d_k(R) :=$ the degree of the $k$-th row of $R$. Finally, consider the *MacMillan degree map* $M : \mathbf{R}^{\bullet \times q}[\xi] \to \mathbf{Z}_+$. This is defined as $M(R) :=$ is the largest degree of all $r(R) \times r(R)$ minors of $R$.

It is easy to prove (using the Smith form for polynomial matrices) that $(R_1 \underset{\pi}{\sim} R_2) \Leftrightarrow$ ($\exists$ unimodular polynomial matrices $U$ and $V$ of suitable dimension such that $R_2 = UR_1$ and $R_1 = VR_2$). From here it immediately follows that $r$ and $M$ are both $\pi$-invariants, and that $\mathbf{R}_{frr}^{\bullet\times q}[\xi]$ is a $\pi$-canonical form, meaning that for all $\Sigma, \pi^{-1}(\Sigma)$ contains a $frr$ polynomial matrix. However, neither $d$ nor $d_{tot}$ are $\pi$-invariants. In order to make the order and the total order of the differential equation describing a $\Sigma \in \mathcal{L}^q$ into invariants, we simply take the smallest order. Thus define the *order* and the *total order maps* $L : \mathcal{L}^q \to \mathbf{Z}_+$ and $L_{tot} : \mathcal{L}^q \to \mathbf{Z}_+$ as

$$L(\Sigma) := \underset{R \in \pi^{-1}(\Sigma)}{\text{minimum}} \; d(R)$$

$$L_{tot}(\Sigma) := \underset{R \in \pi^{-1}(\Sigma)}{\text{minimum}} \; d_{tot}(R)$$

A canonical form which brings into evidence all these invariants consists of the *row proper* polynomial matrices. Consider $R \in \mathbf{R}^{\bullet\times q}[\xi]$. Let $R_L$ be the matrix whose $k$-th row consists of the row vector which is the coefficient of the highest degree power of $\xi$ of the $k$-th row of $R$. $R$ is called *row proper* ($rp$) if $R_L$ has full row rank. Let $\mathbf{R}_{rp}^{\bullet\times q}[\xi]$ denote the set of all row proper elements of $\mathbf{R}_{rp}^{\bullet\times q}[\xi]$. It can be shown that $R_{rp}^{\bullet\times q}[\xi]$ is a $\pi$-canonical form. Moreover if $\Sigma \in \mathcal{L}^q$, $R \in \mathbf{R}_{rp}^{\bullet\times q}[\xi]$, and $\pi(R) = \Sigma$, then $M(R) = d_{tot}(R) = L_{tot}(\Sigma)$.

## 4. Input, Output, and State Dimension

One of the important advantages of the behavioral approach as the foundation of system theory is that there is no need to distinguish *a priori* between inputs and outputs, and certainly that there is no need to start with a state representation.

However, it is possible to prove that any system $\Sigma \in \mathcal{L}^q$ admits an *input/output representation*. That is, a system (1) can be equivalently represented by means of a system of the following special form

$$P\left(\frac{d}{dt}\right)y = Q\left(\frac{d}{dt}\right)u \tag{2}$$

$$w = \Pi \begin{bmatrix} u \\ y \end{bmatrix}$$

with $P \in \mathbf{R}^{p\times p}[\xi]$, $\det P \neq 0$, $Q \in \mathbf{R}^{p\times m}[\xi]$, $p+m = q$, and $\Pi$ a $(p+m)\times(p+m)$ permutation matrix. It is even possible to choose the permutation matrix such that $P^{-1}Q$ is a matrix of proper rational functions.

The question whether properness of $P^{-1}Q$ is a necessary feature for an I/O representation is to some extent a matter of taste: it is the nagging question whether a differentiator is an I/O system. If $P^{-1}Q$ is not proper, then the I/O system will indeed contain a differentiating action. The distinguishing feature of a behavioral equation of the type (2) is that $u$ is *free* and $y$ *bound*. This means that for any $u \in C^\infty(\mathbf{R}, \mathbf{R}^m)$ there exists a $y \in C^\infty(\mathbf{R}, \mathbf{R}^p)$ such that (2) is satisfied. Moreover, the set of all such $y$'s forms a *finite dimensional* linear variety. Properness of $P^{-1}Q$ follows iff for all $u \in \mathcal{L}^{loc}(\mathbf{R}, \mathbf{R}^q)$ there exists a $y \in \mathcal{L}^{loc}(\mathbf{R}, \mathbf{R}^q)$ such that $(u, y)$ belongs to the behavior of (2). In fact, $y$ will be always at least as smooth as $u$.

Also, it is possible to prove that any system $\Sigma \in \mathcal{L}^q$ admits a "*state*" representation

$$E\frac{dx}{dt} + Fx + Gw = 0 \tag{3}$$

with $E, F, G$ matrices of suitable dimension. Let $n$ be the dimension of the vector $x$ in (3). The *external behavior* of (3) is given by

$\{w \in \mathcal{L}^{loc}(\mathbf{R}, \mathbf{R}^q) \mid \exists x : \mathbf{R} \to \mathbf{R}^n$, absolutely continuous, such that $E\frac{dx}{dt}(t) + Fx(t) + Gw(t) = 0$ a.e.$\}$

We will call (3) a *state representation* of $\Sigma$ if this external behavior is equal to the behavior of $\Sigma$. The rationale for calling (3) a state representation is explained in [1].

Finally, it is even possible to combine the representations (2) and (3) yielding an *input/state/output representation*, the usual starting point of the theory of linear systems:

$$\begin{aligned} \frac{dx}{dt} &= Ax + Bu \\ y &= Cx + Du \\ w &= \Pi \begin{bmatrix} u \\ y \end{bmatrix} \end{aligned} \tag{4}$$

We will call the representation (3) or (4) (*state-*) *minimal* if among all such representations of a given $\Sigma \in \mathcal{L}^q$, the dimension of the state space, $n$, is as small as possible. It can be shown that (4) is minimal iff $(A, C)$ is observable.

The number of inputs and outputs in (2) and (4) and the number of states in a minimal representation (3) and (4) are immediately related to the invariants $r$, $M$, $L$, and $L_{tot}$. In fact, with $p$ (the number of outputs), $m$ (the number of inputs), $n$ (the – minimal – number of states), and $\nu$ (the observability index of a state minimal representation) defined in the obvious way, we have

$$r = p = q - m \tag{5}$$

$$M = L_{tot} = n \tag{6}$$

$$L = \nu \tag{7}$$

These invariant play an important role in many system theory problems, for example in the pole placement problem. In particular if $R \in \mathbf{R}_{rp}^{\bullet \times q}[\xi]$ is such that $\pi(R) = \Sigma$, then

$$\begin{array}{lllllll} C(R) & = & r(R) & = & p(\Sigma) & = & q - m(\Sigma) \\ M(R) & = & d_{tot}(R) & = & L_{tot}(\Sigma) & = & n(\Sigma) \\ d(R) & = & L(\Sigma) & = & \nu(A, C) & & \end{array}$$

## 5. Interconnection and Feedback

Let $\Sigma_1, \Sigma_2 \in \mathcal{L}^q$. We view the interconnection of $\Sigma_1$ and $\Sigma_2$ simply as imposing the laws of both $\Sigma_1$ and $\Sigma_2$. Formally the *interconnection* of $\Sigma_1 = (\mathbf{R}, \mathbf{R}^q, \mathfrak{B}_1)$ and $\Sigma_2 = (\mathbf{R}, \mathbf{R}^q, \mathfrak{B}_2)$ is denoted by $\Sigma_1 \wedge \Sigma_2$ and defined as

$$\Sigma_1 \wedge \Sigma_2 := (\mathbf{R}, \mathbf{R}^q, \mathfrak{B}_1 \cap \mathfrak{B}_2) \tag{8}$$

It is easy to prove that $\Sigma_1, \Sigma_2 \in \mathcal{L}^q$ implies $\Sigma_1 \wedge \Sigma_2 \in \mathcal{L}^q$. We will be only interested in interconnections in which the laws of the systems can be viewed as independent. Thus we will call $\Sigma_1 \wedge \Sigma_2$ a *feedback interconnection* if

$$p(\Sigma_1 \wedge \Sigma_2) = p(\Sigma_1) + p(\Sigma_2) \tag{9}$$

We will call a feedback interconnection *regular* if

$$n(\Sigma_1 \wedge \Sigma_2) = n(\Sigma_1) + n(\Sigma_2) \tag{10}$$

Otherwise, it is called *singular*. We shall soon see that the usual feedback interconnections considered in system theory are regular. PID control laws with a differentiating action are examples of singular feedback interconnections.

An interconnection $\Sigma_1 \wedge \Sigma_2$ is a regular feedback interconnection iff $\Sigma_1 = (\mathbf{R}, \mathbf{R}^q, \mathfrak{B}_1)$ and $\Sigma_2 = (\mathbf{R}, \mathbf{R}^q, \mathfrak{B}_2)$ have the following property. For all $w_1 \in \mathfrak{B}_1$ and $w_2 \in \mathfrak{B}_2$ there will exist $w \in \mathfrak{B}_1 \cap \mathfrak{B}_2$ such that the $w_1 \wedge w \in \mathfrak{B}_1$ and $w_2 \wedge w \in \mathfrak{B}_2$. The concatenation $v' \wedge v''$ is defined as follows:

$$(v' \wedge v'')(t) := \begin{cases} v'(t) & \text{for} \quad t < 0 \\ v''(t) & \text{for} \quad t \geq 0 \end{cases}$$

In other words in a regular feedback interconnection a switch can be closed at any time, forcing not yet interconnected systems to proceed in an interconnected symbiosis. Singular feedback interconnections however require that the state of the interconnected components be suitably prepared before one is able to close the switch and proceed.

Let $\Sigma_1, \Sigma_2 \in \mathcal{L}^q$ and assume that $\Sigma_1 \wedge \Sigma_2$ is a feedback interconnection. Then there exists a componentwise partition of $\mathbf{R}^q$ into $\mathbf{R}^{q_1} \times \mathbf{R}_{q_2} \times \mathbf{R}^{q_3}$ such that the corresponding equations for $w = \begin{bmatrix} w_1 \\ w_2 \\ w_3 \end{bmatrix}$ are given by:

for $\Sigma_1$:

$$P_1\left(\frac{d}{dt}\right)w_1 = Q_1\left(\frac{d}{dt}\right)w_2 + Q_3\left(\frac{d}{dt}\right)w_3 \tag{11}$$

for $\Sigma_2$:

$$P_2\left(\frac{d}{dt}\right)w_2 = Q_2\left(\frac{d}{dt}\right)w_1 + Q_4\left(\frac{d}{dt}\right)w_3 \tag{12}$$

with $det\ P_1 \neq 0$ and $det\ P_2 \neq 0$; $det \begin{bmatrix} P_1 & -Q_1 \\ -Q_2 & P_2 \end{bmatrix} \neq 0$; and

$$\begin{bmatrix} P_1 & -Q_1 \\ -Q_2 & P_2 \end{bmatrix}^{-1} \begin{bmatrix} Q_3 \\ Q_4 \end{bmatrix}$$

proper. In other words, in $\Sigma_1$, $w_2, w_3$ is the input and $w_1$ the output; in $\Sigma_2$, $w_1, w_3$ is the input and $w_2$ the output; and in $\Sigma_1 \wedge \Sigma_2$ $w_3$ is the input and $w_1, w_2$ the output. The interconnection thus imposes conditions forcing the input $w_2$ to $\Sigma_1$ to be equal to the output $w_2$ to $\Sigma_2$ and vice-versa for $w_1$. This is the usual looping of the signal in a feedback structure. In principle this looping allows for a differentiating action.

In a regular feedback interconnection we can in addition assume that $P_1^{-1}[Q_1\ Q_3]$ and $P_2^{-1}[Q_2\ Q_3]$ are proper. The interconnection then yields the usual feedback structure in which a differentiating action is not allowed.

From the above it follows that we can view the usual feedback looping simply in terms of regular feedback interconnections, provided we are allowed to choose the input/output structure of the plant and the controller. However, in many applications the input/output structure of the plant will be imposed a priori. In this case we should use the notion of feedback equivalence.

Let $\Sigma_1 \in \mathcal{L}^q$ be given. Two systems $\Sigma_2', \Sigma_2'' \in \mathcal{L}^q$ are called *feedback equivalent relative* to $\Sigma_1$ if $\Sigma_1 \wedge \Sigma_2' = \Sigma_1 \wedge \Sigma_2''$. Now let $\Sigma_1 \wedge \Sigma_2'$ be a regular feedback interconnection and assume that we start with a given partition of $\Sigma_1$ as given in (11) with $P_1^{-1}[Q_2 \vdots Q_3]$ proper. It may then not be true that the interconnection $\Sigma_1 \wedge \Sigma_2'$ allows the usual interpretation in terms of feedback looping. However, it can be proven that there will exist a $\Sigma_2'' \in \mathcal{L}^q$, with $p(\Sigma_2') = p(\Sigma_2'')$, $n(\Sigma_2') = n(\Sigma_2'')$, and $L(\Sigma_2') = L(\Sigma_2'')$, with $\Sigma_2'$ and $\Sigma_2''$ feedback equivalent relative to $\Sigma_1$, and such that $\Sigma_1 \wedge \Sigma_2''$ does allow the usual interpretation in terms of feedback looping.

These considerations allow us to conclude that:

- singular feedback interconnections occur when considering differentiating actions in feedback loops.
- regular feedback interconnections can always be seen as arising by looping back outputs from the plant to the controller and outputs from the controller to the plant.

Consequently, the ordinary feedback design problems can be viewed as the question of finding, for a given $\Sigma_1 \in \mathcal{L}^q$, a desired $\Sigma_2 \in \mathcal{L}^q$ such that $\Sigma_1 \wedge \Sigma_2$ is a regular feedback interconnection. Considerations of what variables are inputs and outputs need hence not enter the formulation of feedback design problems. The desired restrictions on the signal flow graph are captured by the regularity of the feedback interconnection.

## 6. The Pole Placement Question

Let $\Sigma_1, \Sigma_2 \in \mathcal{L}^q$ and assume that $p(\Sigma_1) = m(\Sigma_2)$ (and hence $p(\Sigma_2) = m(\Sigma_1)$). Hence, if $\Sigma_1 \wedge \Sigma_2$ is a feedback interconnection, there will hold $p(\Sigma_1 \wedge \Sigma_2) = q$, equivalently $m(\Sigma_1 \wedge \Sigma_2) = 0$. The interconnected system will hence not have inputs. Systems $\Sigma \in \mathcal{L}^q$ having the property that $m(\Sigma) = 0$ are what we call autonomous systems. Formally, a system $\Sigma = (\mathbb{R}, \mathbb{R}^q, \mathfrak{B}) \in \mathcal{L}^q$ is called *autonomous* if $w_1, w_2 \in \mathfrak{B}$ and $w_1(t) = w_2(t)$ for $t < 0$ imply $w_1 = w_2$. It can be shown that the following conditions are equivalent:

1. $\Sigma = (\mathbb{R}, \mathbb{R}^q, \mathfrak{B}) \in \mathcal{L}^q$ is autonomous
2. $\mathfrak{B}$ is finite dimensional
3. $m(\Sigma) = 0$ (equivalently $p(\Sigma) = 0$)
4. there exists $R \in \mathbb{R}^{q \times q}[\xi]$ with $det\ R \neq 0$ such that $\mathfrak{B} = ker\ R(\frac{d}{dt})$

It can also be shown that $dim\ \mathfrak{B} = n(\Sigma)$ is also equal to the degree of $det\ R$.

Let $\Sigma \in (\mathbb{R}, \mathbb{R}^q, \mathfrak{B}) \in \mathcal{L}^q$ be autonomous and $R \in \mathbb{R}^{q \times q}[\xi]$ be such that $\pi(R) = \Sigma$. Clearly $R' \underset{\pi}{\sim} R$ iff there exists a unimodular polynomial matrix $U \in \mathbb{R}^{q \times q}[\xi]$ such that $R' = UR$. This implies that $det\ R$ is, up to a non-zero multiplicative constant, a $\pi$-invariant. Hence if we restrict attention to $R$'s such that $det\ R$ is a monic polynomial we will obtain a $\pi$-invariant. Thus, for autonomous $\Sigma \in \mathcal{L}^q$, we will denote by $\chi_\Sigma$ the unique monic polynomial such that $\chi_\Sigma$ is equal to $det\ R$ up to a non-zero multiplicative constant. $\chi_\Sigma$ will be called the *characteristic polynomial* of $\Sigma$. Hence $\chi_\Sigma \in \mathbb{R}[\xi]$ is monic and has degree equal to the dimension of $\mathfrak{B}$. If $\frac{dx}{dt} = Ax, w = Cx$ is a state minimal ($\Leftrightarrow$ $(A, C)$ observable) representation of $\Sigma$, then $\chi_\Sigma(\xi) = det\ (I\xi - A)$.

We can now formulate, purely in terms of behaviors, the usual pole placement problem both in the case of singular and regular feedback interconnections.

Let $\Sigma \in \mathcal{L}^q$ (the "*plant*") be given. Define $\mathcal{J}^\Sigma \subseteq \mathcal{L}^q$ (the admissible "*feedback processors*") as follows. $(\Sigma' \in \mathcal{J}^\Sigma) :\Leftrightarrow$ (1. $\Sigma \wedge \Sigma'$ is autonomous, and 2. $\Sigma \wedge \Sigma'$ is a feedback interconnection). Let $\mathcal{C}^\Sigma \subseteq \mathcal{L}^q$ (the "*closed loop systems*") be defined as: $\mathcal{C}^\Sigma := \Sigma \wedge \mathcal{J}^\Sigma$. Denote by $\chi^\Sigma$ the resulting closed loop characteristic polynomials. Thus $\chi^\Sigma \subseteq \mathbf{R}_m[\xi]$ (:= the monic elements of $\mathbf{R}[\xi]$) is defined by $\chi^\Sigma := \{\chi_{\Sigma'}$ for some $\Sigma' \in \mathcal{C}^\Sigma\}$.

We will use a completely analogous notation $\mathcal{J}_R^\Sigma, \mathcal{C}_R^\Sigma, \chi_R^\Sigma$ when we consider only regular feedback interconnections.

The pole placement problem may now be formulated as follows: *Given* $\Sigma \in \mathcal{L}^q$, *identify respectively* $\chi^\Sigma$ *and* $\chi_R^\Sigma$.

## 7. CONTROLLABILITY

The main result regarding pole placement states that pole placement can be achieved for controllable systems [2]. We will now introduce controllability in a behavioral setting.

A dynamical system $\Sigma = (\mathbf{R}, \mathbf{R}^q, \mathfrak{B}) \in \mathcal{L}^q$ is said to be *controllable* if for any $w_1, w_2 \in \mathfrak{B}$ there exists $T > 0$ and $w \in \mathfrak{B}$ such that $w(t) = w_1(t)$ for $t < 0$ and $w(t) = w_1(t - T)$ for $t > T$.

Let $R \in \mathbf{R}^{\bullet \times q}[\xi]$ and consider $\pi(R)$. *For what $R$'s is $\pi(R)$ controllable?* There is a very nice test for this [1]. Indeed, $\pi(R)$ is controllable if and only if the complex matrix $R(\lambda)$ has constant rank for $\lambda \in \mathbf{C}$.

There is many alternative equivalent tests for controllability. A matrix $R \in \mathbf{R}_{frr}^{p\times q}[\xi]$ is said to be *left prime* if any factorization $R = R_1R_2$ with $R_1 \in \mathbf{R}^{p\times p}[\xi]$ is such that $R_1$ is unimodular. If $R \in \mathbf{R}_{frr}^{\bullet\times q}[\xi]$, then $\pi(R)$ is controllable iff $R$ is left prime.

The following is an explicit test in terms of the coefficients of the differential equation (1). Let $R(\xi) = R_0 + R_1\xi + \cdots + R_L\xi^L$. Consider for $k = 0, 1, 2, \cdots$

the following matrices formed from these coefficients:

$$M_k := \begin{bmatrix} R_0 & 0 & \cdots & 0 \\ R_1 & R_0 & \cdots & 0 \\ \vdots & \vdots & \ddots & \vdots \\ R_k & R_{k-1} & \cdots & R_0 \\ \vdots & \vdots & \vdots & \vdots \\ R_L & R_{L-1} & \cdots & R_{L-k} \\ 0 & R_L & \cdots & R_{L-k+1} \\ \vdots & \vdots & \ddots & \vdots \\ 0 & 0 & \cdots & R_L \end{bmatrix} \tag{13}$$

Let $r_k := rank\ M_k$ and define $\rho_k := r_k - r_{k-1}$ with $r_{-1} := 0$. Then it can be shown that $q \geq \rho_1 \geq \cdots \geq \rho_k \geq \cdots \geq 0$. Therefore there will exist a $k^*$ such that $\rho_k = \rho_{k^*}$ for $k \geq k^*$. Actually, it can be shown that $M(R)$ is a possible choice for $k^*$, and that $\rho_{k^*} = q - r(R)$.

It can be shown that $\pi(R)$ is controllable iff $r_{k^*} = M(R) + (q - r(R))(k^* + 1)$.

A final test for controllability which is useful is the following. $\pi(R)$ is controllable iff there exist unimodular polynomial matrices $U$ and $V$ such that

$$R = U \begin{bmatrix} I & 0 \\ 0 & 0 \end{bmatrix} V \tag{14}$$

## 8. POLE PLACEMENT BY SINGULAR FEEDBACK

Using the results of Section 6, it follows that pole placement question may be reformulated as follows in the case of singular feedback. Let $R \in \mathbf{R}^{p \times q}_{frr}[\xi]$ be given. Consider $R' \in \mathbf{R}^{(q-p) \times q}[\xi]$ such that $det \begin{bmatrix} R \\ R' \end{bmatrix} \neq 0$. The question is what polynomials $det \begin{bmatrix} R \\ R' \end{bmatrix} \in \mathbf{R}[\xi]$ are achievable this way. The following constitutes the main result in this area.

**Lemma:** Let $R \in \mathbf{R}^{p \times q}_{frr}[\xi]$ be given. Then there exists, for any $\chi \in \mathbf{R}[\xi]$, a polynomial matrix $R' \in \mathbf{R}^{(q-p) \times q}[\xi]$ such that

$$det \begin{bmatrix} R \\ R' \end{bmatrix} = \chi \tag{15}$$

iff $R$ is left prime.

Translating this lemma into system theoretic language yields the following important result.

**Theorem:** Let $\Sigma \in \mathcal{L}^q$ be given. Then there exist for all $\chi \in \mathbf{R}_m[\xi]$ a $\Sigma' \in \mathcal{L}^q$ such that $p(\Sigma \wedge \Sigma') = p(\Sigma) + p(\Sigma) = q$ and $\chi_{\Sigma \wedge \Sigma'} = \chi$ iff $\Sigma$ is controllable.

Actually, it is easy to classify the achievable polynomials $\chi$ even when $\Sigma$ is not controllable. Let $R \in \mathbf{R}^{\bullet \times q}_{frr}[\xi]$ be such that $\pi(R) = \Sigma$. Then $R$ can be factored as $R = R_1 R_2$ with $R_1$ square and $R_2$ left prime. The achievable polynomials in (15) are precisely those $\chi$'s which have *det* $R_1$ as a factor.

## 9. Pole Placement by Regular Feedback

The pole placement problem may be formulated as follows in the case of regular feedback. Let $R \in \mathbf{R}^{p \times q}_{frr}[\xi]$ be given. Consider $R' \in \mathbf{R}^{(q-p) \times q}[\xi]$ such that $det \begin{bmatrix} R \\ R' \end{bmatrix} \neq 0$, and such that $M(\begin{bmatrix} R \\ R' \end{bmatrix}) = M(R) + M(R')$. The question now is what polynomials $det \begin{bmatrix} R \\ R' \end{bmatrix}$ are achievable this way. The condition of regularity ($M(\begin{bmatrix} R \\ R' \end{bmatrix}) = m(R) + m(R')$) implies, of course, that the degree of the achievable polynomials must be sufficiently large (in particular, at least $m(R)$). The following result may be obtained.

**Lemma:** Let $R \in \mathbf{R}^{p \times q}_{frr}[\xi]$ be given. Then there exists, for each $\chi \in \mathbf{R}[\xi]$ of degree not smaller than $M(R) + L(R) - 1$, a matrix $R' \in \mathbf{R}^{(q-p) \times q}[\xi]$ such that

$$det \begin{bmatrix} R \\ R' \end{bmatrix} = \chi \tag{16}$$

and $M(\begin{bmatrix} R \\ R' \end{bmatrix}) = M(R) + M(R')$ iff $R$ is left prime.

Translating this into theoretic language yields.

**Theorem:** Let $\Sigma \in \mathcal{L}^q$ be given. Then there exists, for all $\chi \in \mathbf{R}_m[\xi]$ with degree $\chi \geq n(\Sigma) + L(\Sigma) - 1$, a $\Sigma' \in \mathcal{L}^q$ such that $p(\Sigma \wedge \Sigma') = p(\Sigma) + p(\Sigma') = q$, $n(\Sigma \wedge \Sigma') = n(\Sigma) + n(\Sigma') = \textit{degree}\ \chi$, and $\chi_{\Sigma \wedge \Sigma'} = \chi$, iff $\Sigma$ is controllable.

Again it is possible to generalize this result to the non-controllable case. It suffices to require that (in addition to degree $\chi \geq n(\Sigma) + L(\Sigma) - 1$) that $\chi$ has *det* $R_1$ as a factor.

## 10. The Single Input/Single Output Case

The obtain some insight in the above results, we will analyse in detail the single input/single output case. The model is then

$$p\left(\frac{d}{dt}\right)y = q\left(\frac{d}{dt}\right)u \tag{17}$$

with $p, q \in \mathbf{R}[\xi]$, and $p$ and $q$ not both zero. Let $\Sigma$ be the resulting dynamical system. We have $p(\Sigma) = m(\Sigma) = 1$, and $n(\Sigma) = L(\Sigma) = max$ (degree $p$, degree $q$). $\Sigma$ will be controllable iff $p$ and $q$ are co-prime polynomials.

Consider the feedback controller

$$a\left(\frac{d}{dt}\right)u + b\left(\frac{d}{dt}\right)y = 0 \tag{18}$$

with $a, b \in \mathbf{R}[\xi]$ and $a$ and $b$ not both zero. Let $\Sigma'$ be the resulting dynamical system. We have $p(\Sigma') = m(\Sigma') = 1$, and $n(\Sigma') = L(\Sigma') = max$ (degree $a$, degree $b$).

The closed loop system is described by

$$\begin{bmatrix} p\left(\frac{d}{dt}\right) & -q\left(\frac{d}{dt}\right) \\ b\left(\frac{d}{dt}\right) & a\left(\frac{d}{dt}\right) \end{bmatrix} \begin{bmatrix} u \\ y \end{bmatrix} = 0 \tag{19}$$

Assuming $pa + qb \neq 0$, then $\Sigma \wedge \Sigma'$ is a feedback interconnection. The question is what polynomials $pa + qb$ are achievable by choosing $a, b$ for given $p, q$. It follows immediately from the Bezout equality that if $p$ and $q$ are co-prime, then for every $\chi \in \mathbf{R}[\xi]$ there exists $a, b$ such that

$$pa + qb = \chi \tag{20}$$

However, in order to understand what happens in the case of regular feedback it is useful to consider the mapping $(a, b) \mapsto (pa + qb)$. Let $\mathbf{R}^k[\xi]$ denote the set of elements in $\mathbf{R}[\xi]$ with degree $\leq k$. Consider the mapping

$$L_k : (a, b) \in \mathbf{R}^k[\xi] \mapsto pa + qb \in \mathbf{R}^{k+n(\Sigma)+1} \tag{21}$$

For $k < n(\Sigma)$ it is easy to see that $p$ and $q$ are co-prime iff $L_k$ is injective. Now, a mapping from $\mathbf{R}^{2n(\Sigma)}$ into $\mathbf{R}^{2n(\Sigma)}$. Therefore, if $p$ and $q$ are co-prime, $L_{n(\Sigma)-1}$ is injective, hence surjective. Moreover $k = n(\Sigma) - 1$ will be the smallest $k$ for which $L_k$ is surjective. Thus $n(\Sigma') = n(\Sigma) - 1$ is the lowest order compensator which yields pole placement in the single input/single output case.

Up to now, we have considered singular feedback, but a similar result holds for regular feedback. Indeed, assume that $p$ and $q$ are co-prime. Then $L_{n(\Sigma)-1}$ is surjective. It immediately follows that if degree $\chi = 2n(\Sigma) - 1$ then degree $\chi = n(\Sigma) + n(\Sigma')$ showing that the feedback connection is indeed regular.

The main open problem concerning pole placement is to generalize this to the multivariable case. The problem can be formulated as follows in terms of polynomial matrices. Let $R \in \mathbb{R}^{\bullet \times q}_{frr}[\xi]$ be given and assume that it is left prime. Define $\mathcal{C}_{n'} := \{R' \in \mathbb{R}^{(q-C(R))\times q}[\xi] \mid M(R') \geq n'\}$. Now consider the mapping $f$ which associates with $R' \in \mathcal{C}_{n'}, det \begin{bmatrix} R \\ R' \end{bmatrix} =: f(R')$. Obviously $f$ maps $\mathcal{C}_{n'}$ into $\mathbb{R}_{n(\Sigma)+n'}[\xi]$ (the polynomials with degree $\leq n(\Sigma) + n'$). The open question is find for a given $R$ what the smallest $n'$ is such that $f$ is surjective.

## 11. Conclusions

Viewing feedback as interconnection of behaviors has a number of advantages. Firstly, in applications where neither the forward nor the feedback loop has an imposed signal flow graph, it corresponds to the physics of the problem. Dynamical systems encountered in mechanics, chemistry, electrical engineering etc. will sedomly have a signal flow restriction and as such the input/output structure may be unnatural. Secondly, as we have seen, we can always associate an imposed signal flow graph with a regular feedback interconnection. In feedback structures with logic devices it is important to respect the imposed signal flow graph. As such, viewing feedback as an interconnection captures both the situation that there is and that there is not an imposed signal flow graph.

In this paper we have considered pole placement (and stabilization) as a specific problem to treat in this setting. Our next step is to treat optimal feedback control from this point of view.

## REFERENCES

[1] J.C. Willems, *Paradigms and Puzzles in the Theory of Dynamical Systems*, IEEE Transactions on Automatic Control, Vol. 36, No. 3, pp. 259-294, 1991.

[2] W.M. Wonham, *Linear Multivariable Control: A Geometric Approach*, Springer Verlag, 1979.

*Author's address:*
Mathematics Institute, University of Groningen,
P.O. Box 800, 9700 AV Groningen, The Netherlands.

# Learning Techniques and Neural Networks for the Solution of N-Stage Nonlinear Nonquadratic Optimal Control Problems

R. ZOPPOLI and T. PARISINI

**Abstract**

This paper deals with the problem of designing closed-loop feed-forward control strategies to drive the state of a dynamic system (in general, nonlinear) so as to track <u>any</u> desired trajectory joining the points of given compact sets, while minimizing a certain cost function (in general, nonquadratic). Due to the generality of the problem, conventional methods (e.g., dynamic programming, maximum principle, etc.) are difficult to apply. Then, an approximate solution is sought by constraining control strategies to take on the structure of multi-layer feed-forward neural networks. After discussing the approximation properties of neural control strategies, a particular neural architecture is presented, which is based on what has been called the "Linear-Structure Preserving Principle" (the LISP principle). The original functional problem is then reduced to a nonlinear programming one, and backpropagation is applied to derive the optimal values of the synaptic weights. Recursive equations to compute the gradient components are presented, which generalize the classical adjoint system equations of $N$-stage optimal control theory. Simulation results related to non-LQ problems show the effectiveness of the proposed method.

## 1. Introduction

In the paper, we address the problem of designing an intelligent device (a controller, a decision maker, or other, depending on the application considered) that, in a fixed number $N$ of temporal stages, drives a dynamic system so as its state vector follows <u>any</u> specified trajectory $\{\underline{x}_1^*, \underline{x}_2^*, \ldots, \underline{x}_N^*\}$, while minimizing a certain cost function. The states of the trajectory to be tracked must belong to given compact sets, but are otherwise completely arbitrary.

The optimal control strategy must have a closed-loop form, even though the problem is posed in a deterministic context (no stochastic variables are acting on the dynamic system, and the current state vector $\underline{x}_i$ can be measured without noise). Moreover, as, at time $i$, the future trajectory $\{\underline{x}^*_{i+1}, \ldots, \underline{x}^*_N\}$ is assumed to be perfectly known to the control device, this can exploit such information to generate the control vector $\underline{u}_i$, using a feed-forward scheme. The optimal control strategy can then be expressed in the form $\underline{u}_i = \underline{\gamma}^\circ_i(\underline{x}_i, \underline{X}^*_{i+1})$ (where $\underline{X}^*_{i+1} \triangleq \text{col}\,(\underline{x}^*_{i+1}, \ldots, \underline{x}^*_N)$), which allows it to react to unpredictable actions that might change both the current state and the remaining part of the trajectory to be tracked.

As is well known, the $N$-stage optimal control problem described above can be solved analytically only in a few cases, typically, under LQ hypotheses (linear dynamic system / quadratic cost). If such assumptions are not verified, and one confines oneself to deriving open-loop control strategies, a variety of numerical techniques are available, like the discrete-time maximum principle, dynamic programming, reduction of the optimal control problem to a nonlinear programming one, etc. (see, for instance, [1],[2] for some general concepts). Obviously, the open-loop case is much simpler than the closed-loop one, as it entails a parametric optimization problem instead of a functional one. In other words, a sequence of optimal control vectors has to be determined (one sequence for any trajectory to be tracked), instead of a sequence of functions $\underline{\gamma}^\circ_i(\underline{x}_i, \underline{X}^*_{i+1})$. As stated before, in our case an optimal control strategy is searched for in closed-loop form and in this case the techniques to find it are not many; in practice, only dynamic programming can be reasonably applied.

Also the dynamic programming approach, however, exhibits some drawbacks. In the most conventional situation, it is not necessary to track one of many possible trajectories. A proper neighbourhood of a given final state $\underline{x}^*_N$ must be reached, while the initial state $\underline{x}_0$ must take any possible value in a given initial set $A_0$. In this case, by following the dynamic programming approach, the optimal closed-loop strategy is derived in the backward phase of the procedure. This phase starts from the final stage $N$, where a suitable terminal cost function penalizing, for example, the Euclidean distance $\|\underline{x}^*_N - \underline{x}_N\|$ is defined, and terminates at the initial stage $i = 0$, when the state trajectories turn out to be optimized for all states $\underline{x}_0 \in A_0$. Such a procedure, however, is not an easy numerical task, for it requires, at each decision stage, the definition of a suitable grid that is obtained by discretizing all the components of the state vector. The optimal control vectors are then determined and stored in the memory, for all the grid points, and, as soon as the dimension of $\underline{x}_i$ increases, this may result in prohibitive requirements for storage capacity. Possible different versions of the conventional dynamic programming technique, to overcome the dimensionality barrier, can be found in [1] and [3].

Suppose now that $\underline{x}^*_N$ is not fixed, but, like $\underline{x}_0$, has to take any possible value in a given final set $A_N$. Then, a set of optimal control problems turn out to be stated, each characterized by the presence of a proper final cost parametrized by $\underline{x}_N^*$. This means that we have to choose a suitable number of vectors $\underline{x}_N^*$ belonging to $A_N$, and solve all the corresponding problems by means of dynamic programming. The problem complexity increases if we introduce some intermediate vectors $\underline{x}_i^*$ to be tracked, each of which has to take on any possible value in a given set $A_i$. The maximum computational complexity is exhibited by the problem we want to face in this paper, where the dynamic system must track any possible trajectory $\{\underline{x}_1^*, \underline{x}_2^*, \ldots, \underline{x}_N^*\}$. Actually, in this case, a cost function must be defined to penalize the distance between the desired and the real trajectories. A finite (possibly very large) number of desired trajectories should be fixed, and all the corresponding state-tracking optimal control problems should be solved via dynamic programming.

Due to the high complexity of calculations, to the huge amount of memory needed to store the closed-loop feed-forward strategy, and to the fact that, in any case, only an approximate solution can however be obtained, we give up dynamic programming and prefer to use another kind of approximation. This consists in assigning the control strategies a given structure, in which the values of a certain number of parameters have to be determined via nonlinear programming in order to minimize the cost function. This approach is not new in control theory; actually, it dates back to the so-called Specific Optimal Control (S.O.C) problem considered by Eisenberg and Sage [4], for the general non-LQ case, and, under LQ assumptions, to the parametric optimal control problem introduced by Kleinman and Athans [5] (see also the survey reported in [6]). Once this kind of approach has been decided, implementing our control strategies on multi-layer feed-forward neural networks appears quite a natural choice, since it has been shown, both experimentally (in the past few years) and theoretically (more recently [12]), that these networks are very well suited to approximating nonlinear functions. Some preliminary results on this approach can be found in [7],[8].

## 2. Statement of the N-Stage Optimal Control Problem

We consider the discrete-time dynamic system (in general, nonlinear)

$$\underline{x}_{i+1} = \underline{f}_i(\underline{x}_i, \underline{u}_i), \quad i = 0, 1, \ldots, N-1 \tag{1}$$

where $\underline{x}_i \in \Re^n$ is the state vector of the time-varying dynamic system and $\underline{u}_i \in \Re^m$ is the control vector. The cost function (in general, nonquadratic) is given by

$$J = \sum_{i=0}^{N-1} \left[ h_i(\underline{x}_i, \underline{u}_i) + \rho_{i+1}\left( \|\underline{x}_{i+1}^* - \underline{x}_{i+1}\| \right) \right] \tag{2}$$

where $\{\underline{x}_1^*, \underline{x}_2^*, \ldots, \underline{x}_N^*\}$ is the state trajectory to be tracked, and $\|\cdot\|$ is the Euclidean norm on $\Re^n$. Moreover, $\underline{f}_i \in C^1[\Re^n \times \Re^m]$, $h_i \in$

$C^1[\Re^n \times \Re^m]$, and $\rho_i \in C^1[\Re^+]$. The transition costs $h_i(\underline{x}_i, \underline{u}_i)$, being in general nonquadratic, can take easily into account the control effort, smoothness conditions for the state trajectory, penalty functions describing constraints (possibly time-varying) on the control and state vectors, etc.. $\rho_i(z)$ are increasing functions for $z \geq 0$, with $\rho_i(0) = 0$. We assume that the initial state $\underline{x}_0$ and the vectors $\underline{x}_1^*, \underline{x}_2^*, \ldots, \underline{x}_N^*$ can take any possible value in the given compact sets $A_0, A_1, \ldots, A_N$, respectively. Then we can state the following

**Problem 1.** *Find the optimal closed-loop feed-forward control strategies* $\left\{\underline{u}_i^\circ = \underline{\gamma}_i^\circ(\underline{x}_i, \underline{X}_{i+1}^*), i = 0, 1, \ldots, N-1\right\}$ *that minimize cost (2) for any* $N+1$*-tuple* $(\underline{x}_0, \underline{x}_1^*, \underline{x}_2^*, \ldots, \underline{x}_N^*) \in A_0 \times A_1 \times A_2 \times \ldots \times A_N$.

We want to remark explicitly that we are looking for closed-loop strategies, although Problem 1 is stated in a deterministic context (no random variables are acting on the system, and all state variables can be measured without noise). Actually, closed-loop strategies are preferable in practical situations, whenever unpredictable input variables affect the dynamic system. Likewise, the control strategies have to be expressed in feed-forward form, as the vector $\underline{X}_{i+1}^*$ characterizing, at time $i$, the remaining part of the desired trajectory can vary in an unpredictable way (also this vector is assumed to be perfectly measurable).

It is worth noting that the explicit introduction of the sets $A_0, A_1, \ldots, A_N$ into the formulation of Problem 1 is due to the fact that this problem is nonlinear and nonquadratic. For its LQ version, we should minimize the cost function

$$J = \sum_{i=0}^{N-1} \left( \|\underline{u}_i\|_{P_i}^2 + \|\underline{x}_{i+1}^* - \underline{x}_{i+1}\|_{V_{i+1}}^2 \right) \tag{3}$$

where $P_i = P_i^T > 0$, and $V_i = V_i^T \geq 0$. As is well known, this LQ "state-tracking" problem can be solved analytically, and the optimal control strategy is given by

$$\underline{u}_i^\circ = -L_i \underline{x}_i + F_i \underline{v}_i \,, \qquad i = 0, 1, \ldots, N-1 \tag{4}$$

where $\underline{v}_i$ is a vector generated backward by means of the recursive equation

$$\begin{aligned} \underline{v}_i &= G_i \underline{v}_{i+1} + V_{i+1} \underline{x}_{i+1}^* \,, \qquad i = 0, 1, \ldots, N-2 \\ \underline{v}_{N-1} &= V_N \underline{x}_N^* \end{aligned} \tag{5}$$

and matrices $L_i, F_i, G_i$ can be computed after solving a discrete-time Riccati equation. Since these matrices are independent of $A_0, A_1, \ldots, A_N$, the introduction of these sets into the formulation of Problem 1 is unnecessary. We want also to remark that in (4) the control vector $\underline{u}_i^\circ$ does not depend explicitly on the remaining part of the desired trajectory described by $\underline{X}_{i+1}^*$ but on the vector $\underline{v}_i$, whose dimension is constant in time. This fact will be exploited in Section 4 to account for the time-invariant structure that we shall choose for the neural strategy approximating the optimal solution of Problem 1.

If the LQ hypotheses are not verified, a possible approach consists in applying dynamic programming, but, as stated in the Introduction, this algorithm entails cumbersome numerical procedures, and, in general, does not enable us to obtain the optimal solution of Problem 1 in analytical form. This leads us to state the terminal control problem in a different, though approximate, way: the control strategies are assigned a given structure in which the values of a certain number of parameters have to be determined via nonlinear programming so as to minimize the cost function.

Multi-layer feed-forward neural networks are very attractive candidates for the design of nonlinear control structures. As will be explained in the next section, such neural networks can involve an arbitrarily large number of parameters (synaptic weights), but are well suited for the use of distributed optimizing algorithms that keep the complexity of calculations substantially independent of the number of parameters to be tuned.

## 3. Reduction of the Functional Optimization Problem 1 to a Nonlinear Programming Problem

Let us constrain the control strategies to take on a fixed structure of the form

$$\underline{u}_i = \hat{\underline{\gamma}}_i\left(\underline{x}_i, \underline{X}^*_{i+1}, \underline{w}_i\right), \quad i = 0, 1, \ldots, N-1 \tag{6}$$

where $\hat{\underline{\gamma}}_i$ is the input/output mapping of a multi-layer feed-forward neural network generating the control vector $\underline{u}_i$ at stage $i$, and $\underline{w}_i$ is the vector of the synaptic weights we have to determine. As is shown in Fig. 1, the control scheme gives rise to a chain of $N$ neural networks, each followed by the dynamic system. This chain is strictly related to the control scheme proposed in [9]. Our scheme differs from the latter in the feed-forward action generated by the vector $\underline{x}^*_N$, and in the fact that the neural networks are allowed to be time-dependent.

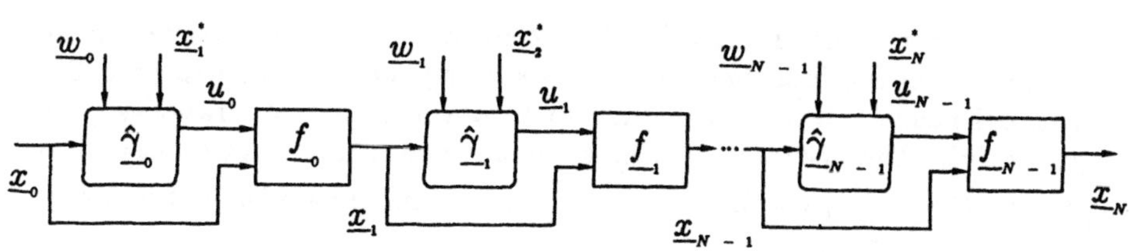


**Fig. 1. Structure of the closed-loop feed-forward neural strategies**

If we now substitute (6) in (1) and (2), and (1) in (2), the cost function takes on the form $J(\underline{w}, \underline{x}_0, \underline{X}_1^*)$, where $\underline{w} \triangleq \mathrm{col}(\underline{w}_i, i = 0, 1, \ldots, N-1)$. As the cost function $J(\underline{w}, \underline{x}_0, \underline{X}_1^*)$ depends on both the vector $\underline{w}$ to be determined and $\underline{x}_0, \underline{X}_1^*$, we eliminate the dependence on $\underline{x}_0, \underline{X}_1^*$ by assuming $\underline{x}_0, \underline{x}_1^*, \ldots, \underline{x}_N^*$ to be mutually independent random vectors uniformly distributed on $A_0, A_1, \ldots, A_N$, respectively, and by averaging $J(\underline{w}, \underline{x}_0, \underline{X}_1^*)$ with respect to these vectors. It is worth noting that this is a typical procedure to solve parametric LQ optimal control problems (see, for instance, [5]). We have now to solve the following

**Problem 2.** *Find the vector* $\underline{w}^\circ$ *that minimizes the expected cost* $\mathrm{E}_{\underline{x}_0, \underline{X}_1^*}[J(\underline{w}, \underline{x}_0, \underline{X}_1^*)]$.

It follows that the functional Problem 1 has been reduced to an unconstrained nonlinear programming problem. Let us now describe in some detail the neural networks that implement our control strategies. Each of the $N$ neural networks implementing the strategies $\hat{\underline{\gamma}}_i$ is composed of $L$ layers, and, in the generic layer $s$, $n_s$ neural units are active. The input/output mapping of the $q$-th neural unit of the $s$-th layer is given by

$$y_q^i(s) = g\left[z_q^i(s)\right], \quad s = 1, \ldots, L; \, q = 1, \ldots, n_s \tag{7}$$

$$z_q^i(s) = \sum_{p=1}^{n_{s-1}} w_{pq}^i(s) y_p^i(s-1) + w_{0q}^i(s) \tag{8}$$

where $y_q^i(s)$ is the output variable of the neural unit, $g(x) = \tanh(x)$ is a shifted sigmoidal activation function, and $w_{pq}^i(s)$ and $w_{0q}^i(s)$ are the weight and bias coefficients, respectively. All these coefficients are the components of the vector $\underline{w}_i$ appearing in the control law $\underline{u}_i = \hat{\underline{\gamma}}_i(\underline{x}_i, \underline{X}_{i+1}^*, \underline{w}_i)$; the variables $y_q^i(0)$ are the components of $\underline{x}_i$ and $\underline{X}_{i+1}^*$, and the variables $y_q^i(L)$ are the components of $\underline{u}_i$.

Clearly, the solution to Problem 2 constitutes an approximation for Problem 1 (in the following, the corresponding control laws will be called "optimal neural control strategies" and "optimal control strategies", respectively). The accuracy to which the optimal neural control strategies $\left\{\hat{\underline{\gamma}}_i(\underline{x}_i, \underline{X}_{i+1}^*, \underline{w}_i^\circ), i = 0, 1, \ldots, N-1\right\}$ can approximate the optimal control strategies $\left\{\underline{\gamma}_i^\circ(\underline{x}_i, \underline{X}_{i+1}^*), i = 0, 1, \ldots, N-1\right\}$ is clearly a crucial point of the method proposed in the paper. By exploiting some results recently appeared in the literature, we are able to provide a quantitative assessment of the approximation properties of the neural strategies $\hat{\underline{\gamma}}_i$.

Toward this end, it is sufficient to assume (as we shall do later on) that the strategies $\hat{\underline{\gamma}}_i$ contain only one hidden layer (i.e., $L = 2$) composed of

$\nu_i$ neural units, and that the output units are linear. Denote then such strategies by $\underline{\hat{\gamma}}_i^{(\nu_i)}, i = 0, 1, \ldots, N-1$. Define now the mappings

$$(9) \quad g_i \triangleq \underline{f}_i[\underline{x}_i, \underline{\gamma}_i^\circ(\underline{x}_i, \underline{X}_{i+1}^*)]: \Re^n \times \Re^{(N-i)n} \to \Re^n, \; i = 0, 1, \ldots, N-1$$

and the sets $\mathcal{B}_i \subset \Re^n \times \Re^{(N-i)n}$ , where

$$(10) \quad \mathcal{B}_i \triangleq \begin{cases} \mathcal{A}_0 \times \mathcal{A}_1 \times \cdots \times \mathcal{A}_N, & \text{for } i = 0 \\ g_{i-1}[\mathcal{B}_{i-1}] \times \mathcal{A}_{i+1} \times \cdots \times \mathcal{A}_N, & \text{for } i = 1, \ldots, N-1 \end{cases}$$

Then we can state the following

**Proposition 1.** *Assume that Problem 1 has only one solution* $\underline{\gamma}_i^\circ(\underline{x}_i, \underline{X}_{i+1}^*) \in \mathcal{C}[\mathcal{B}_i]$, $i = 0, 1, \ldots, N-1$. *Then:* $\forall \varepsilon \in \Re, \varepsilon > 0$, *and* $\forall i: 0 \le i \le N-1$, $\exists \nu_i$ *and* $\underline{w}_i$ *such that*

$$(11) \quad \left\| \underline{\gamma}_i^\circ(\underline{x}_i, \underline{X}_{i+1}^*) - \underline{\hat{\gamma}}_i^{(\nu_i)}(\underline{x}_i, \underline{X}_{i+1}^*, \underline{w}_i) \right\| < \varepsilon, \; \forall \, (\underline{x}_i, \underline{X}_{i+1}^*) \in \mathcal{B}_i$$

Consider now the functions $\gamma_{ij}^\circ(\underline{x}_i, \underline{X}_{i+1}^*)$ and $\hat{\gamma}_{ij}^{(\nu_i)}(\underline{x}_i, \underline{X}_{i+1}^*, \underline{w}_i)$, which represent the $j$-th components of the vector functions $\underline{\gamma}_i^\circ$ and $\underline{\hat{\gamma}}_i^{(\nu_i)}$, respectively. We let also $d_i \triangleq n + (N-i)n$. Then we have the following

**Proposition 2.** *Assume that Problem 1 has only one solution* $\underline{\gamma}_i^\circ(\underline{x}_i, \underline{X}_{i+1}^*) \in \mathcal{C}^1[\mathcal{B}_i]$, $i = 0, 1, \ldots, N-1$, *and that all its derivatives of order* $s_i$ *are square integrable on* $\mathcal{B}_i$, *where* $s_i$ *is the least integer such that* $s_i > 1 + \frac{d_i}{2}$. *Then:* $\forall i: 0 \le i \le N-1$, $\forall j: 1 \le j \le m$, $\forall$ *probability measure* $\mu$, *and* $\forall \nu_i \ge 1$, $\exists c_{ij} \in \Re, c_{ij} > 0$, *and* $\underline{w}_i$ *such that*

$$(12) \quad \int_{\mathcal{B}_i} \left| \gamma_{ij}^\circ(\underline{x}_i, \underline{X}_{i+1}^*) - \hat{\gamma}_{ij}^{(\nu_i)}(\underline{x}_i, \underline{X}_{i+1}^*, \underline{w}_i) \right|^2 \mu\left[d\,(\underline{x}_i, \underline{X}_{i+1}^*)\right] \le \frac{c_{ij}}{\nu_i}$$

Proposition 1 is the direct consequence of the results given in [10],[11], and Proposition 2 has been derived from Barron's recent theorems [12]. The proofs of both propositions are detailed in [13]. As pointed out in [12], Proposition 2 states that the number of parameters required to achieve an integrated square error of order $o(1/\nu_i)$ is $o(\nu_i d_i)$, which grows linearly with $d_i$, i.e., the dimension of the input vector of the neural networks. This result is quite surprising, considering that traditional polynomial, spline and trigonometric expansions require (under suitable assumptions on the smoothness of the function to be approximated) exponentially many parameters $o(\nu_i^{d_i})$ to achieve approximation errors of order $o(1/\nu_i)$.

## 4. Control Strategies Based on Time-Invariant Neural Structures

In the preceding section, we have seen that the structures of the control strategies $\underline{\hat{\gamma}}_i(\underline{x}_i, \underline{X}_{i+1}^*, \underline{w}_i)$, and then the neural networks on which they are implemented, vary in time (hence they are unsuitable for a practical implementation), due to the time-varying dimension of the vector $\underline{X}_i^*$, whose number of components decreases over time. To remove this drawback, it is interesting to note that the optimal solution of the LQ state-tracking problem is given by (4) and (5). These strategies are characterized by a fixed structure, which means that matrices $L_i, F_i, G_i$, even though made up of time-varying elements, are of time-invariant dimensions. This suggests the following reasoning. When dealing with a non-LQ optimal control problem, let us consider its "LQ equivalent structure", consisting of a linear dynamic system with the same input and the same state vectors, and an additive cost containing the same number $N$ of terms (plus the final one). Apart from these very general characteristics, the system and the cost matrices defining the LQ problem are completely arbitrary.

Then we use what we call the "Linear-Structure Preserving Principle" (the LISP principle), which can be stated as follows: *"Once the LQ equivalent structure has been found and one of the infinite LQ optimal control problems has been solved, replace its linear mappings (matrices) with nonlinear mappings (neural networks) so as to preserve the global structure that is correct in the LQ case"*.

According to this principle, the linear mappings (4) and (5) are replaced with the nonlinear mappings

(13) $$\underline{u}_i = \underline{\gamma}_1(\underline{x}_i, \underline{v}_i, \underline{w}_i), \qquad i = 0, 1, \ldots, N-1$$

(14) $$\begin{aligned} \underline{v}_i &= \underline{\gamma}_2(\underline{v}_{i+1}, \underline{x}_{i+1}^*, \underline{\tilde{w}}_i), \qquad i = 0, 1, \ldots, N-2 \\ \underline{v}_{N-1} &= \underline{\gamma}_{2N}(\underline{x}_N^*, \underline{\tilde{w}}_N) \end{aligned}$$

where $\underline{\tilde{w}}_i$ and $\underline{\tilde{w}}_N$ are the vectors of the synaptic weights defined in Section 3. Observe that, in the new control architecture (depicted in Fig. 2), the neural mappings $\underline{\gamma}_1$ constitute the "feedback chain", while the neural mappings $\underline{\gamma}_2$ constitute the "feed-forward chain".

To simplify the notation, the synaptic weights and all the variables of the neural mappings $\underline{\gamma}_1$ are denoted by the same symbols as in Problem 2 (see (7) and (8)), while the weights and the variables of the neural mappings $\underline{\gamma}_2$ are denoted by adding a tilde (˜) to such symbols (for example $\tilde{y}_q^i(s) = g[\tilde{z}_q^i(s)]$). The variables $y_p^i(0)$ are now the components of $\underline{x}_i$ and $\underline{v}_i$, and the variables $y_q^i(L)$ are the components of $\underline{u}_i$. In a similar way, the variables $\tilde{y}_p^i(0)$ are the components of $\underline{v}_{i+1}$ and $\underline{x}_{i+1}^*$, and the variables $\tilde{y}_q^i(\tilde{L})$ are the components of $\underline{v}_i$.

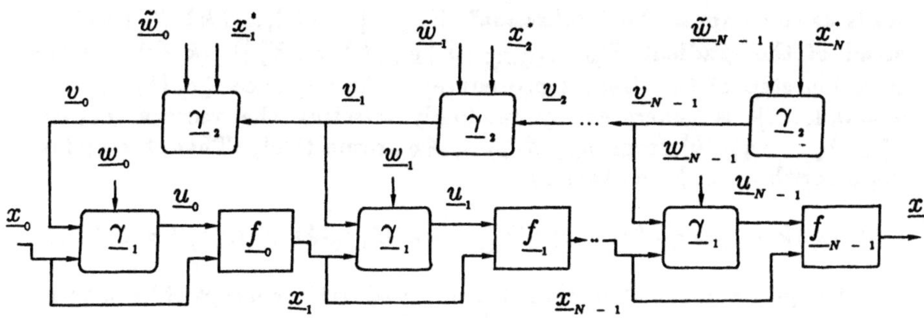

**Fig. 2. Time-invariant structure of the neural controller**

By following the same procedure that led to the statement of Problem 1', we now substitute (14) in (13), (13) in (1) and (2), (1) in (2), and let $\underline{w} \triangleq \text{col}(\underline{w}_i, i = 0, 1, \ldots, N-1)$, $\underline{\tilde{w}} \triangleq \text{col}(\underline{\tilde{w}}_i, i = 0, 1, \ldots, N-1)$, and $\underline{w}_T \triangleq \text{col}(\underline{w}, \underline{\tilde{w}})$. Then the cost function takes on the form $J(\underline{w}_T, \underline{x}_0, \underline{X}_1^*)$. As the cost function $J(\underline{w}_T, \underline{x}_0, \underline{X}_1^*)$ depends on both the vector $\underline{w}_T$ to be determined and $\underline{x}_0, \underline{X}_1^*$, we eliminate again the dependence on $\underline{x}_0$ and $\underline{X}_1^*$ by averaging $J(\underline{w}_T, \underline{x}_0, \underline{X}_1^*)$ with respect to these vectors. Then we have to solve the following

**Problem 2'.** *Find the vector* $\underline{w}_T^\circ$ *that minimizes the expected cost* $\mathrm{E}_{\underline{x}_0, \underline{X}_1^*} [J(\underline{w}_T, \underline{x}_0, \underline{X}_1^*)]$.

It follows that a functional optimization problem has been approximated again to an unconstrained nonlinear programming one. As a final comment about the application of the LISP principle, we want to remark that a given non-LQ optimal control problem does not imply necessarily only one approximating neural control architecture. It is easy to show, for example, that a neural architecture based on the strategies $\hat{\underline{\gamma}}_i(\underline{x}_i, \underline{X}_{i+1}^*, \underline{w}_i)$, instead of strategies (13),(14), can be regarded as another architecture based on the LISP principle. A deeper discussion of this point can be found in [13].

## 5. Application of the Gradient Method to Determine the Optimal Control Strategies

We consider the unconstrained nonlinear programming Problem 2', and solve it by means of some descent algorithm. We focus our attention on methods of the gradient type, as, when applied to neural networks, they are well suited to distributed computation. Due to the very general formulation of the problem, we are unable to express the average cost $\mathrm{E}_{\underline{x}_0, \underline{X}_1^*} [J(\underline{w}_T, \underline{x}_0, \underline{X}_1^*)]$ in explicit form. This leads us to compute the "*realization*" $\nabla_{\underline{w}_T} J[\underline{w}_T(k), \underline{x}_0(k), \underline{X}_1^*(k)]$ instead of the gradient $\nabla_{\underline{w}_T} \mathrm{E}_{\underline{x}_0, \underline{X}_1^*}\{J[\underline{w}_T(k), \underline{x}_0, \underline{X}_1^*]\}$ ($k$ denotes the iteration step of the descent procedure). The sequence $\{\underline{x}_0(k), \underline{X}_1^*(k), k = 0, 1, \dots\}$ is generated by randomly selecting the vectors $\underline{x}_0(k)$, $\underline{x}_1^*(k), \dots, \underline{x}_N^*(k)$ from $A_0, A_1, \dots, A_N$, respectively. Then, the updating algorithm can be written as

(15) $$\underline{w}_T(k+1) = \underline{w}_T(k) - \alpha(k) \nabla_{\underline{w}_T} J[\underline{w}_T(k), \underline{x}_0(k), \underline{X}_1^*(k)], \; k = 0, 1, \dots$$

The probabilistic algorithm (15) is related to the concept of "*stochastic approximation*" (see, for instance, [14]). Observe that the step-size must suitably decrease to ensure convergence, as, at the optimal point $\underline{w}_T^\circ$, $\nabla_{\underline{w}_T} J[\underline{w}_T^\circ, \underline{x}_0(k), \underline{X}_1^*(k)]$ differs in general from zero. In the examples given in the following, we take $\alpha(k) = c_1 / (c_2 + k)$, $c_1, c_2 > 0$. To accelerate convergence, we modify algorithm (15) by using heuristic techniques that have proved effective in training neural networks. Then we add the so-called "momentum" $\eta[\underline{w}_T(k) - \underline{w}_T(k-1)]$ to (15), thus causing the algorithm to behave like a smoothing filter that tends to compute the average gradient of the cost.

We now want to derive the components of $\nabla_{\underline{w}_T} J[\underline{w}_T(k), \underline{x}_0(k), \underline{X}_1^*(k)]$, i.e., the partial derivatives $\partial J[\underline{w}_T(k), \underline{x}_0(k), \underline{X}_1^*(k)] / \partial w_{pq}^i(s)$, $\partial J[\underline{w}_T(k), \underline{x}_0(k), \underline{X}_1^*(k)] / \partial \tilde{w}_{pq}^i(s)$. Toward this end, we define the following three variables, which play a basic role in the development of the proposed algorithm (to simplify the notation, we drop the index $k$)

(16)
$$\delta_q^i(s) \triangleq \frac{\partial J(\underline{w}_T, \underline{x}_0, \underline{X}_1^*)}{\partial z_q^i(s)}, \; i = 0, 1, \dots, N-1; \; s = 1, \dots, L; \; q = 1, \dots, n_s$$

(17)
$$\tilde{\delta}_q^i(s) \triangleq \frac{\partial J(\underline{w}_T, \underline{x}_0, \underline{X}_1^*)}{\partial \tilde{z}_q^i(s)}, \; i = 0, 1, \dots, N-1; \; s = 1, \dots, \tilde{L}; \; q = 1, \dots, \tilde{n}_s$$

(18) $$\underline{\lambda}_i \triangleq \nabla_{\underline{x}_i} J(\underline{w}_T, \underline{x}_0, \underline{X}_1^*), \; i = 0, 1, \dots, N-1$$

Then, the computational procedure to obtain the required gradient can be summarized in the following

**Proposition 3.** *The components of the gradient* $\nabla_{\underline{w}_T} J[\underline{w}_T, \underline{x}_0, \underline{X}_1^*]$ *are given by the partial derivatives*

(19) $$\frac{\partial J(\underline{w}_T, \underline{x}_0, \underline{X}_1^*)}{\partial w_{pq}^i(s)} = \delta_q^i(s)\, y_p^i(s-1)$$

(20) $$\frac{\partial J(\underline{w}_T, \underline{x}_0, \underline{X}_1^*)}{\partial \tilde{w}_{pq}^i(s)} = \tilde{\delta}_q^i(s)\, \tilde{y}_p^i(s-1)$$

*where $\delta_q^i(s)$ and $\tilde{\delta}_q^i(s)$ can be computed recursively by means of the equations*

$$\text{(21a)} \qquad \delta_q^i(s) = g'\left[z_q^i(s)\right] \sum_{h=1}^{n_{s+1}} \delta_h^i(s+1) w_{qh}^i(s+1), \; s = 1, \ldots, L-1$$

$$\text{(21b)} \qquad \delta_q^i(L) = g'\left[z_q^i(L)\right] \left[\frac{\partial}{\partial \underline{u}_i} \tilde{h}_i(\underline{x}_i, \underline{u}_i, \underline{x}_i^*) + \underline{\lambda}_{i+1}^T \frac{\partial}{\partial \underline{u}_i} \underline{f}_i(\underline{x}_i, \underline{u}_i)\right]_q$$

$$\text{(22a)} \qquad \tilde{\delta}_q^i(s) = g'\left[\tilde{z}_q^i(s)\right] \sum_{h=1}^{\tilde{n}_{s+1}} \tilde{\delta}_h^i(s+1) \tilde{w}_{qh}^i(s+1), \; s = 1, \ldots, \tilde{L}-1$$

$$\text{(22b)} \qquad \tilde{\delta}_q^i(\tilde{L}) = g'\left[\tilde{z}_q^i(\tilde{L})\right] \left[\sum_{h=1}^{\tilde{n}_1} \tilde{\delta}_h^{i-1}(1) \tilde{w}_{qh}^{i-1}(1) + \sum_{h=1}^{n_1} \delta_h^i(1) w_{qh}^i(1)\right]$$

*where $g'$ is the derivative of the activation function, $\tilde{h}_i(\underline{x}_i, \underline{u}_i, \underline{x}_i^*) \triangleq h_i(\underline{x}_i, \underline{u}_i) + \rho_i\left(\|\underline{x}_i^* - \underline{x}_i\|\right)$, $i = 1, 2, \ldots, N-1$, and $[\underline{z}]_q$ is the q-th component of the vector $\underline{z}$. The vectors $\underline{\lambda}_i$ appearing in (21b) are generated as follows*

$$\text{(23a)} \qquad \begin{aligned} \underline{\lambda}_i^T = & \frac{\partial}{\partial \underline{x}_i} \tilde{h}_i(\underline{x}_i, \underline{u}_i, \underline{x}_i^*) + \underline{\lambda}_{i+1}^T \frac{\partial}{\partial \underline{x}_i} \underline{f}_i(\underline{x}_i, \underline{u}_i) \\ & + \text{col}\left[\sum_{q=1}^{n_1} \delta_q^i(1) w_{pq}^i(1), \; p = 1, \ldots, n_0\right], \; i = 1, \ldots, N-1 \end{aligned}$$

$$\text{(23b)} \qquad \underline{\lambda}_N^T = \frac{\partial}{\partial \underline{x}_N} h_N\left(\|\underline{x}_N - \underline{x}_N^*\|\right)$$

Equations (19), (20), (21a), and (22a) are the standard ones with the backpropagation technique (see, for instance, [15],[16]). The initializing conditions (21b) and (22b) are peculiar to the present problem. Eqs. (23) are the classical adjoint equations of N-stage optimal control theory, with the addition of a term (the third) to take into account the introduction of neural networks. The computational details for deriving these equations are given in [13] and partially in [8]. It can be seen that the algorithm consists of the following two "passes":

**Forward pass** . The state vectors $\underline{x}_0(k), \underline{x}_1^*(k), \ldots, \underline{x}_N^*(k)$ are randomly generated from $A_0, A_1, \ldots, A_N$, respectively. Then, the control sequence and the state trajectory are computed on the basis of these vectors and of $\underline{w}(k)$ and $\underline{\tilde{w}}(k)$.

**Backward pass .** All the variables $\delta_q^i(s)$, $\tilde{\delta}_q^i(s)$, and $\underline{\lambda}_i$ are computed, and the gradient $\nabla_{\underline{w}_T} J\,[\underline{w}(k), \underline{x}_0(k), \underline{X}_1^*(k)]$ is determined by using (19) and (20). Then, the new weight vector $\underline{w}_T(k+1)$ is generated by means of (15).

In the next section, some examples will be given to illustrate the potentiality of the proposed method.

## 6. Simulation Results

We now present three examples of non-LQ optimal control problems to show the learning properties of the neural control strategies. These strategies were implemented according to the architecture depicted in Fig. 2, i.e., after solving Problem 2'. This was due to the fact that several simulation trials showed that the convergence rates in solving Problems 2 and 2' are essentially equivalent. Therefore, to attain the goal of implementing the control strategies on time-invariant neural structures, we solved the examples within the framework of Problem 2'.

**Example 1.** In this example, we address the problem of driving a dynamic system from any point of a given region $\mathcal{A}_0$ to any point of a given region $\mathcal{A}_N$ so as to minimize a certain cost function, while avoiding an obstacle placed in a known position. The dynamic system is given by

$$\underline{x}_{i+1} = \begin{bmatrix} 0.65 & -0.19 \\ 0 & 0.83 \end{bmatrix} \underline{x}_i + \begin{bmatrix} 7 & 0 \\ 0 & 7 \end{bmatrix} \underline{u}_i$$

The cost function is

$$J = \sum_{i=0}^{N-1} \left[ \|\underline{u}_i\|^2 + v\|\underline{x}_{i+1} - \underline{x}_i\|^2 + \rho \max\left[0, g(\underline{x}_i)^3\right]\right] + v_N \|\underline{x}_N - \underline{x}_N^*\|^2$$

where $v = 10$, $v_N = 90$, $\rho = 200$, and $N = 15$. A differentiable penalty function was introduced to describe a circular obstacle ($g(\underline{x}) = 0$ is the equation for the circle). The initial and final regions are $\mathcal{A}_0 = \left\{(x,y) \in \Re^2 : x = 0, -1.5 \le y \le 3.5\right\}$ and $\mathcal{A}_N = \left\{(x,y) \in \Re^2 : x = 5, -1.5 \le y \le 3.5\right\}$, respectively. Fig. 3 shows the optimal state trajectories for different points of the regions $\mathcal{A}_0$ and $\mathcal{A}_N$. They were learned by the control strategies after some tens of thousands of iterations. The optimal neural strategies $\underline{\gamma}_1(\underline{x}_i, \underline{v}_i, \underline{w}_i^\circ)$, $\underline{\gamma}_2(\underline{v}_{i+1}, \underline{x}_N^*, \underline{\tilde{w}}_i^\circ)$ were implemented on neural networks containing two hidden layers of eight units. A momentum was added to the right-hand side of (15), with $\eta = 0.8$. The constants of the step-size $\alpha(k)$ are $c_1 = 100$, and $c_2 = 100000$.

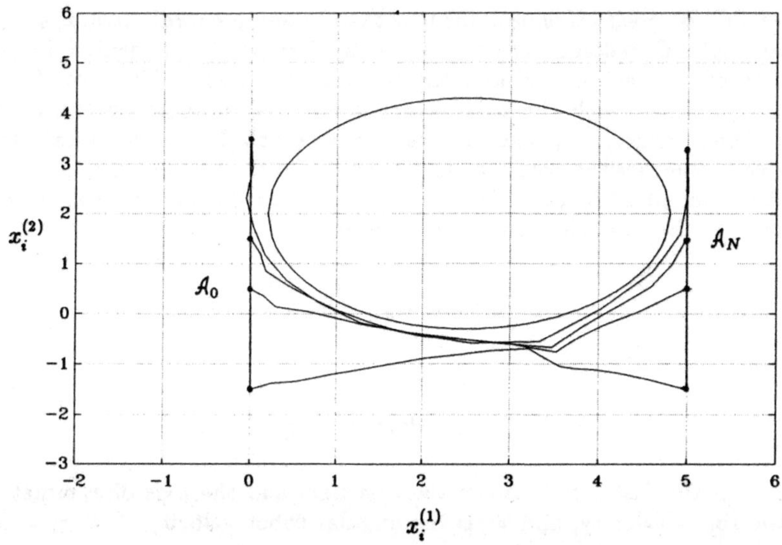


**Fig. 3. Optimal state trajectories**

**Example 2.** Consider the space robot presented in Fig.4; for the sake of simplicity, the robot is assumed to move in the plane of the coordinate system shown in the figure.

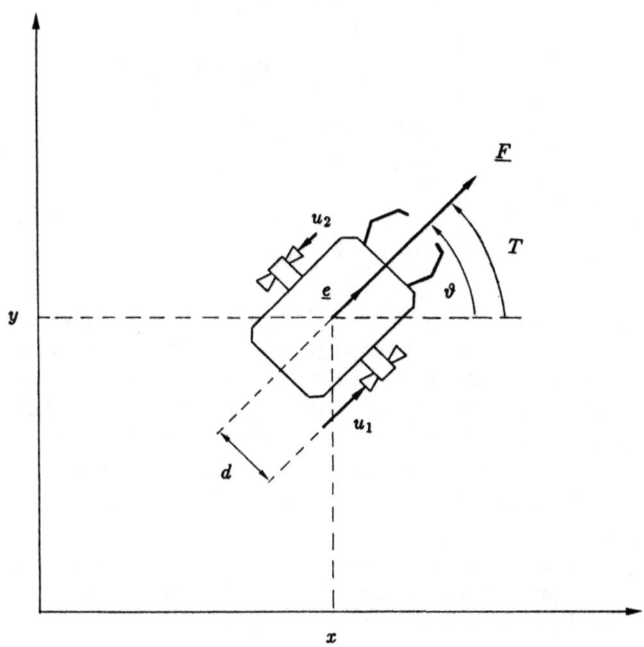


**Fig. 4. The space robot**

The robot position is described by the Cartesian coordinates $x, y$ and by the angle $\vartheta$ between the robot's axis of symmetry (oriented in the direction of the vector $\underline{e}$ of unit length) and the $x$ axis. Two couples of thrusters, aligned with the axis of symmetry, are mounted on the robot sides. Their thrusts, $u_1$ and $u_2$, can be modulated so as to obtain the modulus of the desired force $\underline{F}$ and torque $T$. We assume the mass $m$ and the moment of inertia $J$ to remain constant during the manoeuvre described in the following; then we can write

$$\underline{F} = (u_1 + u_2)\,\underline{e} = m\,\frac{d\underline{v}}{dt} \tag{24}$$

$$T = (u_1 - u_2)\,d = J\,\frac{d\omega}{dt} \tag{25}$$

where $d$ is the distance between the thrusters and the axis of symmetry, $\underline{v}$ is the robot velocity, and $\omega$ is the angular robot velocity. Let $x_1 = x$, $x_2 = \dot{x}$, $x_3 = y$, $x_4 = \dot{y}$, $x_5 = \vartheta$, $x_6 = \dot{\vartheta}$, and $\underline{x} \triangleq \mathrm{col}\,(x_i, i = 1, \ldots, 6)$. Then, from (24) and (25), we obtain the nonlinear differential dynamic system

$$\begin{cases} \dot{x}_i = x_{i+1}\,, \text{ for } i = 1, 3, 5 \\ \dot{x}_2 = \dfrac{1}{m}\,(u_1 + u_2)\,\cos x_5 \\ \dot{x}_4 = \dfrac{1}{m}\,(u_1 + u_2)\,\sin x_5 \\ \dot{x}_6 = \dfrac{d}{J}\,(u_1 - u_2) \end{cases} \tag{26}$$

with the constraints

$$|u_1| \leq U\,, \; |u_2| \leq U \tag{27}$$

where $U$ is the maximum thrust value allowed. The space robot has to start from any given point of the segment $AB$ shown in Fig. 5 (the parking edge of a space platform) and to reach an object moving along the segment $A'B'$ in an unpredictable way. The dashed line shows the path of the object. When the robot reaches the segment $A'B'$, it must stop with the angle $\vartheta = 0$. Then, the initial and final regions are given by $\mathcal{A}_0 = \left\{\underline{x} \in \Re^6 : x_1 = 0, x_2 = 0, 1 \leq x_3 \leq 5, x_4 = 0, x_5 = 0, x_6 = 0\right\}$ and $\mathcal{A}_N = \left\{\underline{x} \in \Re^6 : x_1 = 10, x_2 = 0, 1 \leq x_3 \leq 5, x_4 = 0, x_5 = 0, \; x_6 = 0\right\}$, respectively. The manoeuvre has to be completed at a given time $t_f$, and $N = 10$ control stages are allowed. The fuel consumption has to be minimized, hence the cost function can be expressed as

$$J = \sum_{i=0}^{N-1} \left[ c(u_{1i}) + c(u_{2i}) + \|\underline{x}_N^* - \underline{x}_i\|_V^2 \right] + \|\underline{x}_N^* - \underline{x}_N\|_{V_N}^2$$

where $\underline{x}_i = \underline{x}(i\Delta t)$, $\underline{u}_i = \underline{u}(i\Delta t)$, and $\Delta t = t_f/N$ (for the sake of brevity, we do not write the discretized version of the differential system (26)). Moreover, $V = \mathrm{diag}\,[1, 0.1, 40, 0.1, 40, 0.1]$, $V_N = \mathrm{diag}\,[40, 40, 40, 40, 40, 40]$. The cost of the fuel consumption is taken into account by the functions $c(u_{ji}) = k\left[\frac{1}{\beta}\ln(2 + e^{\beta u_{ji}} + e^{-\beta u_{ji}}) - \frac{1}{\beta}\ln(4)\right]$, $(j = 1, 2)$, which approximate (for large enough values of the parameter $\beta$) the nondifferentiable costs $k\,|u_{ji}|$ (it is realistic to assume the fuel consumption to be proportional to the thrust). Also, observe that the sigmoidal functions generating the control variables $u_{1i}, u_{2i}$ are bounded, thus enabling us to remove constraints (27). For the present example, we took $\beta = 50$, $k = 0.01$.

Using the method proposed in the paper, the optimal neural control strategies $\underline{\gamma}_1\,(\underline{x}_i, \underline{v}_i, \underline{w}_i^\circ)$, $\underline{\gamma}_2\,(\underline{v}_{i+1}, \underline{x}_N^*, \underline{\tilde{w}}_i^\circ)$ were derived by training the 10 feedback neural networks and the 10 feed-forward neural networks on the initial and final regions, $A_0$ and $A_N$. Each neural network had 12 input variables and contained one hidden layer of 40 units. Moreover, $\eta = 0.8$, $c_1 = 100$ and $c_2 = 100000$. Convergence was obtained after some tens of thousands of iterations. The positions of the space robot during its manoeuvre are shown in Fig. 5. The effect of the feed-forward action is clearly revealed by the variation occurring in the robot trajectory when the robot perceives the "right-about turn" of the object to be reached.

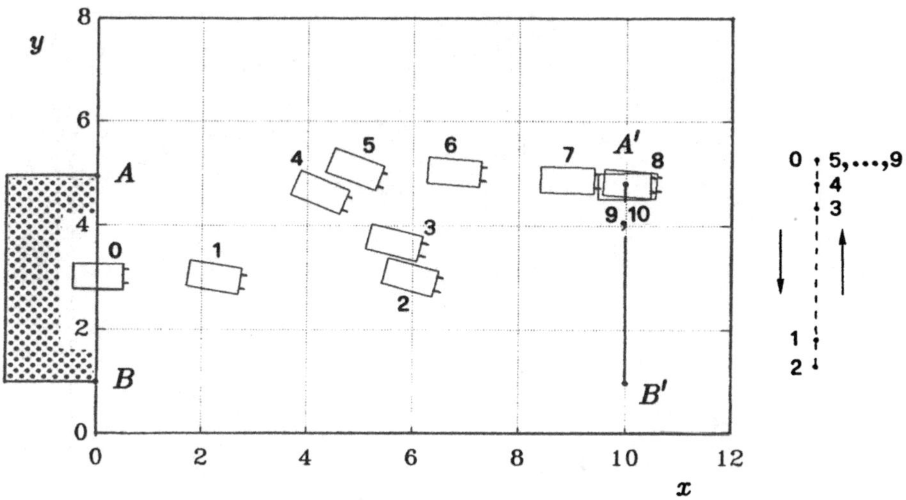


**Fig. 5. Positions of the space robot during its manoeuvre**

**Example 3.** In this example, the same space robot as considered in Example 2 has to start from any given point of the segment $AB$, then to pass through any specified intermediate point of the segment $\overline{A}\,\overline{B}$ at a given time instant $k$, and finally to reach any specified point of the segment $A'B'$ (see Fig. 6). As compared with Example 2, we have the same region $A_0$ and the regions $A_k = \{\underline{x} \in \Re^6 : x_1 = 10, x_2 = 0, 3.5 \le x_3 \le 6.5, x_4 = 0, \; x_5 = 0, x_6 = 0\}$, $A_N = \{\underline{x} \in \Re^6 : x_1 = 20, x_2 = 0, 1 \le x_3 \le 5, \; x_4 = 0, x_5 = 0, x_6 = 0\}$. Hence the present control problem lies midway between a terminal control problem, as in the previous example, and the state-tracking Problem 2'.

The cost function is given by

$$J = \sum_{i=0}^{N-1} [c(u_{1i}) + c(u_{2i})] + \sum_{i=0}^{k} \|\underline{x}_k^* - \underline{x}_i\|_{V_i}^2 + \sum_{i=k+1}^{N} \|\underline{x}_N^* - \underline{x}_i\|_{V_i}^2$$

where $N = 20$, $k = 8$, $V_i = \text{diag}\left[2(i+1)^2, 0.1, 40(i+1)^2, 0.1, 400(i+1), 0.1\right]$, for $i \le k$ and $V_i = \text{diag}\left[2(i+1-k)^2, 0.1, 40(i+1-k)^2, 0.1, 400(i+1-k), 0.1\right]$, for $i > k$.

Two optimal trajectories are shown in Fig. 6: they correspond to the two three-tuples $(\underline{x}_0, \underline{x}_k^*, \underline{x}_N^*)$, whose coordinates $x_3 = y$ are depicted on the three segments $AB$, $\overline{A}\,\overline{B}$, $A'B'$.

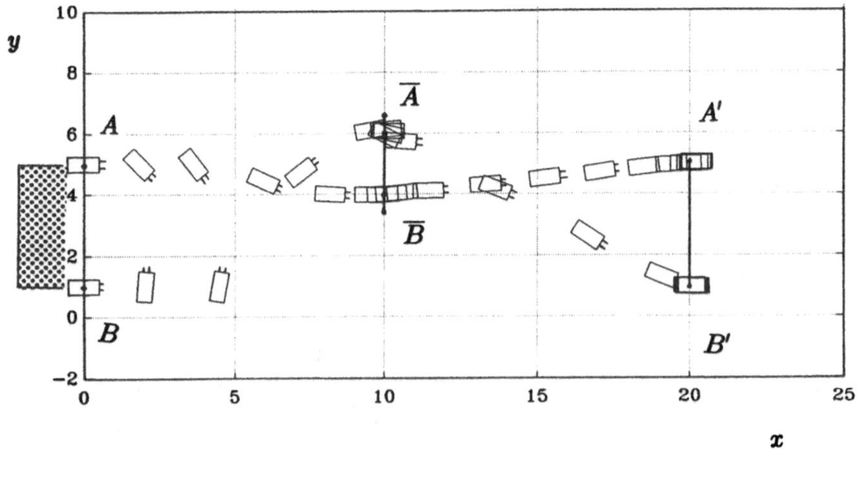


**Fig. 6. Optimal trajectories passing through intermediate points**

The optimal neural strategies $\underline{\gamma}_1(\underline{x}_i, \underline{v}_i, \underline{w}_i^\circ)$, $\underline{\gamma}_2(\underline{v}_{i+1}, \underline{x}_{i+1}^*, \underline{\tilde{w}}_i^\circ)$ (with $\underline{x}_{i+1}^* = \underline{x}_k^*$ for $0 \le i \le k-1$, and $\underline{x}_{i+1}^* = \underline{x}_N^*$ for $k \le i \le N-1$) were implemented on neural networks containing one hidden layer of 60 units. The constants $c_1, c_2, \eta$, were the same as in the previous examples. Convergence was obtained after one hundred of thousands of iterations.

## 7. Conclusions

Non-LQ optimal control problems can be solved by constraining the control strategies to take on the form of a suitable configuration of multilayer feed-forward neural networks connected to one another. Several configurations can be obtained by means of what has been called the "Linear-Structure Preserving principle" (the LISP principle). A particularly interesting configuration, enabling us to implement the control strategies on neural networks characterized by a time-invariant structure, has been presented in the paper. It is worth noting that the LISP principle can be applied to more complex problems, in which, for example, the state vector is not perfectly measurable and must then be reconstructed by means of a state observer. Multi-person (team) optimal control problems can also be faced [17].

Once the structure of the neural control strategy has been fixed, a nonlinear programming problem (instead of a functional one) has to be solved. As the former problem is an approximation of the latter, a suboptimal solution is in general obtained. As regards the accuracy to which neural optimal strategies can approximate the true solution, both experimental and theoretical results, derived from Barron's recent theorems [12], prove that neural networks based on sigmoidal functions exhibit better approximation properties than traditional polynomial, spline and trigonometric expansions. This leads us to believe that the technique of approximating a functional optimization problem by a parametric one, based on the employment of neural networks, may currently be regarded as a very powerful approach to solve "difficult" non-LQ optimal control problems.

## References


[1] A. P. Sage. **Optimum Systems Control**. Prentice-Hall, Englewood Cliffs, 1968.

[2] A. E. Bryson and Y. C. Ho. **Applied Optimal Control**. Blaisdell Publishing Company, 1969.

[3] R. E. Larson. **State Increment Dynamic Programming**. American Elsevier Publishing Company, 1968.

[4] B. R. Eisenberg and A. P. Sage. "Closed-Loop Optimization of Fixed Configuration Systems". **International Journal of Control**, Vol. 3, pp. 183-194, 1966.

[5] D.L. Kleinman and M. Athans. "The Design of Suboptimal Linear Time-Varying Systems". **IEEE Trans. Automatic Control**, Vol. AC-13, pp. 150-159, 1968.

[6] P. M. Mäkilä and H. T. Toivonen. "Computational Methods for Parametric LQ Problems- A survey". **IEEE Trans. Automatic Control**, Vol. AC-32, pp. 658-671, 1987.

[7] T. Parisini and R. Zoppoli. "Neural Networks for the Solution of N-Stage Optimal Control Problems", **Artificial Neural Networks**, T. Kohonen, K. Mäkisara, O. Simula, and J. Kangas, Eds., North-Holland, 1991.

[8] T. Parisini and R. Zoppoli. "Multi-Layer Neural Networks for the Optimal Control of Nonlinear Dynamic Systems", **Proc. First IFAC Symposium on Design Methods of Control Systems**, Zurich, Switzerland, 1991.

[9] D. H. Nguyen and B. Widrow. "Neural Networks for Self-Learning Control Systems". **IEEE Control System Magazine**, Vol. 10, pp. 18-23, 1990.

[10] K. Hornik, M. Stinchombe, and H. White. "Multilayer Feedforward Networks are Universal Approximators". **Neural Networks**, Vol. 2, pp. 359-366, 1989.

[11] R. Hecht-Nielsen. "Theory of the Backpropagation Neural Network". **Proc. Int. Joint Conf. on Neural Networks**, San Diego, CA, 1989, pp. 593-605.

[12] A. R. Barron. "Universal Approximation Bounds for Superpositions of a Sigmoidal Function". **Technical Report No. 58**, Department of Statistics, University of Illinois, 1991.

[13] T. Parisini and R. Zoppoli. "Backpropagation and Dynamic Programming in N-Stage Optimal Control Problems". **DIST Int. Rep. No. 90/4**, University of Genoa, Italy, 1991.

[14] Ya. Z. Tsypkin. **Adaptation and Learning in Automatic Systems**. Academic Press, 1971.

[15] D. E. Rumelhart and J. L. McClelland. **Parallel Distributed Processing**, Vol. 1, Chapter 8, Cambridge, MA, MIT Press, 1986.

[16] P. J. Werbos. "Beyond Regression: New Tools for Prediction and Analysis in the Behavioural Sciences". **Ph.D. Dissertation**, Harvard University, Cambridge, MA, 1974.

[17] G. Frisiani, T. Parisini, L. Siccardi, and R. Zoppoli. "Team Theory and Back-Propagation for Dynamic Routing in Communication Networks". **Proc. Int. Joint Conf. on Neural Networks**, Seattle, WA, 1991, Vol. 1, pp. 325-334.

The authors are with the Department of Communications, Computers and Systems Sciences, University of Genoa, Italy.

# Shock Waves for Riccati Partial Differential Equations Arising in Nonlinear Optimal Control*

CHRISTOPHER I. BYRNES** and ALI JHEMI

**Abstract**

**The derivation and analysis of the Riccati partial differential equation for optimal control has been recently developed in a series of papers and announcements. In addition to finite escape time, which already is an issue for linear quadratic problems, the existence of shock waves for solutions of these Riccati partial differential equations is the principle object of the nonlinear analysis and is known to be the only obstruction to the "off-line" characterization of optimal feedback laws. The purpose of this paper is to illustrate, in a simple optimal control problem, the use of Riccati partial differential equations which, in this case, reduces to the inviscid Burgers' equation. One rather interesting aspect of this analysis is the development of a variational interpretation of the Burgers' equation, quite similar to the variational interpretation of the Riccati ordinary differential equation in linear quadratic theory; viz., that the existence of unique optimal control laws, the existence of global solutions of the Riccati partial differential equation and convexity of the constraint on the terminal state are mutually equivalent. We also illustrating these concepts and the existence of shock waves for a nonconvex terminal constraint.**

## 1. Introduction

One of the most noteworthy aspects of the nonlinear regulator theory introduced in ([8], [9]) is that the feedback laws produced by this design methodology have a somewhat surprisingly familiar form; viz. that of a superposition of both a feedback and a feedforward "gain" which involves a quantity which may be computed "off-line" by solving a system of nonlinear partial differential equations. In the linear case, these equations reduce to the "off-line" Sylvestor equations discovered by Francis [10] in his treatment of the regulator problem for linear systems. On the other hand, the "off- line" characterization of feedback laws is also reminiscent of the solution of linear quadratic problems by nonlinear Riccati

*Dedicated, in homage, to Professor Antonio Ruberti on the occasion of his $65^{th}$ birthday.

**Research supported in part by AFOSR Grant No. 91-0266 and by NSF Grant DMS 89-9008223

equationsand raises the question of whether similar geometric methods might lead to a direct, nonlinear analogue of Riccati equations for optimal control.

Indeed, the derivation and analysis of the Riccati partial differential equation for optimal control, which has antecedents in the methods of dynamic programming and invariant imbedding, has been recently developed in a series of papers ([1], [7]) and announcements ([2], [3]). In addition to finite escape time, which already is an issue for linear quadratic problems, the existence of shock waves for solutions of these Riccati partial differential equations is the principle object of the nonlinear analysis and the only obstruction to the "off-line" characterization of optimal feedback laws. In fact, nonexistence of shocks can be shown to be equivalent([1]) to continuous differentiability of the value function of the optimal control problem. Consequently, the development of a geometric theory of weak solutions to Riccati equations provides an alternative approach to analyzing smoothness properties of value functions. It should also be remarked that other forms of the Riccati equation have been discovered in connection with the Kalman- Yacubovitch-Popov Lemma for positive real nonlinear systems [11] and with the factorization of nonlinear systems [12], so that one might anticipate Riccati partial differential equations playing a role in nonlinear systems similar in scope to that of Riccati equations in linear systems theory.

The purpose of this paper is to illustrate the use of Riccati partial differential equations in a simple optimal control problem, involving an elementary linear system, a quadratic integral performance measure and a nonlinear penalty on the terminal state. For this problem, it turns out that the Riccati partial differential equation reduces to the inviscid Burgers' equation, which is well understood in the theory of nonlinear shock waves [5]. One rather interesting aspect of this analysis is the development of a variational interpretation of the Burgers' equation which is quite similar to the variational interpretation of the Riccati ordinary differential equation. More explicitly, for those Riccati ordinary differential equation arising in variational problems, there is no classical blow-up (finite escape time) for initial data coming from a positive semi- definite constraint on the terminal state. We show that, similarly, for the Burgers' equation viewed in a variational context, there are no blow-ups(finite escape time or shock waves) precisely for initial data coming from a convex constraint on the terminal state. We also illustrate, by simulation, the existence and propagation of shock waves for nonconvex constraints.

We begin in section 2 by recalling the basic set-up for a nonlinear Bolza problem and the specialization of the Pontryagin Maximum Principle to such problems under a suitable set of hypotheses. In section 3, we describe a geometric derivation of and interpretation for the Riccati partial differential equation, illustrating these concepts and the existence of shock waves for a simple optimal control problem. In section 4, we sketch a proof of our main result for this model problem:, viz., that the existence of unique optimal control laws, the existence of global solutions of the Riccati partial differential equation and convexity of the constraint on the terminal state are mutually equivalent. This result is then illustrated for a nonquadratic but convex terminal constraint.

Once again, it is a pleasure to congratulate Il Professore Ruberti for all he has done, throughout his brilliant careeer, for systems engineering and science, for academic institutions and for education and research.

## 2. Preliminaries

Consider a control system having the form

$$\dot{x} = f(x) + g(x)u(t) \tag{2.1}$$

where $x \in \mathbb{R}^n$, and where for each $t$, $u(t) \in \mathbb{R}^m$. We assume that the vector fields $f$, $g_i$ are $C^r$ in $x$, $r \geq 2$ and that the $u_i(t)$ are piecewise continuous functions. In particular, for each pair $(x(0), u(t))$, the system (2.1) has a unique solution for $t << \infty$. Furthermore, we assume that $f(0) = 0$. For the system (2.1) we shall consider the problem of minimizing the cost functional

$$J^T(x(0), u) = \int_0^T L(x, u)dt + Q(x(T)) \tag{2.2}$$

In the next section we shall present some explicit solutions, in feedback form, to such optimal control problem, thereby providing constructive existence results. In this section, we review some well-known consequences of the existence of optimal control laws for (2.1)–(2.2).

Our general assumptions concerning (2.2) are that $Q(x)$ is $C^{q+1}$, $q \geq 1$ and that $L(x, u)$ is $C^{s+1}$, $s \geq 1$, and satisfies:

(H1) for each fixed $x$, $\frac{\partial L}{\partial u}(x, \cdot)$ is a diffeomorphism;

(H2) for each fixed $x$, $L(x, u)$ has a nondegenerate minnimum at $u = 0$.

In particular, we note that $\frac{\partial^2 L}{\partial u^2}(x, u) > 0$ for all $(x, u)$ and therefore that (H1)–(H2) imply that $L(x, u)$ is strictly convex in $u$.

(H3) 0 is a critical point of $Q$.

**Remark 2.1** Without loss of generality, we shall normalize $L(x, u)$ so that $L(0, 0) = 0$. We shall also assume that $Q(0) = 0$.

Finally, we set $k = \min(q, r, s)$.

Fixing notation, we reinterpret (2.1)–(2.2) as a problem in Mayer form by defining

$$\dot{x}_{n+1} = L(x, u), \quad x_{n+1}(0) = 0$$

and considering an augmented state variable

$$\tilde{x} = \begin{pmatrix} x \\ x_{n+1} \end{pmatrix} \in \mathbb{R}^{n+1}$$

The corresponding penalty-constraint function

$$\varphi : \mathbb{R}^{n+1} \times \mathbb{R}^{n+1} \times \mathbb{R}^2 \to \mathbb{R}^{n+4} \tag{2.3}$$

is therefore a function of $e = (\tilde{x}^0, \tilde{x}^1, t^0, t^1)$ where $\tilde{x}^0 = \begin{pmatrix} x^0 \\ x^0_{u+1} \end{pmatrix}$ is an initial state, $\tilde{x}^1 = \begin{pmatrix} x^1 \\ x^1_{n+1} \end{pmatrix}$ a terminal state, $t^0$ an initial time and $t^1$ a terminal time. In particular, the Mayer problem corresponding to (2.1)–(2.2) is to minimize

$$\varphi_1(e) = x^1_{n+1} + Q(x^1)$$

subject to the $n+3$ constraints

$$\begin{aligned}
\varphi_2(e) &= x^0_{n+1} = 0 \\
\varphi_{2+i}(e) &= x^0_i - x_i(0) = 0, \quad i = 1, \ldots, n \\
\varphi_{n+3}(e) &= t^1 - T = 0 \\
\varphi_{n+4}(e) &= t^0 = 0.
\end{aligned}$$

Following the Pontryagin Maximum Principle (see e.g. [4]), we first form an augmented Hamilton function

$$\tilde{H}(\tilde{x}, \tilde{p}, u) = \langle p, f(x) + g(x)u \rangle + p_{n+1} L(x, u)$$

and consider the system on $\mathbb{R}^{2n+2}$

$$\dot{\tilde{x}} = \frac{\partial \tilde{H}}{\partial \tilde{p}} \tag{2.4}$$

$$\dot{\tilde{p}} = \frac{-\partial \tilde{H}}{\partial \tilde{x}} \tag{2.5}$$

According to the Maximum Principle, for $x(0)$ fixed a necessary condition for $u_*(t)$ to be an optimal control is that there exist an initial condition $\tilde{p}(0)$ such that if $\tilde{x}(0) = \begin{pmatrix} x(0) \\ 0 \end{pmatrix}$, the corresponding solution $(\tilde{x}(t), \tilde{p}(t))$ of (2.4)–(2.5) exists and satisfies the following conditions:

$$\max_{u \in \mathbb{R}^{n+1}} \tilde{H}(\tilde{x}(t), \tilde{p}(t), u) = \tilde{H}\left(x(t), p(t), u_*(t)\right) \quad a.e. \tag{2.6}$$

$$\tilde{p}(t) \not\equiv 0. \tag{2.7}$$

Furthermore, there must also exist $\lambda \in \mathbb{R}^{n+4}$ with $\lambda_1 \leq 0$ such that the transversality conditions

$$\tilde{p}(t^0)^T = -\lambda^T \varphi_{\tilde{x}^0}(e) \tag{2.8}$$

$$\tilde{p}(t^1)^T = \lambda^T \varphi_{\tilde{x}^1}(e) \tag{2.9}$$

$$\tilde{H}\left(\tilde{x}(t^0), \tilde{p}(t^0), u_*(t^0)\right) = \lambda^T \varphi_{t^0}(e) \tag{2.10}$$

$$\tilde{H}\left(\tilde{x}(t^1), \tilde{p}(t^1), u_*(t^1)\right) = -\lambda^T \varphi_{t^1}(e) \tag{2.11}$$

are satisfied.

**Definition 2.1** Suppose, for some initial condition $\tilde{x}(0)$, that $u(t)$ is a control for which there exists a solution $\tilde{p}(t)$ of the adjoint system and a "multiplier" $\lambda \in \mathbb{R}^{n+4}$ such that the conditions of the Pontryagin Maximum Principle (2.6)–(2.12) are satisfied. Then, we shall say that $u(\cdot)$ is an extremal control, that the corresponding trajectory $x(\cdot)$ is an extremal trajectory and that the pair $(x(\cdot), p(\cdot))$ is a canonical pair.

In the remainder of this section, we shall simplify (2.4)–(2.12) in the light of assumptions (H1)–(H2) and the form of (2.1)–(2.2). We first note that this minimization problem is "normal"; i.e. that $p_{n+1} < 0$. Since the Maximum Principle is invariant under a scaling of $\tilde{H}$ by any positive constant, we may take $p_{n+1} = -1$, leading to the augmented Hamiltonian

$$H(x,p,u) = \langle p, f(x) + g(x)u \rangle - L(x,u)$$

We note that, by (H1), for each $(x,p)$ there is a unique $u_*(x,p)$, $C^k$ in $(x,p)$, which satisfies

$$0 = \frac{\partial H}{\partial u}|_{u=u_*} = \langle p, g(x) \rangle - \frac{\partial L}{\partial u}(x, u_*) \tag{2.13}$$

Moreover, in the light of (H2), for each fixed $(x,p)$ $u_*(x,p)$ in fact maximizes $H(x,p,u)$, since

$$\frac{\partial^2 H}{\partial u^2}|_{u=u_*} = -\frac{\partial^2 L}{\partial u^2}(x, u_*) < 0. \tag{2.14}$$

In particular, if $u_*(t)$ is an optimal control for an initial condition $x(0)$, giving rise to an optimal trajectory $x_*(t)$ and some nontrivial solution $p(t)$ of the adjoint equation satisfying the conditions of the Maximum Principle, we must have

$$u_*(t) = u_*(x_*(t), p(t)) \tag{2.15}$$

There is an important converse to this conclusion. Define a Hamiltonian function, $H_*(x,p)$, via

$$H_*(x,p) = H(x,p,u_*(x,p))$$

where $u_*(x,p)$ is the unique solution of (2.14). Note that (2.4) can be written as

$$\dot{x} = \frac{\partial H_*}{\partial p} \tag{2.16}$$

$$\dot{p} = -\frac{\partial H_*}{\partial x} \tag{2.17}$$

together with

$$\dot{x}_{n+1} = L(x, u) \qquad x_{n+1}(0) = 0 \tag{2.18}$$

$$\dot{p}_{n+1} = 0 \qquad p_{n+1}(0) = -1 \tag{2.19}$$

We can now refine the observation implicit in (2.16), which follows from a specialization of the Maximum Principle to the current problem.

**Proposition 2.1** Consider the "canonical system" (2.16)–(2.17), with the final value condition

$$p(T) = -\nabla Q(x(T)) \tag{2.20}$$

For any initial condition $x(0)$, consider any extremal control $u(t)$ and a corresponding $p(t)$ for which $(x(t), p(t))$ is a canonical pair. Then, $(x(t), p(t))$ is a solution of the canonical system satisfying the final value condition (2.19). Moreover, $u(t)$ is given by the formula (2.16). Conversely, for any solution of the canonical system satisfying the final value condition (2.19) the control $u(t)$ defined by (2.16) is an extremal control with extremal trajectory $x(t)$ and having $(x(t), p(t))$ as a canonical pair.

## 3. Riccati Partial Differential Equations for the Bolza Problem

According to Proposition 2.1, for the Bolza Problem every extremal control $u_*(t)$ for any initial condition $x(0)$ gives rise to a trajectory, or canonical pair, $(x(t), p(t))$ satisfying

$$\dot{x}(t) = \frac{\partial H_*}{\partial p} \tag{3.1}$$

$$\dot{p}(t) = -\frac{\partial H_*}{\partial x} \tag{3.2}$$

and

$$p(T) = -\nabla Q(x(T)). \tag{3.3}$$

Furthermore, $u_*(t)$ may be expressed as

$$u_*(t) = u_*(x(t), p(t)) \tag{3.4}$$

where $u_*(x, p)$ is the unique solution of (2.14). Conversely, any solution $(x(t), p(t))$ of (3.1)–(3.3) generates, via (3.4), an extremal control $u(t)$ with extremal trajectory $x(t)$. In this section, we are interested in developing a closed-loop expression for extremal controls, deferring a discussion of optimality of extremals to sections 5 and 6. To this end, we develop a nonlinear analogue of the Riccati equation for the "state-costate" equations (3.1)–(3.2) with a nonlinear, but smooth, terminal constraint (3.3). More explicitly, motivated by Dynamic Programming, we shall attempt to find an expression

$$p(t) = \pi(x(t), t) \tag{3.5}$$

by integrating the smooth constraint (3.3) backwards in time. If this were possible, one could then rewrite (3.4) in feedback form

$$u(t) = u_*(x(t), \pi(x(t), t))$$

or more simply as

$$u = u_*(x, \pi(x, t)) \tag{3.6}$$

as in $LQ$ Theory. We shall first impose a further simplifying asumption on (2.1)–(2.2).

(H4) The canonical system (3.1)–(3.2) is complete.

Equivalently, we assume that the flow

$$\Phi\left(t, \begin{pmatrix} x(0) \\ p(0) \end{pmatrix}\right) = \begin{pmatrix} x(t) \\ p(t) \end{pmatrix},$$

which is always defined for small $t$, is in fact defined for all $t$. In particular, we note that

$$\Phi : \mathbb{R} \times \mathbb{R}^{2n} \to \mathbb{R}^{2n}$$

is jointly $C^k$ and, for each $t$, the map

$$\Phi_t(\cdot) = \Phi(t, \cdot)$$

is a $C^k$ diffeomorphism

$$\Phi_t : \mathbb{R}^{2n} \to \mathbb{R}^{2n} \tag{3.7}$$

Now consider the closed, connected $C^k$ submanifold of $\mathbb{R}^{2n}$ defined via

$$M_T = \{(x, p) : p = -\nabla Q(x)\} \tag{3.8}$$

$M_T$ is of course the submanifold of terminal constraints given by the transversality conditions. We note that for $t \in [0, T]$

$$M_t = \Phi_{t-T}(M_T) \tag{3.9}$$

is a closed, connected $C^k$ submanifold of $\mathbb{R}^{2n}$ consisting of those pairs $(x(t), p(t))$ which satisfy (3.1)–(3.3), with initial time $t_0 = t$. In particular, for $s \in [t, T]$

$$u(s) = u_*(x(s), p(s))$$

is an extremal control for the Bolza problem (2.1)–(2.3) with initial time $t_0 = t$. Moreover, to say

$$M_s = \text{graph}(-\pi(x, s)), \quad t \leq s \leq T \tag{3.10}$$

is to say that all extremal controls for $t \leq s \leq T$ can be given in feedback form

$$u(s) = u_*(x(s), -\pi(x(s), s)), \quad t \leq s \leq T.$$

We first derive conditions on $\pi(x,t)$ so that the geometric condition (3.10) will be satisfied.

**Theorem 3.1** Necessary and sufficient conditions for the existence of a $C^\ell$ function $\pi(x,t)$, $1 \leq \ell \leq k$, such that

$$M_t = \text{graph}(-\pi(x,t))$$

for $t \in [t_0, T]$, $x \in \mathbb{R}^n$ is that $\pi(x,t)$ satisfy the following "Riccati" partial differential equation, for $(x,t) \epsilon \mathbb{R}^n \times (t_0, T)$

$$\frac{\partial \pi}{\partial t} = \frac{\partial H_*}{\partial x}(x,-\pi) - \frac{\partial \pi}{\partial x}\frac{\partial H_*}{\partial p}(x,-\pi) \tag{3.11}$$

$$\pi(0,t) = 0, \quad \pi(x,T) = \nabla Q(x) \tag{3.12}$$

In particular, the Riccati partial differential equation has a $C^k$ solution if, and only if, it has a $C^1$ solution.

Theorem 3.1 gives a geometric criterion for the existence of a smooth solution of the Riccati PDE, viz

$$M_t = \text{graph}(-\pi(\cdot,t)) \tag{3.13}$$

where $M_t$ is the set of initial conditions $(x(t), p(t))$ for initial time $t = t_0$, which are feasible or compatible with the conditions of the Maximum Principle. On the other hand, Dynamic Programming would suggest that any function $\pi$, $C^1$ or not, satisfying (3.13) would lead to an optimal control law in feedback form

$$u_*(t) = u_*(x(t), -\pi(x(t),t))$$

Indeed, the paper [1] develops a rigorous treatment of the optimality of such control laws, whenever the Riccati PDE has a $C$ solution on $W \times [t,T]$, for $W$ an open neighborhood of 0 in $\mathbb{R}$. This treatment includes the relationship between classical, or even weak, solutions of the Riccati PDE and smooth solutions to the Hamilton-Jacobi-Bellman equation, also yielding results on smoothness of value functions.

Of course, $\pi$ may fail to $C^1$ in several ways, among these being either a "classical blow-up", or finite escape time, where $\frac{\partial \pi}{\partial t}$ becomes infinite or the existence of "shock waves," where $\frac{\partial \pi}{\partial x}$ becomes infinite. In neither case, would $\pi$ be considered to be a classical solution of the Riccati PDE. Rather, in this sense $\pi$, or even $M_t$ would be a weak or generalized, solution of the Riccati PDE with potential importance for the analysis, design or synthesis of optimal control laws (see [1] for a construction of optimal control laws using weak solutions of the Riccati PDE). We conclude this section by illustrating this point using a simple example.

**Example 3.1** Consider the one-dimensional dynamical system

$$\dot{x}_1 = u(t), \quad x, u \in \mathbb{R}$$

with the following cost function:

$$J(x_1(0), u) = \frac{1}{2}\int_0^T u^2 dt + \frac{1}{2}(x_1^4(T) - x_1^2(T))$$

where $x_1(0)$, $x_1(T)$ are initial and final conditions of the system respectively. Setting the system in Mayer form, we introduce the running cost as a new variable,

$$\dot{x}_2 = \frac{1}{2}u^2(t) \quad \text{with } x_2(0) = 0.$$

As before the augmented Hamiltonian is

$$H_*(x, p_1, u) = p_1 u - \frac{1}{2}u^2$$

and the Pontryagin Maximum Principle implies that

$$u = p_1$$

which must be constant since the canonical system has the form

$$\begin{aligned} \dot{x} &= p \\ \dot{p} &= 0. \end{aligned}$$

In particular, for this system, the Riccati PDE for $p = -\pi(x, t)$ reduces to the well known inviscid Burgers' equation:

$$\frac{\partial \pi}{\partial t} = \frac{\partial \pi}{\partial x}\pi \tag{3.14}$$

with the side conditions

$$\pi(T, x) = 2x^3 - x \text{ and } \pi(t, 0) = 0.$$

The fact that this Burgers' equations, with a non-monotone nondecreasing final condition, does not have a solution for all time $t < T$ is classical (see e.g. [5]). Indeed, it is easy to see, either from the method of characteristics or by integrating the canonical equations, that this problem has a $C^1$ solution on $[0, T]$ if, and only if, $T < 1$ (see [6]). The weak solution for $T = 1$ has an infinite vertical slope at $x - 0$ corresponding to the existence (or development) of a shock wave; i.e. a point when $\frac{\partial \pi}{\partial x}$ becomes infinite. The simulations in Figure 3.1 illustrates the fact that as we extend the time interval $[0, T]$ for $T > 1$, this shock wave propagates, with the vertical tangent bifurcating into two shocks, one travelling to the right and one travelling to the left. These shock waves are depicted for the times: $T = .20,\ .50,\ .95,\ 1.5$, and $2.0$. We also note that, if $T > 1$, $M_t$ is no longer of the form

$$M_t = \text{graph}(-\pi(x, t))$$

Indeed, $M_t$ is the graph of a multivalued function. Nonetheless, $M_t$ remains a smooth curve and the "branches" of this curve as well as the vertical tangents, or shock waves, have an interesting interpretation in terms of the optimal control problem. Briefly, integrating the canonical equations, we can explicitly determine the cost functional, as a function of $p$, yielding

$$J^T(x_0, p) = \frac{p^2T}{2} + \frac{1}{2}(x_0 + pT)^4 - \frac{1}{2}(x_0 + pT)^2$$

The extrema of $J^T(x_0, p)$ occur at solutions of

$$T(p + 2(x_0 + pT)^3 - (x_0 + pT)) = 0$$

which is, of course, just the transversality condition of the Maximum Principle. One can therefore compute the Hessian of $J^T(x_0, \cdot)$ obtaining

$$T(1 + 6T(x_0 + pT)^2 - T)$$

The zeroes of $\frac{\partial^2 J}{\partial p^2}$ are therefore the points at which $M_t$, for $T - t > 1$, has a vertical tangent. That is, the onset of shock waves for solutions of the Riccati PDE coincides with the existence of degenerate critical points of the cost functional. Indeed, explicit calculation, for final times $T > 1$, shows that deleting the two points of vertical tangency from $M_t, T - t > 1$, leaves 3 branches, two semi-infinite branches and one finite branch containing 0. On each of the semi-infinite branches we have $\frac{\partial^2 J}{\partial p^2} > 0$ while on the finite branch we have $\frac{\partial^2 J}{\partial p^2} < 0$, representing a bifurcation of the extremal minimizing controls to a family of 3 extremal controls. One can, of course, develop a synthesis of an optimal control for this problem, using these geometric constructions, (see [7]). We emphasize, however, that smoothness of $M_t$ is crucial in this analysis.

**Remark 3.1** This example also sheds some light on the question of global existence of solutions of the Riccati PDE and its relationship with uniqueness of optimal controls and global existence of optimal state feedback laws; viz., all three coexist in precisely the case when the time horizon is less than or equal to 1. For larger horizon times, none of these three properties persists–a phenomenon which remains true in considerable generality (see [7]).

## 4. Non-existence of shock wave for convex penalty functions

Our study of solutions of (2.1)–(2.2) for the system described in Example 3.1 using Burgers' equation suggests that globally defined, smooth optimal state feedback laws exist only for those problems in which $Q'(x)$ is monotone nondecreasing; i.e. when $Q(x)$ is convex. On the other hand, such a result would give a variational interpretation of Burgers' equation, also yielding a new interpretation of the well-known criterion for existence of global solutions.

**Theorem 4.1** The optimal control problem

$$\begin{aligned}\dot{x}(t) &= -u(t), \quad x(t_0) = x_0 \\ \min_u J(x_0, u) &= \min_u \left[\frac{1}{2}\int_0^T u^2 dt + Q(x(T))\right]\end{aligned}$$

has a unique solution for $t_0 \in (-\infty, T]$ if, and only if, $Q(x)$ is convex. Moreover, if $Q$ is convex then this solution can be expressed as a smooth feedback control law.

**Proof** If $Q(x)$ is convex, then $\pi(x,t)$ exists for all $x \in R$, $t \in (-\infty, T]$. Fixing $t \in (-\infty, T]$ defining, for $t_0 \leq t \leq T$

$$u_*(x(t)) = \pi(x(t), t)$$

We claim $u_*$ is optimal, for each $x(t_0) = x_0$.

Define $\bar{x}(t)$ as the trajectory of the equation

$$\dot{x}(t) = -\pi(x(t), t)$$

$\bar{x}(t)$ exists for $t \in (t_0 - \varepsilon,\ t_0 + \varepsilon)$. Now define $\bar{p}(t)$ via

$$\bar{p}(t) = -\pi(x(t), t).$$

Then $(\bar{x}(t),\ \bar{p}(t))$ is a solution of the canonical system

$$\begin{aligned}\dot{x} &= p \\ \dot{p} &= 0\end{aligned}$$

which satisfies the transversality condition $p(T) = -Q'(x(T))$.

Moreover, $u(t) = u_*(\bar{x}(t), t)$ is an extremal control which generates $\bar{x}(t)$ as an extremal trajectory. In particular,

$$\frac{\partial H}{\partial u}(x, p, u) = 0 \text{ at } u = u*$$

and

$$\frac{\partial^2 H}{\partial u^2} < 0 \text{ at } u = u*$$

so that $u_*$ maximizes $H(x, p, u)$.

Therefore

$$\pi(\bar{x}, t)u - \frac{1}{2}u^2 \leq \pi(\bar{x}, t)u_* - \frac{1}{2}u_*^2$$

Set

$$\pi(x, t) = -\nabla_x W(x, t)$$

where

$$W(0, t) = 0.$$

We define, for the associated Mayer problem,

$$V(x,x_2,t) = x_2 + W(x,t).$$

and we claim that $V(x,x_2,t)$ is the value function. First, note

$$\begin{aligned}\dot{V}(x,x_2,t) &= \frac{\partial V}{\partial t} + \frac{\partial W}{\partial x}\dot{x} + \dot{x}_2 \\ &= \frac{\partial V}{\partial t} + \frac{\partial W}{\partial x}(-u) + \frac{1}{2}u^2\end{aligned}$$

Set $u = u_* + v$

$$\begin{aligned}&= \frac{\partial V}{\partial t} - \pi(x,t)u_* - \pi(x,t)v + \frac{1}{2}u_*^2 + u_*v + \frac{1}{2}v^2 \\ &= \frac{\partial V}{\partial t} - \pi(x,t)u_* + \frac{1}{2}u_*2 + \frac{1}{2}v^2\end{aligned}$$

Note, in particular, that $\frac{\partial V}{\partial t} - \pi(x,t)u_* + \frac{1}{2}u_*^2 = \frac{\partial V}{\partial t} - H(x,p,u_*)$

In particular, it suffices to show that $V$ and the control law $u_*$ satisfy the Hamilton-Jacobi-Bellman equation .

**Lemma 4.2**

$$\frac{\partial V}{\partial t} - H(x,p,u_*) = 0$$

**Proof** Taking the derivative of this quantity with respect to $x$, we obtain

$$\begin{aligned}\frac{\partial}{\partial x}\left(\frac{\partial V}{\partial t} - H(x,p,u_*)\right) &= \frac{\partial \pi}{\partial t} - \frac{\partial H}{\partial x}(x,p,u) + \frac{\partial H}{\partial u}|_{u=u_*}\frac{\partial u}{\partial x} \\ &= \frac{\partial \pi}{\partial t} - \frac{\partial H}{\partial x}(x,-\pi,u) \quad (\text{since } \frac{\partial H}{\partial u}|_{u=u_*} = 0) \\ &= \frac{\partial \pi}{\partial t} - \frac{\partial \pi}{\partial x}u_* \\ &= \frac{\partial \pi}{\partial t} - \frac{\partial \pi}{\partial x}\pi = 0\end{aligned}$$

Therefore

$$\frac{\partial V}{\partial t} - H(x,p,u_*) = F(t)$$

for $F$ a function of time.

Evaluating the left-hand side of this equation at $x = 0$, we obtain

$$\frac{\partial V}{\partial t}(0,t) - \pi(0,t)u_* + \frac{1}{2}u_*^2$$

We first note that

$$V(0,0,t) = x_2(0) + W(0,t) = 0$$

and therefore

$$\frac{\partial V}{\partial t}(0,t) = 0.$$

Since

$$\frac{\partial V}{\partial t}(0,t) = 0, \pi(0,t) = 0, \text{ and } u_*^2(0) = [\pi(0,t)]^2 = 0$$

we conclude

$$F(t) = 0$$

which completes the proof of the lemma.

Summarizing, we have shown that the conditions of the verification principle of Dynamic Programming hold: viz. for any $u$, $\dot{V}(x, x_2, t) \geq 0$, while for $u = u*$, $\dot{V}(x, x_2, t) = 0$. Therefore, $V$ is the value function and $u*$ is an optimal control. Moreover, since the proof shows that the Hamilton-Jacobi-Bellman inequality is strict for $u = u*$, it follows from an elementary comparison argument that $u*$ is the unique optimal control.

Conversely, under hypotheses which apply in the case at hand, it is shown in [7] that global existence of a unique optimal control implies global existence of a solution to the Riccati PDE, which in this case is of course the inviscid Burgers' equation. Matters being so, global existence of a solution implies that $Q^1(x)$ is monotone nondecreasing so that $Q(x)$ must be convex, as claimed. ■

In the following example, a convex penalty function is chosen, and shown to have a unique smooth control for all $T$.

**Example 4.1**

Consider the same one dimensional system given in example 3.1, with a convex penalty function

$$J(x(0), u) = \frac{1}{2}\int_0^T u^2 dt + \frac{1}{2}x_1^4(T)$$

Again, the Hamiltonian is

$$H(x, p_1, u) = < p_1, -u > -\frac{1}{2}u^2 = -p_1 u - \frac{1}{2}u^2$$

Since

$$\dot{p}_1 = -\frac{\partial H}{\partial x_1}(x, p, u) = 0$$

we conclude that

$$p_1 = \text{ constant}$$

Also,

$$\frac{\partial H}{\partial u}(x, p, u) = 0 \text{ at } u = u*$$

implies that

$$u = -p_1 = \text{constant}$$

so that

$$x(t) = x(0) - ut \text{ and } x(T) = x(0) - uT$$

Substituting for $x(T)$ in the cost function, we obtain

$$J(x,u) = \frac{1}{2}\int_0^T u^2 dt + \frac{1}{2}(x(0) - uT)^4$$

so that

$$\frac{\partial J}{\partial u} = 0 \Leftrightarrow uT - 2T(x(0) - uT)^3 = 0$$

Eliminating $T$ gives

$$u - 2(x(0) - uT)^3 = 0$$

and, since $T > 0$,

$$\frac{\partial^2 J}{\partial u^2} = T + 6T^2(x(0) - uT)^2 > 0$$

After some manipulation, we obtain

$$u^3 - \frac{6x(0)}{T}u^2 + \frac{12x^2(0)T + 1}{2T^3}u - \frac{x^3(0)}{T^3} = 0$$

This equation may be solved numerically, yielding only one solution, as it should be. A graph of $u$ versus $x(0)$ is on Figure 4.1.

## REFERENCES


[1] C.I. Byrnes, *The Solution of Nonlinear Lagrange and Bolza Problems via Riccati Partial Differential Equations*, to appear.

[2] C.I. Byrnes, Some Partial Differential Equations Arising in Nonlinear Control and Optimization, Computation and Control, Vol. II, Birkhäuser-Boston, 1991.

[3] C.I. Byrnes, *New Methods in Nonlinear Optimal Control*, Proceedings of 1st ECC, Grenoble, 1991.

[4] W.H. Fleming and R.W. Rishel, *Deterministic and Stochastic Optimal Control*, Springer-Verlag, Berlin 1975.

[5] J. Moller, "Shock Waves and Reaction-Diffusion Equations," Springer-Verlag, New York, 1983.

[6] A.A. Jhemi, *Linear and Nonlinear Regulation*, M.S. Thesis, Washington University, 1991.

[7] C.I. Byrnes and H. Frankowska, *Global Solvability of Riccati Equations for Nonlinear Optimal Control*, to appear.

[8] C.I. Byrnes and A. Isidori, *Règulation asymptotique des systèmes non-lineàres*, C.R. Acad. Sci. Paris **309** (1989), 527–530.

[9] A. Isidori and C.I. Byrnes, *Output Regulation of Nonlinear Systems*, IEEE Trans. Aut. Contr. **AC-35** (1990), 131–140.

[10] B.A. Francis, *The Linear Multivariable Regulator Problems*, SIAM J. Contr. Optimiz. **115** (1977), 486–505.

[11] K. Doll, *Global Stabilization of Nonlinear Systems*, D.Sc. Dissertation, Washington University, in preparation.

[12] A. Ben-Artzi and J.W. Helton, *A Riccati Partial Differential Equation for Factoring Nonlinear Systems*, preprint, 1990.

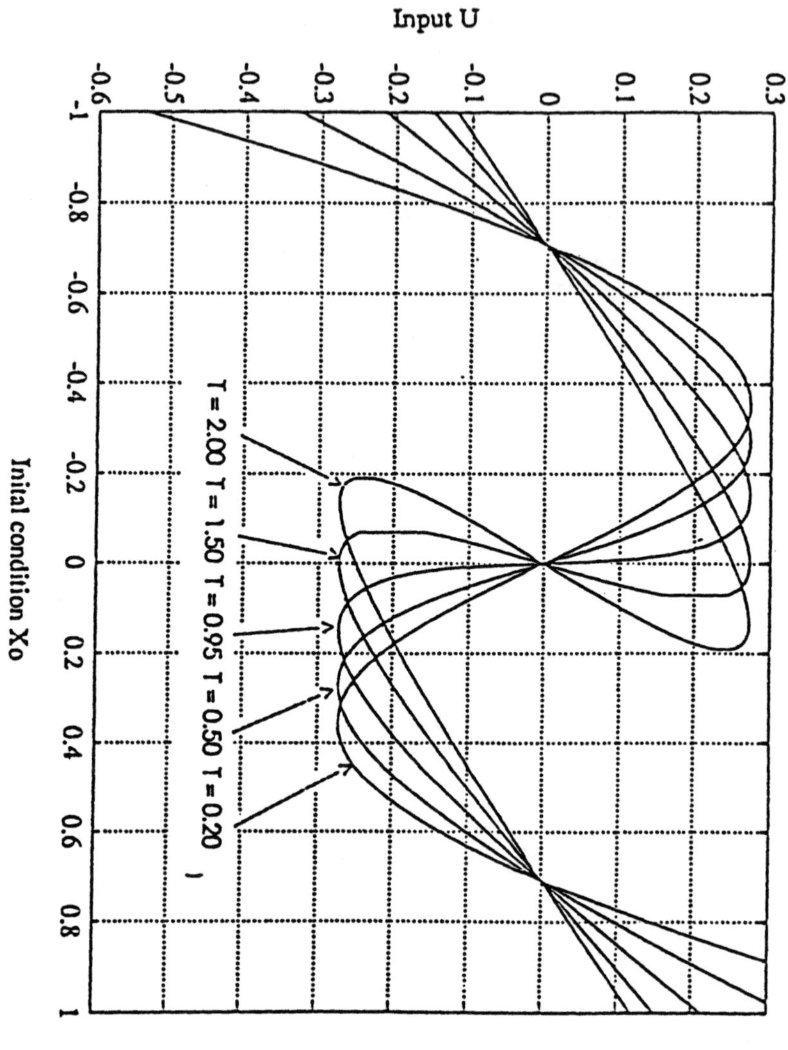


Figure 3.1: $U$ as function of $X_0$

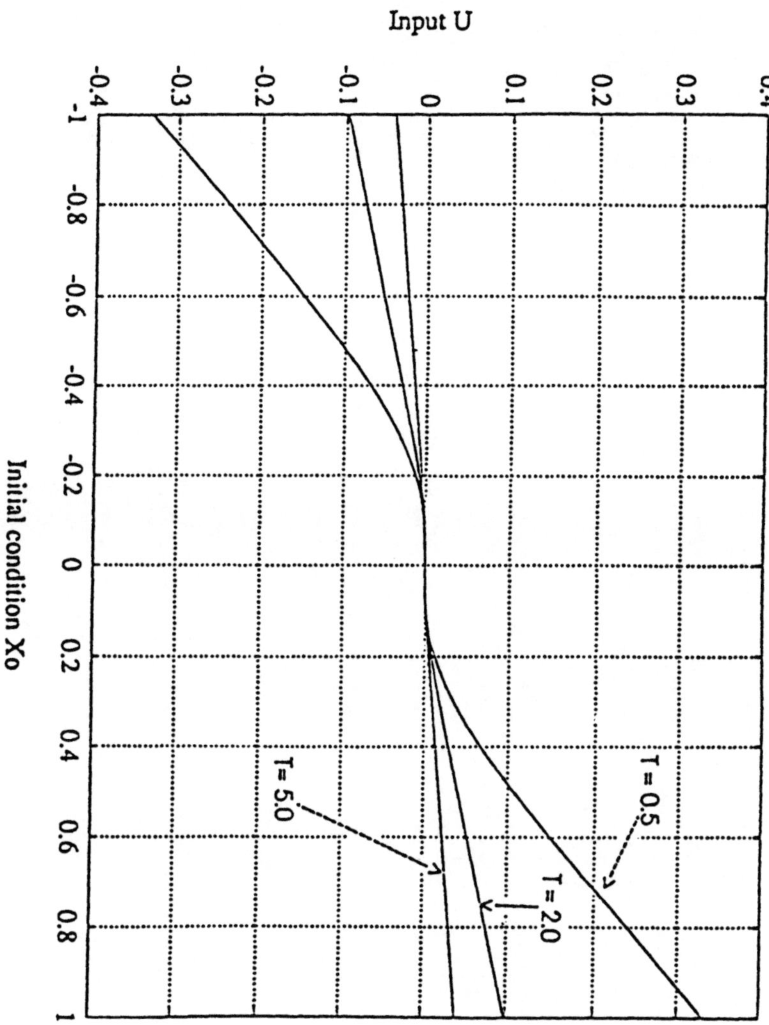


Figure 4.1: $U$ as function of $X_0$

Department of Systems Science and Mathematics
Washington University
St. Louis, Missouri 63130
U.S.A.

# Robust Stability and Performance in the Presence of Parametric and Norm Bounded Uncertainty

G. FIORIO, M. MILANESE and A. VICINO

## 1. Introduction

Control design of actual plants is usually based on mathematical descriptions of the system to be controlled. In practice, approximate models only can be constructed, and the objective of robust design techniques is to take into account the discrepancies between the model and the real system. These discrepancies are often accounted for by complementing the model with perturbations of parametric or non-parametric nature. The former kind of perturbation, often called highly structured perturbation, mainly accounts for approximate knowledge of exact values of system physical parameters or for different operation conditions the system may undergo. The latter type of uncertainty, often referred to as unstructured perturbation, generally accounts for unmodelled dynamics. The common way of representing parametric uncertainty is by assuming an admissible set for model parameters; unstructured uncertainty is generally represented by a frequency dependent tolerance band on the nominal response of the system.

While robust stability and control in the presence of norm bounded perturbations has been studied since long time (see [38,18,17,19,31,30]), the parametric problem has become a key issue in the robust control field only recently, after the appearence in the literature of the Kharitonov Theorem [27] (see e.g. [36,40] for survey papers and [33,25,7] including proceedings of workshops focused on the subject). The aim of most contributions in the parametric area is twofold. First, for a given uncertainty set establish nonconservative stability tests. Second, for a given nominal stable system, find the minimum perturbation destabilizing the system (*stability margin problem*). Most of the results of the first kind regard *interval* feedback systems, i.e. systems where the plant is represented by an interval family of transfer functions [9]. A more recent line of research aimed at reducing discrepancies between the real system and the mathematical model is that of studying systems including both parametric and norm bounded [10] or sector bounded [15,41,11,46] uncertainties. Regarding the stability margin problem, most contributions investigate the case where plant coefficients are linear or polynomial functions of a vector of physical parameters (see e.g. [6,24,36,42] for linear dependence and [16,47,39,48] for the nonlinear case). Also, robust performance of systems in the presence of parametric and/or norm bounded uncertainty is a topic of growing interest in the parametric area of research. Several contributions in this direction have appeared recently. Some of them provide extremal results for interval systems [4,45,14], while others propose algorithms for the case of polynomial dependency on parameters [34].

The aim of the present paper is to review some of the recent results in the robustness area. Particular emphasis is given to uncertainty models incorporating both parametric and nonparametric perturbations. This allows one to extend many classical results, like the small gain theorem or the small $\mu$ theorem, to the case where parametric families of plants are considered.

The paper is structured as follows. Section 2 contains basic notation. Section 3 presents robust stability results for interval feedback systems with norm-bounded perturbations. In Section 4 results on robust performance of interval systems are given, while Section 5 reviews main results on robust strict positive realness for the same class of systems. Section 6 addresses the robust stability and performance problems for systems with nonlinear parametric perturbations.

## 1. Notation

Feedback systems considered in the paper are represented in Figures 1a-1b. We denote by $\mathcal{G}_I$ a family of proper interval plants, i.e.

(1) $$\mathcal{G}_I = \{G(s) : G(s) = \frac{N(s)}{D(s)},\ N(s) \in \mathcal{N}_I,\ D(s) \in \mathcal{D}_I\}$$

where $\mathcal{N}_I$ and $\mathcal{D}_I$ are families of interval polynomials, i.e.

(2) $$\mathcal{N}_I = \{N(s) : N(s) = \sum_{i=0}^{m} b_i s^i,\ b_i \in [b_i^-, b_i^+]\}$$

(3) $$\mathcal{D}_I = \{D(s) : D(s) = s^n + \sum_{i=0}^{n-1} a_i s^i,\ a_i \in [a_i^-, a_i^+]\}\ .$$

The transfer function $C(s) = N_c(s)/D_c(s)$ is the controller and $\delta(s)$ accounts for plant perturbations, assumed bounded according to an $H_\infty$ norm, i.e.

(4) $$|\delta(j\omega)| \leq \epsilon,\ \forall \omega \geq 0$$

where $\epsilon$ is a given positive number. Notice that the norm bounded uncertainty $\delta(s)$ may be located in different points of the loop to represent different uncertainty structures. For example, the multiplicative perturbation of Figure 1b reflects output sensor errors and/or neglected high frequency dynamics. Although we will make explicit reference to the configuration of Figure 1b, most of the relevant results given in the following sections hold also for other structures of perturbations, e.g. additive or coprime factor perturbations ([13]).

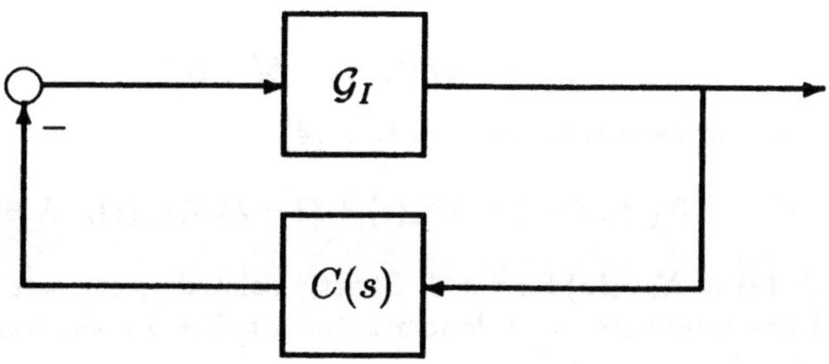


Figure 1a. Interval feedback system.

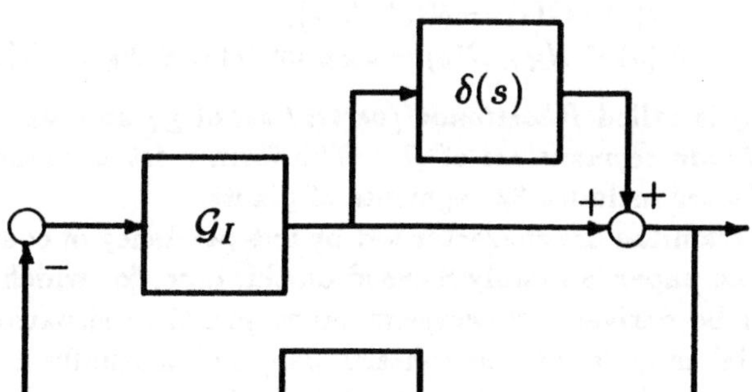


Figure 1b. Interval system with norm bounded perturbation.

For a given interval family of polynomials, we define two subsets playing a key role in robust stability and performance analysis. For a moment, we make specific reference to the numerator family, but similar definitions hold for the denominator family. The first set $\mathcal{N}_K$ includes the 4 *Kharitonov polynomials*, representing four corners of the interval family in coefficient space

(5) $$\mathcal{N}_K = \{N_1(s),\ N_2(s),\ N_3(s),\ N_4(s)\},$$

where

(6) $$N_1(s) = b_0^+ + b_1^+ s + b_2^- s^2 + b_3^- s^3 + b_4^+ s^4 + b_5^+ s^5 + \dots,$$

(7) $$N_2(s) = b_0^+ + b_1^- s + b_2^- s^2 + b_3^+ s^3 + b_4^+ s^4 + b_5^- s^5 + \dots,$$

(8) $$N_3(s) = b_0^- + b_1^- s + b_2^+ s^2 + b_3^+ s^3 + b_4^- s^4 + b_5^- s^5 + \dots,$$

(9) $$N_4(s) = b_0^- + b_1^+ s + b_2^+ s^2 + b_3^- s^3 + b_4^- s^4 + b_5^+ s^5 + \dots .$$

The second set, called $\mathcal{N}_{KS}$, is made of four *segments* of polynomials, i.e. convex combinations of Kharitonov polynomials with coinciding odd or even part

(10) $$\mathcal{N}_{KS} = \{S_1^N,\ S_2^N,\ S_3^N,\ S_4^N\},$$

where the four segments are $(i = 1, \dots, 4)$

(11) $$S_i^N = \{N(s) : N(s) = \lambda N_i(s) + (1-\lambda) N_{i+1}(s),\ \ \lambda \in [0,1]\},$$

where $N_i(s) = N_{i-4}(s)$ for $i > 4$. Strictly related to the sets $\mathcal{N}_K, \mathcal{N}_{KS}, \mathcal{D}_K, \mathcal{D}_{KS}$ of the numerator and denominator families are the following subsets of transfer functions of $\mathcal{G}_I$

(12) $$\mathcal{G}_K = \{G(s) : G(s) = \frac{N(s)}{D(s)},\ N(s) \in \mathcal{N}_K,\ D(s) \in \mathcal{D}_K\},$$

(13) $$\begin{aligned}\mathcal{G}_{KS} = \ & \{G(s) : G(s) = N(s)/D(s),\\ & N(s) \in \mathcal{N}_K,\ D(s) \in \mathcal{D}_{KS} \text{ or } N(s) \in \mathcal{N}_{KS},\ D(s) \in \mathcal{D}_K\}.\end{aligned}$$

The set $\mathcal{G}_K$ is called *Kharitonov (vertex) set* of $\mathcal{G}_I$ and $\mathcal{G}_{KS}$ is denoted as the *Kharitonov segment set* of $\mathcal{G}_I$. The former set is made of 16 plants, while the latter includes 32 segments of plants.

Interval families are characterised by independency of coefficient perturbations. The paper is mainly focused on this case, for which very powerful results can be derived. However,in many practical situations this uncertainty model may result too conservative, and mutually correlated parametric perturbations need be considered. In section 6 some results are reported for the case when coefficients $a_i$, $b_i$ are multilinear or polynomial functions of a vector of parameters $p$.

## 3. Robust stability of linear feedback systems with interval plants and norm bounded uncertainty

In this section we illustrate some extremal results for stability of interval plant-controller families. It will be shown that in general Kharitonov segments have to be checked, while for particular classes of compensators Kharitonov vertices only are needed.

Consider first the case when no norm bounded perturbation affects the interval plant. This is a case which has been widely studied in the recent literature. The following extremal robust stability result given in [9] provides necessary and sufficient conditions for stability of the interval feedback system of Figure 1a.

**Theorem 1.** *(Interval Perturbation). A controller $C(s)$ stabilizes the feedback system for $G(s) \in \mathcal{G}_I$ if and only if it stabilizes the system for $G(s) \in \mathcal{G}_{KS}$.* ■

*Remark.* It can be verified that Theorem 1 holds also if $N_c(s)$ and $D_c(s)$ are real coefficient polynomials each multiplied by a fixed complex number. This fact is exploited in proving many of the subsequent results.

We now consider the case when both an interval plant and norm bounded perturbations are present. The following theorem ([13]) states a parametric version of the well-known small gain theorem for the case of multiplicative perturbations (see Figure 1b).

**Theorem 2.** *(Interval and Norm Bounded Perturbations). A controller $C(s)$ stabilizes the feedback system for $G(s) \in \mathcal{G}_I$ and $\delta$ such that $\|\delta\|_\infty \leq \epsilon$ if and only if it stabilizes $\mathcal{G}_{KS}$ and*

$$\epsilon < 1 / \max_{G(s) \in \mathcal{G}_{KS}} \|G(s)C(s)(1 + G(s)C(s))^{-1}\|_\infty . \tag{14}$$

■

The proof of Theorem 2 is based on the next lemma which allows one to transform an $H_\infty$ problem to an equivalent stability problem. Then, application of Theorem 1 and the relative extension in the remark allows one to complete the proof.

**Lemma 1.** *For a given stable proper transfer function $F(s) = N(s)/D(s)$ and a given positive number $r$, the following statements are equivalent*
*i)* $\|F(s)\|_\infty < r$
*ii)* $|F(0)| < r$ and $rD(j\omega) - e^{j\gamma} N(j\omega) \neq 0, \ \ \forall \omega, \ \forall \gamma \in [0, 2\pi]$ ■

A result similar to Theorem 2 for the case of additive norm bounded perturbation has been given in [10].

In the remaining of this section we examine extremal robust stability results for interval feedback systems with special classes of controllers $C(s)$. As intuition may suggest, by setting hypotheses on the structure of the controller, stronger results than those of the previous theorems are expected. In fact, it turns out that in several cases stability of the endpoints of a Kharitonov segment of systems may be sufficient for stability of the entire segment. In this case, robust stability of the closed loop system may be inferred by that of a finite number of feedback systems, consisting of the Kharitonov vertex plants $\mathcal{G}_K$ of the interval family.

Consider a system with $\delta = 0$ and with a compensator $C(s)$ given as

$$C(s) = k_o \frac{1}{s^\nu} \frac{s+\beta}{s+\alpha} \tag{15}$$

The following result given in [3] states conditions for robust stability of this special class of interval feedback systems.

**Theorem 3.** *(Interval Perturbation). A controller $C(s)$ with structure (15) stabilizes the feedback system for $G(s) \in \mathcal{G}_I$ if and only if it stabilizes the system for $G(s) \in \mathcal{G}_K$.* ∎

*Remark.* In [26] it had been shown that the vertices of the interval family suffice to check robust stability of the interval feedback system. Moreover, for the case when $C(s) \equiv k_o$, it had been shown [22] that a subset of $\mathcal{G}_K$ made of 8 plants is sufficient for the same purpose.

Several vertex type results may be obtained by assuming other special structures for the controller. For example, it is easy to verify that $\mathcal{G}_K$ suffices for checking robust stability of the interval feedback system if the controller $C(s)$ is such that

$$\begin{aligned} N_c(s) &= k_o N_a(s) N_e(s^2) \\ D_c(s) &= s^\nu D_a(s) D_e(s^2) \end{aligned} \tag{16}$$

where $N_a(s)$ and $D_a(s)$ are anti-stable polynomials, i.e. polynomials with all roots located in the right half plane and $N_e(s^2)$ and $D_e(s^2)$ are even polynomials. Some further advances in the direction of proving vertex type results for different or wider classes of perturbations are in [37].
When the structure of the controller is as in (16) and also unstructured uncertainty is present in the feedback loop (configuration of Figure 1b), the following theorem can be proven ([13])

**Theorem 4.** *(Interval an Norm Bounded Perturbations). A proper controller $C(s)$ with structure (16) stabilizes a strictly proper interval plant $\mathcal{G}_I$ for all $\delta$ such that $\|\delta\|_\infty \leq \epsilon$ if and only if it stabilizes $\mathcal{G}_K$ and*

$$\epsilon < 1 / \max_{G(s) \in \mathcal{G}_K} \|G(s)C(s)(1+G(s)C(s))^{-1}\|_\infty . \tag{17}$$

∎

## 4. Robust performance of interval feedback systems

*4.1 Robust performance against parametric and unstructured perturbations*
In this subsection we consider robust performance of an interval feedback system with unstructured perturbation $\delta(s)$ (see Figure 1b). We identify the robust performance problem as the following sensitivity problem.

For given bounds $\epsilon$ and $|S(j\omega)|$, find a controller $C(s)$ such that

1. $C(s)$ stabilizes the feedback system for $G(s) \in \mathcal{G}_I$ and $\|\delta\|_\infty \leq \epsilon$.

2. $\sup_{G(s)\in\mathcal{G}_I,\ \|\delta\|_\infty\leq\epsilon} \|S^{-1}(s)(1+G(s)(1+\delta(s))C(s))^{-1}\|_\infty \leq 1$.

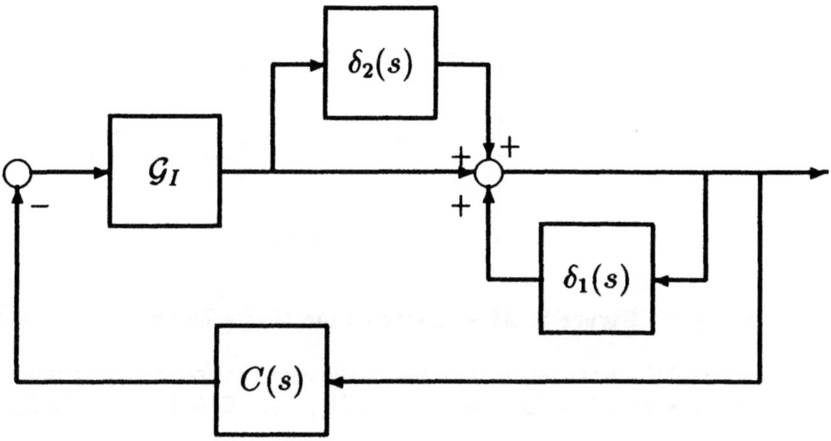


Figure 2. Block scheme for robust performance analysis.

The basic result of this section states that a robustly stabilizing controller $C(s)$ guarantees robust performance if and only if it guarantees robust performance for $G(s) \in \mathcal{G}_{KS}$. To show this result, we use the classical formulation of the robust performance problem [19]. Thus, we make reference to the scheme of Figure 2, where the block $\delta_2(s)$ accounts for unstructured perturbations, while the block $\delta_1(s)$ represents a fictitious perturbation which allows us to impose requirements on the output sensistivity function. By reducing the feedback system of Figure 2 to the configuration of Figure 3, we obtain

$$M(s) = \begin{pmatrix} (1+G(s)C(s))^{-1} & (1+G(s)C(s))^{-1} \\ -C(s)G(s)(1+G(s)C(s))^{-1} & -C(s)G(s)(1+G(s)C(s))^{-1} \end{pmatrix}. \tag{18}$$

Consider now the three classes of matrices $M(s)$ corresponding to the interval plant, the Kharitonov segments and the Kharitonov vertex plants of the family, respectively

$$\begin{aligned} \mathcal{M} &= \{M(s) : G(s) \in \mathcal{G}_I\} \\ \mathcal{M}_{KS} &= \{M(s) : G(s) \in \mathcal{G}_{KS}\} \\ \mathcal{M}_K &= \{M(s) : G(s) \in \mathcal{G}_K\}\,. \end{aligned} \tag{19}$$

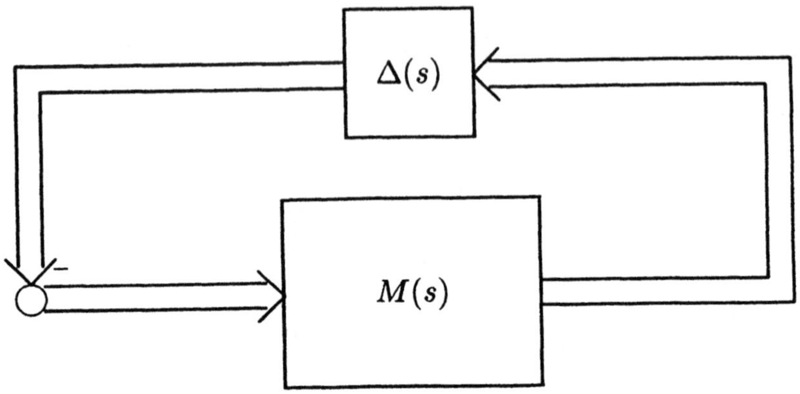


Figure 3. $M - \Delta$ interconnection scheme.

For ease of illustration, we assume in the sequel a frequency independent bound on the fictitious perturbation $\delta_1(s)$, i.e. $S(j\omega) = \alpha$. Define the matrix $\Delta(s)$

$$\Delta(s) \;=\; \text{diag}[\alpha, \epsilon]\Theta(s),$$

where

$$\Theta(s) = \text{diag}[\alpha^{-1}\delta_1(s), \epsilon^{-1}\delta_2(s)].$$

We know [19] that for a given $G(s) \in \mathcal{G}_I$, $C(s)$ solves the robust performance problem if and only if

$$\mu[M'(j\omega)] < 1, \quad \forall\omega \tag{20}$$

where

$$M'(j\omega) = M(j\omega)\text{diag}[\alpha, \epsilon]$$

and

$$\mu[M'(j\omega)] \doteq \left\{ \min_{\bar{\sigma}(\Theta(j\omega))\leq 1} \bar{\sigma}(\Theta(j\omega)) : \; \det(I + M(j\omega)\Delta(j\omega)) = 0 \right\}.^{-1}$$

Hence, robust performance of the feedback system in the face of both unstructured and parametric perturbations is achieved if and only if

$$\sup_{M \in \mathcal{M}} \sup_{\omega} \mu[M'(j\omega)] \; < 1\,. \tag{21}$$

We point out that transfer function coefficient perturbations may be included in the $M - \Delta$ representation of Figure 3. Nevertheless, the reason why here we prefer to keep them separated from unstructured perturbations and performance specifications is that proceeding along this way we

are able to exploit their structure to reduce the dimensionality of the resultant $\mu$ problem.
Now, we can state the following theorem on robust performance ([14])

**Theorem 5.** *A compensator $C(s)$ solves the robust performance problem for $G(s) \in \mathcal{G}_I$ if and only if it solves the problem for $G(s) \in \mathcal{G}_{KS}$.* ■

Theorem 5 holds even if frequency dependent perturbation bounds are considered. On the contrary, the validity of the successive theorem giving a vertex result for the case where the structure (16) is assumed for the controller, is restricted to the case where constant bounds are allowed ([13]).

**Theorem 6.** *A compensator $C(s)$ with structure (16) solves the robust performance problem with $S(j\omega) = \alpha$ for $G(s) \in \mathcal{G}_I$ and $\|\delta\|_\infty \leq \epsilon$, if and only if it solves the problem for $G(s) \in \mathcal{G}_K$.* ■

*4.2 Frequency response of interval plant-controller families*

In this subsection some results on the locus of the Nyquist plots of an interval plant-controller family of transfer functions are illustrated. These results deserve attention because, beyond their own meaning, they provide as byproduct extremal results for some of the most used performance parameters in classical control design. Moreover, they allow us to deduce several robust strict positive realness results and provide alternative views of many of the results stated in the previous sections.

Before stating the main result of this subsection we need to introduce some sets and maps. Consider the family of loop transfer functions

$$\mathcal{L} = \{L(s) : L(s) = C(s)G(s),\ G(s) \in \mathcal{G}_I\}\ . \tag{22}$$

For a given $\omega$, define the map $L(j\omega)$ associating to each $G(s) \in \mathcal{G}_I$ the complex value $C(j\omega)G(j\omega)$. This map is defined only if $D_c(j\omega)D(j\omega) \neq 0$. Denote by $I_L(\omega)$ and $I_L^{ks}(\omega)$ the value sets of the families $\mathcal{G}_I$ and $\mathcal{G}_{KS}$, respectively. $\partial I_L(\omega)$ denotes the boundary of the set $I_L(\omega)$. Let us assume for the moment that $D_c(j\omega)D(j\omega) \neq 0$, $\forall D(s) \in \mathcal{D}_I$. Under this condition, the set $I_L(\omega)$ is bounded and the following theorem can be proven [45]

**Theorem 7.** *Let $\omega$ be such that $D(j\omega)D_c(j\omega) \neq 0$. Then*

$$\partial I_L(\omega) \subseteq I_L^{ks}(\omega)\ . \tag{23}$$

■

Important conclusions can be drawn from Theorem 7. Observe that at each frequency $\omega$ such that no transfer function of the open loop family $\mathcal{L}$ has a pole at $s = j\omega$, the value set boundary $\partial I_L(\omega)$ is determined by the Kharitonov segment set $\mathcal{G}_{KS}$. Hence, the locus of Nyquist plots of transfer functions of the family is bounded by the set of plots corresponding to the

32 Kharitonov segments of plants $\mathcal{G}_{KS}$. An immediate consequence of this fact is that extremal phase and gain margins or maximum loop sensitivity and complementary sensitivity peaks for an interval plant-controller family are achieved for a plant belonging to one of the 32 Kharitonov segments of the interval family.

## 5. Robust strict positive realness

Absolute stability of nonlinear systems as well as convergence and performance analysis of adaptive schemes are typical areas where strict positive realness (SPR) is a key property. In this section we give some extremal results for robust strict positive realness of families of interval plant-controller transfer functions. Strict positive realness of a plant requires that the plant is stable and that its phase does not exceed in magnitude $\pi/2$, i.e. the Nyquist plot lies in the open right half plane.

Consider the interval plant-controller family defined in (22). A problem which is frequently of interest is the real shifted positive realness of transfer functions. By real shift strict positive realness of a transfer function $L(s)$ we mean SPR of the transfer function $k^{-1} + L(s)$, with $k$ a given real number (we will denote this property by kSPR ). One of the main problems motivating the study of robust SPR and kSPR is the presence of nonlinearities in physical plants which cannot be accounted for by parametric linear models. It is well known that a large class of uncertain nonlinear systems may be represented as Lur'e systems (see Figure 4). The only knowledge assumed on the memoryless nonlinear feedback of these systems is that the function $\phi(\cdot)$ lies in a given sector $[0, k]$.

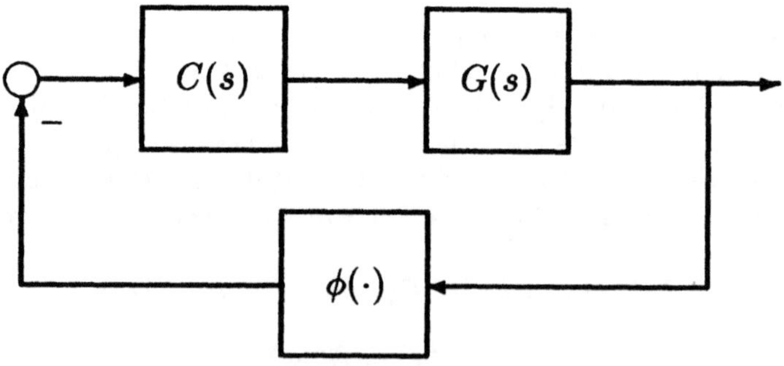


Figure 4. Lur'e control system.

Stability in the large of the system for any nonlinear function in the sector, i.e. *absolute stability* of the Lur'e system, can be checked via a suitable SPR test through any of the classical frequency domain criteria, like circle or Popov. Studying absolute stability for an *interval Lur'e system*, i.e. a Lur'e system with an interval plant in the forward path, leads to a robust strict positive realness problem.

Define two extremal sets of transfer functions of the family $\mathcal{L}$

$$
(24) \qquad \begin{aligned} \mathcal{L}_{KS} &= \{L(s) : L(s) = C(s)G(s),\ G(s) \in \mathcal{G}_{KS}\} \\ \mathcal{L}_K &= \{L(s) : L(s) = C(s)G(s),\ G(s) \in \mathcal{G}_K\}\,. \end{aligned}
$$

In view of the frequency domain criteria for absolute stability, the first issue we are interested in is finding conditions under which the family $\mathcal{L}$ is kSPR. The following theorem [46] answers this question

**Theorem 8.** *The interval plant-controller family $\mathcal{L}$ is kSPR if and only if $\mathcal{L}_{KS}$ is kSPR.* ■

*Remark.* Theorem 8 allows one to build immediately a robust version of the well-known circle criterion for absolute stability. In particular, it ensures that if the Lur'e systems corresponding to the Kharitonov segments of the interval plant satisfy the circle criterion, then the whole interval Lur'e system is absolutely stable in the sector $[0, k]$.

Several types of vertex results can be obtained also for the kSPR property whenever $k = \infty$ and/or special structures of the controller $C(s)$ are assumed. Here, we report some of the most important of them.

First, we consider the SPR property, i.e. the case $k^{-1} = 0$. Regarding SPR of an interval plant, references [15,8] solved the problem, reaching the conclusion that $\mathcal{G}_K$ is sufficient for SPR of $\mathcal{G}_I$. Later references ([11,46]) have shown by different approaches that only 8 out of the 16 Kharitonov plants guarantee SPR of $\mathcal{G}_I$. Moreover, it turns out that $\mathcal{L}_K$ is sufficient for SPR of the whole interval plant-controller family $\mathcal{L}$ ([46]). For the case when $k^{-1} \neq 0$, it can be shown [11] that, if $C(s)$ is a proportional controller, then kSPR of $\mathcal{L}_K$ is necessary and sufficient for kSPR of $\mathcal{L}$. If the sign of the real shift $k^{-1}$ is known, then only a subset of 12 Kharitonov plants suffices for ensuring kSPR of the family [46].

The next theorem shows that a vertex result for the kSPR problem still holds for the more general case when $C(s)$ is a fixed order order polynomial, i.e. $C(s) \equiv N_c(s)$ ([13]).

**Theorem 9.** *Given a strictly proper interval plant $\mathcal{G}_I$ and a fixed polynomial $C(s) \equiv N_c(s)$, a necessary and sufficient condition for kSPR of the interval plant-controller family $\mathcal{L}$ is that $\mathcal{L}_K$ is kSPR.* ■

This theorem allows us to give a robust version of the well-known Popov criterion for absolute stability of interval Lur'e control systems. The Popov criterion for a Lur'e system with a stable linear feedforward path $G(s)$ states that the system is absolutely stable in the sector $[0, k]$, if there exists a real $q$ verifying the following positiveness condition (Popov condition)

$$k^{-1} + \text{Re}[(1+j\omega q)G(j\omega)] > 0, \ \forall\omega \ . \tag{25}$$

Theorem 9, which obviously holds for $C(s) = 1 + sq$, allows us to state a robust version of the Popov criterion

**Corollary 1.** *A Lur'e control system is absolutely stable in the sector* $[0,k]$ *for* $G(s) \in \mathcal{G}_I$ *if there exists a real* $q$ *verifying the Popov condition for* $G(s) \in \mathcal{G}_K$. ∎

## 6. Robust stability and performance with nonlinear perturbations

In this section we consider more general classes of parametric perturbations.

First, we focus on a special nonlinear perturbation structure, arising from the cascade of two interval plants, and we report an extremal result for this case.

Let $\mathcal{G}_I^p$ represent a family of proper plants, defined as follows

$$\mathcal{G}_I^p = \{G(s) : G(s) = \frac{N^1(s)N^2(s)}{D^1(s)D^2(s)}, \ N^i(s) \in \mathcal{N}_I^i, \ D^i(s) \in \mathcal{D}_I^i\} \tag{26}$$

where $\mathcal{N}_I^i$ and $\mathcal{D}_I^i$ $(i \in \{1,2\})$ are families of interval polynomials.

In a fashion similar to Section 2, we can define the sets $\mathcal{N}_K^i$, $\mathcal{D}_K^i$, $\mathcal{N}_{KS}^i$, $\mathcal{D}_{KS}^i$, and

$$\begin{aligned} \mathcal{G}_{KS}^p = \ & \{G(s) : G(s) = N^1(s)N^2(s)/D^1(s)D^2(s), \\ & N^i(s) \in \mathcal{N}_K^i, \ D(s)^i \in \mathcal{D}_{KS} \text{ or } N(s)^i \in \mathcal{N}_{KS}^i, \ D(s)^i \in \mathcal{D}_K^i\}. \end{aligned} \tag{27}$$

The following theorem ([13]) states conditions for robust stability of a feedback system with the class of multilinear perturbations introduced above.

**Theorem 10.**
*i) (Interval Perturbation). A controller* $C(s)$ *stabilizes the feedback system for* $G(s) \in \mathcal{G}_I^p$ *if and only if it stabilizes the system for* $G(s) \in \mathcal{G}_{KS}^p$.
*ii) (Interval and Norm Bounded Perturbations). A controller* $C(s)$ *stabilizes the feedback system for* $G(s) \in \mathcal{G}_I^p$ *and* $\|\delta\|_\infty \le \epsilon$ *if and only if it stabilizes* $\mathcal{G}_{KS}^p$ *and*

$$\epsilon < 1 / \max_{G(s) \in \mathcal{G}_{KS}^p} \|G(s)C(s)(1+G(s)C(s))^{-1}\|_\infty \ . \tag{28}$$

∎

*Remark.* The result i) of Theorem 10 appeared in [12]. Similar results can be obtained for more general classes of systems involving sums and products of independent interval polynomials.

Now we consider the more general case in which uncertainty on the plant is related to uncertainty in a parameter vector $p$ entering (possibly nonlinearly) in the coefficients of $G(s)$. In fact, this case often arises when $G(s)$ represents the transfer function of a model built on the base of internal physical knowledge of the plant. In such cases the model usually depends on some uncertain physical parameters which, even for very simple physical models, enter nonlinearly in tranfer function coefficients.

The parameter vector $p$ usually is not exactly known, but is known to belong to an uncertainty set $\Pi$ represented as

$$\Pi = \{p : \|p - p^o\|_\infty \leq \sigma \} \tag{29}$$

Let us denote by $\mathcal{G}_p$ the family of perturbed plants described as:

$$\mathcal{G}_p = \{G(s;p) : G(s;p) = N(s;p)/D(s;p),\ p \in \Pi \} \tag{30}$$

The answer to the basic robustness questions is clearly much harder than for interval plants. An obvious approach is to reduce to the interval plant case by evaluating the range of variation of the coefficients of $G(s)$. In this way conservative results are obtained.

Several methods have been proposed, in order to obtain nonconservative results ([16,39,50,28,48,49,35,1]. They are essentially related to the positivity of a suitable function $f(\omega;p)$ over the domain $B = \Omega \times \Pi$ ($\Omega$ is the range of active frequencies $\omega$).

Under the assumption that $C(s)$ achieves stability and performance specification for the "nominal plant" ($p = p^0$, $\delta = 0$), specific expressions of $f(\omega;p)$ can be obtained by the following results.

Let $P_{cl}(s;p)$ be the system closed loop characteristic polynomial of degree $\nu$ and $H_\nu(p)$ the corresponding $\nu$-th Hurwitz determinant.

**Proposition 1.** *(Robust Stability for Nonlinear Parametric Perturbation). The following statements are equivalent:*
*i) $C(s)$ stabilizes the feedback system for $G(s) \in \mathcal{G}_p$.*
*ii)* $|P_{cl}(j\omega;p)| > 0 \ \ \forall p \in \Pi,\ \ \omega \in R^+$
*iii)* $H_\nu(p) > 0 \ \ \forall p \in \Pi$ ■

Inequality *ii*) follows from the zero exclusion condition on the closed loop characteristic polynomial. In fact, $\min_{p\in\Pi} |P_{cl}(j\omega;p)|$ represents the euclidean distance of the value set from the origin. Inequality *iii*) follows from Hurwitz criterion of stability and Orlando's formula.

**Proposition 2.** *(Robust Stability for Nonlinear Parametric and Norm Bounded Perturbations). The following statements are equivalent:*
*i) $C(s)$ stabilizes the feedback system for $G(s) \in \mathcal{G}_p$ and $\|\delta\|_\infty \leq \epsilon$.*
*ii)* $1 - \epsilon|G(j\omega;p)C(j\omega)(1 + G(j\omega;p)C(j\omega))^{-1}| > 0 \ \ p \in \Pi,\ \omega \in R^+$. ■

**Proposition 3.** *(Robust Performance for Nonlinear Parametric and Norm Bounded Perturbation). The following statements are equivalent:*
*i)* $C(s)$ *achieves robust performance for* $G(s) \in \mathcal{G}_p$ *and* $\|\delta\|_\infty \leq \epsilon$.
*ii)* $1 - \alpha|G(j\omega;p)C(j\omega)(1+G(j\omega;p)C(j\omega))^{-1}| - \epsilon|(1+G(j\omega;p)C(j\omega))^{-1}| > 0, \quad \forall p \in \Pi, \quad \omega \in R^+$. ■

Several typical performance specifications, other than sensitivity, can be considered, giving rise to similar inequalities ([34]).

Positivity of such functions can be checked by computing their minimum. The key point is that they may be multimodal and their global solutions are needed, in order to have the necessary and sufficient properties previously cited.

Finding global minima of multimodal optimization problems is known to be a formidable task. Methods for dealing with such problems are usually based on random search algorithms (see e.g. [29,23]). However, these methods guarantee convergence to the global minimum only in probability and, more importantly, they do not give any measure on how far the obtained solution is from the true global minimum.

Better results can be obtained if the functional structure of $f(\omega;p)$ is exploited. If parameter vector $p$ enter polynomially in transfer function coefficients, then functions $f(\omega;p)$ are polynomial in $p$ and $\omega$. In such cases, the problem may be still nonconvex, admitting local extrema, but algorithms exist, able to produce a sequence of upper and lower bounds converging with certainty to global extrema, ([16,39,48,49,1]. In particular, the algorithm given in [49], integrating the signomial algorithm proposed in [48] and Bernstein polynomial expansion, proved to be able to solve non trivial robust stability and performance problems [20]. The most interesting feature of Bernstein expansion is that it provides an easy test to check if the polynomial achieves the minimum on one of the vertices of the uncertainty domain.

Let the polynomial $f(\omega;p)$ be expressed as

$$f(x) = \sum_{i_1,\ldots,\, i_q=0}^{r} a_{i_1 i_2 \cdots i_q} \prod_{j=1}^{q} x_j^{i_j}\ . \tag{31}$$

where $x = [\omega, p^t]^t$ has dimension $q$, $r$ is the maximum degree of variables $x_i$ and $a_{i_1 i_2 \cdots i_q}$ are real coefficients. Without loss of generality, we consider a 'unit' hypercube domain (called $B_u$) generated by $q$ cartesian products of the closed interval $[0,1]$ by itself in parameter space. Define the set of $q$-tuples $K = \{(i_1,\ldots,i_q) : i_1 = 0,\ldots,r,\ \ldots,\ \ldots, i_q = 0,\ldots,r\}$ .

The Bernstein coefficients are defined as

$$(32)\quad \beta_{i_1\ldots i_q} = \sum_{l_1=0}^{i_1} \cdots \sum_{l_q=0}^{i_q} \left( \prod_{j=1}^{q} \binom{i_j}{l_j} \right) \rho_{l_1\ldots l_q} \alpha_{l_1\ldots l_q}, \quad (i_1 \cdots i_q) \in K$$

where

$$(33)\qquad \rho_{l_1\ldots l_q} = \left( \prod_{j=1}^{q} \binom{r}{l_j} \right)^{-1}, (l_1 \ldots l_q) \in K \; .$$

and $\alpha_{l_1\ldots l_q} = 0$ if some $l_j$ is such that $l_j > r$.

The next theorem holds ([21, 49].

**Theorem 11.**
*i)* $\min_{(i_1\cdots i_q)\in K} \beta_{i_1\ldots i_q} \leq \min_{x\in B_u} f(x)$.
*ii) Equality holds in i) if and only if the minimum of Bernstein coefficients* $\beta_{i_1\ldots i_q}$ *is achieved for each* $i_j$ *equals* 0 *or* $k$.
*iii) If the minimum of* $\beta_{i_1\ldots i_q}$ *is achieved for each* $i_j$ *equals* 0 *or* $k$*, the global minimum of* $f(x)$ *is achieved at a vertex of* $B_u$. ■

The algorithm proposed in [49], though comparing favourably with other algorithms proposed in the literature, does not fully exploit the properties of Bernstein coefficients. A better way of taking advantage of Theorem 11 is to combine it with a global random search algorithm. These algorithms have the major drawback of being unable to detect if they actually reach the global extremum. Theorem 11 can be used for such a test. In fact, suppose that the algorithm stops at a (possibly local) minimum point $x^*$. Subdivide $B_u$ with hyperplanes parallel to the faces of $B_u$ and containing $x^*$. The point $x^*$ is a global minimum point if and only if it is a minimum point of all splitted hyperrectangles, and this can be checked by applying Theorem 11. If *ii)* does not hold for some subdivision, this can be further subdivided and the procedure iterated.

## ACKNOWLEDGEMENTS

This work was partially supported by funds of Ministero dell'Università e della Ricerca Scientifica e Tecnologica and of CNR Special Project "Algoritmi ed Architetture per l' Identificazione e il Controllo Robusto e Adattivo".

## REFERENCES

[1] V. Balakrishnan, S. Boyd and S. Balemi, "Branch and bound algorithm for computing the minimum stability degree of parameter dependennt linear system", *Proc. International Workshop on Robust Control*, San Antonio (USA), CRC Press, 1991.

[2] B. R. Barmish, "A generalization of Kharitonov's four polynomial concept for robust stability problems with linearly dependent coefficient perturbations," *IEEE Trans. Automat. Contr.*, vol. 34, pp. 157-165, 1989.

[3] B. R. Barmish, C. V. Hollot, F. J. Kraus and R. Tempo, "Extreme point results for robust stabilization of interval plants with first order compensators", *Proc. American Control Conference*, San Diego (USA), 1990.

[4] A. C. Bartlett, "Vertex Results for the Steady State Analysis of Uncertain Systems," *Proceedings of the 29th Conference on Decision and Control*, pp. 436-441, Hawaii, 1990.

[5] A. C. Bartlett, C. V. Hollot and L. Huang, "Root locations of an entire polytope of polynomials: it suffices to check the edges", *Mathematics of Control, Signals and Systems*, vol. 1, pp. 61-71, 1988.

[6] S. P. Bhattacharyya, *Robust Stabilization Against Structured Perturbations*. Lecture Notes in Control and Information Sciences, 99, Springer Verlag, 1987.

[7] S. P. Bhattacharyya and L. Keel Eds. , *Robust Control*, CRC Press, 1991.

[8] N. K. Bose and J. F. Delansky, " Boundary implications for interval positive rational functions" *IEEE Trans. on Circuit and Systems*, vol. CAS-36, pp. 454-458, 1989.

[9] H. Chapellat and S. P. Bhattacharyya, "A Generalization of Kharitonov's Theorem: Robust Stability of Interval Plants," *IEEE Trans. on Automatic Control*, vol. 34, no. 4, March 1989.

[10] H. Chapellat, M. Dahleh, and S. P. Bhattacharyya, "Robust Stability under Structured and Unstructured Perturbations," *IEEE Trans. on Automatic Control*, vol. 35, no. 10, pp. 1100-1108, October 1990.

[11] H. Chapellat, M. Dahleh and S. P. Bhattacharyya, "On Robust Nonlinear Stability of Interval Control Systems," *IEEE Trans. on Automatic Control*, vol. 36, no. 1, January 1991.

[12] H. Chapellat, M. Dahleh, and S. P. Bhattacharyya, " Extremal Manifolds in Robust Stability," *Report UCSB-ME-91-3*, 1991.

[13] M. Dahleh, A. Tesi and A. Vicino, "Extremal properties for robust control of interval plants", *Res. Rep. RT 23/91, DSI, Università di Firenze*, August 1991.

[14] M. Dahleh, A. Tesi and A. Vicino, "Extremal properties for the parametric robust performance problem", *Proc. 30th Conference on Decision and Control*, Brighton (UK), 1991.

[15] S. Dasgupta, "A Kharitonov like theorem for systems under nonlinear passive feedback", *Proc. 26th Conference on Decision and Control*, Los Angeles (USA), pp. 2062-2063, 1987.

[16] R. R. E. de Gaston and M. G. Safonov, "Exact calculation of the multiloop stability margin", *IEEE Trans. on Automat. Contr.*, vol. AC-33, pp. 156-171, 1988.

[17] J. C. Doyle, " Analysis of Feedback Systems with Structured Uncertainties," *IEE Proceedings*, vol. 129, Part D, No. 6, 1982.

[18] J. C. Doyle and G. Stein, "Multivariable feedback design: concepts for a classical/modern synthesis", *IEEE Trans. on Automat. Contr.*, vol. AC-26, pp. 4-16, 1981.

[19] J. C. Doyle, J. E. Wall and G. Stein, "Performance and robustness analysis for structured uncertainty", *Proc. 21-st Conference on Decision an Control*, Orlando (USA), pp. 629-636, 1982.

[20] G. Fiorio, S. Malan, M. Milanese and A. Vicino, "Guaranteed specification control design of inertial platforms for high accuracy calibration device", *Proc. of 29th IEEE Conference on Decision and Control*, Honululu (USA), 1990.

[21] J. Garloff, "Convergent bounds for the range of multivariate polynomials", in *Lecture Notes in Computer science*, vol. 212, *Interval Mathematics 1985*, Springer-Verlag, pp. 37-56, 1985.

[22] B. K. Ghosh, "Some new results on the simultaneous stabilization of a family of single input, single out- put systems", *Systems and Control Letters*, vol. 6, pp. 39-45, 1985.

[23] D.E. Goldberg, *Genetic Algorithms in Search, Optimization and Machine Learning*, Addison Wesley, 1989.

[24] D. Hinrichsen and A. J. Pritchard, "New robustness results for linear systems under real perturbations", *Proc. 27-th Conf. Decision and Control*, Austin (USA), pp. 1375-1379, 1988.

[25] D. Hinrichsen and B. Martensson, Eds., *Control of Uncertain Systems*, Birkhauser, 1990.

[26] C. V. Hollot and F. Yang, "Robust stabilization of interval plants using lead or lag compensators", *Systems and Control Letters*, vol. 14, pp. 9-12, 1990.

[27] V. L. Kharitonov, "Asymptotic Stability of an Equilibrium Position of a Family of Systems of Linear Differential Equations," *Differntial'nye Uraveniya*, vol. 14, no. 11, pp. 1483–1485, 1978.

[28] F.J. Kraus, M. Mansour and B.D.O. Anderson, "Robust stability of polynomials with multilinear parameter dependence", *Robustness in Identification and Control*, M. Milanese, R. Tempo and A. Vicino Eds., Plenum Press, 1989.

[29] P.J.M. van Laaroven and E.H Aarts, *Simulated Annealing: Theory and Applications*. Reidel Publishing Company, 1987.

[30] D. C. McFarlane, K. Glover, *Robust Controller Design Using Normalized Coprime Factor Plant Descriptions*, Lecture Notes in Control and Information Sciences, Springer Verlag, 1990.

[31] G. Menga, M. Milanese and A. Negro, "Min-max quadratic cost control of systems described by approximate models", *IEEE Trans. Automat. Control, AC-21,* pp.651-659, 1976.

[32] R. J. Minnichelli, J. J. Anagnost and C. A. Desoer, "An elementary proof of Kharitonov's stability theorem with extensions", *IEEE Trans. on Autom. Contr.*, vol. 34, pp. 995-998, 1989.

[33] M. Milanese, R. Tempo and A. Vicino Eds., *Robustness in Identification and Control*, Plenum Press, 1989.

[34] M. Milanese, G. Fiorio, S. Malan and A. Vicino, "Robust performance design of nonlinearly perturbed control systems". *Proc. International Workshop on Robust Control*, San Antonio (USA), CRC Press, 1991.

[35] T.M. Murdock, W.E. Schmitendorf and S. Forrest, "Use of genetic algorithm to analyze robust stability problems", Proc. ACC, Boston, pp.886-889, 1991.

[36] M. P. Polis, W. Olbrot and M. Fu, "An overview of recent results on the parametric approach to robust stability", *Proc. 28-th Conf. Decision and Control*, Tampa (USA), pp. 23-29, 1989.

[37] A. Rantzer, "Stability conditions for polytopes of polynomials", *Techn. Rep., Dept. of Mathematics*, Royal Inst. of Techn., Stockolm (S), 1990.

[38] M. G. Safonov, *Stability and Robustness of Multivariable Feedback Systems*, Cambridge, MA: MIT Press, 1980.

[39] A. Sideris and R. S. Sánches Peña, "Fast computation of the multivariable stability margin for real interrelated uncertain parameters", *IEEE Trans. on Automat. Contr.*, vol. AC-34, pp. 1272-1276, 1989..

[40] D. D. Šiljak, "Parameter space methods for robust control design: A guided tour", *IEEE Trans. on Automat. Contr.*, vol. AC-34, pp. 674-688, 1989.

[41] D. D. Šiljak,"Polytopes of nonnegative polynomials", *Proc. ACC*, Pittsburgh, (U.S.A), 1989.

[42] A. Tesi and A.Vicino, "Robustness analysis for uncertain dynamical systems with structured perturbations", *Proc. 27th CDC*, Austin (USA),pp. 519-525, 1988 and *IEEE Trans. on Automat. Contr.*, 2, pp. 186-191, 1990.

[43] A. Tesi and A. Vicino, "Design of optimally robust controllers with few degrees of freedom", *Proc. IFAC Symp. on Design Methods of Control Systems*, Zurich (CH), pp. 264-269, 1991.

[44] A. Tesi and A. Vicino, "Robust absolute stability of Lur'e control systems in parameter space", *Automatica*, Vol. 27, pp.147-151, 1991.

[45] A. Tesi and A. Vicino, "Kharitonov segments suffice for frequency response analysis of interval plant-controller families", *Proc. International Workshop on Robust Control*, San Antonio (USA), CRC Press, 1991.

[46] A. Vicino, and A. Tesi, "Robust strict positive realness: new results for interval plant plus controller families", *Proceedings of the 30th Conference on Decision and Control*, Brighton (UK), 1991.

[47] A. Tesi and A. Vicino, "Robust stability of state space models with structured uncertainties", *IEEE Transactions on Automatic Control*, No. 2, pp. 191-195, 1990.

[48] A. Vicino, A. Tesi and M. Milanese, "Computation of nonconservative stability perturbation bounds for systems with nonlinearly correlated uncertainties", *IEEE Trans. on Automat. Contr.*, vol. AC-35, pp. 835-841, 1990.

[49] A. Vicino, M. Milanese, "Robust stability of linear state space models via Bernstein polynomials" in *Control of Uncertain Systems*, D. Hinrichsen and B. Martensson Eds., Birkhauser, pp. 329-343, 1990.

[50] E. Zeheb, "Necessary and sufficient condition for robust stability of a continuous system- The continuos dependence case illustrated via multinear dependency", *IEEE Trans. on Circuits and Systems*, vol. CAS-37, pp. 47-35, 1990.

Giovanni Fiorio
Dipartimento di Automatica e Informatica, Politecnico di Torino, Italy.

Mario Milanese
Dipartimento di Automatica e Informatica, Politecnico di Torino, Italy.

Antonio Vicino
Dipartimento di Ingegneria Elettrica, Università di L'Aquila, Italy.

# Design and Comparison of Globally Stabilizing Controllers for an Uncertain Nonlinear System

R. A. FREEMAN and P. V. KOKOTOVIC

**Abstract**

We review various integrator backstepping design procedures for generating globally stabilizing controllers for uncertain nonlinear systems. By way of a case study of a simple planar system, we demonstrate how these procedures can be combined to produce new classes of stabilizing controllers. We present eight different backstepping controllers for the planar system and compare their structural properties and simulated performance.

## 1. Introduction

In the past decade, major advances have been made in feedback design of nonlinear systems. To a large extent this progress is due to the contributions of the University of Rome control theory group founded by Professor Ruberti. Among the current research topics of this group is the robust control of nonlinear systems.

In this paper we address the problem of finding globally asymptotically stabilizing controllers for plants having uncertain nonlinearities. Our first goal is to review existing design procedures leading to such controllers, unify their stability proofs, reveal their common qualitative properties, and illustrate how their combinations produce new classes of stabilizing controllers. Our second goal is to compare the performance of these controllers via simulation. For fairly general nonlinear systems, such a task would involve unmanageably large numbers of different controller structures and design parameters. We therefore present a case study, examining eight different stabilizing controllers for the simple planar system

$$\text{(1)} \qquad \begin{cases} \dot{x}_1 &= px_1^2 + x_2 \\ \dot{x}_2 &= u \end{cases}$$

The key feature of this stabilization problem is that the scalar parameter $p$, which multiplies the polynomial nonlinearity $x_1^2$, is unknown. Some of these eight controllers require knowledge of bounds on $p$ and we therefore assume $|p| \leq \bar{p}$ where $\bar{p}$ is known. This planar problem is simple enough to make our second goal feasible, yet rich enough that in presenting controllers for this system we achieve our first goal.

The desired controllers have access to the values of both states and must globally asymptotically stabilize the origin of (1). Such controllers are referred to as "$u$-controllers" to distinguish them from "$v$-controllers" for the scalar system

$$\text{(2)} \qquad \dot{x}_1 = px_1^2 + v$$

The $v$-controllers are designed to globally asymptotically stabilize the origin $x_1 = 0$ of (2). This is a preliminary step which treats the state $x_2$ in the original system (1) as the control signal. Given such $v$-controllers for the scalar system (2) along with their associated Lyapunov functions, the ***integrator backstepping*** procedure (terminology introduced in [6]) is then used to generate the desired stabilizing $u$-controllers for the planar system (1). This two-step design takes advantage of the fact that the unknown parameter $p$ in the $v$-system (2) trivially satisfies the *matching condition* [7,2,1], i.e. it enters the same state equation as the control $v$. Therefore we may apply *Lyapunov redesign* methods [7] to generate stabilizing $v$-controllers for this matched system. In Section 2 we describe three such Lyapunov redesign controllers for (2) based on essentially the same Lyapunov function: the ***continuous v-controller***, the ***switching v-controller***, and the ***adaptive v-controller***. The continuous $v$-controller uses smooth functions to bound the uncertain nonlinearity and make the Lyapunov derivative negative; for our polynomial system this $v$-controller is a polynomial in $x_1$. The switching $v$-controller uses a smooth approximation to a switch in order to make its Lyapunov derivative negative. Finally, the adaptive $v$-controller employs dynamic feedback to make negative the derivative of an augmented Lyapunov function. These by no means exhaust the list of all $v$-controllers available through Lyapunov redesign or other methods, but they represent three fundamentally different approaches to making negative the derivative of the same (or augmented) Lyapunov function.

In Section 3 we employ four versions of backstepping to generate stabilizing $u$-controllers for the original system (1): ***continuous backstepping***, *switching backstepping*, *adaptive backstepping*, and *adaptive double backstepping*. The reader will notice that these names correspond to those of the $v$-controllers above; the reason for this similar terminology will be made clear in Section 3. Continuous backstepping is applied only to the continuous $v$-controller and generates a $u$-controller which we call the CC controller (for Continuous $v$-controller plus Continuous backstepping). Switching and adaptive backstepping, however, are each applied to all three of the above $v$-controllers and thus together generate six different $u$-controllers (respectively labelled CS, SS, AS, CA, SA, and AA). Adaptive double backstepping "steps back" through two integrators at a time, which for our planar system eliminates the need for the preliminary $v$-step and therefore directly generates what we call the DB controller. The four "mixed" $u$-controllers (CS, AS, CA, and SA) appear for the first time in this paper. They represent an approach in which new classes of stabilizing controllers are designed by combining different backstepping techniques. This approach is not restricted to the planar system of our case study.

We compare these eight controllers in Sections 4 and 5. Section 4 examines the different structural properties of the controllers, noting important advantages or disadvantages of each and pointing out any structural similarities. Section 5 attempts to compare their performance via simulation, concentrating on a comparison of control effort required for stabilization.

## 2. $v$-Controllers for the Scalar System

*Continuous $v$-controller* [8]. Given the Lyapunov function $V_1(x_1) = \frac{1}{2}x_1^2$, we will try to make negative its time-derivative with respect to (2) by choosing the smooth static feedback $v$-controller

$$v = -c_1x_1 - \tfrac{1}{2}\bar{p}\left(\alpha x_1 + \frac{1}{\alpha}x_1^3\right) \tag{3}$$

where $c_1$ and $\alpha$ are positive design parameters. Computing the Lyapunov derivative $\dot{V}_1$ we obtain [1]

$$\begin{aligned} \dot{V}_1 &= px_1^3 + vx_1 && (4)\\ &\leq |p|\,|x_1^3| + vx_1 && (5)\\ &\leq \tfrac{1}{2}\bar{p}\left(\alpha x_1^2 + \frac{1}{\alpha}x_1^4\right) + vx_1 && (6)\\ &= -c_1x_1^2 && (7)\end{aligned}$$

[1]Throughout, $\dot{V}_1$ will indicate the time-derivative of $V_1$ with respect to the system (2) and its associated $v$-controller, while $\dot{V}_2$ will indicate the time-derivative of $V_2$ with respect to the system (1) and its associated $u$-controller.

where (6) follows from the algebraic inequality

$$ab \le \tfrac{1}{2}\left(\gamma a^2 + \frac{1}{\gamma} b^2\right) \quad , \quad \gamma > 0 \tag{8}$$

Thus $\dot{V}_1$ is negative-definite in $x_1$ proving that this $v$-controller globally asymptotically stabilizes the system (2).

*Switching v-controller* [3,7,2,1]. Given the same Lyapunov function as above, $V_1(x_1) = \frac{1}{2}x_1^2$, we now try to make its time-derivative negative by choosing the static feedback switching controller

$$v = -c_1 x_1 - \check{p}_1(x_1)\, x_1^2 \tag{9}$$

where the approximate switching function $\check{p}_1(x_1)$ is defined as follows:

$$\check{p}_1(x_1) = \frac{2}{\pi}\bar{p}\tan^{-1}(\gamma_1\sigma_1) \tag{10}$$

$$\gamma_1(x_1) = 2\bar{p}\mu_1 \tan\left(\frac{\pi}{2}(1-\nu_1)\right) \tag{11}$$

$$\nu_1(x_1) = (1+\bar{p}\mu_1 T_1)^{-1} \tag{12}$$

$$T_1(x_1) = \ln(2\cosh(\sigma_1)) \tag{13}$$

$$\mu_1(x_1) = (\eta_1 W_1 + (1-\eta_1)\varepsilon_1)^{-1} \tag{14}$$

$$\sigma_1(x_1) = x_1^3 \tag{15}$$

$$W_1(x_1) = \tfrac{1}{2}c_1 x_1^2 \tag{16}$$

$$\varepsilon_1 = (32\bar{p}^2)^{-1} \tag{17}$$

and $c_1 > 0$, $\eta_1 \in [0,1)$ are design parameters. This function $\check{p}_1(x_1)$ can be thought of as a smooth approximation to the ideal switch $\bar{p}\,\mathrm{sgn}\,\sigma_1$. Taking the time-derivative of $V_1$ we get

$$\dot{V}_1 = -c_1 x_1^2 - x_1^3(\check{p}_1 - p) \tag{18}$$

$$= -2W_1 - \sigma_1(\check{p}_1 - p) \tag{19}$$

$$= -2W_1 + R_1 \tag{20}$$

where $R_1 = -\sigma_1(\check{p}_1 - p)$ denotes the "remainder", i.e. the term in the Lyapunov derivative which appears because $p$ is unknown. Clearly, replacing $\check{p}_1(x_1)$ by the discontinuous function $\bar{p}\,\mathrm{sgn}\,\sigma_1$ in (9) would render $R_1 \le 0$, thus making $\dot{V}_1 \le -2W_1$. However, for backstepping we require smoothness of $v$ thereby disallowing such a replacement. The slope $\gamma_1(x_1)$ of the approximation to the discontinuity depends on $x_1$ in such a way that $\gamma_1$ remains "small" while insuring $R_1 \le W_1$ for all $x_1$ (see [3]). We then obtain $\dot{V}_1 \le -W_1$ which proves that this $v$ is another globally asymptotically stabilizing controller for (2).

*Adaptive v-controller* [5]. In contrast to the previous two $v$-controllers, the adaptive $v$-controller uses dynamic state feedback to cancel (rather than bound) the "remainder" terms in the Lyapunov derivative. Consider the augmented Lyapunov function $V_1(x_1, \hat{p}_1, p) = \frac{1}{2}x_1^2 + \frac{1}{2}m_1(\hat{p}_1 - p)^2$ associated with the dynamic feedback controller

$$v = -c_1 x_1 - \hat{p}_1 x_1^2 \tag{21}$$

$$\dot{\hat{p}}_1 = \frac{1}{m_1} x_1^3 \tag{22}$$

with positive design parameters $c_1$ and $m_1$. The time-derivative of $V_1$ is

$$\dot{V}_1 = p x_1^3 + v x_1 + m_1 \dot{\hat{p}}_1 (\hat{p}_1 - p) \tag{23}$$

$$= -c_1 x_1^2 - x_1^3(\hat{p}_1 - p) + x_1^3(\hat{p}_1 - p) \tag{24}$$

$$= -c_1 x_1^2 \tag{25}$$

Hence this adaptive $v$-controller also globally asymptotically stabilizes (2). Note that in this case the "remainder" term $R_1 = -x_1^3(\hat{p}_1 - p)$ in the Lyapunov derivative is cancelled through the choice of $\dot{\hat{p}}_1$. This approach has been labelled "adaptive" in the literature, although calling the dynamic feedback state $\hat{p}_1$ a "parameter estimate" is justified only by its position in the "certainty equivalence" form of the feedback law (21). $V_1$ being proper in $\hat{p}_1$ guarantees global boundedness of $\hat{p}_1(t)$, but this dynamic feedback state does not necessarily provide any information about the value of the true parameter $p$.

## 3. $u$-Controllers for the Planar System

Integrator backstepping has recently become an effective tool for extending stability results for matched systems to unmatched systems (see [5,6] and the references therein). Given one of the above three $v$-controllers for the scalar matched system (2) along with its associated Lyapunov function $V_1$, we use backstepping to generate a stabilizing $u$-controller for the planar unmatched system (1). This procedure involves constructing a new Lyapunov function $V_2$ from the given functions $v$ and $V_1$ and then designing the $u$-controller so that the Lyapunov derivative $\dot{V}_2$ is negative. As we have seen in the previous section, there are different ways to make a Lyapunov derivative negative, and thus we have different versions of backstepping. The continuous, switching, and adaptive backstepping methods achieve negativity of $\dot{V}_2$ in ways analogous to how their $v$-controller counterparts achieve negativity of $\dot{V}_1$. Adaptive double backstepping achieves such negativity in a manner similar to that of adaptive backstepping, but its $V_2$ is not constructed from the $v$ and $V_1$ of a preliminary step.

*Continuous backstepping* [8]. Given the continuous $v$-controller and its associated Lyapunov function $V_1$, we construct the following CC controller and Lyapunov function $V_2$ for the system (1):

$$u = -\frac{\partial \Phi}{\partial V_1} x_1 + \frac{\partial v}{\partial x_1} x_2 - \mu \left( x_2 - v \right) \tag{26}$$

$$V_2(x_1, x_2) = \Phi(V_1) + \tfrac{1}{2} \left( x_2 - v \right)^2 \tag{27}$$

where

$$\Phi(V_1) = \frac{1}{c_1} \left( \phi V_1 + \tfrac{1}{2} \bar{p} \beta \left( c_1 + \tfrac{1}{2} \bar{p} \alpha \right) V_1^2 + \frac{3 \bar{p}^2 \delta}{2\alpha} V_1^4 \right) \tag{28}$$

and

$$\mu = c_2 + \frac{c_1 \bar{p}}{2\beta} + \frac{\bar{p}^2 \alpha}{4\beta} + \frac{3\bar{p}^2}{4\alpha\delta} \tag{29}$$

and $c_2$, $\phi$, $\beta$, $\delta$ are additional positive design parameters. By using (8) repeatedly Praly has shown that $\dot{V}_2 \leq -\phi x_1^2 - c_2 \left( x_2 - v \right)^2$, thus rendering $u$ a globally asymptotically stabilizing controller for the system (1). The proof seems to rely on the specific polynomial form of the continuous $v$-controller and thus we apply continuous backstepping only to this $v$-controller.

*Switching backstepping* [3]. First suppose we are given either the continuous or the switching $v$-controller along with the associated Lyapunov function $V_1$. Consider the following controller $u$ and Lyapunov function $V_2$:

$$u = -\frac{\partial \Phi}{\partial V_1} x_1 - c_2 \left( x_2 - v \right) + \frac{\partial v}{\partial x_1} \left( \check{p}_2(x_1, x_2)\, x_1^2 + x_2 \right) \tag{30}$$

$$V_2(x_1, x_2) = \Phi(V_1) + \tfrac{1}{2} \left( x_2 - v \right)^2 \tag{31}$$

where $c_2$ is a positive design parameter and $\Phi(\cdot)$ is any smooth function having positive derivative on $[0, \infty)$ such that $\Phi$ is unbounded on $[0, \infty)$ and $\Phi(0) = 0$. Now define the approximate switching function $\check{p}_2(x_1, x_2)$ as follows:

$$\check{p}_2(x_1, x_2) = \frac{2}{\pi} \bar{p} \tan^{-1} \left( \gamma_2 \sigma_2 \right) \tag{32}$$

$$\gamma_2(x_1, x_2) = 2 \bar{p} \mu_2 \tan \left( \frac{\pi}{2} (1 - \nu_2) \right) \tag{33}$$

$$\nu_2(x_1, x_2) = \left( 1 + \bar{p} \mu_2 T_2 \right)^{-1} \tag{34}$$

$$T_2(x_1, x_2) = \ln \left( 2 \cosh \left( \sigma_2 \right) \right) \tag{35}$$

$$\mu_2(x_1, x_2) = \left( \eta_2 W_2 + (1 - \eta_2) \varepsilon_2 \right)^{-1} \tag{36}$$

$$\sigma_2(x_1, x_2) = -\left( x_2 - v \right) \frac{\partial v}{\partial x_1} x_1^2 \tag{37}$$

$$W_2(x_1, x_2) = \tfrac{1}{2} \frac{\partial \Phi}{\partial V_1} W_1(x_1) + \tfrac{1}{2} c_2 \left( x_2 - v \right)^2 \tag{38}$$

where $\eta_2 \in [0,1)$ is a design parameter, $W_1(x_1)$ is any positive-definite function satisfying $\dot{V}_1 \leq -W_1$, and $\varepsilon_2$ is a sufficiently small positive number. Taking the time-derivative of $V_2$ we get

$$\begin{aligned}
(39)\quad \dot{V}_2 &= \frac{\partial\Phi}{\partial V_1}x_1\left(px_1^2+x_2\right)+(x_2-v)\left(u-\frac{\partial v}{\partial x_1}\left(px_1^2+x_2\right)\right)\\
(40)\quad &= \frac{\partial\Phi}{\partial V_1}\dot{V}_1+(x_2-v)\left(\frac{\partial\Phi}{\partial V_1}x_1+u-\frac{\partial v}{\partial x_1}\left(px_1^2+x_2\right)\right)\\
(41)\quad &= \frac{\partial\Phi}{\partial V_1}\dot{V}_1-c_2\left(x_2-v\right)^2-\sigma_2\left(\check{p}_2-p\right)\\
(42)\quad &\leq -2W_2(x_1,x_2)+R_2(x_1,x_2)
\end{aligned}$$

where $R_2(x_1,x_2) = -\sigma_2\left(\check{p}_2 - p\right)$ is the "remainder" term. Following [3], our choice of $\check{p}_2$ is a smooth approximation to the switch $\bar{p}\,\mathrm{sgn}(\sigma_2)$ and guarantees that $R_2 \leq W_2$ for all $x_1$ and $x_2$ and hence $\dot{V}_2 \leq -W_2$. This proves that $u$ is a globally asymptotically stabilizing controller for the system (1). Given the two choices of $v$-controllers we have thus generated two stabilizing $u$-controllers, CS and SS.

If we are instead given the adaptive $v$-controller, minor modifications to the above are needed to accommodate for the additional state $\hat{p}_1$. First of all, $u$ as given in (30) should include the additional term $(\partial v/\partial\hat{p}_1)\dot{\hat{p}}_1$. Next, we require that $\Phi(V_1) = \phi V_1$ where $\phi$ is a positive design parameter (otherwise $\partial\Phi/\partial V_1$ would depend on $p$ and would thus be unavailable to the controller). With this AS $u$-controller, calculations similar to those above again yield $\dot{V}_2 \leq -W_2$ proving global boundedness of $\hat{p}_1(t)$ and global asymptotic stability of the origin of (1).

*Adaptive backstepping* [5]. Given one of the three $v$-controllers and its associated Lyapunov function $V_1$, we construct the following controller $u$ and augmented Lyapunov function $V_2$:

$$\begin{aligned}
(43)\quad u &= -\phi x_1 - c_2\left(x_2-v\right)+\frac{\partial v}{\partial x_1}\left(\hat{p}_2x_1^2+x_2\right)+\frac{\partial v}{\partial\hat{p}_1}\dot{\hat{p}}_1\\
(44)\quad \dot{\hat{p}}_2 &= \frac{1}{m_2}\sigma_2(x_1,x_2,\hat{p}_1)
\end{aligned}$$

$$(45)\quad V_2(x_1,x_2,\hat{p}_1,\hat{p}_2,p) = \phi V_1 + \tfrac{1}{2}\left(x_2-v\right)^2+\tfrac{1}{2}m_2\left(\hat{p}_2-p\right)^2$$

where $c_2$, $\phi$, $m_2$ are positive design parameters and $\sigma_2$ is defined by (37). Note that for the continuous and switching $v$-controllers, the last term in (43) would be zero. Now when we calculate the time-derivative of $V_2$, we will get the "remainder" term $R_2 = -\sigma_2\left(\hat{p}_2 - p\right)$ which we cancel through our choice of $\dot{\hat{p}}_2$. The calculation is otherwise similar to that of the switching backstepping above, and we conclude that with the three $u$-controllers generated by this adaptive backstepping (CA, SA, and AA), the dynamic feedback states $\hat{p}_1(t)$ (if present) and $\hat{p}_2(t)$ are globally bounded while the origin of (1) is globally asymptotically stable.

*Adaptive double backstepping* [4]. The previous three backstepping methods all construct essentially the same Lyapunov function $V_2$ for the planar system (1) from given functions $v$ and $V_1$. As shown in [4], we can eliminate the preliminary $v$-step by directly considering the Lyapunov function

$$V_2 = \tfrac{1}{2}\phi x_1^2 + \tfrac{1}{2}\left(\hat{p}x_1^2 + x_2 + x_1\right)^2 + \tfrac{1}{2}m\left(\hat{p} - p\right)^2 \tag{46}$$

and dynamic feedback controller

$$u = -\phi x_1 - \dot{\hat{p}}x_1^2 - c\left(\hat{p}x_1^2 + x_2 + x_1\right) - \left(\hat{p}x_1^2 + x_2\right)\left(1 + 2\hat{p}x_1\right) \tag{47}$$

$$\dot{\hat{p}} = \frac{1}{m}x_1^2\left[\phi x_1 + \left(\hat{p}x_1^2 + x_2 + x_1\right)\left(1 + 2\hat{p}x_1\right)\right] \tag{48}$$

where $\phi$, $c$, and $m$ are positive design parameters. With this DB controller, the Lyapunov function (46) has time-derivative

$$\dot{V}_2 = -\phi x_1^2 - c\left(\hat{p}x_1^2 + x_2 + x_1\right)^2 \tag{49}$$

thus guaranteeing global boundedness of $\hat{p}(t)$ and global asymptotic stability of the origin of (1).

## 4. Structural Comparison

We first compare the stuctural properties of the above controllers, noting that such comparisons remain valid in settings more general than that of our case study. To begin with, we observe that the above eight $u$-controllers become actually classes of controllers when we take design parameters into account. These classes include static controllers (SS, CS, and CC), dynamic controllers which add one state to the system (SA, AS, CA, and DB), and dynamic controllers which add two states to the system (AA). When applying adaptive backstepping repeatedly in order to obtain controllers for higher-dimensional systems, many such additional states might appear in the feedback law. These would increase both the complexity of the controller and the risk of "parameter drift" instabilities. Using adaptive double backstepping would roughly cut in half the number of additional states needed [4]; using continuous and switching backstepping could provide a further reduction. If, however, the parameter bound $\bar{p}$ is unknown, we have no proof of stability for controllers employing continuous or switching backstepping (note that only the AA and DB controllers make no use of $\bar{p}$).

We next observe that switching and adaptive backstepping are closely related. The outputs of the switches ($\check{p}_1$ or $\check{p}_2$) and the outputs of the integrators ($\hat{p}_1$ or $\hat{p}_2$) appear interchangeably in the control laws, and their inputs are identical (ignoring positive scaling factors which, for the switching case, may vary with the states). Adaptive backstepping has the advantage of not using $\bar{p}$, but switching backstepping adds no extra states and can be applied to more general types of uncertainties [3].

Finally, we emphasize that all of the above classes of $u$-controllers (except the DB controller) are generated by making negative the time-derivatives of the *same* (or perhaps augmented) Lyapunov function. They distinguish themselves from each other by using different techniques for achieving negativity at different stages of backstepping. We see in the next section how this leads to different controller performance.

## 5. Simulation Comparison

A thorough examination of simulation results would be tedious and would interfere with our presentation of ideas in this paper. We therefore point out in this section only the most interesting performance characteristics of the eight classes of controllers described above, concentrating on a comparison of control signals rather than state trajectories.

We first made the problem more specific by setting $\bar{p} = 2$. The next step was to select a representative from each of the eight classes of controllers by fixing design parameters. Choosing appropriate values for the controller design parameters was crucial in making meaningful simulation comparisons. Many of these parameters largely affected the controller performance, and no systematic method was available for obtaining satisfactory values for them. Our choices were based on trying to achieve "optimal" controller performance under a set of nominal operating conditions. In order to simplify this task, we based our sense of "optimality" on the maximum control effort required for stabilization. We first fixed an initial condition in the $x_1x_2$-plane, choosing a large initial value for $x_1$ in order to emphasize the nonlinearity $px_1^2$. We arbitrarily chose the sign of this initial $x_1$-value to be positive and then set $p = \bar{p}$ to create a "worst-case scenario." In this manner we selected the values $(x_1^0, x_2^0, p) = (10, 1, 2)$. Based on these nominal conditions, we simulated our system (1) with each different controller, adjusting design parameters until we believed to have achieved the best possible performance with respect to maximum control effort.

Figures 1–3 show respectively the SS, SA, and DB control signals under the nominal initial conditions $(x_1^0, x_2^0) = (10, 1)$ and for various values of $p$. Figure 1 illustrates a basic property of the SS controller, namely that its control signals exhibit "jumps" corresponding to the switching points.

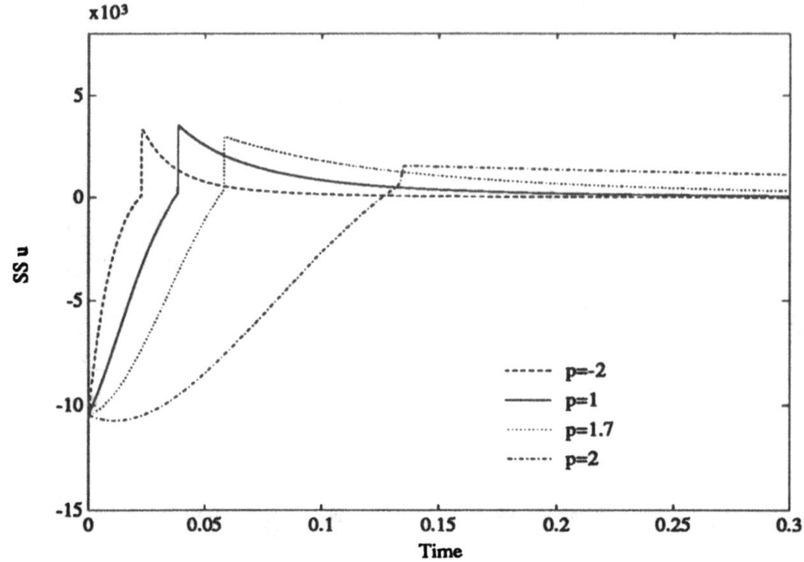


**Figure 1: SS control signals for $(x_1^0, x_2^0) = (10, 1)$**

One should also note that as $p$ approaches $\bar{p}$, the SS controller becomes very slow, allowing the states to grow to large values before forcing them to the origin; in some sense the controller barely stabilizes the system under these conditions. To prevent such behavior, one could simply replace $\bar{p}$ in the SS control law by a larger value, although this would increase the magnitudes of the control signals. The performance of the CS controller under the nominal initial conditions is similar to that of the SS controller, but its behavior as $p$ approaches $\bar{p}$ is more satisfactory.

The AA, AS, and CA control signals under the above nominal initial conditions all look similar to the SA control signals in Figure 2. The peaking evident in the SA signals as $p$ approaches $\bar{p}$ also appears in the other three cases but to a slightly lesser extent. We will soon see how this peaking can lead to large oscillatory behavior under different simulation conditions.

In Figure 3 we observe that the maximum magnitudes of the DB control signals are very large, almost a factor of three larger than those of the SS control signals. The CC control signals also suffer from such large magnitudes. As described above, the controller design parameters were chosen specifically to try to minimize these magnitudes under these initial conditions. We therefore suggest that the DB and CC designs require in general more control effort; simulations run under different conditions support this hypothesis.

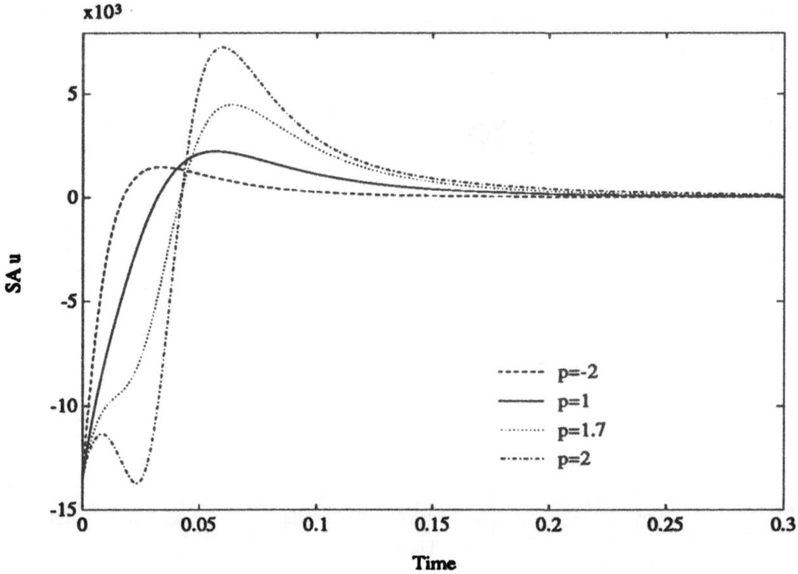


**Figure 2: SA control signals for $(x_1^0, x_2^0) = (10, 1)$**

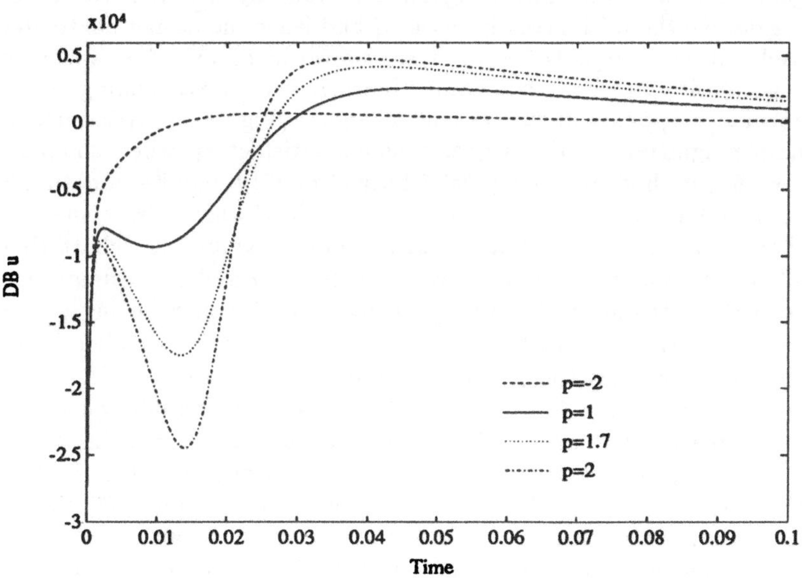


**Figure 3: DB control signals for $(x_1^0, x_2^0) = (10, 1)$**

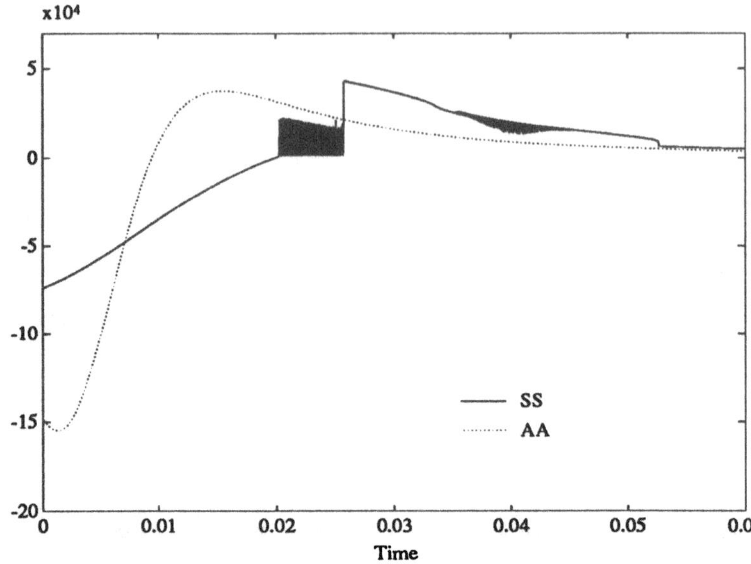


**Figure 4: SS, AA control signals for $(x_1^0, x_2^0, p) = (20, 1, 1)$**

Figures 4–7 show the control signals generated by the eight controllers when we double the initial condition on $x_1$ and leave the design parameters unaltered. All eight signals for $p = 1$ are shown in Figures 4–6, and three of the signals for $p = 2$ are shown in Figure 7. The most striking change in performance appears in the SS control signal (Figure 4). Although the maximum magnitude of this signal remains satisfactory when compared to the others, its high-frequency "chattering" quality may be undesirable. The CS and AS control signals (Figure 5), both of which also come from controllers generated by switching backstepping, exhibit no "chattering" but still have reasonable maximum magnitudes. This and other simulations suggest that switching backstepping might be useful in reducing control effort but best suited for the last backstepping stages in a controller design.

Figure 7 shows that the peaking exhibited by the SA and CA control signals in Figures 5 and 6 respectively leads to large oscillatory behavior when we increase the value of $p$. We noticed through additional simulations that in general the dynamic controllers were more prone to such oscillations than the static ones. As a final note, Figures 6 and 7 show that the DB and CC controllers again generate the largest control signals.

We end this section by pointing out that, as compared to the other controllers, three of the "unmixed" controllers (SS, CC, and DB) exhibited undesirable behavior under certain simulation conditions. From this obser-

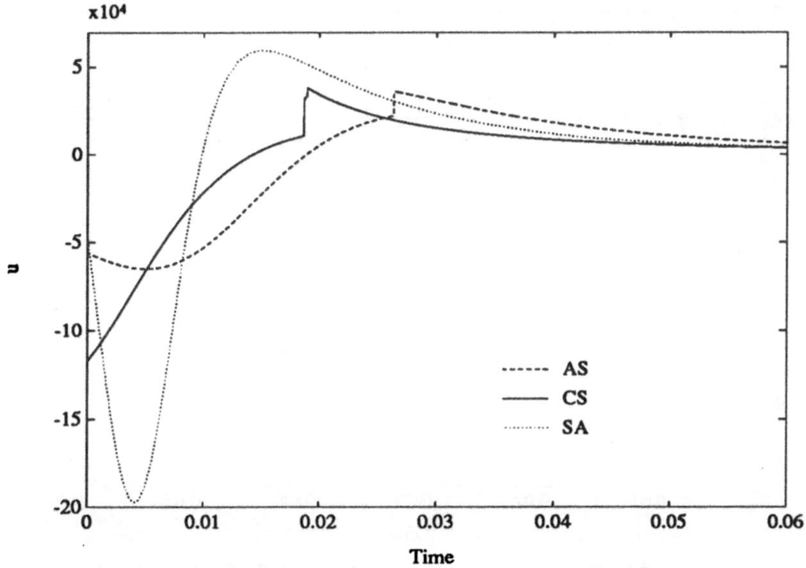


**Figure 5: AS, CS, SA control signals for $(x_1^0, x_2^0, p) = (20, 1, 1)$**

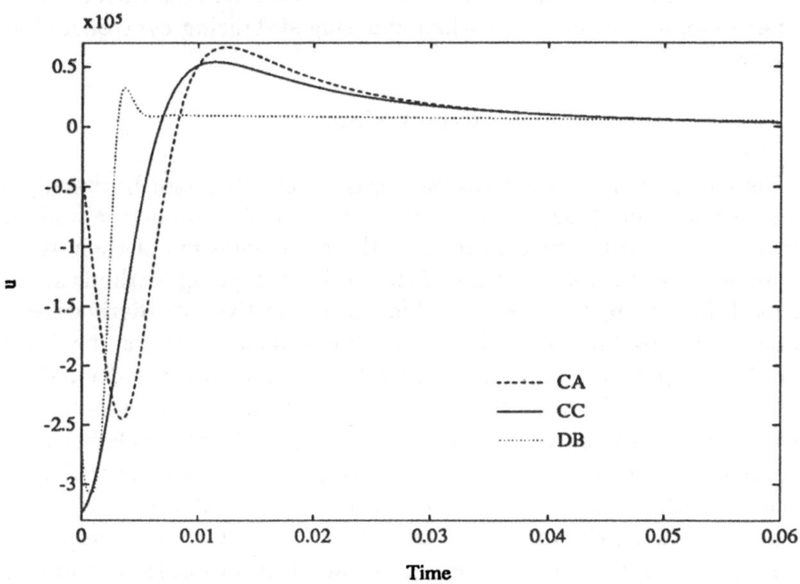


**Figure 6: CA, CC, DB control signals for $(x_1^0, x_2^0, p) = (20, 1, 1)$**

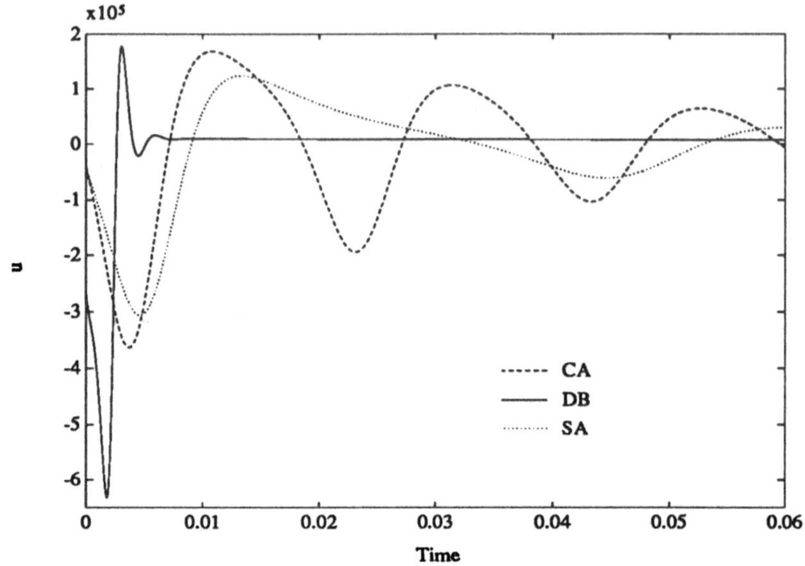


**Figure 7: CA, DB, SA control signals for $(x_1^0, x_2^0, p) = (20, 1, 2)$**

vation we conclude that the "mixed" backstepping designs introduced in this paper deserve consideration when choosing stabilizing controllers for a given system.

## 6. Conclusion

In this paper we have used various versions of integrator backstepping to generate classes of globally stabilizing controllers for an uncertain planar nonlinear system. In the presentation of the main elements of their stability proofs, we have shown how these different backstepping methods simply represent different approaches to making negative the time-derivatives of the same Lyapunov functions. For higher-dimensional systems, backstepping would be applied repeatedly (if possible) to generate the desired stabilizing controllers. We have emphasized that at each backstepping stage the designer is free to choose from more than one of the available backstepping methods. In our case study this led to four new classes of stabilizing controllers for the planar system (the "mixed" controllers). For higher-dimensional systems, the number of such new classes would be very large, and choosing the best controller for a given application would be a difficult task. By presenting some simulation results for the planar system, we have illustrated various performance characteristics of the different versions of backstepping, hopefully providing guidelines for selecting the most appropriate backstepping method at a given stage of the controller design. The simulations suggest that the best controller for a given application could very well come from a combination of *different* backstepping techniques and thus belong to one of the new classes of controllers introduced in this paper.

*Acknowledgements*: This research was supported in part by the U.S. Department of Energy under Grant DE-FG-02-88-ER-13939, in part by the Air Force Office of Scientific Research under Grant AFOSR 90-0011 90-91, and in part by the National Science Foundation under an NSF Graduate Fellowship.

## REFERENCES

[1] M. R. Corless and G. Leitman, *Continuous state feedback guaranteeing uniform ultimate boundedness for uncertain dynamic systems*, IEEE Transactions on Automatic Control, vol. AC-26, no. 5, pp. 1139–1144, 1981.

[2] F. Esfandiari and H. K. Khalil, *Stability analysis of a continuous implementation of variable structure control*, IEEE Transactions on Automatic Control, vol. 36, no. 5, pp. 616–620, 1991.

[3] R. A. Freeman and P. V. Kokotovic, *Backstepping design of robust controllers for a class of nonlinear systems.* In preparation, 1991.

[4] Z. P. Jiang and L. Praly, *Iterative designs of adaptive controllers for systems with nonlinear integrators.* Submitted for presentation at The 30th IEEE Conference on Decision and Control, 1991.

[5] I. Kanellakopoulos, P. V. Kokotovic, and A. S. Morse, *Systematic design of adaptive controllers for feedback linearizable systems*, IEEE Transactions on Automatic Control, vol. 36, no. 11, 1991.

[6] I. Kanellakopoulos, P. V. Kokotovic, and A. S. Morse, *A toolkit for nonlinear feedback design*, Tech. Rep. CCEC-91-0619, Center for Control Engineering and Computation, Electrical and Computer Engineering, University of California, Santa Barbara, 1991.

[7] H. K. Khalil, *Nonlinear Systems*. New York: Macmillan Publishing Company, 1992.

[8] L. Praly, July 1991. Personal communication.

R. A. Freeman is with the Department of Electrical and Computer Engineering, University of Illinois, Urbana, IL 61801, USA.

P. V. Kokotovic is with the Department of Electrical and Computer Engineering, University of California, Santa Barbara, CA 93106, USA.

# Nicely Nonlinear Modelling and Control

GUIDO O. GUARDABASSI

**Abstract.** A system $M$ tangent to a plant $P$ at the nominal equilibrium is a *nicely nonlinear model* of $P$ if it is (properly) linearizable by output feedback. For any nicely nonlinear model $M$, a composite controller can naturally be devised as the cascade connection of a (nonlinear) linearizing compensator $C_L$ and a (possibly linear) controller $R$ the design (or the tuning) of which can be carried out following standard paradigms of any peace of control theory where the plant is by assumption linear and time-invariant. Constructing parametrized families of nicely nonlinear models is therefore of obvious interest, inasmuch as it is also possible to provide, for each one of them, conceivably cheap algorithms able to identify one (at least) of the individual elements in the family, that best fit with (the known and significant behavior of) a given plant. Such a task is here referred to as *nicely nonlinear modelling*. The purpose of the present paper is to discuss, and to illustrate by examples, some aspects of its general (abstract) structure.

## 1 Introduction

The idea of using feedback to linearize the behavior of a given plant in order to make its control easier is an old and fascinating one [1],[2][3][4]. The line of thinking inherent in nicely nonlinear modelling consists of bringing the linearization idea back to the modelling phase [5]. Thus, standard techniques for control system design based on pure (control oriented, possibly uncertain, currently or periodically updated) linear modelling can naturally be used to solve control problems concerning a significantly broader class of basically nonlinear plants. The purpose of the present paper is to introduce in fairly general (abstract) terms the nicely nonlinear modelling concept, and to illustrate some simple instances of it, for which the available theory enables solving nontrivial problems of practical significance.

Consider a plant $P$ described by:

$$\dot{x}_P(t) = F_P(x_P(t), u(t)) \quad , \quad y_P(t) = H_P(x_P(t))$$

Without loss of generality, let $O := (\bar{u} = 0, \bar{x}_P = 0, \bar{y}_P = 0)$ be the "nominal" equilibrium of $P$; whence: $F_P(0,0) = 0, H_P(0) = 0$. Denote, then, by $\delta P$ the *linear time-invariant system tangent to P at the "nominal" equilibrium:*

$$\delta\dot{x}(t) = A_P\delta x(t) + B_P\delta u(t) \quad , \quad \delta y(t) = C_P\delta x(t)$$

where:

$$A_P := \frac{\partial F_P}{\partial x}(0,0) \quad , \quad B_P := \frac{\partial F_P}{\partial u}(0,0) \quad , \quad C_P := \frac{\partial H_P}{\partial x(0)}$$

Any (nonlinear) system $M$ described by:

$$\dot{x}_M(t) = F_M(x_M(t), u(t)) \quad , \quad y_M(t) = H_M(x_M(t))$$

is *tangent to P at O* if $O$ is an equilibrium point of $M$, and $\delta M = \delta P$; i.e.: $\delta M$ is equivalent to $\delta P$ (up to a change of coordinates in the state space).

A system $M$ (tangent to $P$ at $O$) is said to be ***properly linearizable by output-feedback*** if there exists a *linearizing compensator* $C_L$, with input $(v, y_M)$ and output $u$, such that:

i) the interconnected system $(C_L, M)$ of Fig.1-a is equivalent (up to a change of coordinates in the state space) to a system of the form shown in Fig.1-b, where $L$ is linear and time-invariant, $N_1$ and $N_2$ are BIBS and BSBO stable, respectively (see Appendix A);

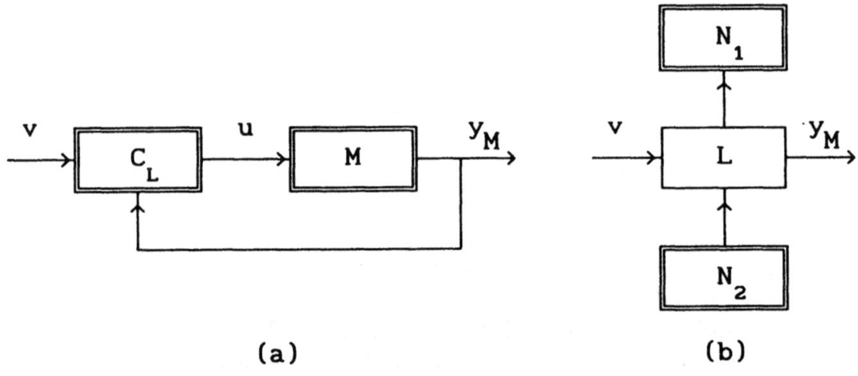


*Fig.1: Rigorously linearizing compensator*

ii) $C_L$ is functionally invertible relative to $v$; i.e.: there exists a causal $C_L^{-1}$, with input $(\hat{u}, y_M)$ and output $v$, such that the tandem connection of $C_L$ with $C_L^{-1}$ (Fig.2-a) is equivalent (up to a change of coordinates in the state space) to a system of the form shown in Fig.2-b, where $N$ is BSBO stable.

The linearizing compensator $C_L$ is said to be *normal* if system $L$ of Fig.1-b is externally equivalent to the linear system $\delta M$ tangent to $M$ at $O$; namely if the reachable and observable parts of $L$ and $\delta M$ do coincide, up to a change of coordinates in the state space.

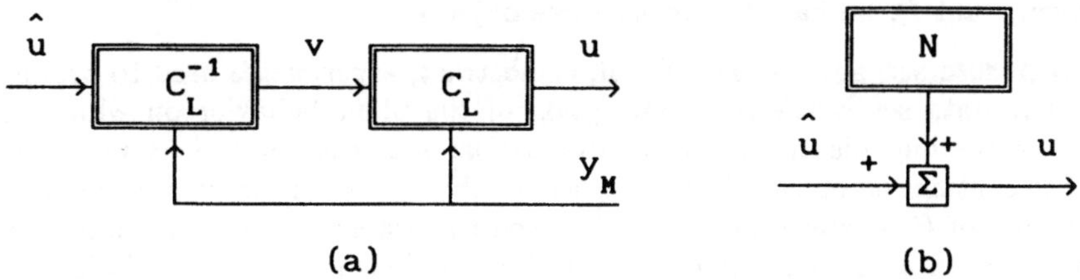


*Fig.2: Functional invertibility*

The system $M$ is a *nicely nonlinear model* of $P$ if it is tangent to $P$ at $O$, and properly linearizable by output-feedback.

*Remark 1.* Any (linear) controller $R$ (Fig.3), designed (or tuned) on the linear time-invariant plant $\delta M = \delta P$, when connected to a nicely nonlinear model $M$ of $P$ through a *normal* linearizing compensator $C_L$, immediately extends its (linear) performance to the whole of M. Therefore, if $R$ is any satisfactory linear or nonlinear controller of $\delta P$, and $C_L$ is any normal linearizing compensator of a (good) nicely nonlinear model of $P$, then the tandem connection of $R$ and $C_L$ (Fig.4) naturally candidates itself as a nonlinear controller of $P$ not worse than $R$ in a neighborhood of $O$, and potentially much better than $R$ alone, outside that neighborhood.

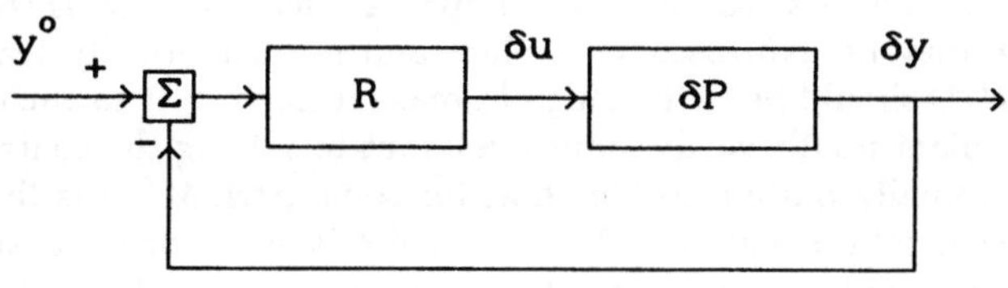


*Fig.3: Linear control*

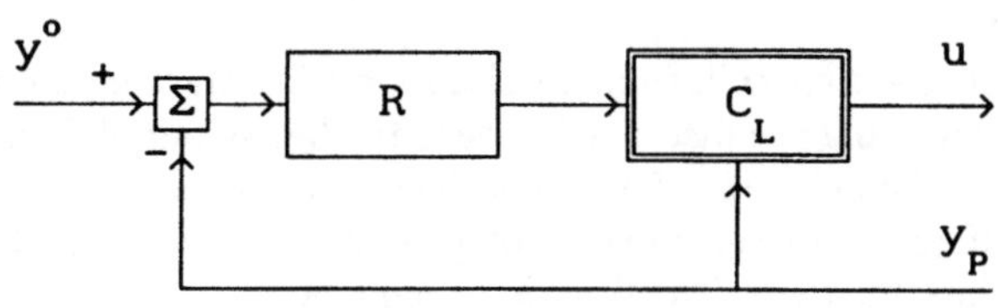


*Fig.4: Nicely nonlinear control*

*Remark 2.* The existence of a smooth linearizing compensator of $M$ implies the existence of a normal linearizing compensator of $M$ (for an outline of the proof, see Appendix B).

## 2 Nicely nonlinear modelling

Generally speaking, in order to state a nicely nonlinear modelling problem for a given plant $P$, we have to specify three objects.

i) A *data set*, and the set of real, or abstract, *experiments* used to produce it. The data set is the sole description of the plant behavior on which the entire modelling/identification process stands: it is then extremely important that it contains as much information as possible about whatever of (nonlinear) dynamics of $P$ is expectedly relevant to control system design; and, conversely, that as much as possible of the useless information be suitably excluded. As in many other modelling/identification processes, the experiments underneath the data set may amount either to nothing but measuring the (zero-initial-state or periodic) response of the plant to standard input functions (e.g. steps of different amplitudes, sinusoids of different amplitudes and frequencies, colored noises of different intensities, etc.), or to measuring the plant input and output in normal operation, or even to leading the plant through one or more transients, or regimes, which are in some sense optimal from an identification point of view. Significantly enough, these experiments can often be carried out simply resorting to accurate and relyable simulation models of $P$, which are of no direct use, though, for (nonlinear) control design, due to their hard, many-sided or anyhow inextricable complexity.

ii) A *parametrized class of nicely nonlinear models.* That is a parametric model $M(\alpha)$, with input $u$ and output $y_M$, the parameter vector of which is restrained to an admissible set $A \subseteq \mathbf{R}^m$ such that $M(\alpha)$ is a nicely nonlinear model of $P$ for all $\alpha \epsilon A$. An obviously important feature of $M$ is that the admissible set A may be expressed in explicit and mathematically tractable terms. Second, $M(A)$ should be rich enough to make it likely of it to encompass a good deal of the plant nonlinear dynamics, relevant to solving the control problem at hand. It is finally quite sensible that, for some $\alpha \epsilon A, M(\alpha)$ is linear and time invariant (whereby equal to $\delta P$) so that $\delta P$ itself belongs to the considered class of nicely nonlinear models, whence the optimal nicely nonlinear model of $P$ in $M(A)$, relative to the considered data set, cannot be worse than its linear competitor $\delta$P.

iii) A *model selection criterion*; a suitable misfit function, to minimize with respect to $\alpha$, expressing the distance between the behaviour of $P$, as described by the data set, and the behavior of $M(\alpha)$ in the very same experimental conditions. For instance, if the data set consists of $N$ experiments, each of length $\tau_i$, an obvious candidate for the misfit function might be

$$J(\alpha) := \sum_{i=1}^{N} \frac{1}{\tau_i} \int_0^{\tau_i} |y_{Pi}(t) - y_{Mi}(t, \alpha)|^2 dt.$$

As is well known, though, it is generally rewarding to look for slightly different yet equally sensible misfit functions endowed with properties (quadratic structure, convexity, etc.) known to significantly simplify the fundamental problem of minimizing $J(\alpha)$, subject to $\alpha \epsilon \mathrm{A}$.

In conclusion, the data set (or, more specifically, the underneath experimental setting), the class of models, and the selection criterion cannot be chosen independently. It should be fairly apparent, in fact, that all these elements strongly interact with each other in determining the complexity of the ultimate nonlinear programming problem to solve. When dealing with real time identification problems, as is usually the case with adaptive control, a further and certainly more subtle aspect has to be taken into account: that is the possibility of giving the basic optimization algorithm a suitable recursive form. Research in this direction, for nicely nonlinear real time identification, is just slowly moving its first steps.

## 3 Examples

### *3.1 A class of single-input single-output nicely nonlinear models*

Consider the class of models described by the block-diagram of Fig.5, where $f$ and $g$ are polynomials of degree $m_f$ and $m_g$ , respectively. For polynomials (of degree $m-1$) we adopt the following notation:

$$p(v;\vartheta,m) := \chi(v;m)\vartheta$$

where $\vartheta \epsilon \mathbf{R}^m$ is a parameter vector, while

$$\chi(v;m) := [v^{m-1} \quad v^{m-2} \quad \dots \quad v \quad 1].$$

Note that:

$$\frac{dp}{dv}(v;\vartheta,m) = \frac{d\chi}{dv}(v;m)\vartheta := \xi(v;m)\vartheta$$

$$\xi(v;m) := [(m-1)v^{m-2} \; (m-2)v^{m-3} \quad \dots \quad 2v \quad 1 \quad 0].$$

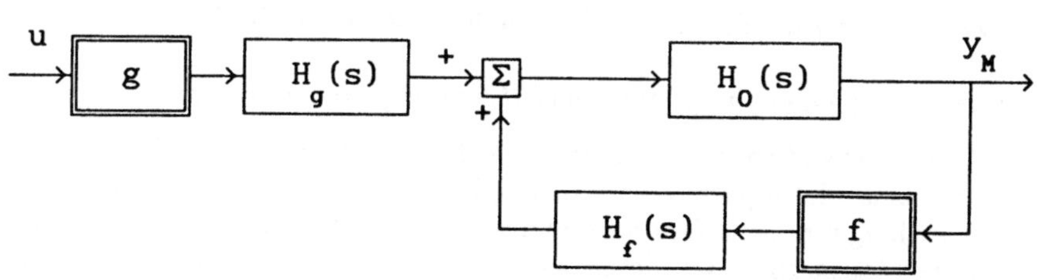


*Fig.5: A class of nicely nonlinear models*

We have, then:

$$f(y_M) = p(y_M; \varphi, m_f) \quad , \quad g(u) = p(u; \gamma, m_g)$$

where $\varphi$ and $\gamma$ are the parameter vectors of $f$ and $g$, respectively.

Assume that $u(t) \epsilon U \subseteq \mathbf{R}$, for all $t$; assume also that $\delta P$ is stabilizable, and denote by $T(s)$ its transfer function. Then, it is fairly easy to check that the following statement is true. If

i)

$$\varphi_{m_f} = \varphi_{m_f - 1} = \gamma_{m_g} = 0 \ , \ \gamma_{m_g - 1} = 1 \ ;$$

ii) $g$ is invertible on $U$; i.e.:

$$\xi(u; m_g)\gamma \geq 0 \ , \ \forall u \epsilon U \ ;$$

iii) $T_f(s)$ is asymptotically stable; $T_g(s)$ is causal and asymptotically stable, with causal and asymptotically stable inverse; $T_0(s)$ is such that: $T_0(s)T_g(s) = T(s)$;

then, model $M$ of Fig.5 is nicely nonlinear (tangent to $P$ at $O$, and properly linearizable by output feedback), for all admissible values of the system parameters. Note that condition (ii) consists of an infinity of linear inequalities; as such, it leads to a semi-infinite mathematical programming problem.

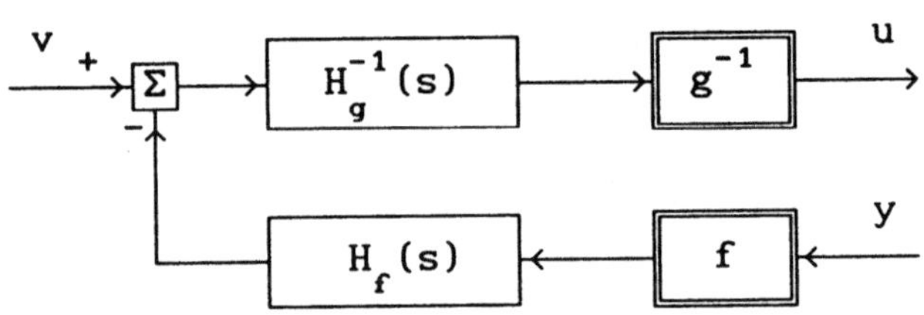


*Fig.6: Linearizing compensator*

Instead of formally proving the statement above, we simply show the structure (Fig.6) of a normal linearizing compensator (Section 1) suited to (properly) linearize any nicely nonlinear model of the form shown in Fig.5.

### *3.2 A describing function approach and the corresponding data set*

A classical way of (partially) describing the periodic response of a nonlinear system to sinusoidal inputs of various amplitudes and frequencies is by the so called describing function of the system itself [6]. The data set may consist, then, of the values taken on by the describing function of the plant over a finite grid of the amplitude-frequency space. Consistently, the misfit function to minimize with respect to the model parameters could be a weighted average, over the considered grid, of the distance between the plant and the model describing functions. If the transfer function $T(s)$ of $\delta P$ is known, such a misfit function can be carried into a convex quadratic function of the parameter vectors $\varphi$ and $\gamma$ [7]. The final problem to solve is then a convex quadratic semi-infinite program [7],[8],[9]. Within this framework, a possible difficulty, here, refers to the significance of the describing function (and the validity of the so called filtering assumption) in a frequency range (much) lower than the inverse of the plant response-time; in particular, the steady-state behaviour of both the plant and the model, as implicitly described by their describing functions, may prove strongly inadequate. A possible remedy currently under investigation to this elusive, sporadic, yet potentially harmful event might be to add in the misfit function an explicit term separately accounting for the distance between the plant and the model steady-state input-output relationships.

## 4 Concluding remarks

Nicely nonlinear modelling has been introduced some years ago as a means to widen the range of satisfactory operation of (industrial) controllers originally designed (or tuned) on a linear time-invariant model of the considered plant (e.g. the linear system tangent to the plant at the nominal equilibrium).

In this paper, the main features of a general nicely nonlinear modelling/ identification problem have been shortly discussed. Many aspects need further investigation, both from a theoretical and from a computational point of view. Of them all, the development of workable recursive algorithms for real-time nicely nonlinear identification to be included into specifically far-sighted adaptive control processes seems the most challenging task.

Acknowledgment: this research has been supported by CNR, Centro di Teoria dei sistemi, and by MURST.

## REFERENCES

[1] H.S.Black, *Stabilized feedback amplifiers*, Bell Sys. Tech. J., pp.1-18, Jan. 1934.

[2] C.A.Desoer, Y.T.Wang and A.N.Payne, *Linearizing effect of feedback on nonlinear dynamical systems*, 16th Annual Allerton Conf. on Commun., Contr. and Comp., 1978.

[3] R.W.Brockett, *Feedback invariants for nonlinear systems*, Proc. 7th IFAC Congress, Vol.6, pp.115-1120, Pergamon 1978.

[4] A.Isidori, *Nonlinear Control Systems*, Springer-Verlag 1989.

[5] G.O.Guardabassi, A.Righettini and I.Ruffoni, *Output linearizable models and nonlinear control of a distillation column*, System Science, Vol.12, pp.43-66, 1986.

[6] D.P.Atherton, *Nonlinear Control Engineering*, Van Nostrand 1975.

[7] D.Abastanotti, P.Colaneri, J.C.Geromel, R.Ghidoni and G.O.Guardabassi, *Nicely nonlinear modelling: ADF approach with application to "linearized process control"*, Proc. IFAC Symp. on Nonlinear Control System Design, Pergamon 1989.

[8] D.Abastanotti, J.C.Geromel, R.Ghidoni and G.O.Guardabassi, *Optimization-based nicely nonlinear modelling*, in System Modelling and Optimization, H.J.Sebastian and K.Tammer (Eds.), pp.885-893, Springer- Verlag 1990.

[9] J.C.Geromel and G.O.Guardabassi, *Computer-aided nicely nonlinear modelling: A set of problem specific MATLAB functions*, IFAC Symp. on Computer Aided Design in Control Systems, Swansea (UK), July 1991.

## Appendix A

In the background of several definitions given in the present paper, a particularly important role is played by the notions of Bounded-Input/ Bounded - State (BIBS) and Bounded - (initial) - State / Bounded-Output (BSBO) stability. They are shortly introduced in this Appendix, for ease of reference. A number of similar or related concepts actually exist in the literature, where even the very same acronyms are occasionally used with more or less sharply different meanings. Reviewing and relating with each other all these concepts is far beyond the scope of this Appendix.

Def.1: *Class K functions*
A function $\varphi : \mathbf{R}^+ \rightarrow \mathbf{R}^+$ belongs to class $K$ if it is continuous, strictly increasing, and $\varphi(0) = 0$.

Note that a function in class $K$ is bijective iff it is unbounded.

Def.2: *BIBS stability*
A system $N$ described by

$$\dot{x}(t) = f(x(t), u(t)) \quad , \quad f(0,0) = 0 \quad , \quad x(0) = x_0$$

is said to be BIBS stable (relative to the norm $\|\cdot\|$ of the input function space) if there exist a bijective function $\varphi_1$ and a function $\varphi_2$, both in class $K$, such that, for any $r \epsilon \mathbf{R}^+$:

$$(\| u \| \leq \varphi_1(r), | x_0 | \leq \varphi_2(r)) \Rightarrow (| x(t) | \leq r, \forall t \epsilon \mathbf{R}^+) \quad ,$$

where $|\cdot|$ denotes the Euclidean norm in $\mathbf{R}^k$.

Note that, by definition, it must be: $\varphi_2(r) \leq r, \forall r \epsilon \mathbf{R}^+$. If, in addition to $\varphi_1$, also $\varphi_2$ is bijective (unbounded), then system $N$ is said to be globally BIBS stable.

Def.3: *Perturbed system*
To any autonomous system $N$ described by

$$\dot{x}(t) = f(x(t)) \quad , \quad f(0) = 0 \quad , \quad x(0) = x_0$$

a perturbed system $N_p$ is associated as follows:

$$\dot{x}_p(t) = f(x_p(t)) + u(t) \quad , \quad x_p(0) = x_0 .$$

Def.4: *BSBO stability*
The autonomous system $N$ of Definition 3 is said to be BSBO stable (relative to $\|\cdot\|$) if there exist a function $\varphi_1$ and a bijective function $\varphi_2$, both in class $K$, such that, for any $r \epsilon \mathbf{R}^+$:

$$(\| u \| \leq \varphi_1(r), | x_0 | \leq \varphi_2(r)) \Rightarrow (| x_p(t) | \leq r, \forall t \epsilon \mathbf{R}^+).$$

Note that, according to the above definition, the BSBO stability of an autonomous system does not depend on the output mapping.

## Appendix B

Assume that $M$ is a model tangent to $P$ at $O$, and that $C_L$ is a smooth (continuosly differentiable) linearizing compensator of $M$ (Fig.1-a). Then, there exists a normal linearizing compensator of M.

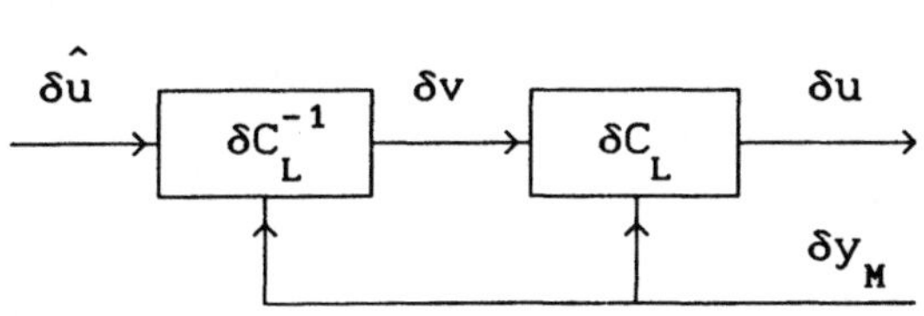


*Fig. 7: Linear system tangent at 0 to that of Fig: 2-a*

Functional invertibility of $C_L$ implies that the origins of its input, state, and output spaces do represent, in fact, an equilibrium condition. Consider then the linear (time-invariant) system (Fig.7) tangent at the origin to that of Fig.2-a, and isolate the block denoted $\delta C_L^{-1}$.A little thought reveals that the system of Fig.8 is a normal linearizing compensator of M.

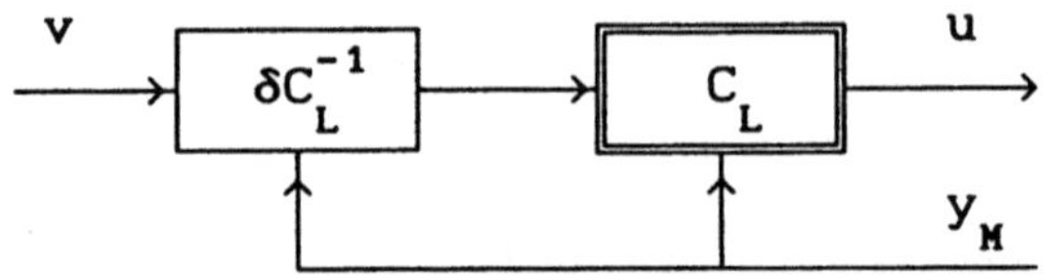


*Fig.8: Normal linearizing compensator*

Guido O. Guardabassi
*Politecnico di Milano*
*Dipartimento di Elettronica*
*Piazza Leonardo da Vinci 32*
*20133 Milano (Italy)*
*FAX ++39.2.23993587*

# Attenuation of Disturbances in Nonlinear Control Systems

ALBERTO ISIDORI

*Dedicato al maestro e amico Antonio Ruberti, in occasione del suo 65esimo compleanno, con profonda gratitudine per essere stato introdotto e autorevolmente guidato nella Teoria dei Sistemi e del Controllo e in particolare per essere stato indirizzato, con felice preveggenza, allo studio dei sistemi nonlineari.*

## 1 Introduction

In the last few years, the solution of the $H_\infty$ (sub)optimal control problem via state-space methods was developed by several authors (for a rather comprehensive coverage of this subject the reader may consult the paper [1] and the theses [2] [3]). In the state-space formulation, the problem of minimizing the $H_\infty$ norm (or, equivalently, the $L_2$ gain) of a closed loop system is viewed as a two-person, zero sum, differential game and, thus, the existence of the desired controller can be related to the existence of a solution of the algebraic Riccati equations arising in linear quadratic differential game theory (see, e.g. [4], [5] and [6]).

The advantage of the state-space approach and of the related game-theoretic interpretation is that the analysis of the $H_\infty$ control problem for a linear system (namely, the characterization of necessary and sufficient conditions for the existence of a solution, the construction of one particular solution and the description of the set of all possible solutions) can be rendered exclusively dependent on *time domain* arguments, thus laying the basis for appropriate extension to broader classes of systems.

The analysis of the "nonlinear equivalent" of the $H_\infty$ control problem has been addressed in a few recent papers (see [7], [8], [9], [10], [11]). In this domain, the Riccati equation of the linear quadratic differential game is replaced by a particular type of Hamilton-Jacobi equation, known as *Isaacs equation*. As a matter of fact, it has been rather straightforward to realize that the existence of a solution of the appropriate Isaacs equation is a sufficient condition for the existence of a *full-information* feedback law providing disturbance attenuation, in the sense of the $L_2$ gain [8], [9]. In addition, it has also been shown that, under appropriate assumptions, the existence of such a solution is a necessary condition for the solvability of the problem [10] and that the design of a control law based on *measurement* feedback is also possible [11].

A basic system-theoretic issue underlying the analysis of such a design problem is the existence a state-space characterization of the $L_2$ gain of a system. Illustrated in [12], [13], [14], for the case of linear systems and originally in [15], [16] for the more general case of nonlinear systems affine in the input, this characterization essentially provides the arguments to understand that the existence of solutions to certain Riccati, or Hamilton-Jacobi-Isaacs, equations is a necessary condition for the solvability of the problem under consideration.

The purpose of this paper is to present some of the basic concepts and properties underlying the solution of the nonlinear equivalent of the $H_\infty$ control problem and the related issue of characterizing the $L_2$ gain of a system. In particular, the purpose is to analyze and estabilsh new results for those nonlinear systems which possibly have a "critical" linear approximation. The existing results, in fact, say very little about this interesting and challenging class of problems.

## 2 Problem statement

Consider a nonlinear system modeled by equations of the form

$$\begin{aligned} \dot{x} &= f(x) + g_1(x)w + g_2(x)u \\ z &= h_1(x) + k_{11}(x)w + k_{12}(x)u \\ y &= h_2(x) + k_{21}(x)w \end{aligned} \tag{1}$$

The first equation of this system describes a plant with *state* $x$, defined on a neighborhood $X$ of the origin in $\mathbb{R}^n$ with *control input* $u \in \mathbb{R}^m$ and subject to a set of *exogenous input* variables $w \in \mathbb{R}^r$ which includes *disturbances* (to be rejected) and/or *references* (to be tracked).

The second equation defines a *penalty* variable $z \in \mathbb{R}^s$, which may include a *tracking error*, for instance the difference between the actual plant output and its desired reference behavior, expressed as a function of some of the exogenous variables $w$, as well as a *cost* of the input $u$ needed to achieve the prescribed control goal.

The third equation defines a set of *measured variables* $y \in \mathbb{R}^p$, which also are functions of the state plant $x$ and the exogenous input $w$.

The mappings $f(x), g_1(x), g_2(x), h_1(x), h_2(x), k_{11}(x), k_{12}(x), k_{21}(x)$, are smooth (i.e. $C^\infty$) mappings defined in a neighborhood of the origin in $\mathbb{R}^n$. We assume also that $f(0) = 0$, $h_1(0) = 0$ and $h_2(0) = 0$.

The control action to (1) is to be provided by a *controller*, which processes the measured variable $y$ and generates the appropriate control input $u$, and is modeled by equations of the form

$$\begin{aligned} \dot{\xi} &= \eta(\xi, y) \\ u &= \theta(\xi, y), \end{aligned} \tag{2}$$

in which $\xi$ is defined on a neighborhood $\Xi$ of the origin in $\mathbb{R}^\nu$ and $\eta : \Xi \times \mathbb{R}^p \to \mathbb{R}^\nu$, $\theta : \Xi \times \mathbb{R}^p \to \mathbb{R}^m$ are $C^k$ functions (for some $k \geq 1$), satisfying $\eta(0,0) = 0$ and $\theta(0,0) = 0$.

The purpose of the control is twofold: *to achieve closed loop stability* and *to attenuate the influence* of the exogenous input $w$ on the penalty variable $z$. A controller which locally asymptotically stabilizes the equilibrium $(x, \xi) = (0,0)$ of the closed loop system is said to be an *admissible controller*. The requirement of disturbance attenuation - on the other hand - may be dealt with in several different manners, depending on the specific class of exogenous signals to be considered and /or the performance criteria chosen to evaluate the penalty variable. In the present paper we consider a characterization based on the so-called $L_2$ gain, which is usually defined as follows. Consider a nonlinear system $\Sigma$ described by equations of the form

$$\begin{aligned} \dot{x} &= f(x) + g(x)u \\ y &= h(x). \end{aligned}$$

with $x \in \mathbb{R}^n$, $u \in \mathbb{R}^m$ and $y \in \mathbb{R}^p$. Let $U_\rho$ denote the open ball in $\mathbb{R}^n$ of radius $\rho$ centered at $x = 0$ and $\mathcal{U}_\delta$ the set of all piecewise continuous functions $u : \mathbb{R} \to \mathbb{R}^m$ satisfying $\|u(t)\| < \delta$ for all $t \in \mathbb{R}$. Suppose $x = 0$ is a locally asymptotically stable equilibrium of $\Sigma$. Then, for each $\epsilon > 0$ there exists $\delta(\epsilon) > 0$ such that for every input $u(\cdot) \in \mathcal{U}_{\delta(\epsilon)}$ the state response $x(\cdot)$ of $\Sigma$ from the initial state $x(0) = 0$ exists and satisfies $x(t) \in U_\epsilon$ for all $t \geq 0$. $\Sigma$ is said to have $L_2$ gain less than or equal to $\gamma$ if for each $u(\cdot) \in \mathcal{U}_{\delta(\epsilon)}$ the response $y(\cdot)$ of $\Sigma$ to $u(\cdot)$ from the initial state $x(0) = 0$ satisfies

$$\int_0^t \|y(s)\|^2 ds \leq \gamma^2 \int_0^t \|u(s)\|^2 ds.$$

for all $t > 0$. In what follows, by *problem of local disturbance attenuation with internal stability* we mean the problem of finding an admissible controller yielding a closed loop system with an $L_2$ gain which less than or equal to a prescribed number $\gamma$.

## 3 Computing the $L_2$ gain

### 3.1 *The scenario in the case of linear systems*

In this subsection, we recall some well-known results related to the characterization of the $L_2$ gain of a linear system (see [13], [14]).

**Lemma 3.1** *Consider a linear system* $\Sigma$

$$\begin{aligned} \dot{x} &= Ax + Bu \\ y &= Cx. \end{aligned}$$

*Suppose that $\sigma(A) \subset \mathbb{C}^-$ and that $(A, B)$ is reachable.*

*The response of $\Sigma$ from the initial state $x(0) = 0$ satisfies*

$$\int_0^t (\gamma^2\|u(s)\|^2 - \|y(s)\|^2)ds \geq 0 \tag{3}$$

*for every piecewise continuous input and every $t > 0$, if and only if the algebraic Riccati equation*

$$A^T X + XA + \frac{1}{\gamma^2} XBB^T X + C^T C = 0 \tag{4}$$

*has a real symmetric solution. If this is the case, then there exists precisely one real symmetric solution $X^-$ of (4) having the additional property that $\sigma(A + BB^T X^-) \subset cl(\mathbb{C}^-)$ and one real symmetric solution $X^+$ of (4) having the additional property that $\sigma(A + BB^T X^+) \subset cl(\mathbb{C}^+)$. Moreover, $X^+ \geq X^- \geq 0$ and every real symmetric solution of (4) satisfies $X^+ \geq X \geq X^-$. If $(C, A)$ is observable, $X^- > 0$. The strict inequalities:*
$X^+ - X^- > 0$,
$\sigma(A + BB^T X^-) \subset \mathbb{C}^-$,
$\sigma(A + BB^T X^+) \subset \mathbb{C}^+$,
$\int_0^t (\tilde{\gamma}^2\|u(s)\|^2 - \|y(s)\|^2)ds \geq 0$ *for some* $0 \leq \tilde{\gamma} < \gamma$,
*all hold simultaneously.*

*Remark.* If $(A, B)$ is not reachable, some of the properties described in this statement still hold. As a matter of fact (see [12]), provided that $\sigma(A) \subset \mathbb{C}^-$, the inequality

$$\int_0^t (\tilde{\gamma}^2\|u(s)\|^2 - \|y(s)\|^2)ds \geq 0$$

holds for some $0 \leq \tilde{\gamma} < 0$ if and only if the hamiltonian matrix

$$H = \begin{pmatrix} A & \frac{1}{\gamma^2}BB^T \\ -C^T C & -A^T \end{pmatrix}$$

has no eigenvalue on the imaginary axis. If this is the case, there exists precisely one real symmetric solution $X^-$ of (4) having the additional property that $\sigma(A + BB^T X^-) \subset \mathbb{C}^-$. $X^- \geq 0$ and every real symmetric solution of (4) satisfies $X \geq X^-$. If $(C, A)$ is observable, $X^- > 0$. ◁

### 3.2 *The Hamilton-Jacobi equation of the $L_2$-gain*

In what follows, we discuss possible extensions of the results described above to the case of a nonlinear system $\Sigma$ of the form

$$\begin{aligned} \dot{x} &= f(x) + g(x)u \\ y &= h(x). \end{aligned} \tag{5}$$

In this case, the algebraic Riccati equation (4) is replaced by a Hamilton-Jacobi equation, which can be motivated in the following way. Consider the function $F : \mathbb{R}^n \times \mathbb{R}^n \times \mathbb{R}^m \to \mathbb{R}$ defined by

$$H(x,p,u) = p^{\mathrm{T}}(f(x) + g(x)u) - \gamma^2 u^{\mathrm{T}} u + h^{\mathrm{T}}(x)h(x).$$

which is a polynomial of degree two in $u$. This function has a unique maximum in $u$, at

$$u^*(x,p) = \frac{1}{2\gamma^2} g^{\mathrm{T}}(x)p,$$

which is given by

$$H^*(x,p) = p^{\mathrm{T}} f(x) + \frac{1}{4\gamma^2} p^{\mathrm{T}} g(x)^{\mathrm{T}} g(x)p + h^{\mathrm{T}}(x)h(x).$$

Therefore,

$$H(x,p,u) = H^*(x,p) - \gamma^2(u - u^*(x,p))^{\mathrm{T}}(u - u^*(x,p)). \tag{6}$$

Suppose $V : U_\rho \to \mathbb{R}$ is a $C^2$ function satisfying the Hamilton-Jacobi equation

$$H^*(x, V_x^{\mathrm{T}}) = 0$$

with $V(0) = 0$. If such a function exists, then the identity (6) implies the inequality

$$V_x(f(x) + g(x)u) \leq \gamma^2 u^{\mathrm{T}} u - h^{\mathrm{T}}(x)h(x).$$

If $x = 0$ is a locally asymptotically stable equilibrium of $f(x)$, for each $\epsilon > 0$ there exists $\delta(\epsilon) > 0$ such that for every input $u(\cdot) \in \mathcal{U}_{\delta(\epsilon)}$ the response $x(\cdot)$ of $\Sigma$ from the initial state $x(0) = 0$ exists and satisfies $x(t) \in U_\epsilon$ for all $t \geq 0$. Integrating the previous inequality along this response yields

$$V(x(t)) \leq \int_0^t (\gamma^2 \|u(s)\|^2 - \|y(s)\|^2) ds.$$

The asymptotic stability of $f(x)$ implies also that $V(x)$ is positive semidefinite (see Proposition 3.1 below). Therefore, it is concluded that the response $y(\cdot)$ of $\Sigma$ from the initial state $x(0) = 0$ satisfies

$$\int_0^t (\gamma^2 \|u(s)\|^2 - \|y(s)\|^2) ds \geq 0$$

for every $u(\cdot) \in \mathcal{U}_{\delta(\epsilon)}$ and for every $t > 0$.

### 3.3 *Preliminary properties of the solution set*

The previous conclusions motivate the importance, in the study of the $L_2$ gain of a nonlinear affine system, of the Hamilton-Jacobi equation $H^*(x, V_x^{\mathrm{T}}) = 0$, that is, of the equation

$$V_x f(x) + \frac{1}{4\gamma^2} V_x g(x)(V_x g(x))^{\mathrm{T}} + h^{\mathrm{T}}(x)h(x) = 0 \tag{7}$$

whose properties are reviewed in the sequel.

One of the fundamental features of this equation is that, if we set

$$\alpha(x) = \frac{1}{2\gamma^2}(V_x g(x))^{\mathrm{T}} \tag{8}$$

the basic identity (6) yields

$$V_x(f(x) + g(x)u) = \gamma^2 u^{\mathrm{T}} u - h^{\mathrm{T}}(x)h(x) - \gamma^2(u - \alpha(x))^{\mathrm{T}}(u - \alpha(x)). \tag{9}$$

In fact, this identity is the point of departure of most of the properties described below.

For obvious reasons, we consider solutions of the Hamilton-Jacobi equation (7) satisfying $V(0) = 0$. Since we are - in particular - interested in positive semidefinite solutions of this equation, $x = 0$ is a point of minimum for any of such functions, and therefore $V_x(0) = 0$. It is also useful observe that, under appropriate additional assumptions on the system, *any* function $V(x)$ which solves the Hamilton-Jacobi equation (7) is *positive semidefinite.*

**Proposition 3.1** (see [15]) *If $V(x)$ is any solution of the Hamilton-Jacobi equation (7), and $f(x)$ is locally asymptotically stable, then $V(x)$ is positive semidefinite in a neighborhood of $x = 0$.*

*Proof.* By construction, along any trajectory of $\dot{x} = f(x)$ we have

$$\begin{aligned} V(x(t)) - V(x) &= \int_0^t V_x f(x(s))ds \\ &= -\tfrac{1}{4\gamma^2} \int_0^t \|V_x g(x(s))\|^2 ds - \int_0^t \|h(x(s))\|^2 ds \leq 0 \end{aligned}$$

where $x = x(0)$. Asymptotic stability of $f(x)$ implies $\lim_{t\to\infty} V(x(t)) = 0$, and therefore

$$V(x) \geq 0$$

for all $x$ in a neighborhood of $x = 0$. ◁

Moreover, it is also easy to show that if $\Sigma$ satisfies a suitable observability property, *any* function $V(x)$ which solves the Hamilton-Jacobi equation (7) is even positive *definite*. To this end, recall that $(h(x), f(x))$ is said to be *locally zero-state observable* if there exists a neighborhood $U$ of $x = 0$ such that, for each $x \in U$, the integral curve $x(\cdot)$ of $\dot{x} = f(x)$ satisfying $x(0) = 0$ exists for all $t \geq 0$ and, moreover,

$$h(x(t)) = 0 \quad \text{for all} \quad t \geq 0 \quad \Rightarrow \quad x = 0.$$

**Proposition 3.2** (see [16]) *If $V(x)$ is any solution of the Hamilton-Jacobi equation (7), $f(x)$ is locally asymptotically stable and $(h(x), f(x))$ is locally zero-sate observable, then $V(x)$ is positive definite in a neighborhood of $x = 0$.*

*Proof.* From the proof of Proposition 3.1, we see that if $V(x) = 0$ for some $x$, then

$$0 \leq V(x(t)) \leq -\int_0^t \|h(x(s))\|^2 ds \leq 0.$$

Thus necessarily $h(x(t)) = 0$ for all $t > 0$ and zero-state observability implies $x = 0$. ◁

Finally, we recall the following well-known property of the solutions of a Hamilton Jacobi equation.

**Proposition 3.3** *Let $H : I\!R^n \times I\!R^n \to I\!R$ be a smooth function, with $H(0,0) = 0$. Consider the Hamiltonian system*

$$\begin{aligned} \dot{x} &= \left(\frac{\partial H^*(x,p)}{\partial p}\right)^{\mathrm{T}} \\ \dot{p} &= -\left(\frac{\partial H^*(x,p)}{\partial x}\right)^{\mathrm{T}}. \end{aligned} \tag{10}$$

*The Hamilton-Jacobi equation $H^*(x, V_x^{\mathrm{T}}) = 0$ has a $C^2$ solution $V : U_\rho \to I\!R$, with $V(0) = 0$ and $V_x(0) = 0$ if and only if the manifold*

$$M = \{(x,p) \in U_\rho \times I\!R^n : p = V_x^{\mathrm{T}}(x)\}$$

*is an invariant manifold of the corresponding Hamiltonian system (10). In particular, if $V(x)$ is a solution of the Hamilton-Jacobi equation (7) then the restriction of the Hamiltonian system to $M$ is given by*

$$\dot{x} = f(x) + g(x)\alpha(x).$$

3.4 *Basic properties of the solution set*

As observed in subsection 3.2, the existence of a solution of the Hamilton-Jacobi equation (7) implies that the $L_2$ gain of $\Sigma$ is less than on equal to $\gamma$.

**Proposition 3.4** (see [16]) *Suppose $f(x)$ is locally asymptotically stable. Suppose the Hamilton-Jacobi equation (7) has a solution $V : U_\rho \to I\!R$. Then there* *exists $\delta > 0$ such that the response $y(\cdot)$ of $\Sigma$ from the initial state $x(0) = 0$ satisfies*

$$\int_0^t (\gamma^2 \|u(s)\|^2 - \|y(s)\|^2) ds \geq 0 \tag{11}$$

*for every $u(\cdot) \in \mathcal{U}_\delta$ and every $t > 0$.*

*Proof.* By Proposition 3.1, there is $\rho' \leq \rho$ such that the restriction of $V(x)$ to $U_{\rho'}$ is positive semidefinite. Thus, the result follows from the analysis described in subsection 3.2. ◁

In the following statements, we assume the existence of two solutions $V^-(x)$ and $V^+(x)$ of the Hamilton-Jacobi equation (7). For conveniencey, we refrain from explicitly indicating the open neighborhood of $x = 0$ on which both are defined and, in case the hypotheses of Propositions 3.1 (respectively 3.2) hold, positive semidefinite (respectively, definite). With each one of these solutions, we associate a feedback law of the form (8), that is

$$\alpha^-(x) = \frac{1}{2\gamma^2}(V_x^- g(x))^{\mathrm{T}}$$

and, respectively,

$$\alpha^+(x) = \frac{1}{2\gamma^2}(V_x^+ g(x))^{\mathrm{T}}.$$

Finally, motivated by the properties described in Proposition 3.3 , we consider the two systems

$$\dot{x} = f(x) + g(x)\alpha^-(x) \tag{12}$$

$$\dot{x} = f(x) + g(x)\alpha^+(x) \tag{13}$$

**Proposition 3.5** *Let $V^-(x)$ be a solution of (7) with the additional property that $f(x) + g(x)\alpha^-(x)$ is asymptotically stable. Then any other solution $V(x)$ of (7) satisfies $V(x) \geq V^-(x)$. As a consequence, there exists at most one such solution. Let $V^+(x)$ be a solution of (7) with the additional property that $f(x) + g(x)\alpha^+(x)$ is asymptotically antistable. Then any other solution $V(x)$ of (7) satisfies $V(x) \leq V^+(x)$. As a consequence, there exists at most one such solution.*

*Proof.* Consider the basic identity (9) for $V^-(x)$, for any other solution $V(x)$, and set $u = \alpha^-(x)$, to obtain

$$V_x^-(f(x) + g(x)\alpha^-(x)) = \gamma^2\|\alpha^-(x)\|^2 - \|h(x)\|^2$$
$$V_x(f(x) + g(x)\alpha^-(x)) = \gamma^2\|\alpha^-(x)\|^2 - \|h(x)\|^2 - \gamma^2\|\alpha^-(x) - \alpha(x)\|^2.$$

Thus, the difference $V(x) - V^-(x)$ satisfies

$$(V - V^-)_x(f(x) + g(x)\alpha^-(x)) \leq 0$$

that is

$$V(x(t)) - V^-(x(t)) \leq V(x) - V^-(x)$$

along the integral curve $x(\cdot)$ of (12) satisfying $x(0) = x$. Then, taking the limit as $t \to \infty$, the asymptotic stability of (12) implies $V(x) \geq V^-(x)$. The other part of the Proposition is proven in the same way. ◁

If $V(x)$ is a solution of the Hamilton-Jacobi equation (7) and $f(x)$ is locally asymptotically stable, the inequality (11) holds and the system in question is *dissipative* (see [17]) with respect to the *supply rate*

$$w(y,u) = \gamma^2 u^{\mathrm{T}} u - y^{\mathrm{T}} y.$$

In fact, $V(x)$ is a *storage function* for $\Sigma$. In the following theorem, we compare the solutions of (7) with two special storage functions of $\Sigma$: the available storage $V^a(x)$ and the required supply $V^r(x)$. To this end, recall that the *available storage* of a dissipative system is the function

$$V^a(x) = \sup_{\substack{x(0)=x \\ u(\cdot) \\ t_1 \geq 0}} \left( -\int_0^{t_1} w(y(t), u(t))dt \right),$$

while the *required supply*, which is well-defined under the additional assumption that $(f(x), g(x))$ is reachable from the zero-state, is the function

$$V^r(x) = \inf_{\substack{x(t_{-1})=0, x(0)=x \\ u(\cdot) \\ t_{-1} \leq 0}} \left( \int_{t_{-1}}^{0} w(y(t), u(t))dt \right).$$

*Reachability from the zero state* means, as usual, that for every $x \in \mathbb{R}^n$ there exist a time $t_1 \leq 0$ and a piecewise continuos input which transfers the state $x = 0$ at time $t = t_1$ into the state $x = x$ at time $t = 0$.

**Proposition 3.6** *Suppose $f(x)$ is asymptotically stable. Let $V^-(x)$ be a solution of (7) with the additional property that $f(x) + g(x)\alpha^-(x)$ is locally asymptotically stable and $V^+(x)$ a solution of (7) with the additional property that $f(x) + g(x)\alpha^+(x)$ is locally asymptotically antistable. Then $V^+(x) \geq V^-(x) \geq 0$ and any other solution $V(x)$ of (7) satisfies $V^+(x) \geq V(x) \geq V^-(x)$. If $(h(x), f(x))$ is zero-state observable, $V^-(x) > 0$. Moreover, $V^-(x) = V^a(x)$, the available storage of $\Sigma$ and, if $(f(x), g(x))$ is reachable from the zero-state, $V^+(x) = V^r(x)$, the required supply of $\Sigma$.*

*Proof.* We only have to show that $V^-(x) = V^a(x)$ and $V^+(x) = V^r(x)$. Consider the basic identity (9) for $V^-(x)$, with $u = \alpha^-(x)$. Then, integrating along the trajectory of (12) satisfying $x(0) = x$, we obtain

$$-\int_0^t w(y(s), \alpha^-(x(s)))ds = V^-(x) - V^-(x(t))$$

Form the definition of $V^a(x)$ and from the fact that system (12) is locally asympototically stable, it is deduced that

$$V^a(x) \geq \sup_{t\geq 0}\Big(V^-(x) - V^-(x(t))\Big) = V^-(x).$$

On the other hand, $V^a(x) \leq V^-(x)$, because $V^-(x)$ is a storage function for $\Sigma$ (see [17]). Thus, the result for $V^-(x)$ follows. The other part of the Proposition is proven by similar arguments. ◃

Finally, in the following statement we describe a result which in some sense extends the characterization of the strict inequalities in Lemma 3.1.

**Proposition 3.7** *Suppose $f(x)$ is asymptotically stable. Let $V^-(x)$ and $V^+(x)$ be two solutions of (7). Suppose that no trajectory of either (12) or (13), other than the trivial trajectory $x(t) = 0$, is such that*

$$(V_x^+ - V_x^-)g(x(t)) = 0$$

*for all $t$. Then, the following are equivalent:*
*(a) $V^+(x) - V^-(x) > 0$*
*(b) system (12) is locally asymptotically stable, system (13) is asymptotically antistable, and there exists $\delta > 0$ such that the response $y(\cdot)$ of $\Sigma$ from the initial state $x(0) = 0$ satisfies the strict inequality*

$$\int_0^t (\gamma^2\|u(s)\|^2 - \|y(s)\|^2)ds > 0 \tag{14}$$

*for every nonzero $u(\cdot) \in \mathcal{U}_\delta$ and every $t > 0$.*

*Proof.* (a) ⇒ (b). Set $W(x) = V^+(x) - V^-(x)$. Then, the basic identity (9) yields (see the beginning of the proof of Proposition 3.5)

$$W_x(f(x) + g(x)\alpha^-(x)) = -\gamma^2\|\alpha^-(x) - \alpha^+(x)\|^2.$$

Therefore, $W(x)$ is nonicreasing along the trajectories of (12), and the equilibrium $x = 0$ is stable in the sense of Lyapunov. By Lasalle's invariance principle, for every sufficiently small $\epsilon > 0$ there is $\delta > 0$ such that all trajectories of (12) satisfying $\|x(0)\| < \delta$ converge, as $t \to \infty$, to the largest invariant subset of

$$N = \{x \in U_\epsilon : \alpha^-(x) - \alpha^+(x) = 0\}$$

Observe now that

$$\alpha^-(x) - \alpha^+(x) = -\frac{1}{\gamma^2}(W_x g(x))^{\mathrm{T}}.$$

Thus,

$$N = \{x \in U_\epsilon : W_x(x)g(x) = 0\}.$$

Since by hypothesis the only trajectory of (12) entirely contained in this set is the trivial trajectory, it is concluded that (12) is locally asymptotically stable. The antistability of (13) is proved by identical arguments.

Finally, to prove the strict inequality (14), observe (see again the basic identity (9)) that by definition $V^+(x)$ and $V^-(x)$ satisfy an inequality of the form

$$V(x(t_2)) - V(x(t_1)) \leq \int_{t_1}^{t_2} (\gamma^2\|u(s)\|^2 - \|y(s)\|^2)ds$$

for every $t_2 \geq t_1$. Suppose - by contradiction - that the left-hand side of (14) is 0 for some nonzero $u(\cdot)$, along a trajectory satisfying $x(0) = 0$. Then necessarily $y(\cdot)$ is not identically zero and this in turn implies that $x(\cdot)$ is not identically zero. Let $t_0$ be a time $0 \leq t_0 \leq t$ such that $x = x(t_0) \neq 0$. Then,

$$\begin{aligned} 0 &= \int_0^{t_0} (\gamma^2\|u(s)\|^2 - \|y(s)\|^2)ds + \int_{t_0}^{t} (\gamma^2\|u(s)\|^2 - \|y(s)\|^2)ds \\ &\geq V^+(x(t_0)) - V^+(x(0)) + V^-(x(t)) - V^-(x(t_0)) \\ &= W(x) + V^+(x(t)). \end{aligned}$$

Since $V^-(x(t)) \geq 0$, this implies $W(x) \leq 0$, which contradicts the assumption.

(a) $\Leftarrow$ (b) We already know from Proposition 3.6 that $W(x) \geq 0$. Suppose - by contradiction - that $W(x) = 0$ for some nonzero $x$. Let $x^-(t)$, with $t \geq 0$, denote the solution of (12) satisfying $x^-(0) = x$ and $x^+(t)$, with $t \leq 0$, denote the solution of (13) satisfying $x^+(0) = x$. Consider the input $u^*(\cdot)$ defined by

$$\begin{aligned} u^*(t) &= \alpha^+(x^+(t)) \qquad t < 0 \\ &= \alpha^-(x^-(t)) \qquad t > 0. \end{aligned}$$

Finally let, for convenience, $I^*(x^0, T)$ denote the integral

$$I^*(x^0, T) = \int_{-T}^{T} (\gamma^2\|u^*(s)\|^2 - \|y^*(s)\|^2)ds$$

where $y^*(\cdot)$ is the response of $\Sigma$ to the input $u^*(\cdot)$ starting from the initial state $x^0$ at time $t = -T$.

By construction

$$\begin{aligned} &I^*(x^+(-T), T) = \\ &= \int_{-T}^{0} (\gamma^2\|u^*(s)\|^2 - \|y^*(s)\|^2)ds + \int_0^{T} (\gamma^2\|u^*(s)\|^2 - \|y^*(s)\|^2)ds \\ &= V^+(x) - V^+(x^+(-T)) + V^-(x^-(T)) - V^-(x) \\ &= V^-(x^-(T)) - V^+(x^+(-T)). \end{aligned}$$

Since $\lim_{T\to\infty} x^+(-T) = 0$, then

$$\lim_{T\to\infty} |I^*(0, T) - I^*(x^+(-T), T)| = 0.$$

Moreover, since also $\lim_{T\to\infty} x^-(T) = 0$,

$$\lim_{T\to\infty} I^*(x^+(-T), T) = 0.$$

Therefore

$$\lim_{T\to\infty} I^*(0, T) = 0.$$

This contradicts the strict inequality (14), because the input $u^*(\cdot)$ is such that

$$\inf_{T\geq T_0>0}\Big(\int_{-T}^{T} \|u^*(s)\|^2 ds\Big) > 0$$

for any $T_0 > 0$.◁

*Remark.* Note that the hypothesis of the Proposition, namely the condition that no trajectory of either (12) or (13), other than the trivial trajectory $x(t) = 0$, is such that

$$(V_x^+ - V_x^-)g(x(t)) = 0$$

for all $t$, can be replaced by a somewhat stronger, but nevertheless much more explicit, hypothesis involving only the function $W(x) = V^+(x) - V^-(x)$ and repeated Lie Brackets of the vector fields $f(x) + g(x)\alpha^-(x)$, $f(x) + g(x)\alpha^+(x)$, and the columns of $g(x)$. To this purpose, consider the collection of vector fields

$$F^- = \{ad^k_{f+g\alpha^-} g_i(x) : k \geq 0, 1 \leq i \leq m\}$$

and define, for sufficiently small $\epsilon$, the set

$$S^- = \{x \in U_\epsilon : W_x X(x) = 0 \text{ for all } X \in F^-\}.$$

Then, it is easy to show that the condition

$$S^- = \{0\} \tag{15}$$

implies that no trajectory of (12), other than the trivial trajectory $x(t) = 0$, renders $W_x g(x(t)) = 0$ for all $t$.

In fact, let $x(t)$ be a trajectory of (12) yielding $W_x g(x(t)) = L_g W(x(t)) = 0$ and note that this trajectory also satisfies, as shown in the proof of the Proposition, $L_{\tilde{f}} W(x(t)) = 0$, where

$$\tilde{f}(x) = f(x) + g(x)\alpha^-(x).$$

Moreover, since $L_{\tilde{f}} W(x)$ is negative semidefinite, $L_{\tilde{f}} W(x)$ is maximal along $x(t)$ and

$$\Big(\frac{\partial L_{\tilde{f}} W(x)}{\partial x}\Big)_{x=x(t)} = 0.$$

From these conditions, it is immediate to deduce that

$$\begin{aligned} L_{[\tilde{f},g]}W(x(t)) &= L_{\tilde{f}}L_gW(x(t)) - L_gL_{\tilde{f}}W(x(t)) \\ &= \frac{dL_gW(x(t))}{dt} - \frac{\partial L_{\tilde{f}}W}{\partial x}g(x(t)) = 0. \end{aligned}$$

An simple iteration yields

$$L_X W(x(t)) = 0$$

for all $X \in F^-$. Thus, the condition $S^- = \{0\}$ clearly implies $x(t) = 0$. A similar argument can be used for (13).

The condition (15), for system (12), and the corresponding condition for (13) are special forms of the well-know algebraic conditions for controllability. In fact, it can be shown that if the system $\Sigma$ is linear, and the hypothesis (a) is fulfilled, both conditions are implied by the standard controllability condition, provided $V^-(x)$ and $V^+(x)$ are quadratic functions (which is indeed the case). In this case, the set $S^-$ reduces for example to the set

$$S = \{x \in I\!R^n : x^T P(A + BF)^k B = 0, k \geq 0\}$$

where $P$ is such that $W(x) = \frac{1}{2}x^T Px$ and $F$ such that $Fx = \alpha^-(x)$. $P$ being nonsingular ($W(x)$, by hypothesis (a), is positive definite), it follows that $S = \{0\}$ if and only if $(A, B)$ is controllable. ◁

We wish to stress that the occurrence of the strict inequality (14) is related to the asymptotic stability of (12), to the asymptotic instability of (13), and to the fact that the difference $V^+(x) - V^-(x)$ is positive definite, and it may well occurr if the equilibrium $x = 0$ of (12) and (13) is *not* hyperbolic. In other words, the strict inequality may well occur in cases in which the jacobian matrices of $f(x) + g(x)\alpha^-(x)$ and $f(x) + g(x)\alpha^+(x)$ at the equilibrium $x = 0$ (and consequently the jacobian matrix of the Hamiltonian system (10) at the equilibrium $(x, p) = (0, 0)$) have eigenvalues with zero real part. This fact is clarified in the following simple example.

*Example.* Consider the system

$$\dot{x} = -4x^3 + u$$
$$y = 2\sqrt{3}x^3$$

and let $\gamma = 1$. The Hamilton-Jacobi equation (7) is this case is

$$-V_x 4x^3 + \frac{1}{4}(V_x)^2 + 12x^6 = 0.$$

The two functions

$$V^-(x) = x^4 \qquad V^+(x) = 3x^4$$

are solutions of this equation and

$$W(x) = V^+(x) - V^-(x) = 2x^4$$

is positive definite, i.e. condition (a) holds. Since $W_x g(x) = 8x^3$ the hypothesis of Proposition 3.7 holds as well, and therefore the properties indicated in (b) hold. In fact,

$$\alpha^-(x) = 2x^3 \qquad \alpha^+(x) = 6x^3$$

are such that

$$f(x) + g(x)\alpha^-(x) = -2x^3$$

is asymptocically stable and

$$f(x) + g(x)\alpha^+(x) = 2x^3$$

is asymptotically antistable. Moreover, to double-check that the strictly inequality (14) also holds, consider the convex combination $V(x) = \frac{1}{2}V^+(x) + \frac{1}{2}V^-(x) = 2x^4$, which satisfies

$$V_x(f(x) + g(x))u - u^2 + y^2 = -4x^6 - (u - 4x^3)^2 \leq -4x^6$$

for all $u$. Thus, along any trajectory of the system starting from $x(0) = 0$,

$$\int_0^t (u^2(s) - y^2(s))ds \geq 2x^4(t) + \int_0^t 4x^6(s)ds > 0$$

prodided $u(\cdot)$, and therefore $x(\cdot)$, is not identically zero.

Note that the Hamiltonian system associated with the equation (7) in the present case is

$$\begin{aligned} \dot{x} &= -4x^3 + \tfrac{1}{2}p \\ \dot{p} &= -72x^5 + 12x^2 p. \end{aligned}$$

The jacobian matrix of the right-hand side at $(x,p) = (0,0)$, which has the form

$$H = \begin{pmatrix} 0 & \frac{1}{2} \\ 0 & 0 \end{pmatrix},$$

has only one one-dimensional invariant subspace. Nevertheless, the Hamiltonian systems has two one-dimensional invariant manifolds,

$$\begin{aligned} M^- &= \{(x,p) : p = 4x^3\} \\ M^+ &= \{(x,p) : p = 12x^3\} \end{aligned}$$

The restriction of the hamiltonian system to $M^-$ is a asymptotically stable, while its restriction to $M^+$ is asymptotically antistable. ◁

### 3.5 *Existence of solutions*

Van der Schaft [10] has shown that under appropriate hypotheses on $\Sigma$, the sufficient condition described in Proposition 3.4 is also necessary. The hypoteses in question are expressed in terms of the parameters of the linear approximation of $\Sigma$ at the equilibrium $x = 0$, namely of the linear system

$$\begin{aligned} \dot{x} &= Fx + Gu \\ y &= Hx, \end{aligned} \tag{16}$$

characterized by the triplet of matrices $(F, G, H)$ defined by

$$F = \left(\frac{\partial f}{\partial x}\right)_{x=0} \qquad G = g(0) \qquad H = \left(\frac{\partial h}{\partial x}\right)_{x=0}$$

**Lemma 3.2** (see [8]) *Suppose $X^-$ (respectively $X^+$) is a solution of the algebraic Riccati equation*

$$F^{\mathrm{T}}X + XF + \frac{1}{\gamma^2}XGG^{\mathrm{T}}X + H^{\mathrm{T}}H = 0$$

*having the additional property that $\sigma(F + GG^{\mathrm{T}}X^-) \subset \mathbb{C}^-$ (respectively $\sigma(F + GG^{\mathrm{T}}X^+) \subset \mathbb{C}^+$). Then, there exists $\rho > 0$ and a smooth solution $V^- : U_\rho \to \mathbb{R}$ (respectively $V^+ : U_\rho \to \mathbb{R}$) of the Hamilton-Jacobi equation (7).*

**Lemma 3.3** (see [10]) *Suppose that, for some $\delta > 0$, the response $y(\cdot)$ of (5) from the initial state $x(0) = 0$ is well-defined and satisfies the inequality*

$$\int_0^t (\tilde{\gamma}^2\|u(s)\|^2 - \|y(s)\|^2)ds \geq 0. \tag{17}$$

*for every input $u(\cdot) \in \mathcal{U}_\delta$ and every $t > 0$. Then also the response $y(\cdot)$ of (16) from the initial state $x(0) = 0$ satisfies the inequality (17) for every piecewise continuous input and every $t > 0$.*

These two lemmas can be used, in conjunction with the results illustrated in subsection 3.1, to deduce a partial converse of Proposition 3.4.

**Proposition 3.8** (see [10]) *Consider the nonlinear system (5) and let (16) denote its linear approximation at the equilibrium $x = 0$. Suppose that $\sigma(F) \subset \mathbb{C}^-$ and that $(F, G)$ is reachable. Suppose that, for some $\delta > 0$, the response $y(\cdot)$ of $\Sigma$ from the initial state $x(0) = 0$ is well-defined and satisfies the inequality (17), with $0 \leq \tilde{\gamma} < \gamma$, for every input $u(\cdot) \in \mathcal{U}_\delta$ and every $t > 0$. Then there exists $\rho > 0$ and two smooth solutions $V^- : U_\rho \to \mathbb{R}$ and $V^+ : U_\rho \to \mathbb{R}$ of the Hamilton-Jacobi equation (7), with the additional property that system (12) is asymptotically stable in the first approximation, and system (13) is asymptotically antistable in the first approximation.*

*Remark.* If $(F, G)$ is not reachable, some of the properties described in this statement still hold. As a matter of fact, provided that $\sigma(F) \subset \mathbb{C}^-$, if the inequality

$$\int_0^t (\tilde{\gamma}^2 \|u(s)\|^2 - \|y(s)\|^2) ds \geq 0$$

holds, there exists precisely one solution $V^- : U_\rho \to \mathbb{R}$ of (7) having the additional property that system (12) is asymptotically stable in the first approximation. ◃

## 4 The $H_\infty$ gain of a nonlinear system

A fundamental property of stable linear systems is that the response $y(\cdot)$ starting from the initial state $x(0) = 0$ satisfies the inequality (3) for every piecewise continuous input and every $t > 0$, if and only if

$$\sup_{\omega \in \mathbb{R}} \|G(j\omega)\|_2 \leq \gamma,$$

that is the $H_\infty$ norm of the transfer matrix $G(s)$ of the system is less than or equal to $\gamma$.

In what follows we describe a similar equivalence for nonlinear systems. To this end, it is convenient to observe that the $H_\infty$ norm of a linear system has a simple time-domain interpretation which can easily be extended to more general classes of nonlinear systems. As a matter of fact, recall that in a stable linear system the response to a sinousidal input function (or, more in general, to any periodic and bounded input function) converges, as time tends to infinity, towards a uniquely defined response which is indepenent of the initial state and is itself a sinusoidal function, of the same frequency, with amplitude and phase which are uniquely determined by $G(j\omega)$. This response - which is called the *steady state response* - can be identified with the actual response - to the specific input - obtained from a properly chosen initial state.

If we let $y_{ss}(t)$ denote the steady state response of the system to any input of the form

$$u_{ss}(t) = cos(\omega t) u$$

(where $u \in \mathbb{R}^m$), then the $H_\infty$ norm of the transfer matrix $G(s)$ of the system can be expressed as

$$\|G\|_\infty = \sup_{\substack{\omega \in \mathbb{R} \\ \|u\|=1}} \frac{\frac{1}{T}\left(\int_t^{t+T} \|y_{ss}(s)\|^2 ds\right)^{\frac{1}{2}}}{\frac{1}{T}\left(\int_t^{t+T} \|u_{ss}(s)\|^2 ds\right)^{\frac{1}{2}}}.$$

where $T = 2\pi/\omega$ (and $t$ is any arbitrary time).

Note that this caracterization involves only *time-domain* defined quantities. Since the notion of steady-state response is not necessarily restricted to linear systems (see [18]), such an interpretation can be used to extend the the notion of $H_\infty$ norm to more general classes of systems. To this end, recall that in [18] it was observed that if in a nonlinear system

$$\dot{x} = f(x, w)$$

the equilibrium $(x, w) = (0, 0)$ is stable in the first approximation, and if the exogenous inputs are "generated" by a dynamical system

$$\dot{w} = s(w)$$

in which all trajectories are *periodic* functions with a given period $T$ (like, for instance, in the case on a harmonic oscillator), the composed system

$$\dot{x} = f(x, w) \quad (18)$$

$$\dot{w} = s(w) \quad (19)$$

has a *center manifold*, the graph

$$S = \{(x, w) \in W \times \mathbb{R}^n : x = \pi(w)\}$$

of a $C^1$ mapping $x = \pi(w)$, defined in a neighborhood of $W$ of the point $w = 0$. Since such a center manifold is locally attractive, if the exogenous input $w(\cdot)$ is such that $w(t) \in W$ for all $t$ and if the initial condition $x(0)$ is sufficiently close to that manifold (i.e. if $\|x(0) - \pi(w(0))\|$ is sufficiently small), then the integral curve $x(\cdot)$ of the system (18) converges, as time tends to infinity, to the (uniquely defined) integral curve $x_{ss}(\cdot)$ which passes through the initial state $\pi(w(0))$. The trajectory $x_{ss}(\cdot)$ is therefore the *steady state response* of the system to the (periodic) input $w(\cdot)$. Since the manifold $S$ is invariant, the integral curve $x_{ss}(\cdot)$ is such that $x_{ss}(t) = \pi(w(t))$ for all $t$ and, therefore, since $w(\cdot)$ is periodic with a given period $T$, so is also the corresponding steady state response $x_{ss}(\cdot)$.

Consider now a nonlinear system satisfying the assumptions of Proposition 3.8 (or the weaker assumptions in the subsequent Remark). This system is stable in the first approximation and therefore, for some $\delta > 0$, every *periodic* input $u_{ss}(\cdot) \in \mathcal{U}_\delta$ has a steady state response $x_{ss}(\cdot)$, which is a periodic function (with the same period $T$ as that of $u_{ss}(\cdot)$). Moreover, there exists a smooth function $V(x)$ such that

$$V(x(t_2)) - V(x(t_1)) \leq \int_{t_1}^{t_2} (\gamma^2 \|u(s)\|^2 - \|y(s)\|^2) ds$$

along any trajectory of the system. Evaluating this inequality along the trajectory of the steady state response and choosing $t_1 = t$ and $t_2 = t + T$, since

$x_{ss}(t_2) = x_{ss}(t_1)$, yields

$$\int_t^{t+T} ||y_{ss}(s)||^2 ds \le \gamma^2 \int_t^{t+T} ||u_{ss}(s)||^2 ds.$$

where $y_{ss}(t) = h(x_{ss}(t))$.

Thus

$$\sup_{u_{ss}(\cdot)\in\mathcal{U}_\delta} \frac{\frac{1}{T}\left(\int_t^{t+T} ||y_{ss}(s)||^2 ds\right)^{\frac{1}{2}}}{\frac{1}{T}\left(\int_t^{t+T} ||u_{ss}(s)||^2 ds\right)^{\frac{1}{2}}} \le \gamma$$

Note that the denominator and the numerator of this fraction, which are indeed independent of $t$, are the *r.m.s. values* of the periodic input $u_{ss}(\cdot)$ and that of the corresponding (periodic) steady state response $y_{ss}(\cdot)$.

## 5 Disturbance attenuation in the case of full information

For reasons of space, we review in this paper the solution of the problem of disturbance attenuation under the so called *full information* pattern, i.e. when the full state $x$ of the plant and the exogenous input $w$ are available for feedback. The more general case of arbitrary measurement feedback is discussed in [11].

In view of the results illustrated in Section 3, in order to solve the problem of disturbance attenuation, it is natural to seek a feedback law $\alpha(x,w)$ which - for some positive definite function $V : U_\epsilon \to \mathbb{R}$ - renders the inequality

$$(20) \qquad V(x(t_1)) - V(x(t_0)) + \int_{t_0}^{t_1} L(x(s), w(s), \alpha(x(s), w(s)))ds \le 0,$$

satisfied for all $t_1 \ge t_0$, where

$$L(x,w,u) = ||h_1(x) + k_{11}(x)w + k_{12}(x)u||^2 - \gamma^2||w||^2$$

and $x(t)$ is the integral curve of the closed loop system

$$\dot{x} = f(x) + g_1(x)w + g_2(x)\alpha(x,w).$$

Passing from the *dissipation* inequality (20) to its infinitesimal version, one is led to study the existence of a feedback $\alpha(x,w)$ and a positive definite function $V$ satisfying

$$(21) \qquad V_x(f(x) + g_1(x)w + g_2(x)\alpha(x,w)) + L(x,w,\alpha(x,w)) \le 0.$$

To this end, observe that the quantity

$$H(x,p,w,u) = p^{\mathrm{T}}(f(x) + g_1(x)w + g_2(x)u) + L(x,w,u)$$

is *quadratic* in $(w,u)$. If $(\alpha_1,\alpha_2)$ is a pair such that

$$\left(\frac{\partial H}{\partial w}\right)_{(w,u)=(\alpha_1,\alpha_2)} = 0, \qquad \left(\frac{\partial H}{\partial u}\right)_{(w,u)=(\alpha_1,\alpha_2)} = 0 \tag{22}$$

then clearly $H(x,p,w,u)$ can be expressed as

$$H(x,p,w,u) = H(x,p,\alpha_1,\alpha_2) + \begin{bmatrix}(w-\alpha_1)\\(u-\alpha_2)\end{bmatrix}^{\mathrm{T}} R(x) \begin{bmatrix}(w-\alpha_1)\\(u-\alpha_2)\end{bmatrix}$$

where

$$R(x) = \begin{bmatrix} r_{11}(x) & r_{12}(x)\\ r_{21}(x) & r_{22}(x)\end{bmatrix} = \begin{bmatrix} k_{11}^{\mathrm{T}}(x)k_{11}(x) - \gamma^2 I & k_{11}^{\mathrm{T}}(x)k_{12}(x)\\ k_{12}^{\mathrm{T}}(x)k_{11}(x) & k_{12}^{\mathrm{T}}(x)k_{12}(x)\end{bmatrix}.$$

The solution $(\alpha_1,\alpha_2)$ of (22) exists and is unique if and only if the matrix $R(x)$ is invertible. If this is the case, then $\alpha_1$ and $\alpha_2$ have the expression

$$\begin{bmatrix}\alpha_1\\ \alpha_2\end{bmatrix} = \begin{bmatrix}\alpha_1(x,p)\\ \alpha_2(x,p)\end{bmatrix} = -R^{-1}(x)\begin{bmatrix}(1/2)g_1^{\mathrm{T}}(x)p + k_{11}^{\mathrm{T}}(x)h_1(x)\\ (1/2)g_2^{\mathrm{T}}(x)p + k_{12}^{\mathrm{T}}(x)h_1(x)\end{bmatrix}.$$

In particular, $R(x)$ is nonsingular if

$$\begin{aligned} &r_{22}(x) > 0\\ &r_{11}(x) - r_{12}(x)r_{22}^{-1}(x)r_{21}(x) < 0\end{aligned}$$

for each $x$. Suppose these conditions hold and that $\alpha_1$ and $\alpha_2$ are such that the following equation is satisfied

$$H(x, V_x^{\mathrm{T}}, \alpha_1(x,V_x^{\mathrm{T}}), \alpha_2(x,V_x^{\mathrm{T}})) = 0. \tag{23}$$

If this is the case, it is immediate to check that

$$H(x,V_x^{\mathrm{T}},w,u) = \tag{24}$$

$$\begin{bmatrix}(w-\alpha_1)\\(u-\alpha_2)\end{bmatrix}^{\mathrm{T}} \begin{bmatrix} I & r_{12}r_{22}^{-1}\\ 0 & I\end{bmatrix} \begin{bmatrix} r_{11} - r_{12}r_{22}^{-1}r_{21} & 0\\ 0 & r_{22}\end{bmatrix} \begin{bmatrix} I & 0\\ r_{22}^{-1}r_{21} & I\end{bmatrix} \begin{bmatrix}(w-\alpha_1)\\(u-\alpha_2)\end{bmatrix}$$

and therefore the control law

$$u = \alpha(x,w) = \alpha_2(x,V_x^{\mathrm{T}}) - r_{22}^{-1}(x)r_{21}(x)(w - \alpha_1(x,V_x^{\mathrm{T}})) \tag{25}$$

renders the infinitesimal version (21) of the dissipation inequality (20) *satisfied*.

The identity (23) is a Hamilton-Jacobi-Isaacs equation. Note that it can be rewritten as

$$V_x f(x) + h_1^{\mathrm{T}}(x)h_1(x) - \begin{pmatrix}\alpha_1(x,V_x^{\mathrm{T}})\\ \alpha_2(x,V_x^{\mathrm{T}})\end{pmatrix}^{\mathrm{T}} R(x) \begin{pmatrix}\alpha_1(x,V_x^{\mathrm{T}})\\ \alpha_2(x,V_x^{\mathrm{T}})\end{pmatrix} = 0. \tag{26}$$

In the following statement it is shown that, under an appropriate additional assumption, from the knowledge of a solution of this equation it is possible to find a solution of the problem of disturbance attenuation.

**Theorem 5.1** (see [9]) *Consider the nonlinear system (1). Suppose*

$$r_{22}(x) > 0$$
$$r_{11}(x) - r_{12}(x)r_{22}^{-1}(x)r_{21}(x) < 0$$

*at each $x$. Suppose there exists an open neighborhood $U_\epsilon$ of $x = 0$ in $\mathbb{R}^n$ and a positive definite solution $V : U_\epsilon \to \mathbb{R}$ of the Hamilton-Iacobi-Isaacs equation (26). Moreover, suppose equilibrium $x = 0$ of the system*

$$\dot{x} = f(x) + g_1(x)\alpha_1(x, V_x^{\mathrm{T}}) + g_2(x)\alpha_2(x, V_x^{\mathrm{T}}) \tag{27}$$

*is locally asymptotically stable. Then the feedback law (25) locally asymptotically stabilizes the equilibrium $x = 0$ of the corresponding closed-loop system (with $w = 0$). Moreover, there exists $\delta > 0$ such that the response $z(\cdot)$ from the initial state $x(0) = 0$ satisfies*

$$\int_0^t \|z(s)\|^2 ds \leq \gamma^2 \int_0^t \|w(s)\|^2 ds. \tag{28}$$

*for every exogenous input $w(\cdot) \in \mathcal{W}_\delta$ and every $t > 0$.*

*Proof.* Set $w = 0$ in the closed loop system (1)-(25), to obtain

$$\dot{x} = f(x) + g_2(x)\alpha(x, 0) \tag{29}$$

where

$$\alpha(x, 0) = \alpha_2(x, V_x^{\mathrm{T}}) + r_{22}^{-1}(x)r_{21}(x)\alpha_1(x, V_x^{\mathrm{T}}).$$

Choosing $w = 0$ and $u = \alpha(x, 0)$ in the identity (24) yields

$$\begin{aligned} &V_x(f(x) + g_2(x)\alpha(x, 0)) = \\ &\quad -\|h_1(x) + k_{12}(x)\alpha(x, 0)\|^2 + \alpha_1^{\mathrm{T}}(x, V_x^{\mathrm{T}})P(x)\alpha_1(x, V_x^{\mathrm{T}}) \end{aligned} \tag{30}$$

where $P(x) = r_{11}(x) - r_{12}(x)r_{22}^{-1}(x)r_{21}(x)$. The left-hand side of this expression is exactly the derivative of the function $V(x(t))$ along the trajectories of the closed loop system (29). Since $P(x)$ is by hypothesis negative definite, (30) shows that $V(x(t))$ is nonincreasing along the trajectories of the closed loop system and therefore the equilibrium $x = 0$ is Lyapunov stable. By Lasalle's invariance principle, for every sufficiently small $\epsilon > 0$ there is $\delta > 0$ such that all trajectories of (29) satisfying $\|x(0)\| < \delta$ converge, as $t \to \infty$, to the largest subset of

$$N = \{x \in U_\epsilon : h_1(x) + k_{12}(x)\alpha(x, 0) = 0 \quad \text{and} \quad \alpha_1(x, V_x^{\mathrm{T}}) = 0\}$$

which is invariant for (29). But, for each $x \in N$, $\alpha_1(x, V_x^{\mathrm{T}}) = 0$ and therefore

$$g_2(x)\alpha(0, x) = g_1(x)\alpha_1(x, V_x^{\mathrm{T}}) + g_2(x)\alpha_2(x, V_x^{\mathrm{T}}).$$

Therefore, any trajectory of (29) which is entirely contained into $N$ is also a trajectory of the system (27). Since the latter is by assumption locally asymptotically stable, so is system (29). The disturbance attenuation property is a straightforward consequence of the analysis which has preceded the Theorem. ◁

*Remark.* Recall (see Proposition 3.3) that $V(x)$ is a solution of the Hamilton-Jacobi-Isaacs equation (26) if and only if

$$M = \{(x,p) \in U_\epsilon \times \mathbb{R}^n : p = V_x^{\mathrm{T}}(x)\} \tag{31}$$

is an invariant manifold of the Hamiltonian system

$$\begin{aligned} \dot{x} &= \Big(\frac{\partial H(x,p,\alpha_1(x,p),\alpha_2(x,p))}{\partial p}\Big)^{\mathrm{T}} \\ \dot{p} &= -\Big(\frac{\partial H(x,p,\alpha_1(x,p),\alpha_2(x,p))}{\partial x}\Big)^{\mathrm{T}}. \end{aligned} \tag{32}$$

If this is the case, then system (27) coincides with the restriction of the Hamiltonian system (32) to $M$. Thus, the condition requested by the previous Theorem is the existence, for the Hamiltonian system (32), of an invariant manifold of the form (31) with the additional property that the restriction of this Hamiltonian system to $M$ has an asymptotically stable equilibrium at $x = 0$. Note also that the manifold $M$ is not necessarily the stable manifold of this Hamiltonian system, because the restriction of (32) to $M$, namely the system (27), needs not to have an exponentially stable equilibrium at $x = 0$. ◁

If in the controlled system (1)

$$k_{11}(x) = 0,$$

the feedback law (25) which solves the problem of disturbance attenuation reduces to

$$u = \alpha(x) = \alpha_2(x, V_x^{\mathrm{T}}) \tag{33}$$

that is independent of the exogenous input $w$. Moreover, if the following *orthogonality condition* is satified

$$h_1^{\mathrm{T}}(x)k_{12}(x) = 0,$$

then in order to make sure that the feedback law (33) locally asymptotically stabilizes the closed loop system, it suffices to require that $(h_1(x), f(x))$ is locally zero-state observable.

**Theorem 5.2** *Consider the nonlinear system (1). Suppose*

$$\begin{aligned} &k_{11}(x) = 0 \\ &h_1^{\mathrm{T}}(x)k_{12}(x) = 0 \\ &k_{12}^{\mathrm{T}}(x)k_{12}(x) > 0 \end{aligned}$$

*at each $x$. Suppose $(h_1(x), f(x))$ is locally zero state-observable. Suppose there exists an open neighborhood $U_\epsilon$ of $x = 0$ in $\mathbb{R}^n$ and a positive definite solution $V : U_\epsilon \to \mathbb{R}$ of the Hamilton-Iacobi-Isaacs equation (26). Then the feedback law (33) locally asymptotically stabilizes the equilibrium $x = 0$ of the corresponding closed-loop system (with $w = 0$). Moreover, there exists $\delta > 0$ such that the response $y(\cdot)$ of the closed loop system from the initial state $x(0) = 0$ satisfies (28) for every exogenous input $w(\cdot) \in \mathcal{W}_\delta$ and every $t > 0$.*

*Proof.* Under the specific assumptions of the Proposition, the set $N$ introduced in the proof of Theorem 5.1 becomes

$$N = \{x \in U_\epsilon : h_1(x) = 0 \quad \text{and} \quad \alpha_1(x, V_x^{\mathrm{T}}) = 0 \quad \text{and} \quad \alpha_2(x, V_x^{\mathrm{T}}) = 0\}$$

Therefore, any trajectory of (29) which is entirely contained into $N$ is also a trajectory of $\dot{x} = f(x)$ entirely contained in the set

$$N' = \{x \in U_\epsilon : h_1(x) = 0\}.$$

By zero-state observabilty, the only trajectory of $\dot{x} = f(x)$ contained in $N'$ is the trivial trajectory, and thus the closed-loop system is locally asymptotically stable. ◃

The previous results have (partial) converse. For simplicity, we limit ourselves to state the converse of Theorem 5.2.

**Theorem 5.3** (see [10]) *Consider the nonlinear system (1). Suppose*

$$\begin{aligned} k_{11}(x) &= 0 \\ h_1^{\mathrm{T}}(x)k_{12}(x) &= 0 \\ k_{12}^{\mathrm{T}}(x)k_{12}(x) &> 0 \end{aligned}$$

*at each $x$. Suppose the linear approximation of $(h_1(x), f(x))$ at the equilibium $x = 0$ is observable. Suppose there exists a feedback law $u = \alpha(x)$ such that the closed loop system*

$$\begin{aligned} \dot{x} &= f(x) + g_1(x)w + g_2(x)\alpha(x) \\ z &= h_1(x) + k_{12}(x)\alpha(x) \end{aligned}$$

*is stable in the first approximation at the equilibrium $x = 0$ (with $w = 0$) and such that, for some $\delta > 0$, the response $z(\cdot)$ from the initial state $x(0) = 0$ satisfies the inequality*

$$\int_0^t \|z(s)\|^2 ds \le \tilde{\gamma}^2 \int_0^t \|w(s)\|^2 ds,$$

*with $0 \le \tilde{\gamma} < \gamma$, for every exogenous input $w(\cdot) \in \mathcal{W}_\delta$ and every $t > 0$. Then there exists a positive definite solution $V(x)$ of the Hamilton-Jacobi-Isaacs equation (26). In particular, there exists a positive definite solution such that the equilibrium $x = 0$ of (27) is stable in the first approximation.*

## 6 A family of controllers

In this section, we show how to build a family of controllers which solve the problem of local disturbance attenuation with internal stability. The interesting feature of this family is that, in case both the controlled plant and the controller are linear systems, it reduces exactly to the family of all controllers which solve the $H_\infty$-(sub)optimal control problem [1].

Assume the hypotheses of Theorem 5.2 hold and, for the sake of notational simplicity, also that $k_{12}^{\mathrm{T}}(x)k_{12}(x) = I$. Then, as shown in the proof of Theorem 5.2, the feedback law (33) locally stabilizes the closed loop system and renders its $L_2$ gain less than or equal to $\gamma$. Define now a "new control input"

$$v = u - \alpha_2(x)$$

where $\alpha_2(x) = \alpha_2(x, V_x^{\mathrm{T}})$, and a "new measured output"

$$r = w - \alpha_1(x)$$

where $\alpha_1(x) = \alpha_1(x, V_x^{\mathrm{T}})$. Motivated by the results which hold in the case of linear systems, it is reasonable to expect that the interconnection of the system thus defined, namely

$$\begin{aligned} \dot{x} &= f(x) + g_2(x)\alpha_2(x) + g_1(x)w + g_2(x)v \\ z &= h_1(x) + k_{12}(x)\alpha_2(x) + k_{12}(x)v \\ r &= -\alpha_1(x) + w \end{aligned}$$

with a system of the form

$$\begin{aligned} \dot{\xi} &= a(\xi) + b(\xi)r \\ v &= c(\xi) \end{aligned} \tag{34}$$

is still asymptotically stable and has a $L_2$ gain less than or equal to $\gamma$ if the latter, that is system (34), is such.

In the following Proposition we show that this is indeed the case, provided that the hypothesis that system (34) has a $L_2$ gain less than or equal to $\gamma$ is replaced by the stronger assumption that there exists a solution of a Hamilton-Jacobi equation of the form (7).

**Proposition 6.1** *Suppose the hypotheses of Theorem 5.2 hold and, in particular, $k_{12}^{\mathrm{T}}(x)k_{12}(x) = I$. Let*

$$\begin{aligned} \dot{\xi} &= a(\xi) + b(\xi)r \\ v &= c(\xi) \end{aligned}$$

*be any smooth nonlinear system, with $a(0) = 0$ and $c(0) = 0$, whose equilibrium $\xi = 0$ is locally asymptotically stable and such that the Hamilton-Jacobi equation*

$$U_\xi a(\xi) + c^{\mathrm{T}}(\xi)c(\xi) + \frac{1}{\gamma^2}U_\xi b(\xi)b^{\mathrm{T}}(\xi)U_\xi^{\mathrm{T}} = 0. \tag{35}$$

*has a positive definite solution $U(\xi)$ in the neighborhood of $\xi = 0$.*
*Then, the feedback law*

$$\begin{aligned} \dot{\xi} &= a(\xi) + b(\xi)(w - \alpha_1(x)) \\ u &= c(\xi) + \alpha_2(x) \end{aligned} \tag{36}$$

*is such that the closed loop system (1)-(36) is locally asympototically stable and, for some $\delta > 0$, the response $y(\cdot)$ from the initial state $(x(0), \xi(0)) = (0,0)$ satisfies (28) for every exogenous input $w(\cdot) \in \mathcal{W}_\delta$ and every $t > 0$.*

*Proof.* Hamilton-Jacobi equation (26) implies the identity

$$\begin{aligned} &V_x(f(x) + g_1(x)w + g_2(x)u) = \\ &= \|u - \alpha_2(x)\|^2 - \gamma^2\|w - \alpha_1(x)\|^2 - \|h_1(x)\|^2 - \|u\|^2 + \gamma^2\|w\|^2. \end{aligned}$$

In addition, Hamilton-Jacobi equation (35) implies the identity

$$U_\xi(a(\xi) + b(\xi)r) \leq -\|c(\xi)\|^2 + \gamma^2\|r\|^2 \tag{37}$$

which, in the present setup, can be rewritten as

$$U_\xi(a(\xi) + b(\xi)(w - \alpha_1(x))) \leq -\|u - \alpha_2(x)\|^2 + \gamma^2\|w - \alpha_1(x)\|^2.$$

The positive definite function $W(x,\xi) = V(x) + U(\xi)$ satisfies, along the trajectories of the closed loop system

$$\frac{dW}{dt} \leq -\|h_1(x)\|^2 - \|u\|^2 + \gamma^2\|w\|^2. \tag{38}$$

Setting $w = 0$ we see that the system is stable in the sense of Lyapunov. Moreover, $\dot{W}(t) = 0$ implies $h_1(x(t)) = 0$ for all $t \geq 0$ and this, by the hypothesis of observability, shows that $\lim_{t\to\infty} x(t) = 0$. This, in turn, implies $\lim_{t\to\infty} \xi(t) = 0$ because system (36) is by hypothesis locally asymptotically stable.

Finally, (38) shows that, on the trajectories starting from $(x,\xi) = (0,0)$,

$$\int_0^T (\gamma^2\|w(t)\|^2 - \|z(t)\|^2)dt \geq 0$$

thus completing the proof. ◁

## REFERENCES


[1] J.C. Doyle, K. Glover, P.P. Khargonekar, and B.A. Francis, State space solutions to standard $H_2$ and $H_\infty$ control problems, *IEEE Trans. Autom. Control*, **AC-34**:831–846, 1989.

[2] A.A. Stoorvogel, *The $H_\infty$ control problem: a state space approach*, PhD thesis, Technical University Eindhoven, 1990.

[3] C. Scherer, *The Riccati inequality and state-space $H_\infty$-optimal control*, PhD thesis, University of Würzburg, 1990.

[4] E.F. Mageirou and Y.C. Ho, Decentralized stabilization via game theoretic methods, *Automatica*, **13**:888–896, 1977.

[5] G. Tadmor, Worst case design in time domain, *Math. Control, Signals and Systems*, **3**:301–324, 1990.

[6] T. Basar and P. Bernhard, *$H_\infty$-optimal control and related Minimax problems*, Birkhauser, 1990.

[7] J.A. Ball and J.W. Helton, $H_\infty$ control for nonlinear plants: connection with differential games, In *Proc. of 28th Conf. Decision and Control*, pages 956–962, Tampa, FL, December 1989.

[8] A.J. Van der Schaft, A state-space approach to nonlinear $H_\infty$ control, *Syst. and Contr. Lett.*, **16**:1–8, 1991.

[9] A. Isidori, Feedback control of nonlinear systems, In *Proc. of 1st European Control Conf.*, Grenoble, France, July 1991.

[10] A.J. Van der Schaft, $L_2$-gain analysis of nonlinear systems and nonlinear $H_\infty$ control, Tech. Memorandum 969, Universiteit Twente, 1991.

[11] A. Isidori and A. Astolfi, Nonlinear $H_\infty$ control via measurement feedback, *J. Math. Systems, Estimation and Control*, **2**:to appear, 1992.

[12] B.D. Anderson, An algebraic solution to the spectral factorization problem, *IEEE Trans. Autom. Control*, **AC-12**:410–414, 1967.

[13] R.W. Brockett, *Finite dimensional linear systems*, Wiley, 1970.

[14] J.C. Willems, Least square optimal control and the algebraic riccati equation, *IEEE Trans. Autom. Control*, **AC-16**:621–634, 1971.

[15] P.J. Moylan, Implications of passivity in a class of nonlinear systems, *IEEE Trans. Autom. Control*, **AC-19**:373–381, 1974.

[16] D. Hill and P.J. Moylan, The stability of nonlinear dissipative systems, *IEEE Trans. Autom. Control*, **AC-21**:708–711, 1976.

[17] J.C. Willems, Dissipative dynamical systems, *Arch. Rational Mechanics and Analysis*, **45**:321–693, 1972.

[18] C.I. Byrnes and A. Isidori, Steady state response, separation principle and the output regulation of nonlinear systems, In *Proc. of 28th Conf. Decision and Control*, pages 2247–2251, Tampa, FL, December 1989.

Research supported in part by MURST and by AFOSR under grant 91-0266

Dipartimento di Informatica e Sistemistica
Università di Roma "La Sapienza"
Roma
*e*
Department of Systems Science and Mathematics
Washington University
St. Louis

# The Construction of Optimal Linear and Nonlinear Regulators

ARTHUR J. KRENER

## 1 Introduction

The regulator servomechanism problem is to design a feedforward and feedback control law to make the output of a given system called the plant, track a signal from a given class. There are various ways to make this precise, we shall follows Francis [5] and Isidori and Byrnes [11]. The class of signals to be tracked are described as the output of a second system called the signal generator (or exosystem). The control law consists of feedforward terms involving the state of signal generator, feedback terms involving the state of the plant and mixed terms involving both.

Francis [5] posed and solved the linear regulator problem and Isidori and Byrnes [11] generalized this to the nonlinear case. The former showed that the linear regulator problem is solvable only if a certain system of linear equation is solvable. The latter showed that the nonlinar regulator problem is solvable only if a certain system of first order partial differential equations is solvable. Huang and Rugh [9], [10] did a formal (term by term) analysis of these PDEs and gave sufficient conditions for its solvability. The degree one terms of the Isidori and Byrnes PDE yield the linear equations of Francis, so we refer to the system as the Francis-Byrnes -Isidori (FBI) equations.

In this paper, we shall show necessary and sufficient conditions for the term by term solvability of the FBI equations when either the signal generator has a semisimple pole structure or the plant has a semisimple zero structure. We present a proof of Huang-Rugh sufficient conditions.

We also give optimal methods for constructing the nonlinear regulator based on term by term analysis of the Hamilton-Jacobi Bellman HJB equations in the spirit of Al'brecht. This optimal approach to constructing a regulator may be novel even in the linear case.

## 2 Linear Regulation

Following Francis [5] and Isidori and Byrnes [11], we review the solution of the linear regulator problem. We are given a plant

$$
\begin{aligned}
\dot{x} &= Fx + Gu \\
y &= Hx + Ju
\end{aligned} \tag{1}
$$

where $x \in \mathbb{R}^{n\times 1}$, $u \in \mathbb{R}^{m\times 1}$ and $y \in \mathbb{R}^{p\times 1}$. Throughout we shall assume $(F, G)$ is stabilizable, $(H, F)$ is detectable, $\begin{bmatrix} G \\ J \end{bmatrix}$ of full column rank and $[H \quad J]$ of full row rank. We are also given a signal generator

$$\dot{w} = Aw \tag{2}$$

$$\bar{y} = Cw \tag{3}$$

where $w \in \mathbb{R}^{d\times 1}$ and $\bar{y} \in \mathbb{R}^{p\times 1}$. Without loss of generality, we assume $(C, A)$ is observable. The initial state of the signal generator is nonzero and the state and output evolve in a smooth fashion. The goal is to find a feedforward and feedback law from $(x, w)$ to $u$ which both stabilizes the plant (1) and drives the output error

$$\tilde{y} = y - \bar{y} = Hx + Ju - Cw \tag{4}$$

to zero as $t \to \infty$. Alternatively, one can occasionally reinitialize the state of the signal generator and obtain piecewise smooth signals. The goal then is to achieve stabilization and tracking on a significantly faster time scale then the times between reinitialization of the signal generator. A third alternative is to include an input channel in the dynamics of the signal generator

$$\dot{w} = Aw + Bv \tag{5}$$

where $v$ is a piecewise constant input. For example, if

$$A = \begin{bmatrix} 0\ 1 & & \\ & \ddots & \\ & & 1 \\ & & 0 \end{bmatrix} \qquad B = \begin{bmatrix} 0 \\ \vdots \\ 1 \end{bmatrix} \qquad C = [1\ 0\ \cdots\ 0]$$

then the output $\bar{y}$ of (5, 3) is a polynomial spline of degree $d$. Again the goal is to achieve stabilization and tracking on a significantly faster time scale than the times between steps of the input. Throughout, we shall consider signal generators of the form (2, 3), the extension to (5, 3) is straightforward.

There are several other interesting extensions of the regulator problem. One is to assume that the plant dynamics is directly disturbed by the signal generator

$$\dot{x} = Fx + Gu - Bw \tag{6}$$

and then the goal of the feedback can be viewed as stabilization, regulation and disturbance rejection. Another is to assume that the state of signal generator is not directly measurable, only the output error $\tilde{y}$ is measurable. Dynamic feedback can be used to solve this problem. We shall not discuss these extensions, they are treated in [11] and the methods that we shall present are generalizable to them. The dynamic feedback problem requires the construction of nonlinear observers.

Francis [5] showed that the original problem is solvable iff $(F, G)$ is stabilizable and there exists mappings

$$\begin{aligned} x &= Pw \\ u &= Kw \end{aligned} \tag{7}$$

which solve the Francis equations,

$$\begin{aligned} FP + GK &= PA \\ HP + JK &= C \end{aligned} \tag{8}$$

Hautus [8] has shown that solvability of these equations is connected with the transmission polynomials of the original plant (1) and the transmission polynomials of the combined plant, signal generator and output error (1, 2, 3, 4). This is also related to the Internal Model Principle of Francis and Wonham [6].

We now offer a necessary and sufficient condition for the solvability of the Francis equations that is slightly different from these and is valid regardless of the relative sizes of $m$ and $p$.

Following standard terminology, a pole $\lambda$ of the signal generator is an eigenvalue of $A$. Introducing new terminology, an output pole triple $(\lambda, w, y) \in \mathbb{C} \times \mathbb{C}^{d\times 1} \times \mathbb{C}^{p\times 1}$ satisfies $w \neq 0$ and

$$\begin{aligned} Aw &= \lambda w \\ Cw &= y \end{aligned} \tag{9}$$

Of course $w$ is a right eigenvector of $A$ corresponding to $\lambda$. If $A$ is not semisimple, i.e., $A$ has nontrivial Jordan blocks, then there are sequences of generalized output pole triples, $(\lambda, w^j, y^j)$ for $j = 1, \cdots, r$ satisfying $w^1, \ldots, w^r$ linearly independent and

$$\begin{bmatrix} A - \lambda I \\ C \end{bmatrix} [w^1 w^2 \cdots w^r] = \begin{bmatrix} 0 & w^1 & \cdots & w^{r-1} \\ y^1 & y^2 & \cdots & y^r \end{bmatrix} \tag{10}$$

The $w^j$'s are generalized right eigenvectors of $A$ and it is always possible to choose a basis for $\mathbb{C}^{d\times 1}$ consisting of right eigenvectors and generalized right eigenvectors of $A$.

The terminology for zeros of a multivariable system (1) varies among authors, we adopt the following. An $s \in \mathbb{C}$ is an output zero if

$$rank \begin{bmatrix} F - sI & G \\ H & J \end{bmatrix} < n + p. \tag{11}$$

This matrix is called the system matrix of (1). An output zero triple $(s, \zeta, \psi) \in \mathbb{C} \times \mathbb{C}^{1\times m} \times \mathbb{C}^{1\times p}$ satisfies $\zeta \neq 0$ and

$$(12) \qquad (\zeta \quad \psi)\begin{bmatrix} F - sI & G \\ H & J \end{bmatrix} = (0 \quad 0).$$

Notice that if $m < p$ then every $s \in \mathbb{C}$ is an output zero.

The output zero structure need not be semisimple. There may exist sequences of generalized output zero triples $(s, \zeta_i, \psi_i)$ for $i = 1, \cdots, r$ satisfying $\zeta_1, \cdots, \zeta_r$ linearly independent and

$$(13) \qquad \begin{bmatrix} \zeta_1 & \psi_1 \\ \zeta_2 & \psi_2 \\ \vdots & \vdots \\ \zeta_r & \psi_r \end{bmatrix} \begin{bmatrix} F - sI & G \\ H & J \end{bmatrix} = \begin{bmatrix} 0 & 0 \\ \zeta_1 & 0 \\ \vdots & \vdots \\ \zeta_{r-1} & 0 \end{bmatrix}.$$

Intuitively the output pole triples describe the modal directions of the state and output of the signal generator. The output zero triples describe the components of the plant's state and output which cannot be excited at a given frequency. There are dual definitions of input pole triples and input zero triples which we omit as we have no need of them. The intuitive content of the next theorem is that the Francis equations are not solvable iff an output pole triple of the signal generator excites an output zero triple of the plant.

**Theorem 2.1** *The Francis equations (8) are not solvable iff*

*(i) there exists an output pole triple $(\lambda, w, y)$ of the signal generator (2, 3) and an output zero triple $(s, \zeta, \psi)$ of the plant (1) such that*

$$(14) \qquad \lambda = s$$

*and*

$$(15) \qquad \psi y \neq 0$$

*or*

*(ii) there exists a sequence of generalized output pole triples $(\lambda, w^j, y^j)$, $j = 1, \cdots, r$, of the signal generator (2, 3) and a sequence of generalized output zero triples $(s, \zeta_i, \psi_i)$, $i = 1, \cdots, \rho$, of the plant (1) such that (14) and for some $j \leq \min(r, \rho)$*

$$(16) \qquad \psi_1 y^j + \psi_2 y^{j-1} + \cdots + \psi_j y^1 \neq 0$$

*Proof.* We choose a basis for $\mathbb{C}^{d\times 1}$ consisting of eigenvectors and generalized eigenvectors of $A$. Multiplying the Francis equations on the right by an eigenvector $w$ corresponding to the eigenvalue $\lambda$ and $y = Cw$, we obtain

$$(17) \qquad \begin{bmatrix} F - \lambda I & G \\ H & J \end{bmatrix} \begin{bmatrix} Pw \\ Kw \end{bmatrix} = \begin{bmatrix} 0 \\ y \end{bmatrix}.$$

This is an equation for $Pw$ and $Kw$. It is solvable iff every $(\zeta \quad \psi) \in \mathbb{C}^{1\times n} \times \mathbb{C}^{1\times p}$ annihilating the system matrix on the left annihilates the vector on the right. There is such a $(\zeta \quad \psi)$ annihilating the system matrix iff $\lambda = s$, where $s$ is an output zero. If $\lambda = s$ and $(s, \zeta, \psi)$ is an output zero triple then (17) is solvable iff $\psi y = 0$. If $A$ is semisimple, its eigenvectors span $\mathbb{C}^{d\times 1}$ so we are done.

Suppose $A$ is not semisimple, say there exists a sequence of generalized output pole triples $(\lambda, w^j, y^j)$, $j = 1, \cdots, r$. Multiplying the Francis equations on the right by $w^j$ yields

$$\begin{bmatrix} F - \lambda I & G \\ H & J \end{bmatrix} \begin{bmatrix} Pw^j \\ Kw^j \end{bmatrix} = \begin{bmatrix} Pw^{j-1} \\ y^j \end{bmatrix} \tag{18}$$

and we have a coupled set of equations for $Pw^j$, $Kw^j$ as $j = 1, \cdots, r$. This can be rewritten as

$$\begin{bmatrix} D(\lambda) & 0 & \dots & 0 & 0 \\ E & D(\lambda) & \dots & 0 & 0 \\ & & \ddots & & \\ 0 & 0 & \dots & D(\lambda) & 0 \\ 0 & 0 & \dots & E & D(\lambda) \end{bmatrix} \begin{bmatrix} \theta^1 \\ \vdots \\ \theta^r \end{bmatrix} = \begin{bmatrix} \gamma^1 \\ \vdots \\ \gamma^r \end{bmatrix} \tag{19}$$

where

$$D(s) = \begin{bmatrix} F - sI & G \\ H & J \end{bmatrix} \qquad E = \begin{bmatrix} -I & 0 \\ 0 & 0 \end{bmatrix}$$

$$\theta^i = \begin{bmatrix} Pw^i \\ Kw^i \end{bmatrix} \qquad \gamma^i = \begin{bmatrix} 0 \\ y^i \end{bmatrix}$$

If there is a sequence $(s, \zeta_i, \psi_i)$, $i = 1, \cdots, \rho$, of generalized output zero triples of (1) where $s = \lambda$ then the matrix

$$\begin{bmatrix} \xi_1 & 0 & \dots & 0 \\ \xi_2 & \xi_1 & \dots & 0 \\ & & \ddots & \\ \xi_\mu & \xi_{\mu-1} & \dots & \xi_1 \end{bmatrix} \tag{20}$$

where $\mu = \min(r, \rho)$ and

$$\xi_i = [\zeta_i \ \psi_i]$$

annihilates the large matrix on the left side of (19). Moreover, it is not hard to see that any vector annihilating the left side of (19) must define a sequence of generalized output zero triples and hence a matrix of the form (20). Therefore (19) is solvable iff

$$\psi_1 y^1 = 0$$
$$\psi_2 y^1 + \psi_1 y^2 = 0$$
$$\psi_\mu y^1 + \cdots + \psi_1 y^\mu = 0$$

QED

If the Francis equations are solvable then the desired feedback law to enable the output of the plant (1) to track the output of the signal generator is

(21) $$u = Kw + L(x - Pw)$$

where $L$ is any feedback matrix such that $F+GL$ is asymptotically stable. Under this feedback the combined system (1, 2) is driven to the subspace $x = Pw$, which is an invariant subspace for the closed loop dynamics [5], [11]. On this subspace the closed loop dynamics is the same as that of the signal generator (2) and this is referred to as the Internal Model Principle [6].

We would like to suggest an optimal way to choose $L$ and also to choose $K$ when the Francis equations admit many solutions, e.g. $m > p$. Suppose we wish to minimize the infinite time quadratic cost

(22) $$\frac{1}{2}\int_0^\infty x^* Qx + 2x^* Su + u^* Ru \, dt$$

where

(23) $$R > 0$$

and

(24) $$\begin{bmatrix} Q & S \\ S^* & R \end{bmatrix} \geq 0.$$

There has been some discussion in the literature about how to suitably interpret this because the cost (22) could very well be infinite, see for example [2]. We believe it should be interpreted in two stages. If the initial condition of the combined system (1, 2) lies in the subspace defined by $x = Pw$ and we use the feedback $u = Kw + v$ where $P$ and $K$ solve the Francis equations and $v$ will be chosen later, then the closed loop dynamics with $v = 0$ will evolve on this subspace. If there are many solutions to the Francis equations then to minimize the integrand of the cost (22) whenever we are on the subspace, we should choose one that minimizes

(25) $$trace \begin{bmatrix} P \\ K \end{bmatrix}^* \begin{bmatrix} Q & S \\ S^* & R \end{bmatrix} \begin{bmatrix} P \\ K \end{bmatrix}.$$

If (24) is positive definite then the minimizing solution is unique.

It should be noted that this criterion is not invariant under change of coordinates in the state space of the signal generator. Therefore it is important that the state coordinates of the signal generator are chosen in such a way that the expected states of the signal generator are uniformly distributed over the ellipsoids.

$$w^*\Gamma w = constant$$

where $\Gamma > 0$. Then we should choose a solution of the Francis equations which minimizes

$$trace\left[\begin{bmatrix} P \\ K \end{bmatrix}^* \begin{bmatrix} Q & S \\ S^* & R \end{bmatrix} \begin{bmatrix} P \\ K \end{bmatrix} \Gamma^{-1}\right] \tag{26}$$

If the initial condition of the combined system is off of the subspace $x = Pw$ then the goal is to drive the solution onto the subspace in an optimal fashion. We use $z = x - Pw$ as coordinates transverse to the subspace $x = Pw$. A simple calculation shows that under the feedback $u = Kw + v$,

$$\dot{z} = Fz + Gv \tag{27}$$

$$\tilde{y} = Hz + Jv \tag{28}$$

If we replace the cost (22) with

$$\frac{1}{2}\int_0^\infty z^*Qz + 2z^*Sv + v^*Rv \; dt \tag{29}$$

then we get a standard LQR problem for the optimal feedback

$$v = Lz = L(x - Pw). \tag{30}$$

Under the standard assumptions, $(F, G)$ stabilizable, $(Q^{1/2}, F)$ detectable and (23) we have the cost (29) is finite and the feedback (30) asymptotically drives the system (27) to zero and hence (20) under the closed loop dynamics remains on the invariant subspace given by $z = x - Pw = 0$.

Of course there is no reason why the cost (29) must use the $Q,\ R, S$ of (22). One could use instead

$$\frac{1}{2}\int_0^\infty \|\tilde{y}\|^2 + \|v\|^2 \; dt \tag{31}$$

or any other cost that satisfies the standard assumptions, as long as the cost is zero whenever $z = 0$ and $v = 0$. This will ensure that the $LQR$ problem will have a finite solution.

## 3 The FBI Equations

Isidori and Byrnes [11] have posed and solved a nonlinear extension of Francis' linear regulator [5]. We are given a plant

$$\begin{aligned} \dot{x} &= f(x,u) \\ y &= h(x,u) \\ 0 &= f(0,0) \end{aligned} \tag{32}$$

and a signal generator

$$\dot{w} = a(w)$$
$$(33) \qquad \bar{y} = c(w)$$
$$0 = a(0).$$

As before, the goal is to find a feedback law from $(x, w)$ to $u$ which stabilizes the plant (1) and drives the output error

$$(34) \qquad \tilde{y} = y - \bar{y} = h(x, u) - c(w)$$

to zero as $t \to \infty$ for all initial conditions. Actually Isidori and Byrnes treated a slightly different problem in which the plant dynamics (32) is affine in both $u$ and $y$ and where the output does not depend explicitly on $u$. Their methods and results extend immediately to the above. They also treat the problem where $(x, w)$ is not directly measurable.

Isidori and Byrnes assume that the signal generator dynamics (33) is Poisson stable around the origin, $w = 0$ and the linear approximation around the origin of the plant dynamics (32) is stabilizable. Under these assumptions they show that the nonlinear regulator problem is solvable iff there exists

$$(35) \qquad \begin{aligned} x &= \pi(w) \\ u &= \kappa(w) \end{aligned}$$

satisfying the first order P.D.E.s

$$(36) \qquad \begin{aligned} f\left(\pi(w), \kappa(w)\right) - \tfrac{\partial \pi}{\partial w}(w)a(w) &= 0 \\ h\left(\pi(w), \kappa(w)\right) - c(w) &= 0. \end{aligned}$$

We expand (32), (33) and (35) in Taylor series

$$(37) \qquad \begin{aligned} \dot{x} &= Fx + Gu + F^{[2]}(x,\ u) + f^{[3]}(x,\ u) + \cdots \\ y &= Hx + Ju + h^{[2]}(x,\ u) + h^{[3]}(x,\ u) + \cdots \end{aligned}$$

$$(38) \qquad \begin{aligned} \dot{w} &= Aw + a^{[2]}(s) + a^{[3]}(x,\ u) + \cdots \\ \bar{y} &= Cw + c^{[2]}(w) + c^{[3]}(w) + \cdots \end{aligned}$$

$$(39) \qquad \begin{aligned} x &= Pw + \pi^{P}[2](w) + \pi^{[3]} + \cdots \\ u &= Kw + \kappa^{[2]}(w) + \kappa^{[3]}(w) + \cdots \end{aligned}$$

where superscripts [2], [3], $\cdots$ denote homogeneous polynomial function of degree $2, 3, \ldots$ respectively. If we insert these series expansions into (36) and truncate after degree one terms, we obtain the Francis equations (8). For this reason we refer to (36) as the Francis-Byrnes-Isidori equations of the FBI equations.

Huang and Rugh [9] [10] have stated sufficient conditions for the formal solvability of the FBI equations. Formal solvability means term by term solvability of the FBI equations without regard to convergence.

To simplify the exposition we restrict our criteria to signal generators which are linear, but the methods can easily be extended to nonlinear signal generators. The methods can also be easily extended to the case where dynamics depends explicitly on the state of the signal generator, i.e., the disturbance rejection and regulation problem. We shall give necessary and sufficient conditions for the formal solvability of the FBI equations when either the signal generator's linear pole structure is semisimples or the plant's linear output zero structure is semisimple. We shall also give a proof of the Huang-Rugh sufficient condition.

We start by considering the degree two part of the FBI equations assuming the degree one part admits the solution $P,\ K$,

$$\begin{aligned} F\pi^{[2]}(w) + G\kappa^{[2]}(w) - \tfrac{\partial\pi^{[2]}}{\partial w}(w)Aw &= -f^{[2]}\left(Pw, Kw\right) \\ H\pi^{[2]}(w) + J\kappa^{[2]}(w) &= -h^{[2]}\left(Pw,\ Kw\right). \end{aligned} \tag{40}$$

Suppose $(\omega_1, \ldots,\ \omega_d)$ is a basis of $\mathbb{C}^{1\times d}$ consisting of left eigenvectors of $A$,

$$\omega_i A = \lambda_i \omega_i, \tag{41}$$

or generalized left eigenvectors of $A$,

$$\omega_i A = \lambda_i \omega_i + \omega_{i+1}. \tag{42}$$

We can expand $\pi^{[2]}(w), \kappa^{[2]}(w), f^{[2]}(Pw,\ Kw)$ and $h^{[2]}(Pw,\ Kw)$ in terms of the $\omega_i$ as

$$\begin{aligned} \pi^{[2]}(w) &= \textstyle\sum \pi^{ij}\left(\omega_i w\right)\left(\omega_j w\right) \\ \kappa^{[2]}(w) &= \textstyle\sum \kappa^{ij}\left(\omega_i w\right)\left(\omega_j w\right) \\ f^{[2]}\left(Pw,\ Kw\right) &= \textstyle\sum f^{ij}\left(\omega_i w\right)\left(\omega_j w\right) \\ h^{[2]}\left(Pw, Kw\right) &= \textstyle\sum h^{ij}\left(\omega_i w\right)\left(\omega_j w\right) \end{aligned} \tag{43}$$

where $\pi^{ij} \in \mathbb{C}^{n\times 1}, \kappa^{ij} \in \mathbb{C}^{m\times 1}, f^{ij} \in \mathbb{C}^{n\times 1},\ h^{ij} \in \mathbb{C}^{p\times 1}$ and the sums range over $1 \le i \le j \le d$.

**Theorem 3.1** *Assume the FBI equations of degree one are solvable and the pole structure of the signal generator (2, 3) is semisimple. The FBI equations of degree two are not solvable iff there exist left eigenpairs $(\lambda_i, \omega_i)$, $(\lambda_j, \omega_j)$ of $A$ and an output zero triple $(s,\ \zeta,\ \psi)$ of the linear part of the plant (1) such that*

$$s = \lambda_i + \lambda_j \tag{44}$$

*and*

$$\left(\zeta \quad \psi\right)\begin{bmatrix} f^{ij} \\ h^{ij} \end{bmatrix} \neq 0. \tag{45}$$

*Proof.* Plugging (43) into the FBI equations of degree two (40), we obtain $(d+1)$ choose 2 linear equations for the unknown $\pi^{ij}, \kappa^{ij}$ in terms of the known $f^{ij}$, $h^{ij}$ and the plant system matrix. Because $A$ is semisimple, these equations are in diagonal from

$$\begin{bmatrix} F-(\lambda_i+\lambda_j) & G \\ H & J \end{bmatrix} \begin{bmatrix} \pi^{ij} \\ \kappa^{ij} \end{bmatrix} = - \begin{bmatrix} f^{ij} \\ h^{ij} \end{bmatrix}. \tag{46}$$

for $1 \leq i \leq j \leq d$.

If $s = \lambda_i + \lambda_j$ is not an output zero of the linear part of the plant then (46) is solvable for any $f^{ij}$, $h^{ij}$. If there exist an output zero triple $(s, \zeta, \psi)$ satisfying (44) then $(\zeta, \psi)$ annihilates the system matrix on the left side of (46). Hence (46) is solvable iff

$$(\zeta \quad \psi) \begin{bmatrix} f^{ij} \\ h^{ij} \end{bmatrix} = 0.$$

QED

Suppose the FBI equations are formally solvable up to degree $k-1$ then the degree $k$ equations are

$$\begin{aligned} F\pi^{[k]}(w) + G\kappa^{[k]}(w) - \tfrac{\partial \pi^{[k]}}{\partial w}(w)Aw &= -f^{[k]}(w) \\ H\pi^{[k]}(w) + J\kappa^{[k]}(w) &= -h^{[k]}(w) \end{aligned} \tag{47}$$

where $f^{[k]}(w)$ and $h^{[k]}(w)$ are the degree $k$ parts of the composition of $f(x,u)$ and $h(x,u)$ with the expansion of (39) up to degree $k-1$.

As before we can expand $\pi^{[k]}(w)$, $\kappa^{[k]}(w)$, $f^{[k]}(w)$ and $h^{[k]}(w)$ in terms of a basis of $\mathbb{C}^{1\times d}$ of left eigenvectors and generalized eigenvectors of $A$,

$$\begin{aligned} \pi^{[k]}(w) &= \sum \pi^{i_1\cdots i_k} (\omega_{i_1} w) \cdots (\omega_{i_k} w) \\ \kappa^{[k]}(w) &= \sum \kappa^{i_1\cdots i_k} (\omega_{i_1} w) \cdots (\omega_{i_k} w) \\ f^{[k]}(w) &= \sum f^{i_1\cdots i_k} (\omega_{i_1} w) \cdots (\omega_{i_k} w) \\ h^{[k]}(w) &= \sum h^{i_1\cdots i_k} (\omega_{i_1} w) \cdots (\omega_{i_k} w) \end{aligned} \tag{48}$$

where the sum ranges over $1 \leq i_1 \leq i_2 \leq \cdots \leq i_k \leq d$.

**Theorem 3.2** *Assume the FBI equations are formally solvable up to degree $k-1$ and the pole structure of the signal generator (2) is semisimple. The FBI equations of degree $k$ are not solvable iff there exist left eigenpairs $(\lambda_{i_1}, \omega_{i_1}), \ldots, (\lambda_{i_k}, \omega_{i_k})$ of $A$ and an output zero triple $(s, \zeta, \psi)$ of the linear part of the plant (32) such that*

$$s = \lambda_{i_1} + \cdots + \lambda_{i_k} \tag{49}$$

$$(\zeta \quad \psi) \begin{bmatrix} f^{i_1\cdots i_k} \\ h^{i_1\cdots i_k} \end{bmatrix} \neq 0. \tag{50}$$

*Proof.* Plugging (48) into the degree $k$ FBI equations (47), we obtain $(d+k-1)$ choose $k$ decoupled sets of linear equations of the form

$$\begin{bmatrix} F-(\lambda_{i_1}+\cdots+\lambda_{i_k})I & G \\ H & J \end{bmatrix} \begin{bmatrix} \pi^{i_1\cdots i_k} \\ \kappa^{i_1\cdots i_k} \end{bmatrix} = - \begin{bmatrix} f^{i_1\cdots i_k} \\ h^{i_1\cdots i_k} \end{bmatrix} \tag{51}$$

If $s = \lambda_{i_1} + \cdots + \lambda_{i_k}$ where $(s, \zeta, \psi)$ is an output zero triple then (51) is solvable iff the left side of (50) is zero,

QED

Conditions like (44) and (49) are called resonance conditions in the literature, see [3], [7]. The intuitive meaning is that the signal generator excites the nonlinearities of the plant so as to create harmonics which the linear part of the plant has difficulty tracking and canceling.

The above theorem excludes many interesting signal generators, for example, those that generate polynomial splines as was discussed in Section 2. Such a signal generator is not semisimple. Unfortunately, a general analysis of regulator problems with nonsemisimple signal generators can be quite complicated. We illustrate this by a relatively simple example.

Suppose $p = 1$, $d = 3$ and the Jordan form of $A$ consist of one block. In other words there exists $\lambda, \omega_1, \omega_2, \omega_3$ such that

$$\begin{bmatrix} \omega_1 \\ \omega_2 \\ \omega_3 \end{bmatrix} (A - \lambda I) = \begin{bmatrix} \omega_2 \\ \omega_3 \\ 0 \end{bmatrix}$$

Notice that $\omega_3$ is a left eigenvector of $A$ and $\omega_1$ and $\omega_2$ are generalized left eigenvectors. Assume the degree one FBI equations have a solution $P, K$. To solve the degree two equations(40) we expand as before (43) and obtain as the coefficients $(\omega_i w)(\omega_j w)$

$$\begin{bmatrix} F-2\lambda I & G \\ H & J \end{bmatrix} \begin{bmatrix} \pi^{ij} \\ \kappa^{ij} \end{bmatrix} - \begin{bmatrix} \pi^{i-1,j} + \pi^{i,j-1} \\ 0 \end{bmatrix} = \begin{bmatrix} f^{ij} \\ h^{ij} \end{bmatrix}$$

where $\pi^{ij} = 0$, $\kappa^{ij} = 0$ if $i$ or $j = 0$. Because the pole structure of the signal generator is not semisimple, the equations are not decoupled. We rewrite them as

$$\begin{bmatrix} D(2\lambda) & 0 & 0 & 0 & 0 & 0 \\ E & D(2\lambda) & 0 & 0 & 0 & 0 \\ 0 & E & D(2\lambda) & 0 & 0 & 0 \\ 0 & 2E & 0 & D(2\lambda) & 0 & 0 \\ 0 & 0 & E & E & D(2\lambda) & 0 \\ 0 & 0 & 0 & 0 & 2E & D(2\lambda) \end{bmatrix} \begin{bmatrix} \theta^{11} \\ \theta^{12} \\ \theta^{13} \\ \theta^{22} \\ \theta^{23} \\ \theta^{33} \end{bmatrix} = - \begin{bmatrix} \gamma^{11} \\ \gamma^{12} \\ \gamma^{13} \\ \gamma^{22} \\ \gamma^{23} \\ \gamma^{33} \end{bmatrix}$$

(52)

where

$$D(s) = \begin{bmatrix} F - sI & G \\ H & J \end{bmatrix} \qquad E = \begin{bmatrix} -I & 0 \\ 0 & 0 \end{bmatrix}$$

$$\theta^{ij} = \begin{bmatrix} \pi^{ij} \\ \kappa^{ij} \end{bmatrix} \qquad \gamma^{ij} = \begin{bmatrix} f^{ij} \\ h^{ij} \end{bmatrix}$$

Clearly if $s = 2\lambda$ is not an output zero of the linear part of the plant then (52) is always solvable. Suppose $s = 2\lambda$ is an output zero frequency and there exist a sequence of generalized output zero triples satisfying (13) for $r \geq 5$. It is not hard to see that that $6 \times 6(n+p)$ matrix

$$(53) \qquad \begin{bmatrix} \xi_1 & 0 & 0 & 0 & 0 & 0 \\ \xi_2 & \xi_1 & 0 & 0 & 0 & 0 \\ \xi_3 & \xi_2 & \xi_1 & 0 & 0 & 0 \\ 2\xi_3 & 2\xi_2 & 0 & \xi_1 & 0 & 0 \\ 3\xi_4 & 3\xi_3 & \xi_2 & \xi_2 & \xi_1 & 0 \\ 6\xi_5 & 6\xi_4 & 2\xi_3 & 2\xi_3 & 2\xi_2 & \xi_1 \end{bmatrix}$$

where $\xi_i = (\zeta_i \quad \psi_i)$ annihilates the $6(n+p) \times 6(n+m)$ matrix on the left side of (52). Moreover, any $1 \times 6(n+p)$ vector annihilating this matrix must be in the form of a row of (53) for some sequence of generalized output zero triples where $s = 2\lambda$. Hence (52) is solvable iff every matrix of the form (53) constructed from sequences of five generalized output zero triples at $s = 2\lambda$ annihilates the right side of (52). If there are several linearly independent sequences of output zero triples at $s = 2\lambda$, then these yield additional conditions for the solvability of (52).

If the sequence of generalized output zero triples is shorter than five triples, then there are fewer conditions and they are obtained by deleting rows of (53) containing the nonexistent triples. In particular, if the output zero structure of the plant is semisimple, i.e., there are no generalized output zero triples, then the degree two FBI equations are solvable iff whenever $s = 2\lambda$ then

$$(54) \qquad (\zeta \quad \psi) \begin{bmatrix} f^{11} \\ h^{11} \end{bmatrix} = 0.$$

This last remark can be generalized to the following theorem.

**Theorem 3.3** *Assume the FBI equations are formally solvable up to degree $k-1$ and the output zero structure of the linear part of the plant (32) is semisimple. The degree $k$ FBI equations are not solvable iff there exists left eigenpairs $(\lambda_{i_j}, \omega_{i_j})$ where $\omega_{i_j}$ is a left eigenvector corresponding to $\lambda_{i_j}$ or the first generalized left eigenvector of a sequence corresponding to $\lambda_{i_j}$ for $j = 1, \ldots, n$ and an output zero triple $(s, \zeta, \psi)$ of the linear part of the plant such that*

$$(55) \qquad s = \lambda_{i_1} + \cdots + \lambda_{i_k}$$

$$(56) \qquad (\zeta \quad \psi)\begin{bmatrix} f^{i_1\cdots i_k} \\ h^{i_1\cdots i_k} \end{bmatrix} \neq 0$$

*Proof.* Plugging (48) into the degree $k$ FBI equations (47), we obtain $(d+k-1)$ choose $k$ set of linear equations. A set of equations corresponding to all simple eigenpairs $(\lambda_{i_1}, \omega_{i_1})\cdots(\lambda_{i_k}, \omega_{i_k})$ takes the form (51) and is decoupled from all the rest. Hence it is not solvable iff (55, 56) hold.

If one of the $(\lambda_{i_j}, \omega_{i_j})$ is part of a sequence of generalized eigenpairs, then this set of equations is coupled to several other sets. All these coupled sets can be written in a large system like (52). If the sets of equations are written in lexographic order with $i_1 \leq i_2 \leq \cdots \leq i_k$ and the rightmost indices moving fastest then this large system is lower triangular and the diagonal consists of $D(\lambda_{i_1} + \cdots + \lambda_{i_k})$ as in (52). If $\lambda_{i_1} + \cdots + \lambda_{i_k}$ is not an output zero of the plant then this matrix is left invertible and the large system is solvable. If (55) holds, where $(s, \zeta, \psi)$ is an output triple of the plant, then the vector

$$(\zeta \; \psi \; 0 \; 0 \quad 0 \; 0 \; \cdots \; 0 \; 0)$$

annihilates the matrix on the left side of the large equations. Hence the system is not solvable iff (56). QED

The Huang-Rugh sufficient condition is proved in a similar fashion.

**Theorem 3.4** [9], [10] *Assume the FBI equations are formally solvable up to degree $k-1$. The degree $k$ FBI equations are solvable if for all output zeros $s$ of linear part of the plant (3.1) and all poles $\lambda_{i_1}, \ldots, \lambda_{i_k}$ of the signal generator,*

$$(57) \qquad s \neq \lambda_{i_1} + \cdots + \lambda_{i_k}.$$

*Proof.* Plugging (48) into the degree $k$ FBI equations (47), we obtain $(d+k-1)$ choose $k$ sets of linear equations. A set of equations corresponding to all simple eigenpairs $(\lambda_{i_1}, \omega_{i_1})\cdots(\lambda_{i_k}, \omega_{i_k})$ takes the form (51) and is decoupled from the rest. If (57) holds then it is solvable.

If one of the $(\lambda_{i_j}, \omega_{i_j})$ is part of a sequence of generalized eigenpairs then this set of equations is coupled to several other sets. All these coupled sets can be written in lexographic order as a large lower triangular system with $D(\lambda_{i_1} + \cdots + \lambda_{i_k})$ on the diagonal. If (57) holds then this large matrix is left invertible, hence the system is solvable.

QED

## 4 Nonlinear Regulation

Now suppose the FBI equations are solvable to degree 2. There are two approaches to constructing the degree two regulator. The first might be called the pole placement approach. For the linear part of the regulator we proceed as in Section 2 and choose a $P$, $K$ satisfying the Francis equations and a $L$ such that $F + GL$ is asymptotically stable. For the quadratic part of the regulator we choose any

$$\bar{\pi}^{[2]}(z,w) \qquad \bar{\kappa}^{[2]}(z,\ w,\ v) \tag{58}$$

such that the restrictions

$$\pi^{[2]}(w) = \bar{\pi}^{[2]}(0,w) \qquad \kappa^{[2]}(w) = \bar{\kappa}^{[2]}(0,w,0) \tag{59}$$

satisfy the FBI equations of degree 2. We consider the effect of the change of state coordinates

$$x = z + Pw + \bar{\pi}^{[2]}(z,w) \tag{60}$$

and feedback

$$u = v + Kw + Lz + \bar{\kappa}^{[2]}(z,w,v) \tag{61}$$

on the combined system (32, 33, 34) neglecting cubic and higher terms. The result is

$$\dot{z} = (F+GL)z + Gv + \bar{f}^{[2]}(z,w,v) + O(z,w,v)^3 \tag{62}$$

$$\dot{w} = Aw \tag{63}$$

$$\tilde{y} = (H+JL)z + Jv + \bar{h}^{[2]}(z,w,v) + O(z,w,v)^3 \tag{64}$$

where

$$\begin{aligned} \bar{f}^{[2]}(z,w,v) &= f^{[2]}(z+Pw,\ v+Kw+Lz) + F\bar{\pi}^{[2]}(z,w) \\ &\quad - \frac{\partial\bar{\pi}^{[2]}}{\partial(z,w)}(z,w)\begin{bmatrix} (F+GL)z+Gv \\ Aw \end{bmatrix} \\ \bar{h}^{[2]}(z,w,v) &= h^{[2]}(z+Pw,\ v+Kw+Lz) + H\bar{\pi}^{[2]}(z,w). \end{aligned} \tag{65}$$

Because the restrictions (59) of $\bar{\pi}^{[2]}$ and $\bar{\kappa}^{[2]}$ satisfy the FBI equations of degree two, it follows that

$$\bar{f}^{[2]}(0,\ w,\ 0) = 0 \qquad \bar{h}^{[2]}(0,\ w,\ 0) = 0. \tag{66}$$

If we ignore the $O(z,w,v)^3$ terms in (62, 63, 64), the submanifold $z = 0$ is an invariant manifold of the dynamics (62, 63) when $v = 0$ and on this submanifold the tracking error (64) is zero. The linear part of the dynamics is asymptotically stable so in a neighborhood of $z = 0,\ w = 0$, the transverse state coordinates

$z(t)$ and tracking error $\tilde{y}(t)$ will get smaller as $t \to \infty$ and $v(t) = 0$. Because of the presence of quadratic and higher degree terms in (62, 63, 64) and because $w(t) \neq 0$, these will not necessarily converge to zero. To make them smaller it would be desirable to choose (58) satisfying (59) so that

$$\begin{aligned} \bar{f}^{[2]}(z,\ w,\ v) &= 0 \\ \bar{h}^{[2]}(z,\ w,\ v) &= 0. \end{aligned} \tag{67}$$

In general this is not possible and one may have to settle for a least squares approximate solution to (67). The extra freedom in (58) as opposed to (59) allows a better approximate solution to (67).

If the FBI equations are solvable to degree $k$, then this approach can be used to build regulator of degree $k$, see [10] for details.

Now we discuss an optimal approach to constructing a degree two regulator. Assume that the state coordinates of the signal generator (33) have been suitably chosen so that the distribution of the state is isotropic and assume that we have a quadratic cost criterion (22). We proceed as in Section 2 and choose a solution $P,\ K$ of the Francis equations which minimizes (25). Next we solve the FBI equation of degree two in a similar fashion. We choose a basis $(\omega_1, \ldots, \omega_d)$ for $\mathbb{C}^{1\times d}$ of left eigenvectors (41) and generalized left eigenvectors (42) such that the $\omega_i$'s are unit vectors and orthogonal whenever possible. It is always possible to choose this basis so that all the eigenvectors and generalized eigenvectors corresponding to the same eigenvalue are orthogonal. We seek a solution (43) of the FBI equations of degree two (40) which minimizes

$$\frac{1}{2} \sum_{1 \le i \le j \le d} \pi^{ij^*} Q \pi^{ij} + 2\pi^{ij^*} S \kappa^{ij} + \kappa^{ij^*} R \kappa^{ij} \tag{68}$$

The reason for doing this is that we wish to make cost of tracking small. Of course (68) is not the only criterion one can employ and there may be better choices which take into account the cost due to the interaction between the linear and quadratic solutions of the FBI equations.

Now we apply the change of coordinates

$$x = z + Pw + \pi^{[2]}(w) \tag{69}$$

and feedback

$$u = v + Kw + \kappa^{[2]}(w) \tag{70}$$

to the combined system (32, 33, 34) neglecting cubic and higher terms. The result is

$$\dot{z} = Fz + Gv + \bar{f}^{[2]}(z,w,v) + O(z,w,v)^3 \tag{71}$$

$$\dot{w} = Aw \tag{72}$$

$$\tilde{y} = Hz + Jv + \bar{h}^{[2]}(z,w,v) + O(z,w,v)^3 \tag{73}$$

where

$$(74)\quad \bar{f}^{[2]}(z,w,v) = f^{[2]}(z+Pw,\ v+Kw)+F\pi^{[2]}(w)-\frac{\partial \pi^{[2]}}{\partial w}(w)Aw$$

$$(75)\quad \bar{h}^{[2]}(z,w,v) = h^{[2]}(z+Pw,\ v+Kw)+H\pi^{[2]}(w).$$

Since (69, 70) is constructed from solutions of the FBI equation of degree one and two, it follows that

$$(76)\quad \begin{aligned} \bar{f}^{[2]}(0,w,0)&=0\\ \bar{h}^{[2]}(0,w,0)&=0. \end{aligned}$$

Next we choose a cost criterion, say (29) and consider the nonlinear quadratic regulator problem (NQR) of minimizing the cost subject to (71, 72, 73). Let $\theta(z,\ w)$ denote the minimial value of (29) starting (71, 72, 73) at $(z,w)$ at $t=0$. If $\theta(z,w)$ is finite and a smooth function, then it satisfies the Hamilton-Jacobi-Bellman (HJB) system of partial differential equations,

$$(77)\quad \frac{\partial\theta(z,w)}{\partial(z,w)}\begin{bmatrix} f(z,\ w,\ v)\\ Aw\end{bmatrix}+\ell(z,\ w)=0$$

$$(78)\quad \frac{\partial\theta(z,w)}{\partial z}\frac{\partial f}{\partial v}(z,w,v)+\frac{\partial \ell}{\partial v}(z,w)=0$$

where $f(z,\ w,\ v)$ is the right side of (71) and $\ell(z,\ w)$ is the integrand of (29).

Al'brecht [1] has analyzed the formal solvability of the HJB equations. We expect the optimal cost $\theta(z,\ w)$ to begin with a quadratic term in $z$

$$(79)\quad \theta(z,\ w)=\frac{1}{2}z^*Tz+\theta^{[3]}(z,\ w)+\cdots$$

and the optimal feedback to begin with a linear term in $z$,

$$(80)\quad v=Lz+\bar{\kappa}^{[2]}(z,w)+\cdots.$$

The lowest degree terms in (77) are the familiar equations of a LQR,

$$(81)\quad \frac{1}{2}z^*\left(TF+F^*T+Q-(TG+S)R^{-1}(TG+S)^*\right)z=0$$

$$(82)\quad z^*(TG+S)+v^*R=0.$$

We solve the Riccati equation (81) for $T$ and from (82) obtain the linear feedback gain,

$$(83)\quad L=-R^{-1}(TG+S)^*.$$

Next we look at the cubic part of (77) and the quadratic part of (78)

$$(84)\qquad \frac{\partial\theta^{[3]}}{\partial(z,w)}(z,\ w)\begin{bmatrix}(F+GL)z\\ Aw\end{bmatrix}+z^*T\bar{f}^{[2]}(z,\ w,\ Lz)=0$$

$$(85)\qquad \frac{\partial\theta^{[3]}}{\partial z}(z,\ w)G+z^*T\frac{\partial\bar{f}^{[2]}}{\partial v}(z,\ w,\ Lz)+\bar{\kappa}^{[2]^*}(z,\ w)R=0$$

where $\bar{f}^{[2]}(z,\ w,\ v)$ is given by (74).

This (84,85) is a system of linear equations for the unknowns, $\theta^{[2]}$ and $\bar{\kappa}^{[2]}$. If we can solve (84) for $\theta^{[3]}(z,\ w)$ then the quadratic feedback $\bar{\kappa}^{[2]}(z,w)$ is given by

$$(86)\qquad \bar{\kappa}^{[2]}(z,\ w)=-R^{-1}\left(\frac{\partial\theta^{[3]}}{\partial z}(z,w)G+z^*T\frac{\partial\bar{f}^{[2]}}{\partial v}(z,\ w,\ Lz)\right)^*$$

Let $(\omega_1,\ldots,\omega_d)$ be a basis of $\mathbb{C}^{1\times d}$ of left eigenvectors and generalized eigenvectors of $A$ as in (41, 42). Let $(\zeta_1,\ \ldots,\zeta_n)$ be a basis for $\mathbb{C}^{n\times 1}$ of left eigenvectors or generalized left eigenvectors of $(F+GL)$

$$(87)\qquad \zeta_i(F+GL)=\mu_i\zeta_i$$

or

$$(88)\qquad \zeta_i(F+GL)=\mu_i\zeta_i+\zeta_{i+1}$$

where $\mu_1,\ldots,\mu_n$ are the eigenvalues of $(F+GL)$. Then $\theta^{[3]}(z,w)$ can be expanded

$$(89)\qquad \begin{aligned}\theta^{[3]}(z,\ w)\ &=\ \textstyle\sum_{1\le i\le j\le k\le n}\theta^{ijk}_{zzz}\,(\zeta_i z)\,(\zeta_j z)\,(\zeta_k z)\\ &+\ \textstyle\sum_{\substack{1\le i\le j\le n\\ 1\le k\le d}}\theta^{ijk}_{zzw}\,(\zeta_i z)\,(\zeta_j z)\,(\omega_k w)\\ &+\ \textstyle\sum_{\substack{1\le i\le n\\ 1\le j\le k\le d}}\theta^{ijk}_{zww}\,(\zeta_i z)\,(\omega_j w)\,(\omega_k w)\\ &+\ \textstyle\sum_{1\le i\le j\le k\le d}\theta^{ijk}_{www}\,(\omega_i w)\,(\omega_j w)\,(\omega_k w)\,.\end{aligned}$$

Because of (76), $z^*T\bar{f}^{[2]}(z,\ w,\ Lz)$ is at least quadratic in $z$ so

$$(90)\qquad \begin{aligned}z^*T\bar{f}^{[2]}(z,\ w,\ Lz)\ &=\ \textstyle\sum_{1\le i\le j\le k\le n}f^{ijk}_{zzz}\,(\zeta_i z)\,(\zeta_j z)\,(\zeta_k z)\\ &+\ \textstyle\sum_{\substack{1\le i\le j\le n\\ 1\le k\le d}}f^{ijk}_{zzw}\,(\zeta_i z)\,(\zeta_j z)\,(\omega_k w)\,.\end{aligned}$$

If $F+GL$ and $A$ are both semisimple then when we substitute (89, 90) into (84) we obtain a diagonal system of equations

$$(91)\qquad (\mu_i+\mu_j+\mu_k)\,\theta^{ijk}_{zzz}=-f^{ijk}_{zzz}$$

for $1\le i\le j\le k\le n$ and

$$(92)\qquad (\mu_i+\mu_j+\lambda_k)\,\theta^{ijk}_{zzw}=-f^{ijk}_{zzw}$$

for $1 \leq i \leq j \leq n$ and $1 \leq k \leq d$.

If $(F+GL)$ or $A$ is not semisimple, then when the terms of the expansions (89, 90) are written in lexographic order we obtain a lower triangular system with (91, 92) on the diagonal. Hence if there are no reasonances, i.e.,

$$\begin{aligned} &\mu_i + \mu_j + \mu_k \neq 0 \qquad 1 \leq i \leq j \leq k \leq n \\ &\mu_i + \mu_j + \mu_k \neq o \qquad 1 \leq i \leq j \leq k \leq n, 1 \leq k \leq d \end{aligned} \tag{93}$$

then (84, 85) is solvable.

Under standard assumptions, the spectrum of $(F+GL)$ will always lie in the open left half of the complex plane. The spectrum of a typical signal generator is in the closed left half of the complex plane. Therefore, under standard assumptions the nonresonance conditions (93) will be satisfied. Notice that the total feedback up to degree two is

$$u = Kw + Lz + \kappa^{[2]}(w) + \bar{\kappa}^{[2]}(z,\ w). \tag{94}$$

Note also that

$$\begin{aligned} &\theta^{[3]}(0, w) = 0 \\ &\frac{\partial \theta^{[3]}}{\partial z}(0,\ w) = 0 \\ &\bar{\kappa}^{[2]}(0,\ w) = 0. \end{aligned} \tag{95}$$

Hence when $z = 0$ then the optimal cost up to degree three is zero and the optimal feedback up to degree two is

$$u = Kw + \kappa^{[2]}(w) \tag{96}$$

which nearly keeps the system (37 , 38) on the manifold

$$z = x - Pw - \pi^{[2]}(w) = 0 \tag{97}$$

The extra terms in feedback (94) when $z \neq 0$ drive the system toward (97) in an optimal way up to degree two. This generalizes to higher degrees as follows.

**Theorem 4.1** *Assume the following:*

*(i) The linear part of the plant (37) is stabilizable.*

*(ii) The poles of the linear signal generator (2) lie in the closed half of the complex plane.*

*(iii) A solution (39) exists of the FBI equations up to degree $k$.*

*(iv) The quadratic cost (29) satisfies (23, 24) and $(Q^{1/2},\ F)$ is detectable.*

*Then the HJB equations for the NQR of minimizing (29) where*

$$z = x - Pw - \pi^{[2]}(w) - \cdots - \pi^{k}(w) \tag{98}$$

$$v = u - Kw - \kappa^{[2]}(w) - \cdots - \kappa^{[k]}(w) \tag{99}$$

*are formally solvable up to degree $k+1$ in the optimal cost $\theta(z,w)$ and up to degree $k$ in the optimal feedback $\bar{\kappa}(z, w)$. The unique solution*

$$\theta(z,w) = \frac{1}{2}z^*Tz + \theta^{[3]}(z,w) + \cdots + \theta^{[k+1]}(z,w) \tag{100}$$

$$\bar{\kappa}(z,w) = Lz + \bar{\kappa}^{[2]}(z,w) + \cdots + \bar{\kappa}^{[k]}(z,w) \tag{101}$$

*satisfies*

$$\begin{aligned} \theta(0,w) &= 0 \\ \frac{\partial\theta}{\partial z}(0,w) &= 0 \\ \bar{\kappa}(0,w) &= 0. \end{aligned} \tag{102}$$

*Proof* (By induction on $k$). The above analysis showed that theorem is true for $k=1$ and 2. Now suppose the HJB equations are solvable up to degree $k$ in the cost and degree $k-1$ in the feedback. The degrees $k+1$, $k$ parts of the H-J-B equations (84, 85) are

$$\begin{aligned} &\frac{\partial\theta^{[k+1]}}{\partial(z,w)}(z,w)\begin{bmatrix} (F+GL)z \\ Aw \end{bmatrix} + \frac{\partial\theta^{[k]}}{\partial z}(z,w)\bar{f}^{[2]}(z,w,0) \\ &\quad + \cdots + z^*T\left(\bar{f}^{[k]}(z,w,0) + G\bar{\kappa}^{[k]}(z,w)\right) \\ &\quad + z^*\left(S + L^*R\right)\bar{\kappa}^{[k]}(z,w) + \sum_{i=2}^{k-1} \kappa^{[i]^*}(z,w)R\kappa^{[k+1-i]}(z,w) = 0 \end{aligned} \tag{103}$$

$$\begin{aligned} &\frac{\partial\theta^{[k+1]}}{\partial z}(z,w)G + \frac{\partial\theta^{[k]}}{\partial z}(z,w)\frac{\partial\bar{f}^{[2]}}{\partial v}(z,w,0) \\ &\quad + \cdots + z^*T\frac{\partial\bar{f}^{[k]}}{\partial v}(z,w,0) \\ &\quad + \bar{\kappa}^{[k]^*}(z,w)R = 0 \end{aligned} \tag{104}$$

where $\bar{f}^{[j]}(z,w,v)$ is the degree $j$ part of composition of $f(x,u)$ appearing in (32) with

$$\begin{aligned} x &= z + Pw + \pi^{[2]}(w) + \cdots + \pi^{[k]}(w) \\ u &= v + Kw + Lz + \kappa^{[2]}(w) + \bar{\kappa}^{[2]}(z,w) + \cdots + \bar{\kappa}^{[k-1]}(z, w) \end{aligned} \tag{105}$$

Because of (83), one of the unknowns $\bar{\kappa}^{[k]}(z,w)$ drops out of (103), so we try to solve (103) for the other unknown, $\theta^{[k+1]}(z,w)$. It is possible, then (104) is solvable for $\bar{\kappa}^{[k]}(z,w)$,

$$(106) \quad \bar{\kappa}^{[k]}(z,w) = -R^{-1}\left[\frac{\partial\theta}{\partial z}^{[k+1]}(z,w)G + \cdots + z^{*}T\frac{\partial \bar{f}}{\partial v}^{[k]}(z,w,0)\right]^{*}$$

Since (98, 99) solve the FBI equations up to degree $k$ and (102) holds by induction it follows that

$$(107) \quad \bar{f}^{[j]}(0,w,0) = 0$$

for $j = 2,\ldots,k$. Therefore all the terms but the first in (103) are at least quadratic in $z$. When we expand $\theta^{[k+1]}(z,w)$ and the rest of the terms in (103) in monomials consisting $k+1$ fold products of $(\zeta_i z)$ and $(\omega_j w)$, we obtain a diagonal or at least lower triangular system, which is solvable if the nonresonance condition hold

$$(108) \quad \begin{aligned} &\mu_{i_1} + \cdots + \mu_{i_{k+1}} \neq 0 \\ &\mu_{i_1} + \cdots + \mu_{i_k} + \lambda_{i_{k+1}} \neq 0 \\ &\cdots \\ &\mu_{i_1} + \mu_{i_2} + \lambda_{i_3} + \cdots + \lambda_{i_{k+1}} \neq 0. \end{aligned}$$

These hold since $Re(\mu_i) < 0$ and $Re(\lambda_i) \leq 0$. Hence there eixts a unique $\theta^{[k+1]}(z,\ w)$ satisfying (103) and it is least quadratic in $z$. It follows that $\bar{\kappa}^{[k]}(z,\ w)$ exists (106) and is at least linear in $z$.

QED

We note that if there are several solution (37) to the FBI equations of degree $k$, we can choose that which minimizes the generalization of (68). Of course there are other reasonable choices.

The term by term approach to the HJB equations can also be used with cost criteria like (31) instead of (29). The key points are that the closed loop linear part of the plant $(F+GL)$ be asymptotically stable and that cost vanish up to order $k$ when $z=0$ and $v=0$ so that all but the first term of (103) is at least quadratic in $z$.

Suppose the hypothesis of Theorem 4.1 hold so that we can solve the FBI and HJB equations up to degree $k$. Then we have the total feedback up to degree $k$,

$$(109) \quad u(z,\ w) = \kappa(w) + \bar{\kappa}\left(z\left(x,w\right),w\right)$$

when $z(x,w)$ is given by (98), $\kappa(w)$ is (39) to degree $k$ and $\bar{\kappa}(z,w)$ is (101). The optimal cost is

$$\theta(x,w) = \theta\left(z\left(x,w\right),w\right) \tag{110}$$

where $\theta(z,w)$ is (100) and the running cost $\ell(x,w)$ is

$$\begin{aligned}\ell(x,w) \quad &= \tfrac{1}{2}\left(z^*(x,w)Qz(x,w)_2 z^*(x,w)Sv(x,w)\right) \\ &+v(x,w)Rv(x,w)\end{aligned} \tag{111}$$

where $z(x,w)$ is given by (98), $v(x,w)$ is given by (99) with $u = u(x,w)$ as in (109). From the HJB equations (77) we have

$$\frac{d}{dt}\theta\left(x(t),w(t)\right) = -\ell\left(x(t),w(t)\right) + O\left(x(t),w(t)\right)^{k+2}. \tag{112}$$

This suggests that $\theta(x,w)$ is a candidate Lyapunov function for the tracking problem as formalized in the following.

**Theorem 4.2** *In addition to the hypothesis of Theorem 4.1, assume there exists a region $\Omega$ in $\mathbb{R}^n \times \mathbb{R}^d$ containing $(0,0)$ such that*
*(i) $\Omega$ is positively invariant under the closed loop dynamics (32) (2) and (109).*
*(ii) $\theta(x,w) \geq 0$ on $\Omega$ and $\theta(x,w) = 0$ only on $\{z(x,w) = 0\} \cap \Omega$.*
*(iii) $\frac{d}{dt}\theta(x,w) \leq 0$ on $\Omega$ and the maximal positively invariant subset in $\left\{\frac{d}{dt}\theta(x,w) = 0\right\} \cap \Omega$ is contained in $\{z(x,w) = 0\}$.*
*Then for any initial condition $(x(0), w(0) \in \Omega)$ the solution $(x(t), w(t))$ of the closed loop dynamics converge to $\{z(x,w) = 0\} \cap \Omega$ where the tracking error is $O(w)^{k+1}$.*

The proof of this theorem follows immediately from LaSalle's extension of Lyapunov's direct method.

## REFERENCES

[1] E.G. Al'brect, *On the optimal stabilization of nonlinear systems*, PMM-J. Appl. Math. Mech, vol. 25, pp. 1254–1266, 1961.

[2] B. D. O. Anderson, *Optimal control –linear quadratic methods*, Prentice Hall, Englewood Cliffs, 1990.

[3] V. I. Arnold, *Geometrical methods in the theory of ordinary differential equations*, Springer–Verlag, N.Y., 1983.

[4] J. Carr, *Applications of centre manifold thoery*, Springer–Verlag, N.Y., 1981.

[5] B. A. Francis, *The linear multivariable regulator probem*, SIAM J. Contr. Optimiz. vol. 15, 1977, pp. 486-505.

[6] B. A. Francis, and W. M. Wonham, *The internal model principle for linear multivarable regulators*, J. Appl. Math. Optimiz., vol 2 pp. 170–195, 1975.

[7] J. Guckenheimer and P. J. Holmes, *Nonlinear oscillations, dynamical systems and bifurcations of vector fields*, Springer–Verlag, N.Y., 1983.

[8] M. Hautus, *Linear matrix equations with applications to the regulator problem*, in *Outils and modèles math ḿatique pur l'automatique*, I.D. Landau, Ed., Paris, C.N.R.S., pp. 399–412, 1983.

[9] J. Huang and W. J. Rugh, *Stabilization on zero-error manifolds and the nonlinar servomechanism problem*, Proceedings of the 29th CDC, Honolulu, pp. 1262–1267, 1990. (Also to appear, IEEE TAC 1992.)

[10] J. Huang and W. J. Rugh, *An approximation method for the nonlinear servomechansim problem*, Technical Report 90/08.2, Johns Hopkins University, Electrical and Computer Engineering, 1990. (Also to appear, IEEE TAC 1992.)

[11] A. Isidori and C. I. Byrnes, *Output regulation of nonlinear systems*, IEEE Trans. on Auto. Cont. vol. 35, pp. 131–140, 1990.

Research supported in part byAFOSR-91-0228.

Department of Mathematics
*and*
Institute of Theoretical Dynamics
University of California
Davis, CA 95616-8633
ajkrener@ucdavis.edu

# Feedback Systems Stabilizability in Terms of Invariant Zeros

G. MARRO and A. PIAZZI

## 1. Introduction

It is well known that zeros play a very basic role in the analysis and synthesis of multivariable systems: although the term "zero" clearly comes from the polynomial (transfer matrix) approach, in the early 70's it was extended to the state-space description through geometric tools and techniques [1,2]. In particular, the invariant zeros of a triple $(A, B, C)$ are defined both as the roots with multiplicity of the invariant polynomials of the Rosenbrock system matrix [3] and as the internal unassignable eigenvalues with multiplicity of the maximal $(A, \text{im}B)$-controlled invariant contained in $\ker C$ [4,5].

In this chapter some interesting connections between the invariant zeros layout and the stabilizability features of multivariable feedback systems are investigated. Starting from a very general statement of the problem (the meaning of invariant zeros in the context of dual-lattice structures of controlled and conditioned invariants), conditions for the existence of a stable feedback solution will be stated and proved in the following cases:
(i) achievement of structural changes through algebraic state feedback and output injection, with possible applications in the classical elementary disturbance localization problem and its dual, namely the unknown-input asymptotic state observation;
(ii) achievement of structural changes through dynamic output feedback, as, for instance, in the general disturbance localization problem;
(iii) general feedback regulation in presence of both completely unaccessible disturbances modelled by piecewise-continuous functions and inputs (accessible or not) modelled by an exosystem.

The following notation will be used. Sets, vector spaces and subspaces are denoted by script capitals like $\mathcal{X}$, $\mathcal{V}$, etc.; matrices and linear maps by slanted capitals like $A$, $B$, etc.

Given a linear map $A : \mathcal{X} \rightarrow \mathcal{X}$ and any subspace $\mathcal{B} \subseteq \mathcal{X}$ we say that

$\mathcal{V} \subseteq \mathcal{X}$ is an $(A,\mathcal{B})$-controlled invariant if $A\mathcal{V} \subseteq \mathcal{V}+\mathcal{B}$ and that $\mathcal{S} \subseteq \mathcal{X}$ is an $(A,\mathcal{C})$-conditioned invariant if $A(\mathcal{S}\cap\mathcal{C}) \subseteq \mathcal{S}$. The maximum $(A,\mathcal{B})$-controlled invariant contained in a given subspace $\mathcal{E} \subseteq \mathcal{X}$ will be denoted by $\max \mathcal{V}(A,\mathcal{B};\mathcal{E})$ and the minimum $(A,\mathcal{C})$-conditioned invariant containing a given subspace $\mathcal{D} \subseteq \mathcal{X}$ will be denoted by $\min \mathcal{S}(A,\mathcal{C};\mathcal{D})$; these extremals can easily be determined by means of two recursion algorithms, reciprocally dual, which form the basic tools of the geometric approach. Symbols $\mathcal{B}$, $\mathcal{C}$ $\mathcal{D}$ and $\mathcal{E}$ are used to denote generic subspaces of $\mathcal{X}$, since general properties of controlled and conditioned invariants are first presented without any connection with dynamic system analysis. They will then assume a precise meaning in connection to the state space representation of dynamic systems whose invariant zeros will be considered and discussed.

The restriction of map $A$ to the $A$-invariant subspace $\mathcal{I}$ is denoted by $A|_{\mathcal{I}}$. Given two $A$-invariants $\mathcal{I}_1$ and $\mathcal{I}_2$ such that $\mathcal{I}_1 \subseteq \mathcal{I}_2$, with $A|_{\mathcal{I}_2/\mathcal{I}_1}$ we denote the map induced by $A$ on the quotient space $\mathcal{I}_2/\mathcal{I}_1$. The spectrum of $A$ is denoted by $\sigma(A)$. With reference to finite sets $\mathcal{Z}_1$ and $\mathcal{Z}_2$, that may contain repeated elements, notations $\mathcal{Z}_1 + \mathcal{Z}_2$ and $\mathcal{Z}_1 - \mathcal{Z}_2$ will be used for the union and the difference *with repetition count*.

## 2. The Basic Dual-Lattice Structure and Invariant Zeros

First, let us recall some properties of controlled and conditioned invariants that will be proved to be strongly related to interesting features of invariant zeros in multivariable systems. These properties were the object of previous research work by the authors [6, 8–12].

Given any $(A,\mathcal{B})$-controlled invariant $\bar{\mathcal{V}} \subseteq \mathcal{X}$, all $(A,\mathcal{B})$-controlled invariants $\mathcal{V} \subseteq \bar{\mathcal{V}}$ such that $\mathcal{V} \supseteq \bar{\mathcal{V}}\cap\mathcal{B}$ are called *self-bounded* with respect to $\bar{\mathcal{V}}$. Dually, given any $(A,\mathcal{C})$-conditioned invariant $\bar{\mathcal{S}}$, all $(A,\mathcal{C})$-conditioned invariants $\mathcal{S} \supseteq \bar{\mathcal{S}}$ such that $\mathcal{S} \subseteq \bar{\mathcal{S}}+\mathcal{C}$ are called *self-hidden* with respect to $\bar{\mathcal{S}}$. The sets of all controlled invariants self-bounded with respect to $\bar{\mathcal{V}}$ and of all conditioned invariants self-hidden with respect to $\bar{\mathcal{S}}$ are lattices with respect to $\subseteq, +, \cap$.

We shall now introduce the basic dual-lattice structure. Define

(1) $$\mathcal{V}_0^* := \max \mathcal{V}(A,\mathcal{B};\mathcal{C}),$$

(2) $$\mathcal{S}_0^* := \min \mathcal{S}(A,\mathcal{C};\mathcal{B}),$$

and consider the lattices $\mathcal{M}_0$ and $\mathcal{N}_0$ of all $(A,\mathcal{B})$-controlled invariants self-bounded with respect to $\mathcal{V}_0^*$ and of all $(A,\mathcal{C})$-conditioned invariants self-hidden with respect to $\mathcal{S}_0^*$, defined by

(3) $$\mathcal{M}_0 := \{\mathcal{V} : A\mathcal{V} \subseteq \mathcal{V}+\mathcal{B},\ \mathcal{V} \subseteq \mathcal{C},\ \mathcal{V} \supseteq \mathcal{V}_0^*\cap\mathcal{B}\},$$

(4) $$\mathcal{N}_0 := \{\mathcal{S} : A(\mathcal{S}\cap\mathcal{C}) \subseteq \mathcal{S},\ \mathcal{S} \supseteq \mathcal{B},\ \mathcal{S} \subseteq \mathcal{S}_0^*+\mathcal{C}\}.$$

The supremum and the infimum of $\mathcal{M}_0$ are respectively $\mathcal{V}^*$ itself and

(5) $$\mathcal{R}_0 := \mathcal{V}_0^* \cap \mathcal{S}_0^*,$$

while the infimum and the supremum of $\mathcal{N}_0$ are $\mathcal{S}_0^*$ and

$$\mathcal{Q}_0 := \mathcal{S}_0^* + \mathcal{V}_0^* \,. \tag{6}$$

Consider now the three-map system $(A, B, C)$ defined by

$$\dot{x}(t) = A\,x(t) + B\,u(t)\,, \tag{7}$$

$$y(t) = C\,x(t)\,, \tag{8}$$

where $x \in \mathcal{X}$ $(= \mathbf{R}^n)$, $u \in \mathcal{U}$ $(= \mathbf{R}^p)$ and $y \in \mathcal{Y}$ $(= \mathbf{R}^q)$ denote respectively the state, input and output. Let $\mathcal{B} := \mathrm{im} B$ and $\mathcal{C} := \ker C$.

It is well known that for any $(A, \mathcal{B})$-controlled invariant $\mathcal{V}$ there exists at least one matrix $F$ such that $(A{+}BF)\mathcal{V} \subseteq \mathcal{V}$ and for any $(A, \mathcal{C})$-controlled invariant $\mathcal{S}$ there exists at least one matrix $G$ such that $(A{+}GC)\mathcal{S} \subseteq \mathcal{S}$: these properties are basic in synthesis problems to obtain changes of structure through state feedback and output injection. $\mathcal{V}$ is said to be *internally stabilizable* or *externally stabilizable* if there exists at least one matrix $F$ such that $(A{+}BF)|_{\mathcal{V}}$ is stable or $(A{+}BF)|_{\mathcal{X}/\mathcal{V}}$ is stable.

The main features of dual lattices $\mathcal{M}_0$ and $\mathcal{N}_0$ defined in (3) and (4) are stated as follows.

**Property 1.** *Let $F$ be any matrix such that $(A{+}BF)\mathcal{V}_0^* \subseteq \mathcal{V}_0^*$: then $(A{+}BF)\mathcal{V} \subseteq \mathcal{V}$ for all $\mathcal{V} \in \mathcal{M}_0$. Let $G$ be any matrix such that $(A{+}GC)\mathcal{S}_0^* \subseteq \mathcal{S}_0^*$: then $(A{+}GC)\mathcal{S} \subseteq \mathcal{S}$ for all $\mathcal{S} \in \mathcal{N}_0$.*

**Property 2.** *Let $F$ be any matrix such that $(A{+}BF)\mathcal{V}_0^* \subseteq \mathcal{V}_0^*$: then the eigenstructure of $(A{+}BF)|_{\mathcal{V}/\mathcal{R}_0}$ is independent of $F$ for all $\mathcal{V} \in \mathcal{M}_0$. Let $G$ be any matrix such that $(A{+}GC)\mathcal{S}_0^* \subseteq \mathcal{S}_0^*$: then the eigenstructure of $(A{+}GC)|_{\mathcal{Q}_0/\mathcal{S}}$ is independent of $G$ for all $\mathcal{S} \in \mathcal{N}_0$.*

The eigenstructures referred to in the above property are called respectively *the internal unassignable eigenstructure of $\mathcal{V}$ with respect to $\mathcal{V}^*$* and *the external unassignable eigenstructure of $\mathcal{S}$ with respect to $\mathcal{S}^*$*. Since any $(A, \mathcal{B})$-controlled invariant $\mathcal{V}$ is self-bounded with respect to $\mathcal{V}$ itself, the plain internal unassignable eigenstructure of $\mathcal{V}$ is understood to be with respect to $\mathcal{V}$ itself. A similar assumption is introduced for $(A, \mathcal{C})$-conditioned invariants. It is easily seen (see the proof of Properties 1, 2 and 3 reported below) that any controlled invariant is internally stabilizable if and only if its internal unassignable eigenvalues are all stable and that any conditioned invariant is externally stabilizable if and only if its external unassignable eigenvalues are all stable.

**Property 3.** *Relations*

$$\begin{aligned} \mathcal{S} &= \mathcal{V} + \mathcal{S}_0^*\,, \\ \mathcal{V} &= \mathcal{S} \cap \mathcal{V}_0^* \end{aligned} \tag{9}$$

*establish a one-to-one map and its inverse between $\mathcal{M}_0$ and $\mathcal{N}_0$ such that the internal and external unassignable eigenstructures of each corresponding pair $\mathcal{V}, \mathcal{S}$ are equal.*

**Proof.** Properties 1, 2 and 3 are proved together by using a suitable change of basis in the state space $\mathcal{X}$. This is defined by the similarity transformation $T := [T_1\ T_2\ T_3\ T_4]$ such that $\mathrm{im} T_1 = \mathcal{R}_0 = \mathcal{V}_0^* \cap \mathcal{S}_0^*$, $\mathrm{im}[T_1\ T_2] = \mathcal{V}_0^*$, $\mathrm{im}[T_1\ T_3] = \mathcal{S}_0^*$; the system matrices in the new basis are $A' := T^{-1}AT$, $B' := T^{-1}B$ and $C' := CT$ and exhibit the following structures:

$$A' = \begin{bmatrix} A'_{11} & A'_{12} & A'_{13} & A'_{14} \\ O & A'_{22} & A'_{23} & A'_{24} \\ A'_{31} & A'_{32} & A'_{33} & A'_{34} \\ O & O & A'_{43} & A'_{44} \end{bmatrix}, \quad B' = \begin{bmatrix} B'_1 \\ O \\ B'_3 \\ O \end{bmatrix},$$

$$C' = [\ O \quad O \quad C'_3 \quad C'_4\ ] .$$

The zero submatrices of $B'$ and $C'$ are due to inclusions $\mathrm{im} B \subseteq \mathcal{S}_0^*$ and $\ker C \supseteq \mathcal{V}_0^*$; that in the second row of $A'$ follows from $\mathcal{R}_0$ being an $(A, \mathcal{B})$-controlled invariant and the second row of $B'$ being zero, those in the fourth row from the whole $\mathcal{V}_0^*$ being an $(A, \mathcal{B})$-controlled invariant and the fourth row of $B'$ being zero.

Since $\mathcal{R}_0 = \mathcal{V}_0^* \cap \mathcal{S}_0^*$ is the reachable set on $\mathcal{V}^*$, pair $(A'_{11}, B'_1)$ is controllable and the eigenvalues of $A'_{11} + B'_1 F'_1$ are completely assignable by a suitable choice of $F'_1$.

To prove Property 1, note that, $\mathcal{V}_0^*$ being a controlled invariant, submatrices $A'_{31}$ and $A'_{32}$ can be zeroed by means of a suitable state feedback $F'$ and that, $\mathcal{S}_0^*$ being a conditioned invariant, submatrices $A'_{23}$ and $A'_{43}$ can be zeroed by means of a suitable output injection $G'$. Clearly these feedbacks cause $\mathcal{V}_0^*$ to be an $(A+BF)$-invariant, with $F := F'T^{-1}$, and $\mathcal{S}_0^*$ an $(A+GC)$-invariant, with $G := TG'$. Since $\mathcal{R}_0 \subseteq \mathcal{V} \subseteq \mathcal{V}_0^*$ for all $\mathcal{V} \in \mathcal{M}_0$ and $\mathcal{S}_0^* \subseteq \mathcal{S} \subseteq \mathcal{Q}_0$ for all $\mathcal{S} \in \mathcal{N}_0$, by virtue of the particular structures of $A'$, $B'$ and $C'$ the same state feedback $F$ causes all elements of $\mathcal{M}_0$ to be $(A+BF)$-invariants and the same output injection $G$ causes all elements of $\mathcal{N}_0$ to be $(A+GC)$-invariants. This means that these elements, being controlled invariants, are necessarily expressed as

$$\text{(10)} \qquad \mathcal{V} = \mathrm{im}\left(T \begin{bmatrix} I_1 & O \\ O & X'_2 \\ O & O \\ O & O \end{bmatrix}\right), \quad \mathcal{S} = \mathrm{im}\left(T \begin{bmatrix} I_1 & O & O \\ O & X''_2 & O \\ O & O & I_3 \\ O & O & O \end{bmatrix}\right),$$

where $I_1$ and $I_3$ are properly dimensioned identity matrices, while $X'_2$ and $X''_2$ are basis matrices of $A'_{22}$-invariants.

Property 2 is proved by noting that the internal eigenstrucure of the $A'_{22}$-invariant $\mathrm{im} X'_2$ and the external eigenstructure of the $A'_{22}$-invariant $\mathrm{im} X''_2$ are independent respectively of $F'$ and $G'$.

Property 3 is proved as follows. Assume $X'_2 = X''_2 = X_2$ in (10): since the first two groups of coordinate axes in the new basis span $\mathcal{V}_0^*$, and the first and third span $\mathcal{S}_0^*$, subpaces $\mathcal{V}$ and $\mathcal{S}$ defined by (10) satisfy (9). Referring to the matrix structures obtained through the state feedback $F'$

and the output injection $G'$ (respectively with $A'_{31}, A'_{32}$ and $A'_{23}, A'_{43}$ set to zero), it is easily shown that the eigenstructures of $(A+BF)|_{\mathcal{V}/\mathcal{R}_0}$ and of $(A+GC)|_{\mathcal{Q}_0/\mathcal{S}}$ are equal to each other, and coincide with the internal eigenstructure of the $A'_{22}$-invariant im$X_2$. Q.E.D.

Still referring to a three-map system $(A, B, C)$ we define the invariant zeros $\mathcal{Z}(u;y)$ in completely geometric terms. It has been proved in the literature [1, 4] that the invariant zeros as defined below are completely equivalent to the transmission zeros (the zeros of the matrix transfer function in the Smith-McMillan form).

**Definition 1.** *The invariant zeros of triple $(A, B, C)$ are the internal unassignable eigenvalues of $\mathcal{V}_0^*$, the maximal $(A, \mathcal{B})$-controlled invariant contained in $\mathcal{C}$.*

In other terms, invariant zeros are defined by:

$$\mathcal{Z}(u;y) := \sigma\big((A+BF)|_{\mathcal{V}_0^*/\mathcal{R}_0}\big)\,, \tag{11}$$

where $F$ is any matrix such that $(A+BF)\mathcal{V}_0^* \subseteq \mathcal{V}_0^*$ and $\mathcal{R}_0$ is defined as in (5). The dual, completely equivalent, definition of invariant zeros can be derived as the following corollary of Property 3.

**Corollary 1.** *The invariant zeros of triple $(A, B, C)$ are the external unassignable eigenvalues of $\mathcal{S}_0^*$, the minimal $(A, \mathcal{C})$-controlled invariant containing $\mathcal{B}$.*

In other terms, the following equality holds:

$$\mathcal{Z}(u;y) = \sigma\big((A+GC)|_{\mathcal{Q}_0/\mathcal{S}_0^*}\big)\,, \tag{12}$$

where $G$ is any matrix such that $(A+GC)\mathcal{S}_0^* \subseteq \mathcal{S}_0^*$ and $\mathcal{Q}_0$ is defined as in (6).

## 3. Stabilizability of Some Algebraic Feedback Connections

The previously presented background will be applied to express stabilizability conditions in some basic synthesis problems of the geometric approach in terms of invariant zeros. We shall consider first the well-known *disturbance localization problem*, both with accessible and unaccessible disturbance. A new approach to stabilizability in this problem and its dual without any use of eigenspaces (i.e., by using the dual-lattice structures), but not in terms of invariant zeros, was presented by the authors in [10].

Refer to the four-map system $(A, B, D, E)$ described by

$$\dot{x}(t) = A\,x(t) + B\,u(t) + D\,d(t)\,, \tag{13}$$

$$e(t) = E\,x(t)\,, \tag{14}$$

which is assumed to be subject to the state feedback (state feedback with algebraic feedforward correction)

$$\dot{u}(t) = F\,x(t)\,, \tag{15}$$

$$\big(\,u(t) = F\,x(t) + S\,d(t)\big)\,. \tag{16}$$

In the above equations $x \in \mathcal{X}$ $(= \mathbb{R}^n)$, $u \in \mathcal{U}$ $(= \mathbb{R}^p)$, $d \in \mathcal{U}_d$ $(= \mathbb{R}^r)$ and $e \in \mathcal{Y}_e$ $(= \mathbb{R}^s)$ denote respectively the state, manipulable input, disturbance input and the output. Pair $(A, B)$ is assumed to be stabilizable. Let $\mathcal{B} := \mathrm{im}B$, $\mathcal{D} := \mathrm{im}D$ and $\mathcal{E} := \ker E$.

The disturbance localization problems which are herein considered are geometrically stated in the following terms.

**Problem 1.** (Unaccessible Disturbance Localization) *Find, if possible, a state feedback matrix $F$ such that $A+BF$ is stable and there exists at least one $(A+BF)$-invariant contained in $\mathcal{E}$ and containing $\mathcal{D}$.*

**Problem 2.** (Accessible Disturbance Localization) *Find, if possible, a state feedback matrix $F$ such that $A+BF$ is stable and there exists at least one $(A+BF)$-invariant contained in $\mathcal{E}$ whose sum with $\mathcal{B}$ contains $\mathcal{D}$.*

We shall derive for both problems completely constructive solvability conditions which consist in a structural necessary condition involving the basic $(A, \mathcal{B})$-controlled invariant

$$\mathcal{V}^* := \max \mathcal{V}(A, \mathcal{B}; \mathcal{E}) \tag{17}$$

and a stabilizability condition stated in terms of invariant zeros. All together, these conditions are necessary and sufficient. Due to availability of computer programs providing multivariable invariant or transmission zeros (if $(A, B, E)$ is completely controllable and observable, invariant and transmission zeros coincide), in most cases the sole stabilizability condition, which is in any case necessary, is useful for a quick feasibility screening.

**Theorem 1.** *The unaccessible disturbance localization problem has a solution if and only if:*
(A.1) $\mathcal{D} \subseteq \mathcal{V}^*$,
(A.2) $\mathcal{Z}(u;e) - \mathcal{Z}(u,d;e)$ *has all its elements stable,*
*while the accessible disturbance localization problem has a solution if and only if:*
(A.3) $\mathcal{D} \subseteq \mathcal{V}^* + \mathcal{B}$,
(A.4) $\mathcal{Z}(u;e) - \mathcal{Z}(u,d;e)$ *has all its elements stable.*

**Proof.** An immediate, but nonconstructive, necessary and sufficient condition is the existence of an internally stabilizable[1] $(A, \mathcal{B})$-controlled invariant $\mathcal{V}$ such that

---

[1] The external stabilizability of any $(A, \mathcal{B})$-controlled invariant, hence of $\mathcal{V}$, is implied by the assumption that $(A, B)$ is stabilizable.

(i) $\mathcal{V} \subseteq \mathcal{E}$ and $\mathcal{D} \subseteq \mathcal{V}$ in the case of Problem 1;
(ii) $\mathcal{V} \subseteq \mathcal{E}$ and $\mathcal{D} \subseteq \mathcal{V}+\mathcal{B}$ in the case of Problem 2.

Hence conditions (A.1) and (A.3) are necessary. On the other hand, if they are satisfied it follows that

$$\max \mathcal{V}(A,\mathcal{B};\mathcal{E}) = \max \mathcal{V}(A,\mathcal{B}+\mathcal{D};\mathcal{E}) .$$

In fact the subspace on the left, which is $\mathcal{V}^*$ defined by (17), is the maximal subspace which satisfies $\mathcal{V}^* \subseteq \mathcal{E} \cap A^{-1}(\mathcal{V}^* + \mathcal{B})$, that on the right is the maximal subspace $\mathcal{V}'$ which satisfies $\mathcal{V}' \subseteq \mathcal{E} \cap A^{-1}(\mathcal{V}' + \mathcal{B} + \mathcal{D})$, which is clearly equivalent to the previous one if (A.1) or (A.3) holds.

Consider the subspace

$$\mathcal{V}_m := \mathcal{V}^* \cap \min \mathcal{S}(A,\mathcal{E};\mathcal{B}+\mathcal{D}) , \tag{18}$$

which, if (A.1) or (A.3) holds, is an $(A,\mathcal{B})$-controlled invariant, self-bounded with respect to $\mathcal{V}^*$, satisfying (i) or (ii). Recalling (5), it is seen that $\mathcal{V}_m$ is the infimum of the lattice of all $(A,\mathcal{B}+\mathcal{D})$-controlled invariants self-bounded with respect to $\mathcal{V}^*$, while

$$\mathcal{R}_1 := \mathcal{V}^* \cap \min \mathcal{S}(A,\mathcal{E};\mathcal{B}) \tag{19}$$

is the infimum of the lattice of all $(A,\mathcal{B})$-controlled invariants self-bounded with respect to $\mathcal{V}^*$.

The following property, which was conjectured in [6] and first proved in [7], is basic to prove necessity: *if there exists an internally stabilizable $(A,\mathcal{B})$-controlled invariant $\mathcal{V}$ satisfying* (i) *or* (ii), *$\mathcal{V}_m$ is internally stabilizable*. A simple proof of this property, which uses the particular features of self-bounded controlled invariants, is sketched as follows. Consider the subspace $\bar{\mathcal{V}} := \mathcal{V}+\mathcal{R}_1$, which is an internally stabilizable $(A,\mathcal{B})$-controlled invariant, as the sum of two internally stabilizable $(A,\mathcal{B})$-controlled invariants. It satisfies (i) or (ii) since $\mathcal{V}$ does, and is self-bounded with respect to $\mathcal{V}^*$ since it contains $\mathcal{V}^* \cap \mathcal{B}$: by summing $\mathcal{B}$ to both members of $\mathcal{D} \subseteq \bar{\mathcal{V}}$ or $\mathcal{D} \subseteq \bar{\mathcal{V}}+\mathcal{B}$ we obtain the inclusion $\mathcal{V}^* \cap (\mathcal{B}+\mathcal{D}) \subseteq \bar{\mathcal{V}}$, which means that $\bar{\mathcal{V}}$ is also an $(A,\mathcal{B}+\mathcal{D})$-controlled invariant self-bounded with respect to $\mathcal{V}^*$. Hence it contains $\mathcal{V}_m$: this implies that the internal unassignable eigenvalues of $\mathcal{V}_m$ as an $(A,\mathcal{B})$-controlled invariant self-bounded with respect to $\mathcal{V}^*$, which are a part of those of $\bar{\mathcal{V}}$, are all stable.

Thus $\mathcal{V}_m$ being internally stabilizable is a necessary condition; on the other hand it is also sufficient because, if it holds, a solution of the disturbance localization problem is feasible by synthesizing a state feedback which achieves both $(A+BF)$-invariance of $\mathcal{V}_m$ and stability of $A+BF$.

Subspace (18) is internally stabilizable if and only if its internal unassignable eigenvalues are all stable. Clearly $\mathcal{R}_1 \subseteq \mathcal{V}_m \subseteq \mathcal{V}^*$: all these subspaces are $(A,\mathcal{B})$-controlled invariants self-bounded with respect to $\mathcal{V}^*$.

Let $F$ be such that $(A+BF)\mathcal{V}^* \subseteq \mathcal{V}^*$: the internal unassignable eigenvalues set of $\mathcal{V}_m$ is

$$\sigma\big((A+BF)|_{\mathcal{V}_m/\mathcal{R}_1}\big) = \sigma\big((A+BF)|_{\mathcal{V}^*/\mathcal{R}_1}\big) - \sigma\big((A+BF)|_{\mathcal{V}^*/\mathcal{V}_m}\big)$$
$$= \mathcal{Z}(u;e) - \mathcal{Z}(u,d;e) \;:$$

if (A.1) or (A.3) holds, the condition expressed in terms of invariant zeros (A.2) or (A.4) is equivalent to internal stabilizability of $\mathcal{V}_m$, hence it is necessary and sufficient for solvability of the disturbance localization problem with stability. Q.E.D.

We shall now consider the duals of Problems 1 and 2, namely the ***unknown-input asymptotic state estimation problem***, both with a purely dynamic and a nonpurely dynamic linear observer. When input is unknown, the use of a nonpurely dynamic device in general improves asymptotic state estimation in the sense that the corresponding unobservability subspace is contained in that of a purely dynamic device.

For these problems we shall refer to the four-map system $(A,B,C,E)$ described by

(20) $$\dot{x}(t) = A\,x(t) + D\,d(t)\,,$$
(21) $$y(t) = C\,x(t)\,,$$
(22) $$e(t) = E\,x(t)\,,$$

which is assumed to be connected with the purely (nonpurely) dynamic observer

(23) $$\dot{w}(t) = N\,w(t) + M\,y(t)\,,$$
(24) $$\hat{e}(t) = L\,w(t)\,,$$
(25) $$\big(\hat{e}(t) = L\,w(t) + K\,y(t)\big)\,.$$

In the above equations $d \in \mathcal{U}_d$ $(=\mathrm{R}^r)$ denotes the unknown disturbance, $y \in \mathcal{Y}$ $(=\mathrm{R}^q)$ the accessible output, $e \in \mathcal{Y}_e$ $(=\mathrm{R}^s)$ the variable to be estimated and $\hat{e} \in \mathcal{Y}_e$ the estimate provided by the observer. The vector $x \in \mathcal{X}$ $(=\mathrm{R}^n)$ still denotes the system state and $w \in \mathcal{W}$ $(=\mathrm{R}^m)$ denotes the observer state. Pair $(A,C)$ is assumed to be detectable. Let $\mathcal{D} := \mathrm{im}D$, $\mathcal{C} := \ker C$ and $\mathcal{E} := \ker E$.

The unknown-input asymptotic estimation problems which are herein considered are geometrically stated in the following terms.

**Problem 3.** (Unknown-Input Purely Dynamic Estimation) *Find, if possible, an output injection matrix $G$ such that $A+GC$ is stable and there exists at least one $(A+GC)$-invariant containing $\mathcal{D}$ and contained in $\mathcal{E}$.*

**Problem 4.** (Unknown-Input Nonpurely Dynamic Estimation) *Find, if possible, an output injection matrix $G$ such that $A+GC$ is stable and there* *exists at least one* $(A+GC)$*-invariant containing* $\mathcal{D}$ *whose intersection with* $\mathcal{C}$ *is contained in* $\mathcal{E}$.

Define the basic conditioned invariant

$$\mathcal{S}^* := \min \mathcal{S}(A, \mathcal{C}; \mathcal{D}), \tag{26}$$

which is the dual of (17). The following theorem gives necessary and sufficient conditions for solvability of the stated problems, expressed in terms of invariant zeros. It can be derived by duality from Theorem 1.

**Theorem 2.** *The unknown-input purely dynamic estimation problem has a solution if and only if:*
(B.1) $\mathcal{E} \supseteq \mathcal{S}^*$,
(B.2) $\mathcal{Z}(d;y) - \mathcal{Z}(d;y,e)$ *has all its elements stable,*
*while the unknown-input nonpurely dynamic estimation problem has a solution if and only if:*
(B.3) $\mathcal{E} \supseteq \mathcal{S}^* \cap \mathcal{C}$,
(B.4) $\mathcal{Z}(d;y) - \mathcal{Z}(d,y;e)$ *has all its elements stable.*

## 4. Stabilizability of Dynamic Output Feedback

We shall now consider dynamic output feedback. We refer to a five-map *controlled system* $(A, B, C, D, E)$ described by

$$\dot{x}(t) = A\,x(t) + B\,u(t) + D\,d(t), \tag{27}$$

$$y(t) = C\,x(t), \tag{28}$$

$$e(t) = E\,x(t), \tag{29}$$

connected to a *controller* described by

$$\dot{w}(t) = N\,w(t) + M\,y(t), \tag{30}$$

$$u(t) = L\,w(t) + K\,y(t). \tag{31}$$

Vector $x \in \mathcal{X}$ $(= \mathbb{R}^n)$ represents the controlled system state and $w \in \mathcal{W}$ $(= \mathbb{R}^m)$ the controller state. Vectors $u \in \mathcal{U}$ $(= \mathbb{R}^p)$ and $d \in \mathcal{U}_d$ $(= \mathbb{R}^s)$ are respectively the control input and the external disturbance, while $y \in \mathcal{Y}$ $(= \mathbb{R}^q)$ and $e \in \mathcal{Y}_e$ $(= \mathbb{R}^r)$ are the informative output and the regulated output. Pair $(A, B)$ and $(A, C)$ are assumed to be respectively stabilizable and detectable. As before, let $\mathcal{B} := \text{im}B$, $\mathcal{C} := \ker C$, $\mathcal{D} := \text{im}D$ and $\mathcal{E} := \ker E$.

A significant application of dynamic output feedback is set as follows.

**Problem 5.** (Dynamic Disturbance Localization with Stability) *Given the five-map system* (27–29), *determine, if possible, a controller of the type* (30, 31) *such that the overall system is stable and output* $e$ *is not influenced by input* $d$.

Nonconstructive necessary and sufficient conditions for solvability of Problem 5 are immediately deductable. They can be stated in very simple and intuitive geometric terms: the dynamic disturbance localization problem with stability is solvable if and only if there exist both an internally stabilizable $(A,\mathcal{B})$-controlled invariant $\mathcal{V}$ and an externally stabilizable $(A,\mathcal{C})$-conditioned invariant $\mathcal{S}$ such that $\mathcal{D}\subseteq\mathcal{S}\subseteq\mathcal{V}\subseteq\mathcal{E}$. This implies the basic structural condition

$$(32)\qquad \mathcal{S}^*\subseteq\mathcal{V}^*,$$

where $\mathcal{S}^*$ and $\mathcal{V}^*$ are defined as in (26) and (17). Inclusion (32) is a necessary condition where the stability requirement is not taken into account.

Constructive necessary and sufficient conditions for the existence of a stable solution that use the dual-lattice structures were presented in [11]. They are recalled below in Proposition 1 since they form the background for deriving sufficient (but unfortunately, in this case, not necessary) conditions in terms of invariant zeros which will be presented in Theorem 3. In the statement two further geometric objects are needed besides $\mathcal{S}^*$ and $\mathcal{V}^*$. One is $\mathcal{V}_m$, defined as in (18) and the other is the dual subspace $\mathcal{S}_M$ which is defined by

$$(33)\qquad \mathcal{S}_M := \mathcal{S}^* + \max\mathcal{V}(A,\mathcal{D};\mathcal{C}\cap\mathcal{E});$$

if (32) holds $\mathcal{V}_m$ and $\mathcal{S}_M$ are respectively the infimum of the lattice of all $(A,\mathcal{B}+\mathcal{D})$-controlled invariants self-bounded with respect to $\mathcal{V}^*$ and the supremum of the lattice of all $(A,\mathcal{C}\cap\mathcal{E})$-conditioned invariants self-hidden with respect to $\mathcal{S}^*$.

**Proposition 1.** *The dynamic disturbance localization problem with stability has a solution if and only if*
(C.1) $\mathcal{S}^*\subseteq\mathcal{V}^*$,
(C.2) $\mathcal{V}_M := \mathcal{V}_m + \mathcal{S}_M$ *is internally stabilizable,*
(C.3) $\mathcal{S}_M$ *is externally stabilizable.*

Note that condition (C.2) has a meaning only if $\mathcal{V}_M$ is an $(A,\mathcal{B})$-controlled invariant. We show that this is true provided (C.1) holds. To this end, define

$$(34)\qquad \mathcal{V}_1^* := \max\mathcal{V}(A,\mathcal{D};\mathcal{C}\cap\mathcal{E}),$$
$$(35)\qquad \mathcal{S}_1^* := \min\mathcal{S}(A,\mathcal{E};\mathcal{B}+\mathcal{D}).$$

It is easily seen that $\mathcal{V}_1^*\subseteq\mathcal{V}^*$ if (C.1) holds. In fact, if (C.1) holds, $\mathcal{V}^*$ is the maximum $(A,\mathcal{B}+\mathcal{D})$-controlled invariant contained in $\mathcal{E}$ (see the proof of Theorem 1), which clearly contains the maximum $(A,\mathcal{D})$-controlled invariant contained in $\mathcal{C}\cap\mathcal{E}$. Inclusion $\mathcal{S}_1^*\supseteq\mathcal{S}^*$ also holds by duality. Hence

$$\begin{aligned}\mathcal{V}_M = \mathcal{V}_m + \mathcal{S}_M &= (\mathcal{V}^*\cap\mathcal{S}_1^*) + \mathcal{S}^* + \mathcal{V}_1^*\\ &= (\mathcal{V}^*+\mathcal{S}^*)\cap\mathcal{S}_1^* + \mathcal{V}_1^*\\ &= (\mathcal{V}^*\cap\mathcal{S}_1^*) + \mathcal{V}_1^* = \mathcal{V}_m + \mathcal{V}_1^*;\end{aligned}$$

the equality in the second row derives from the sum with $\mathcal{S}^*$ being distributive with respect to the intersection $\mathcal{V}^* \cap \mathcal{S}_1^*$ (for two reasons: $\mathcal{S}^* \subseteq \mathcal{V}^*$ and $\mathcal{S}^* \subseteq \mathcal{S}_1^*$), while the first equality in the third row derives from the intersection with $\mathcal{S}_1^*$ being distributive with respect to the sum $\mathcal{V}^* + \mathcal{S}^*$ (for the same two reasons). Note that, if the basic inclusion (C.1) holds, $\mathcal{V}_m$ is an $(A, \mathcal{B}+\mathcal{D})$-controlled invariant containing $\mathcal{D}$, and $\mathcal{V}_1^*$, as an $(A, \mathcal{D})$-controlled invariant, is also an $(A, \mathcal{B}+\mathcal{D})$-controlled invariant; hence their sum $\mathcal{V}_M$ is an $(A, \mathcal{B}+\mathcal{D})$-controlled invariant and, since it contains $\mathcal{D}$, an $(A, \mathcal{B})$-controlled invariant. Furthermore, $\mathcal{V}_M$ is contained in $\mathcal{E}$ since both $\mathcal{V}_m$ and $\mathcal{V}_1^*$ are, and contains $\mathcal{V}^* \cap \mathcal{B}$ since $\mathcal{V}_m$ does; thus $\mathcal{V}_M$ is self-bounded with respect to $\mathcal{V}^*$. We can now proceed to state the sufficient conditions in terms of invariant zeros.

**Theorem 3.** *The dynamic disturbance localization problem with stability has a solution if*
(D.1) $\mathcal{S}^* \subseteq \mathcal{V}^*$,
(D.2) $\mathcal{Z}(u;e) - \mathcal{Z}(u,d;e)$ *has all its elements stable,*
(D.3) $\mathcal{Z}(d;y) - \mathcal{Z}(d;y,e)$ *has all its elements stable,*
(D.4) $\mathcal{Z}(u,d;e) \cap \mathcal{Z}(d;y,e)$ *has all its elements stable.*

**Proof.** We shall show that the above conditions imply those stated in Property 1, which are necessary and sufficient. The structural requirement (D.1) is the same as (C.1). On the other hand, in the proof of Theorem 1 we have already shown that (D.2) is necessary and sufficient for $\mathcal{V}_m$ to be internally stabilizable; by duality (D.3) is necessary and sufficient for $\mathcal{S}_M$ to be externally stabilizable, so that (D.3) is equivalent to (C.3).

Consider the dual subspace of $\mathcal{V}_M$, i.e. $\mathcal{S}_m := \mathcal{S}_M \cap \mathcal{V}_m$, which by an argument perfectly dual to that developed above for $\mathcal{V}_M$ can be proved to be an $(A, \mathcal{C})$-conditioned invariant self-bounded with respect to $\mathcal{S}^*$. By a change of basis similar to that previously used to prove the basic dual-lattice properties we can show that

$$\sigma((A+BF)|_{\mathcal{V}_M/\mathcal{V}_m}) = \sigma((A+GC)|_{\mathcal{S}_M/\mathcal{S}_m}), \tag{36}$$

where $F$ and $G$ are any matrices such that $(A+BF)\mathcal{V}^* \subseteq \mathcal{V}^*$ and $(A+GC)\mathcal{S}^* \subseteq \mathcal{S}^*$. Consider the similarity transformation $T := [T_1\ T_2\ T_3\ T_4]$ such that $\mathrm{im}T_1 = \mathcal{S}_m = \mathcal{S}_M \cap \mathcal{V}_m$, $\mathrm{im}[T_1\ T_2] = \mathcal{V}_m$, $\mathrm{im}[T_1\ T_3] = \mathcal{S}_M$ (hence $\mathrm{im}[T_1\ T_2\ T_3] = \mathcal{V}_M$). Since $\mathcal{S}_M \subseteq \mathcal{V}^* + \mathcal{C} = \mathcal{S}_m + \mathcal{C}$, it is possible to choose $T_3$ in such a way that $\mathrm{im}T_3 \subseteq \mathcal{C}$ and, by duality, since $\mathcal{V}_m \supseteq \mathcal{V}^* \cap \mathcal{B} = \mathcal{V}_m \cap \mathcal{B}$, it is possible to choose $T_4$ in such a way that $\mathrm{im}[T_1\ T_2\ T_4] \supseteq \mathcal{B}$. Matrices $A' := T^{-1}AT$, $B' := T^{-1}B$ and $C' := CT$ exhibit the following structures:

$$A' = \begin{bmatrix} A'_{11} & A'_{12} & A'_{13} & A'_{14} \\ A'_{21} & A'_{22} & O & A'_{24} \\ O & O & A'_{33} & A'_{34} \\ A'_{41} & A'_{42} & O & A'_{44} \end{bmatrix}, \quad B' = \begin{bmatrix} B'_1 \\ B'_2 \\ O \\ B'_4 \end{bmatrix},$$

$$C' = [\ C'_1 \quad C'_2 \quad O \quad C'_4\ ].$$

The zero submatrices of $B'$ and $C'$ are due to the previously considered conditions on $T_3$ and $T_4$, while those in the third row of $A'$ are due to $\mathcal{V}_m$ being an $(A,\mathcal{B})$-controlled invariant and to the particular structure of $B'$, and those in the fourth column to the whole $\mathcal{S}_M$ being an $(A,\mathcal{C})$-controlled invariant and to the particular structure of $C'$.

The new basis allows an immediate check of equality (36): in fact, it is easily seen that both spectra considered in (36) coincide with $\sigma(A'_{33})$. On the other hand,

$$\sigma((A+BF)|_{\mathcal{V}_M/\mathcal{V}_m}) \subseteq \sigma((A+BF)|_{\mathcal{V}^*/\mathcal{V}_m}) = \mathcal{Z}(u,d;e)\,, \tag{37}$$

$$\sigma((A+GC)|_{\mathcal{S}_M/\mathcal{S}_m}) \subseteq \sigma((A+GC)|_{\mathcal{S}_M/\mathcal{S}^*}) = \mathcal{Z}(d;y,e)\,, \tag{38}$$

hence (D.4) implies that the spectrum on the left of (36) is stable. Since $\mathcal{V}_m$ is internally stabilizable by virtue of (D.2), we conclude that (C.2) holds if both (D.2) and (D.4) hold. Q.E.D.

## 5. Stabilizability in the General Regulator Problem

The regulator problem is another application of dynamic output feedback. We refer again to the five-map controlled system $(A,B,C,D,E)$ described by (27–29) connected to the controller described by (30, 31), but the assumption that $(A,B)$ is stabilizable is removed in order to introduce an uncontrollable *exosystem* which models disturbances to be asymtotically neutralized and signals to be asymptotically tracked, such as steps, ramps, sinusoids. To be more precise, the controlled system state will be partitioned as

$$x = \begin{bmatrix} x_1 \\ x_2 \end{bmatrix}\,,$$

where $x_1 \in \mathbb{R}^{n_1}$ denotes the state of the *plant* and $x_2 \in \mathbb{R}^{n_2}$ that of the exosystem. Matrices $A,B,C,D,E$ are accordingly partitioned as

$$A = \begin{bmatrix} A_1 & A_3 \\ O & A_2 \end{bmatrix}\,, \quad B = \begin{bmatrix} B_1 \\ O \end{bmatrix}\,, \quad D = \begin{bmatrix} D_1 \\ O \end{bmatrix}\,,$$

$$C = [\,C_1 \quad C_2\,]\,, \quad E = [\,E_1 \quad E_2\,]\,.$$

The controlled system is no longer stabilizable, since the exosystem is unstable and inputs can act only on the plant, but it is assumed to be detectable through the informative output. In strict mathematical terms our assumptions are: (i) the eigenvalues of the exosystem, i.e. the elements of $\sigma(A_2)$, are all unstable; (ii) the plant, i.e. pair $(A_1,B_1)$, is stabilizable; (iii) the overall regulated system, i.e. pair $(A,C)$, is detectable. The overall system with the exosystem disconnected (i.e. taking into account only the differential equations corresponding to $x_1$ and $w$ with $x_2=0$) will be referred to as the *regulated plant*. The regulator problem is stated as follows.

**Problem 6.** (The Regulator Problem) *Given the five-map system* (27–29) *including an exosystem with all eigenvalues unstable determine, if possible, a controller of the type* (30, 31) *such that the regulated plant is stable and output e is both insensitive to input d and asymptotically moving to zero as time approaches infinity.*

As for disturbance localization by dynamic output feedback, nonconstructive necessary and sufficient conditions for solvability of Problem 6, stated in plain geometric terms, are easily derived. In this case, besides a controlled invariant and a conditioned invariant, they involve another geometric object, the *plant* $\mathcal{P}\subseteq\mathcal{X}$, which is the $A$-invariant defined by

$$\mathcal{P} := \{x \,:\, x_2 = 0\}\,. \tag{39}$$

These nonconstructive conditions are: the regulator problem is solvable if and only if there exist both an $(A,\mathcal{B})$-controlled invariant $\mathcal{V}$ such that $\mathcal{V}\cap\mathcal{P}$ is internally stabilizable and an externally stabilizable $(A,\mathcal{C})$-conditioned invariant $\mathcal{S}$ which satisfy $\mathcal{D}\subseteq\mathcal{S}\subseteq\mathcal{V}\subseteq\mathcal{E}$. Note that subspace $\mathcal{V}\cap\mathcal{P}$ is an $(A,\mathcal{B})$-controlled invariant (hence the word "stabilizable" referred to it has a meaning) as the intersection of an $(A,\mathcal{B})$-controlled invariant and an $A$-invariant containing $\mathcal{B}$, since $A(\mathcal{V}\cap\mathcal{P})\subseteq A\mathcal{V}\cap A\mathcal{P}\subseteq(\mathcal{V}+\mathcal{B})\cap\mathcal{P}=\mathcal{V}\cap\mathcal{P}+\mathcal{B}$.

Constructive necessary and sufficient conditions for solvability of the regulator problem were presented in [12] and are recalled in the following proposition.

**Proposition 2.** *The regulator problem has a solution if and only if*
(E.1) $\mathcal{S}^*\subseteq\mathcal{V}^*$,
(E.2) $\mathcal{V}^*$ *is externally stabilizable,*
(E.3) $\mathcal{S}_M$ *is externally stabilizable,*
(E.4) $\mathcal{V}_M\cap\mathcal{P}$ *is internally stabilizable,*
(E.5) $\mathcal{V}_M+\mathcal{V}^*\cap\mathcal{P}$ *is complementable with respect to* $(\mathcal{V}_M,\mathcal{V}^*)$.

Condition (E.5) expresses the requirement that an $(A,\mathcal{B})$-controlled invariant is complementable with respect to a pair of $(A,\mathcal{B})$-controlled invariants. Complementability of controlled invariants is formally defined as follows: given three $(A,\mathcal{B})$-controlled invariants $\mathcal{V}_1$, $\mathcal{V}$ and $\mathcal{V}_2$ such that $\mathcal{V}_1\subseteq\mathcal{V}\subseteq\mathcal{V}_2$, $\mathcal{V}$ is said to be *complementable with respect to pair* $(\mathcal{V}_1,\mathcal{V}_2)$ if there exists an $(A,\mathcal{B})$-controlled invariant $\mathcal{V}'$ such that $\mathcal{V}\cap\mathcal{V}'=\mathcal{V}_1$, $\mathcal{V}+\mathcal{V}'=\mathcal{V}_2$.

In Proposition 2 subspaces $\mathcal{V}^*$, $\mathcal{S}^*$, $\mathcal{S}_M$ and $\mathcal{V}_M$ are those previously defined in (17), (26), (33) and (C.2) of Proposition 1. Let us denote with $\mathcal{Z}_P(u,d\,;e)$ the invariant zeros referred only to the plant, i.e. to triple $(A_1,[B_1\;D_1],E_1)$. The sufficient conditions in terms of invariant zeros are set as follows.

**Theorem 4.** *The regulator problem has a solution if*
(F.1) $\mathcal{S}^*\subseteq\mathcal{V}^*$,
(F.2) $\mathcal{Z}(u;e)$ *contains all the eigenvalues of the exosystem,*
(F.3) $\mathcal{Z}(u;e)-\mathcal{Z}(u,d;e)$ *has all its elements stable,*

(F.4) $\mathcal{Z}(d;y) - \mathcal{Z}(d;y,e)$ *has all its elements stable,*
(F.5) $\mathcal{Z}_P(u,d;e) \cap \mathcal{Z}(d;y,e)$ *has all its elements stable,*
(F.6) $\mathcal{Z}_P(u,d;e)$ *does not contain any eigenvalue of the exosystem.*

**Proof.** A part of the proof is similar to that of Theorem 3: (F.1) is the same as (E.1) and (F.4) has already been proved to be equivalent to (E.3). Although the set $\mathcal{Z}_P(u,d;e)$ is defined in a reduced state space (that of the controlled plant, without the exosystem), it is easily shown that it can also be handled in the regular state space $\mathcal{X}$ as

$$\mathcal{Z}_P(u,d;e) = \sigma\big((A+BF)|_{(\mathcal{V}^*\cap\mathcal{P})/\mathcal{V}_m}\big)\,. \tag{40}$$

In fact, by intersecting with $\mathcal{P}$ any $(A,\mathcal{B})$-controlled invariant contained in $\mathcal{E}$ we still obtain an $(A,\mathcal{B})$-controlled invariant contained in $\mathcal{E}$ whose elements have the first $n_1$ components belonging to an $(A_1,\mathrm{im}B_1)$-controlled invariant contained in $\ker E_1$ and the last $n_2$ components equal to zero. In this correspondence, which clearly is a projection, in particular $\mathcal{V}_P^* := \max\mathcal{V}(A_1,\mathrm{im}B_1;\ker E_1)$ corresponds to $\mathcal{V}^*\cap\mathcal{P}$. Note that the maximal $(A,\mathcal{B})$-controlled invariant or the maximal $(A,\mathcal{B}+\mathcal{D})$-controlled invariant contained in $\mathcal{E}\cap\mathcal{P}$ is $\mathcal{V}^*\cap\mathcal{P}$, self-bounded with respect to $\mathcal{V}^*$, and that $\mathcal{V}_m$ is the infimum of both lattices of all $(A,\mathcal{B}+\mathcal{D})$-controlled invariants self-bounded with respect to $\mathcal{V}^*\cap\mathcal{P}$ and self-bounded with respect to $\mathcal{V}^*$. Hence (40) holds. Also note that $\mathcal{V}_M\cap\mathcal{P}$ – which contains $\mathcal{V}^*\cap(\mathcal{B}+\mathcal{D})$ since both $\mathcal{V}_m$ and $\mathcal{P}$ do – is both an $(A,\mathcal{B})$-controlled invariant and an $(A,\mathcal{B}+\mathcal{D})$-controlled invariant self-bounded both with respect to $\mathcal{V}^*\cap\mathcal{P}$ and $\mathcal{V}^*$. In the change of basis introduced to prove Theorem 3 we can split matrix $T_3$ as $[T_3'\ T_3'']$, with $T_3'$ such that $\mathrm{im}[T_1\ T_2\ T_3'] = \mathcal{V}_M\cap\mathcal{P}$, so that submatrix $A_{33}'$ assumes the structure

$$\begin{bmatrix} P & Q \\ O & R \end{bmatrix},$$

and, denoting by $F$ any matrix such that $(A+BF)\mathcal{V}^* \subseteq \mathcal{V}^*$, it follows that

$$\sigma\big((A+BF)|_{(\mathcal{V}_M\cap\mathcal{P})/\mathcal{V}_m}\big) = \sigma(P)\,.$$

Equation (37) in this case is replaced by

$$\sigma\big((A+BF)|_{(\mathcal{V}_M\cap\mathcal{P})/\mathcal{V}_m}\big) \subseteq \sigma\big((A+BF)|_{(\mathcal{V}^*\cap\mathcal{P})/\mathcal{V}_m}\big) = \mathcal{Z}_P(u,d;e)\,, \tag{41}$$

while (36) and (38) are still valid. Since

$$\sigma\big((A+BF)|_{(\mathcal{V}_M\cap\mathcal{P})/\mathcal{V}_m}\big) = \sigma(P) \subseteq \sigma\big((A+BF)|_{\mathcal{V}_M/\mathcal{V}_m}\big) = \sigma(A_{33}')\,,$$

and $\mathcal{V}_m$ is internally stabilizable because of (F.3), we conclude that (F.3) and (F.5) together imply (E.4).

It is well known that any $(A,\mathcal{B})$-controlled invariant $\mathcal{V}$ is externally stabilizable if and only if the $A$-invariant

$$\mathcal{V} + \mathcal{R} \tag{42}$$

is externally stable. In (42) $\mathcal{R}$ denotes the controllability subspace of the controlled system, i.e. the minimal $A$-invariant containing $\mathcal{B}$. Conditions (F.2) and (F.3) imply $\mathcal{Z}(u,d;e)\supseteq\sigma(A_2)$; on the other hand, since $\mathcal{Z}(u,d;e)$ and $\mathcal{Z}_P(u,d;e)$ contain respectively the internal unassignable eigenvalues of $\mathcal{V}^*$ and of $\mathcal{V}^*\cap\mathcal{P}$, which are $(A,\mathcal{B}+\mathcal{D})$-controlled invariants self-bounded with respect to $\mathcal{V}^*$ (whose dimensions cannot be different by a number greater than $n_2$ because the dimension of $\mathcal{P}$ is $n_1=n-n_2$), and $\mathcal{Z}_P(u,d;e)\cap\sigma(A_2)=\emptyset$ by virtue of (F.6), it follows that

$$\mathcal{Z}(u,d;e)=\mathcal{Z}_P(u,d;e)+\sigma(A_2)\,; \tag{43}$$

hence the difference in dimensions of $\mathcal{V}^*$ and $\mathcal{V}^*\cap\mathcal{P}$ is equal to $n_2$, so that

$$\mathcal{V}^*+\mathcal{P}=\mathcal{X}\,. \tag{44}$$

Note that if $\mathcal{V}$ is any $(A,\mathcal{B})$-controlled invariant, also $\mathcal{V}+\mathcal{P}$ is. Since the assumption that $(A_1,B_1)$ is stabilizable means that the $A$-invariant $\mathcal{R}$ is externally stable with respect to the $A$-invariant $\mathcal{P}$, (44) implies (E.2).

Consider now subspace $\mathcal{V}_M+\mathcal{V}^*\cap\mathcal{P}$, which is also an $(A,\mathcal{B})$-controlled invariant self-bounded with respect to $\mathcal{V}^*$, and perform the similarity transformation $T:=[T_1\ T_2\ T_3\ T_4]$ such that $\mathrm{im}T_1=\mathcal{V}_M$, $\mathrm{im}[T_1\ T_2]=\mathcal{V}_M+\mathcal{V}^*\cap\mathcal{P}$, $\mathrm{im}[T_1\ T_2\ T_3]=\mathcal{V}^*$. Since $\mathcal{V}_M\supseteq\mathcal{V}^*\cap\mathcal{B}=\mathcal{V}_M\cap\mathcal{B}$, it is possible to choose $T_4$ in such a way that $\mathrm{im}[T_1\ T_4]\supseteq\mathcal{B}$. Matrices $A':=T^{-1}AT$ and $B':=T^{-1}B$ have the structures

$$A'=\begin{bmatrix}A'_{11} & A'_{12} & A'_{13} & A'_{14}\\ O & A'_{22} & A'_{23} & A'_{24}\\ O & O & A'_{33} & A'_{34}\\ A'_{41} & A'_{42} & A'_{43} & A'_{44}\end{bmatrix},\qquad B'=\begin{bmatrix}B'_1\\ O\\ O\\ B'_4\end{bmatrix}.$$

Complementability condition (E.5) is satisfied if and only if the Sylvester equation $A'_{22}X-X\,A'_{33}=-A'_{23}$ admits a solution. It is known that it admits a (unique) solution if $\sigma(A'_{22})$ and $\sigma(A'_{33})$ have no common elements. Condition (E.5) is satisfied if both (F.2) and (F.6) hold since $\sigma(A'_{22})\subseteq\mathcal{Z}_P(u,d;e)$ and $\sigma(A'_{33})\subseteq\mathcal{Z}(u,d;e)-\mathcal{Z}_P(u,d;e)=\sigma(A_2)$. Q.E.D.

## 6. Conclusions

All the presented theorems (Theorems 1–4), related to various feedback stabilizability problems, are characterized by the presence of only one structural condition that does not require any eigenspace computation, and other conditions strictly expressing stability requirements for sets of invariant zeros.

Theorems 3 and 4 give only sufficient, but "tight" conditions, which do not require the minimum phase hypothesis for the controlled plant, as appears from the proofs offered.

## REFERENCES

[1] A.S. Morse, *Structural invariants of linear multivariable systems*, in SIAM J. Control, 11(1973), 446–465.

[2] B.P. Molinari, *Zeros of the system matrix*, in IEEE Trans. Autom. Contr., AC-21(1976), 795–797.

[3] H.H. Rosenbrock, *State space and multivariable theory*, Wiley, New York, 1970.

[4] A.G.J. Macfarlane and K. Karcanias, *Poles and zeros of linear multivariable systems: a survey of the algebraic, geometric, and complex-variable theory*, in International J. Control, 24(1976), 33–74.

[5] C.B. Schrader and M.K. Sain, *Research on system zeros: a survey* in International J. Control, 50(1989), 1407–1433.

[6] G. Basile and G. Marro, *Self-bounded controlled invariants: a straightforward approach to constrained controllability*, in J. Optimiz. Th. Applic., 38(1982), 71–81.

[7] J.M. Schumacher, *On a conjecture of Basile and Marro*, in J. Optimiz. Th. Applic., 41(1983), 371–376.

[8] G. Basile and G. Marro, *Self-bounded controlled invariants versus stabilizability*, in J. Optimiz. Th. Applic., 48(1986), 245–263.

[9] G. Basile and G. Marro, *Controlled and conditioned invariants in linear system theory*, Prentice Hall, Englewood Cliffs, 1991.

[10] G. Basile, G. Marro and A. Piazzi, *A new solution to the disturbance localization problem with stability and its dual*, in Proceedings of the '84 International AMSE Conference on Modelling and Simulation, 1-2(1984), 19–27, Athens.

[11] G. Basile, G. Marro and A. Piazzi, *Stability without eigenspaces in the geometric approach: some new results*, in Frequency domain and state space methods for linear systems, C.A. Byrnes and A. Lindquist, 1986, North-Holland, 441–450.

[12] G. Basile, G. Marro and A. Piazzi, *Stability without eigenspaces in the geometric approach: the regulator problem*, in in J. Optimiz. Th. Applic., 64(1990), 29–42.

Giovanni Marro, Aurelio Piazzi
D.E.I.S., Università di Bologna,
Viale Risorgimento, 2,
40136 Bologna, Italy.

This research was supported by the Ministero dell' Università e della Ricerca Scientifica e Tecnologica.

# High-Order Parameter Tuners for the Adaptive Control of Linear and Nonlinear Systems*

A. S. MORSE

**Abstract**

**A new method of parameter tuning is introduced which generates as outputs not only a set of tuned parameters, but also the first $\bar{n}$ time derivatives of each parameter, $\bar{n}$ being a prespecified positive integer. Using a modification of the idea of certainty-equivalence control, called "surrogate model control," it is explained how to construct an "identifier-based parameterized controller" which, when tuned using the aforementioned method, is capable of adaptively stabilizing any member of a specially structured family of nonlinear systems. An analysis is made of a model reference adaptive control algorithm consisting of a simplified high-order tuner and an identifier-based parameterized controller. It is shown that this algorithm can adaptively stabilize siso processes admitting linear, minimum phase models with different relative degrees.**

## 1. Introduction

In the past few years there have been a number of attempts to extend existing adaptive control techniques to processes modelled by restricted classes of nonlinear dynamical systems {e.g., [1]-[3]}. Such efforts usually consist of showing that a particular algorithm is capable of stabilizing a specially structured linearly parameterized, nonlinear design model $\Sigma_D(p)$ for each fixed but unknown value of $p$. Techniques from nonlinear system theory are then used to characterize, in coordinate independent terms, those dynamical process models whose input-output descriptions match or are close to that of $\Sigma_D$ for particular values of $p$ in $\Sigma_D$'s parameter space $\mathcal{P}$. This then is the class of admissible process models $\mathcal{C}_P$ to which the particular algorithm can be successfully applied.

*This research was supported by the National Science Foundation under grant number ECS-9012551

With the preceding in mind it is natural to wonder just how far existing adaptive techniques {e.g., [4]} can be extended into the world of nonlinear systems without having to introduce significant modifications. For example, for those cases in which parameters enter the model in "just the right way" and nonlinearities are such that with knowledge of the parameters nonadaptive stabilization is possible, one might expect that without knowledge of the parameters and notwithstanding the nonlinear dynamics, adaptive stabilization could be achieved using familiar methods. Interestingly, this has proved *not* to be the case, even for very simply structured examples, when the nonlinearities are not globally Lipschitz. Within the very recent past, several new adaptive algorithms have been discovered {e.g., see [5]-[9]} which overcome some of these difficulties. One of the key features of these algorithms which sharply distinguishes them from more familiar algorithms described in [4] and elsewhere, is the absence of tuning error normalization. The primary objective of this paper is to introduce a new method of parameter tuning, also employing an unnormalized tuning error, which can be used to obtain adaptive controllers with the same capabilities as those reported in [6].

The new tuner is of "high-order". By a tuner of *order* $\bar{n}$, is meant a tuning algorithm capable of generating as outputs not only a set of tuned parameters, but also the first $\bar{n}$ time-derivatives of each parameter in the set. Most tuning algorithms are of order one. The only previous studies of high-order tuners we know of can be found in [10] and [11]. What distinguishes the algorithm introduced here from those considered in [10] and [11], is the absence of tuning error normalization.

The new high-order tuner $\Sigma_T$ is described in §2. Then, in §3.1 we define one possible class $\mathcal{C}_P$ of nonlinear process models $\Sigma_P$ to which high-order tuning might be applied. In §3.2, using a modification of the idea of certainty-equivalence control, called "surrogate model control," we explain how to construct an "identifier-based parameterized controller" $\Sigma_C$ {cf. [12]}, which together with $\Sigma_T$ , is capable of adaptively stabilizing any process admitting a model in $\mathcal{C}_P$. A brief stability analysis is carried out in §3.3.

The second objective of this paper is to introduce a model reference adaptive controller, employing a simplified form of high-order tuning. In §4.1 we define a class $\mathcal{C}_P$ of siso, linear, minimum phase process models, to which this type of adaptive controller might be applied. Unlike the process models considered in classical model reference adaptive control, the process models in $\mathcal{C}_P$ are not presumed to all have the same relative degree. Instead, $\mathcal{C}_P$ is defined to include models of two different relative degrees, $n^*$ and $n^* + 1$, where $n^*$ is a prespecified positive integer. In §4.2, again using the concept of surrogate model control, we explain how to construct an identifier-based parameterized controller $\Sigma_C$ which together with a simplified high-order $\Sigma_T$, is capable of causing the output of any process admitting a model in $\mathcal{C}_P$, to approach and track a suitably defined reference signal. A brief argument which establishes this, is given out in §4.3.

## 2. High-Order Tuners

Consider the "error equation"

$$\dot{e} = -\lambda e + q_0(k - q_P)'w + \epsilon \tag{1}$$

where $\lambda$ is a positive constant, $q_0$ is a nonzero, scalar, constant of unknown magnitude but known sign, $q_P$ is an unknown constant $m$-vector, $w$ is an $m$-vector of known time functions, $e$ is a known scalar tuning error, $\epsilon$ is an exponentially decaying time function, and $k$ is a vector of $m$ parameters to be tuned. Equations of this form arise in connection with certain types of adaptive control problems, as will be shown in the sequel. Our objective here is to explain how to tune $k$ in such a way as to cause both $k$ and $e$ to be bounded wherever they exist and also to cause $e$ to have a finite $\mathcal{L}^2$ norm. The algorithm we propose will generate as outputs not only $k$, but also the first $\bar{n}$ derivatives of $k$, $\bar{n}$ being a prespecified positive integer. These derivatives can thus be used elsewhere in the construction of an overall adaptive controller. In order for the algorithm to have this property we shall have to make the following assumption.

**Assumption 1:** The first $\bar{n} - 1$ derivatives of $w$ are available signals.

**High-Order Tuning Algorithm:**

- If $\bar{n} = 1$, set
$$\dot{k} = -\text{sign}(q_0)we \tag{2}$$
- If $\bar{n} > 1$ pick a monic, stable polynomial $\bar{\alpha}(s)$ of degree $\bar{n} - 1$ and let $(\bar{c}, \bar{A}, \bar{b})$ be a minimal realization of $\bar{\alpha}(0)/\bar{\alpha}(s)$; set
$$\begin{aligned} \bar{e} &= e - v && (3)\\ \dot{v} &= -\lambda v + h_0(k - h)'w && (4)\\ \dot{h}_0 &= (k - h)'w\bar{e} && (5)\\ \dot{h} &= -\text{sign}(q_0)w\bar{e} && (6)\\ \dot{X} &= \bar{A}X(I + D) + \bar{b}h'(I + D) && (7)\\ k' &= \bar{c}X && (8) \end{aligned}$$
where $D$ is the $m \times m$ diagonal matrix whose ith diagonal entry is $w_i^2$, $w_i$ being the ith component of $w$.

*Remark 1:* Note that for $\bar{n} > 1$, the definition of $(\bar{c}, \bar{A}, \bar{b})$ together with (6) to (8) ensure that the first $\bar{n}$ derivatives of $k$ can be expressed as functions of $X, h, w, \bar{e}$ and the first $\bar{n} - 1$ derivatives of $w$. Therefore, in view of Assumption 1, the first $\bar{n}$ derivatives of $k$ are realizable from available signals.

*Remark 2:* The significant properties of the preceding algorithm will remain unchanged if $D$ is replaced with any positive definite matrix $\bar{D}$ satisfying $\bar{D} \geq D$. One such matrix is $\bar{D} = w'wI$.

Our first result is as follows.

**Theorem 1 - High-Order Tuning:** *Let* $[0, T)$ *be any subinterval of* $[0, \infty)$ *on which* $w$ *is defined and continuous.*

- *For* $\bar{n} = 1$, *each solution* $\{e, k\}$ *to (1) and (2) is bounded on* $[0, T)$ *and* $e$ *has a finite* $\mathcal{L}^2$ *norm.*
- *For* $\bar{n} > 1$, *each solution* $\{e, v, h_0, h, X\}$ *to (1), (3) - (8) is bounded on* $[0, T)$ *and* $e$ *and* $v$ *have finite* $\mathcal{L}^2$ *norms.*

A proof of this theorem can be found in the appendix.

If an upper bound $\bar{q}$ for $|q_0|$ is known, the preceding algorithm can be simplified by setting $\bar{e}$ equal to $e$, eliminating (4) and (5), and replacing $D$ with $\mu D$ where $\mu$ is a positive scale factor satisfying

$$\mu > \frac{2m\bar{q}||\bar{c}'||||P\bar{A}^{-1}\bar{b}||}{\lambda} \tag{9}$$

and $P$ is the unique positive definite solution to the Lyapunov equation

$$P\bar{A} + \bar{A}'P + I = 0 \tag{10}$$

In this case, the high-order tuning algorithm reduces to the following.

**Simplified High-Order Tuning Algorithm:**

- If $\bar{n} = 1$, set

$$\dot{k} = -\text{sign}(q_0)we \tag{11}$$

- If $\bar{n} > 1$, set

$$\begin{aligned} \dot{h} &= -\text{sign}(q_0)we && (12)\\ \dot{X} &= \bar{A}X(I + \mu D) + \bar{b}h'(I + \mu D) && (13)\\ k' &= \bar{c}X && (14) \end{aligned}$$

In the sequel it will be shown that this algorithm has exactly the same capabilities as its predecessor -to bound $e$, $k$, and the $\mathcal{L}^2$ norm of $e$ - if $e$ satisfies the error equation

$$\dot{e} = -\lambda e + q_0(k - q_P)'w + \epsilon \tag{15}$$

It will also be shown that if $e$ satisfies the error equation

$$\lambda e = q_0(k - q_P)'w + \epsilon \tag{16}$$

rather than (15), the simplified high-order tuning algorithm will still cause $k$ and the $\mathcal{L}^2$ norm of $e$ to be bounded. In §4 this additional capability will be exploited in the development of adaptive controllers for linear process models whose relative degrees are not known exactly.

Note that (15) and(16) are particular cases of the parameterized error equation

$$q_*\dot{e} = -\lambda e + q_0(k - q_P)'w + \epsilon \tag{17}$$

for which $q_* = 1$ and $q_* = 0$ respectively. In the sequel we shall analyze the effect of high order tuning on (17) with the understanding that $q_*$ might be either of these values.

**Theorem 2 - Simplified High-Order Tuning:** *Let $[0, T)$ be any subinterval of $[0, \infty)$ on which $w$ is defined and continuous.*

- *Assume $\bar{n} = 1$. For each $q_* \in \{0, 1\}$, each solution $\{q_*e, k\}$ to (11) and (17) is bounded on $[0, T)$ and $e$ has a finite $\mathcal{L}^2$ norm.*
- *Assume $\bar{n} > 1$. Suppose $|q_0| \leq \bar{q}$ and let $\mu$ satisfy (9). For each $q_* \in \{0, 1\}$, each solution $\{q_*e, h, X\}$ to (12) - (14) and (17) is bounded on $[0, T)$ and $e$ has a finite $\mathcal{L}^2$ norm.*

A proof of this theorem can be found in the appendix.

## 3. Nonlinear Process Models

The aim of this section is to explain how the first high-order tuning algorithm $\Sigma_T$ discussed in §2, might be used in conjunction with a suitably defined, identifier-based parameterized controller $\Sigma_C$ , to adaptively stabilize a process model $\Sigma_P$ containing a single, locally Lipschitz nonlinearity. We begin by defining the class of admissible process models $\mathcal{C}_P$.

### 3.1 Process Model $\Sigma_P$

Let $n$ and $n^*$ be fixed {i.e., known} integers satisfying $n \geq n^* > 0$ and let $N : \Re \rightarrow \Re$ be a memoryless, locally Lipschitz, nonlinear function satisfying $N(0) = 0$. The class $\mathcal{C}_P$ of process models $\Sigma_P$ to be considered consists of siso systems of the form

$$\begin{aligned} \dot{x}_P &= A_P x_P + b_P u + b_N N(y) \\ y &= c_P x_P \end{aligned} \tag{18}$$

where $(A_P, b_P, c_P)$ is a $n_P$–dimensional, stabilizable, detectable linear subsystem. Each $\Sigma_P \in \mathcal{C}_P$ admits a block diagram representation of the form shown in Figure 1 where $t_P(s)$ and $t_N(s)$ are the strictly proper transfer functions $g_P\frac{\beta_P(s)}{\alpha_P(s)}$ and $\frac{\beta_N(s)}{\alpha_P(s)}$ respectively; here $\alpha_P(s)$ is the characteristic polynomial of $A_P$, $\beta_P(s)$ is monic and $g_P$ is a nonzero constant called the *high frequency gain.*

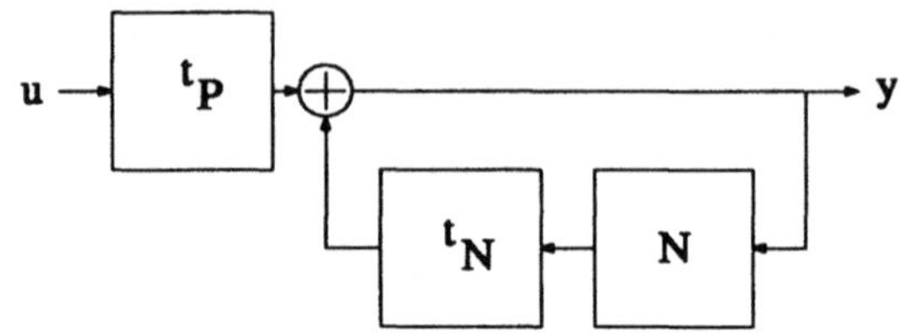


Figure 1: Process Model $\Sigma_P$

We shall make the following

**Process Model Assumptions:** For each $\Sigma_P \in \mathcal{C}_P$

1. $n_P \leq n$
2. relative degree $t_P(s) = n^*$
3. $g_P > 0$
4. $\beta_P(s)$ is stable
5. $N$ is known
6. relative degree $t_P(s) \leq$ relative degree $t_N(s)$

For the case when $t_N(s) = 0$ the preceding reduce to the process model assumptions made in the well-known classical model reference adaptive control problem originally solved in [13]. A more general version of the nonlinear problem considered here, with multiple nonlinear feedback loops, has recently been solved in [6], but without the benefit of high-order tuning.

Assumption 3 is not crucial and can easily be replaced with the weaker assumption that $g_P$ is nonzero but of known sign; and the requirement that $g_P$'s sign be known can be dispensed with by using using a Nussbaum gain [14]. Assumption 6 appears also not to be crucial and can probably be replaced, as in [7, 8, 9], with the much milder requirement that $N$ have a continuous $\mu$th derivative, $\mu$ being either relative degree $\frac{t_P(s)}{t_N(s)}$ or zero, whichever is larger . The additional techniques needed for handling this case, which involve the introduction of "nonlinear damping" can be found in [9].

### 3.2 Parameterized Controller $\Sigma_C$

Our aim here is to construct an "identifier-based parameterized controller" $\Sigma_C$ suitable for high-order tuning. In general, identifier-based parameterized controllers consist of two subsystems, one an "identifier" $\Sigma_I$ whose function is to generate an identification error $e$ and the other an "internal regulator" $\Sigma_R$ whose output serves as the closed-loop control to the process {cf. [12]}.

**Identifier $\Sigma_I$ :**

As the first step in the construction of $\Sigma_C$, we select an appropriate "parameterized design model" $\Sigma_D$ upon which the structure of $\Sigma_I$ and the synthesis of $\Sigma_R$ are to be based. For this, choose a positive number $\lambda$, a monic stable polynomial $\alpha_*(s)$ of degree $n^* - 1$, a controllable, observable realization $(A_*, b_*, c_*, d_*)$ of $\frac{1}{\alpha_*(s)}$, and a $n$-dimensional, single-output, observable pair $(c, A)$ with $A$ stable. These selections determine the "direct control, design model" $\Sigma_D(p)$ {cf. [12]}, parameterized by the vector $p = [\ p_0 \ \ p_1' \ \ p_2' \ \ p_3' \ \ p_4\ ]' \in \mathcal{P} = \Re \oplus \Re^n \oplus \Re^n \oplus \Re^n \oplus \Re$ and described by the equations

$$\begin{aligned} \dot{x}_{D1} &= Ax_{D1} + p_1 y_D + p_2 u_D + p_3 N(y_D) \\ \dot{x}_{D2} &= A_* x_{D2} + b_*(u_D + cx_{D1} + p_4 N(y_D)) \\ \sigma_D &= c_* x_{D2} + d_*(u_D + cx_{D1} + p_4 N(y_D)) \\ \dot{y}_D &= -\lambda y_D + p_0 \sigma_D \end{aligned} \tag{19}$$

$\Sigma_D$ admits the following block diagram representation where $\frac{\pi(s,p_i)}{\alpha(s)}$ is the transfer function of $(A, p_i, c)$.

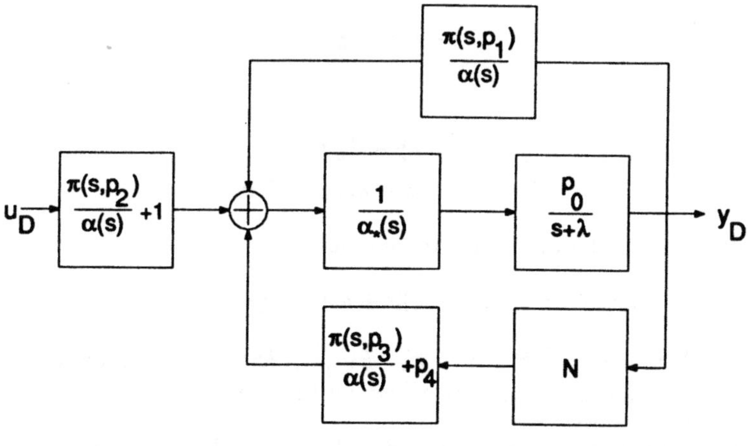


Figure 2: Direct Control Design Model

$\Sigma_D$ has three properties which make it especially useful for the problem at hand.

**Properties of $\Sigma_D$ :**

1. *Detectability:* For each fixed $p \in \mathcal{P}$, each solution $\begin{bmatrix} x'_{D1} & x'_{D2} \end{bmatrix}'$ to (19), along which $y_D(t) = u_D(t) = 0$, goes to zero exponentially fast.
2. *Output Stabilizability:* For each fixed $p \in \mathcal{P}$, the control $u_D = -cx_{D1} - p_4 N(y_D)$ "output stabilizes" $\Sigma_D$ in that $y_D(t) = e^{-\lambda t} y_D(0)$.
3. *Matchability:* For each $\Sigma_P \in \mathcal{C}_P$, there exists a point[1] $q \in \mathcal{P}$ at which $\Sigma_D$ "matches" $\Sigma_P$ in the sense that $\Sigma_D(q)$ admits the same block diagram representation as $\Sigma_P$ which is shown in Figure 1.

Because of matchability and the structure of $\Sigma_D$ in (19), it is possible to write

$$\dot{y} = -\lambda y + q_0(q'_P w + c_* x_* + d_* u) + \epsilon_P \tag{20}$$

where $w$ is a weighting vector generated by the identifier equations

$$\begin{aligned} w' &= c_* H + d_* \begin{bmatrix} z' & N(y) \end{bmatrix} \\ \dot{H} &= A_* H + b_* \begin{bmatrix} z' & N(y) \end{bmatrix} \\ \dot{x}_* &= A_* x_* + b_* u \\ \dot{z} &= \begin{bmatrix} A' & 0 & 0 \\ 0 & A' & 0 \\ 0 & 0 & A' \end{bmatrix} z + \begin{bmatrix} c' & 0 & 0 \\ 0 & c' & 0 \\ 0 & 0 & c' \end{bmatrix} \begin{bmatrix} y \\ u \\ N(y) \end{bmatrix} \end{aligned} \tag{21}$$

Here $\epsilon_P$ is a linear combination of decaying exponentials depending on the eigenvalues of $A$ and $A_*$, and $q_P = \begin{bmatrix} q'_1 & q'_2 & q'_3 & q_4 \end{bmatrix}'$ {cf, [13, 12]}. In view of (20) , it makes sense to use as an identification error

$$e = \hat{y} - y \tag{22}$$

where

$$\dot{\hat{y}} = -\lambda \hat{y} + k_0 \hat{\sigma} \tag{23}$$

$$\hat{\sigma} = k' w + c_* x_* + d_* u \tag{24}$$

since this leads to the familiar error equation

$$\dot{e} = -\lambda e + (k_0 - q_0)\hat{\sigma} + q_0(k - q_P)' w - \epsilon_P \tag{25}$$

Here $k_0 \in \Re$, and $k = \begin{bmatrix} k'_1 & k'_2 & k'_3 & k_4 \end{bmatrix}' \in \Re^{3n+1}$ are parameters to be tuned. $\Sigma_I$ is thus completely described by (21) to (24).

[1] To calculate such a value $q = \begin{bmatrix} q_0 & q'_1 & q'_2 & q'_3 & q_4 \end{bmatrix}'$, write $\alpha(s)$ for the characteristic polynomial of $A$, and let $\gamma(s)$ and $\rho(s)$ denote the unique quotient and remainder respectively of $(s+\lambda)\alpha(s)\alpha_*(s)$ divided by $\alpha_P(s)$. It is easy to verify that with $q_0 = g_P$ and $q_1, q_2, q_3$, and $q_4$ defined so that $(A, b, q'_1)$, $(A, b, q'_2)$, and $(A, b, q'_3, q_4)$ realize $\frac{\beta_P(s)\gamma(s) - \alpha(s)}{\alpha(s)}$, $\frac{\rho(s)}{g_P \alpha(s)}$ and $\frac{\beta_N(s)\gamma(s)}{\alpha(s)}$ respectively, $q$ will have what's required.

**Internal Regulator $\Sigma_R$ :**

To define internal regulator $\Sigma_R$ as a certainty equivalence control, consider first the "surrogate" design model substate

$$\widehat{x}_{D1} = \left[ \begin{array}{ccc} S(k_1) & S(k_2) & s(k_3) \end{array} \right] z \tag{26}$$

where $S(\zeta) = \left[ \begin{array}{cccc} (M^{-1}M_1)'\zeta & (M^{-1}M_2)'\zeta & \cdots & (M^{-1}M_n)'\zeta \end{array} \right]' \ \forall \zeta \in \Re^n$, $M$ is the observability matrix of $(c, A)$, $M_i$ is the observability matrix of $(e_i', A)$ and $e_i$ is the ith unit vector in $\Re^n$ [12]. If $N$ were linear, then in accordance with the Certainty Equivalence Output Stabilization Theorem [12], one would set[2]

$$u = -c\widehat{x}_{D1} - k_4 N(y) \tag{27}$$

since $u_D = -cx_{D1} - p_4 N(y_D)$ output stabilizes $\Sigma_D$ at $p$. This control would then define $\Sigma_R$ .

Unfortunately, problems arise as soon as one tries to use (27) when $N$ is nonlinear. The problems appear to stem from the use of certainty equivalence itself , since this is, in essence, a "static" or "frozen" parameter approach to the design of $\Sigma_R$ [12] . The difficulty is that certainty equivalence promotes control laws such as (27), and does so without directly taking into account the actual time dependence of tuned parameters. Because of this, for certainty equivalence control to work, parameters must be tuned "slowly," at least on the average. This is typically accomplished using some form of tuning error normalization. Simple examples suggest that it is the slow tuning necessitated by the certainty equivalence approach which enables strongly nonlinear functions {i.e., functions which are locally but not globally Lipschitz} to foil adaptive controllers configured in this way [15].

What recent research has shown [5]-[9], is that there are alternative identifier-based methods, not using tuning error normalization and not of the certainty equivalence type which can be used with success, even if $N$ is not globally Lipschitz. Underlying these methods is a useful concept - a modification of the idea of certainty equivalence - to guide one in the selection of $u$. It is this concept which we will now explain in the context of the problem under consideration.

To begin, let us note that the substitution of $\widehat{y} - e$ for $y$ in the equations which define identifier $\Sigma_I(k_0, k)$, uniquely determines a parameter-dependent dynamical system with inputs $u$ and $e$, state $\widehat{x} \sim \{\widehat{y}, H, x_*, z\}$ and output $\widehat{y}$, which is described by

$$\begin{aligned} w' &= c_* H + d_* \left[ z' \quad N(\widehat{y} - e) \right] \\ \dot{H} &= A_* H + b_* \left[ z' \quad N(\widehat{y} - e) \right] \end{aligned}$$

[2] Using (26) and the easily derived identity $cS(\zeta) = \zeta'$, (27) can be rewritten in the more familiar form $u = -\left[ \begin{array}{ccc} k_1' & k_2' & k_3' \end{array} \right] z - k_4 N(y)$.

$$
\begin{aligned}
\dot{x}_* &= A_* x_* + b_* u \\
\dot{z} &= \begin{bmatrix} A' & 0 & 0 \\ 0 & A' & 0 \\ 0 & 0 & A' \end{bmatrix} z + \begin{bmatrix} c' & 0 & 0 \\ 0 & c' & 0 \\ 0 & 0 & c' \end{bmatrix} \begin{bmatrix} \widehat{y} - e \\ u \\ N(\widehat{y} - e) \end{bmatrix} \\
\dot{\widehat{y}} &= -\lambda \widehat{y} + k_0 \widehat{\sigma} \\
\widehat{\sigma} &= k' w + c_* x_* + d_* u
\end{aligned} \tag{28}
$$

Call this system a *surrogate model* and denote it by $\widehat{\Sigma}(\widehat{k})$ where $\widehat{k} = [\,k_0 \quad k'\,]'$. A block diagram of the feedback interconnection of $\widehat{\Sigma}(\widehat{k})$ with a tuner $\Sigma_T$ would then look like this.

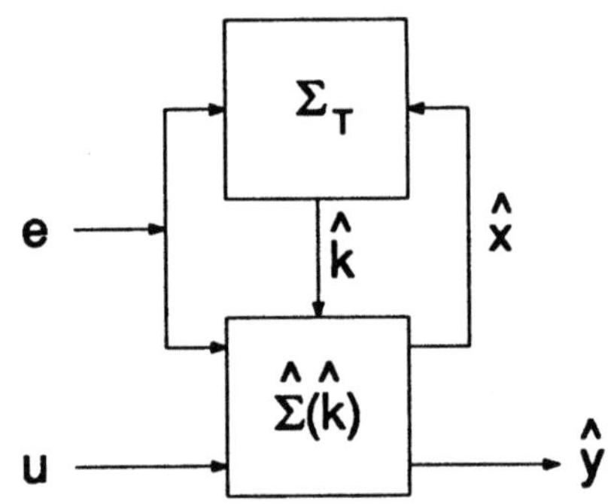


Figure 3: Feedback Interconnection of $\widehat{\Sigma}(\widehat{k})$ and $\Sigma_T$

The following informally stated concept, is a useful alternative to the idea of certainty equivalence control:

**Surrogate Model Control:** *Choose $u$ to "control" $\widehat{\Sigma}(\widehat{k})$, viewing $e$ as an exogenous input to $\widehat{\Sigma}(\widehat{k})$ and taking into account that $\widehat{k}$ is a vector of time-varying parameter generated by $\Sigma_T$.*

One difference between a certainty equivalence control and a surrogate model control is that the former is selected on the basis of a design model {c.f., [12]}, whereas the latter is picked to control a surrogate model, which is not quite the same thing. But what really distinguishes certainty equivalence control from surrogate model control is that the latter takes into account the time-dependence of tuned parameters, whereas the former does not.

The justification for using this type of control is roughly as follows. Many tuning algorithms, including those discussed in this paper, are capable of making $e$ "small" and at the same time bounding $\widehat{k}$, regardless of how $u$ is defined. With this in mind, suppose, for example, that $\Sigma_R$ is a control law of the form

$$u = f(\widehat{x}, \widehat{k}, e) \tag{29}$$

where $f(\cdot)$ is a smooth function satisfying $f(0, \widehat{k}, e) = 0$. Suppose, in addition, that $f$ is chosen to make "small", the resulting closed-loop input to state mapping of $\widehat{\Sigma}$, $e \longmapsto \widehat{x}$. Then smallness of $e$ and boundedness of $\widehat{k}$ would imply smallness of $\widehat{x}$. This , in turn, would imply smallness of both $u$ and $y$, since $u$ satisfies (29), since

$$y = \widehat{y} - e \tag{30}$$

and since $\widehat{y}$ is a component of $\widehat{x}$. From this and the assumed detectability and stabilizability of the process model subsystem $(c_P, A_P, b_P)$ it would therefore follow that $x_P$ is small which is what is desired. This then explains why it makes sense to try to choose u to bound the closed-loop, input to state mapping of $\widehat{\Sigma}$.

For the specific problem under consideration here, $\Sigma_P$ is "minimum phase" {i.e. $\beta_P$ is stable}. For this reason choosing feedback law (29) to bound the closed-loop input to *state* mapping of $\widehat{\Sigma}$ is more than is necessary. Instead it is enough use a feedback law which makes small only the closed-loop input to *output* mapping of $\widehat{\Sigma}$, $e \longmapsto \widehat{y}$. For if (29) were so defined, then smallness of $e$ would imply smallness of $\widehat{y}$, which in turn would imply smallness of $y$ because of (30). And since $\Sigma_P$ is minimum phase, smallness of $y$ would {roughly speaking} imply smallness of $x_P$.

*Remark 3:* With robustness in mind, it is worth noting that the preceding arguments, given to justify surrogate model control, do not depend on $\Sigma_D$'s ability to match $\Sigma_P$ . All that's required for the reasoning to go through is that tuning be done in such a way as to make $e$ small and to bound tuned parameters.

To proceed, suppose $u$ is chosen, to make the closed-loop mapping $e \longmapsto \widehat{y}$ small in the sense that for any input $e$ and initial state $\widehat{x}(0)$, the output response $\widehat{y}$ of $\widehat{\Sigma}$ satisfies

$$||\widehat{y}(t)|| \leq d(||\widehat{x}(0)||) + c\int_0^t ||e||^2 d\tau \tag{31}$$

where $c$ is a constant and $d(\cdot)$ is continuous. Then, if $\Sigma_T$ causes $e$ to have a finite $\mathcal{L}^2$ norm, $\widehat{y}$ will be bounded. If, in addition, $\Sigma_T$ bounds $e$, then $\widehat{y}$ will also be bounded[3].

In the light of the preceding discussion, it makes sense to try to pick $u$ so that (31) holds. There are at least two fundamentally different ways to do this. Examples of the first, which involve the introduction of additional dynamics and the use of "nonlinear damping" can be found in [13] and [6][4] The second, which was outlined in [16], will now be explained.

[3] Note that this conclusion remains valid even if c and $d(\cdot)$ depend continuously on $\widehat{k}$.

[4] Although the internal regulators proposed in these papers all prove to be forms of surrogate model control, they are developed in [13] and [6] without explicitly appealing to the concept.

One way to make (31) hold is to pick $u$ so that $\widehat{y}$ goes to zero, no matter what the behavior of $e$. This, in fact can be achieved with the control law

$$u = -\alpha_*(\delta)(k'w) \tag{32}$$

where $\delta$ is the differential operator. To understand why this is so, note from (28) and the fact that $(A_*, b_*, c_*, d_*)$ realizes $\frac{1}{\alpha_*(s)}$ that

$$\alpha_*(\delta)\widehat{\sigma} = u + \alpha_*(\delta)(k'w) \tag{33}$$

Thus if $u$ is chosen according to (32), then $\widehat{\sigma}$ would satisfy

$$\alpha_*(\delta)\widehat{\sigma} = 0 \tag{34}$$

which means that

$$(\delta + \lambda)\alpha_*(\delta)\widehat{y} = 0$$

Since $(s + \lambda)\alpha_*(s)$ is a stable polynomial, both $\widehat{y}$ and $\widehat{\sigma}$ must go to zero exponentially fast.

Turning to the realizability of the control law given by (32), recall that $\alpha_*(s)$ is of degree $n^* - 1$. Thus control (32) can be realized without time function differentiation, provided the first $n^*-1$ derivatives of both $k$ and $w$ are available signals. In view of (21), $w$ clearly has this property. Therefore, if things can be arranged so that error equation (25) takes on the form of (1), then a tuner of order $n^* - 1$ can be constructed to endow $k$ with the required properties. To get (25) into this form, its enough to simply set $k_0$ equal to any bounded time function {eg. $k_0 = 0$}, since if this is done (25) becomes

$$\dot{e} = -\lambda e + q_0(k - q_P)'w + \epsilon \tag{35}$$

where

$$\epsilon = (k_0 - q_0)\widehat{\sigma} - \epsilon_P \tag{36}$$

In view of (34), $\epsilon$ must go to zero exponentially fast. Henceforth, therefore we shall take (32) as the definition of $\Sigma_R$ and we shall assume that $\Sigma_T$ is a tuner of order $\bar{n} = n^* - 1$ defined by (2) to (7)[5]. Since there is nothing to prevent us from setting both $k_0$ and $\widehat{y}$ equal to zero, for the sake of simplicity, we do. As a consequence, the definition of $e$ reduces to

$$e = -y \tag{37}$$

### 3.3 Stability Analysis

The overall adaptive control system to be examined consists of the high-order tuner $\Sigma_T$ described by (2) to (8), process model $\Sigma_P$ defined by (18), identifier $\Sigma_I$ modeled by (21), (24), (37), and internal regulator $\Sigma_R$ given by (32). The steps involved in establishing global boundedness are as follows:

[5] If $n^* = 1$, set $\bar{n} = 1$.

1. Using (21) it is easy to see that the component subvectors $w_1, w_2, w_3$, and $w_4$ of $w$ satisfy the differential equations

$$\begin{aligned} \alpha_*(\delta)(\delta I - A')w_1 &= c'y & (38)\\ g_P\beta_P(\delta)\alpha_*(\delta)(\delta I - A')w_2 &= c'\alpha_P(\delta)y \\ &\quad - \beta_N(\delta)\alpha_*(\delta)(\delta I - A')w_3 & (39)\\ \alpha_*(\delta)(\delta I - A')w_3 &= c'N(y) & (40)\\ \alpha_*(\delta)w_4 &= N(y) & (41)\end{aligned}$$

Since the process model assumptions imply that the transfer matrices

$$\frac{1}{\alpha_*(s)}(sI - A')^{-1}c', \quad \frac{\alpha_P(s)}{g_P\beta_P(s)\alpha_*(s)}(sI - A')^{-1}c', \quad \frac{\beta_N(s)}{g_P\beta_P(s)}$$

are each proper and stable, in the light of (38) and (39), $w$ can be viewed as the output of a stable linear system $\Sigma_w$ with state $x_w$ and inputs $y, w_3$, and $w_4$. The interconnection of $\Sigma_w$ with $\Sigma_T$ determined by (35) and (37), is thus a dynamical system $\Sigma$ of the form

$$\begin{aligned} \dot{x}_\Sigma &= f(x_\Sigma, w_3, w_4, \epsilon) \\ w &= g(x_\Sigma, w_3, w_4, \epsilon) \end{aligned} \qquad (42)$$

where $x_\Sigma = \{y, v, h_0, h, X, x_w\}$ and $f$ and $g$ are analytic functions[6].

2. Now invoke Theorem 1 and (37), thereby establishing the boundedness of all components of $x_\Sigma$ , except for $x_w$.

3. From (34) and the definition of $\epsilon$ in (36), it follows that $\epsilon$ is in $\mathcal{D}^{n^*-1}$, the space of all vector-valued, bounded functions with bounded time derivatives up to and including those of degree $n^* - 1$. From the boundedness of $y$, (40) and (41) it is clear that $w_3$ and $w_4$ are also in $\mathcal{D}^{n^*-1}$. The definition of $\Sigma_w$ together with the boundedness of its inputs imply that $x_w$ is bounded as well.

4. The preceding characterizes $\Sigma$ as an analytic dynamical system with a bounded state and an input $\{\epsilon, w_3, w_4\}$ in $\mathcal{D}^{n^*-1}$. From this it follows that $w$ must be in $\mathcal{D}^{n^*-1}$. The equations defining $\Sigma_T$ show that $k$ must be in $\mathcal{D}^{n^*-1}$ as well. From these observations, and (32) it follows that $u$ is bounded.

5. The boundedness of $u$ and $y$ and the assumption that $(c_P, A_P)$ is detectable, imply that $x_P$ is bounded[7]. The boundedness of $u$ and $y$ also implies that the identifier state components $H, x_*$ and $z$ defined by (21) are bounded.

---

[6] If $\bar{n} = 1$, $x_\Sigma = \{y, k, x_w\}$

[7] This is true because detectability ensures the existence of a vector $d$ which stabilizes $A_P + dc_P$, because $\dot{x}_P = (A_P + dc_P)x_P - dy + b_P u$, and because $b_P u - dy$ is bounded.

6. The preceding establishes the boundedness of complete adaptive system state $\{v, h_0, h, X, x_P, H, x_*, z\}$ )or $\{k, x_P, H, x_*, z\}$ for the case $\bar{n} = 1$), wherever it exists. From this and the smoothness of the overall adaptive system it follows that its state must exist and be bounded on $[0, \infty)$. It also follows from the boundedness of $\dot{y}$ and the finiteness of the $\mathcal{L}^2$-norm of $y$ established by the Theorem 1 and (37), that $y$ goes to zero as $t \to \infty$ [17].

## 4. Linear Process Models

The aim of this section is to explain how the simplified high-order tuning algorithm $\Sigma_T$ discussed in §2, might be used in conjunction with a suitably defined, identifier-based parameterized controller $\Sigma_C$ of the model reference type, to adaptively control a siso process admitting a minimum phase linear model $\Sigma_P$ whose relative degree is not known exactly. We begin by defining the class of admissible process models $C_P$.

### 4.1 Process Model $\Sigma_P$

Let $\bar{q}, n, n^*$ be fixed positive numbers where $n$ and $n^*$ are integers satisfying $n \geq n^*$. The class $C_P$ of process models $\Sigma_P$ to be considered consists of siso systems of the form

$$\begin{aligned} \dot{x}_P &= A_P x_P + b_P u \\ y &= c_P x_P \end{aligned} \tag{43}$$

where $(A_P, b_P, c_P)$ is a $n_P$—dimensional, stabilizable, detectable linear system. Each $\Sigma_P \in C_P$ has a transfer function $t_P(s)$ of the form

$$t_P(s) = g_P \frac{\beta_P(s)}{\alpha_P(s)}$$

where $\alpha_P(s)$ is the characteristic polynomial of $A_P$, $\beta_P(s)$ is monic and $g_P$ is a nonzero constant.

We shall make the following

**Process Model Assumptions:** For each $\Sigma_P \in C_P$

1. $n_P \leq n$
2. relative degree $t_P(s)$ is either $n^*$ or $n^* + 1$
3. $0 < g_P \leq \bar{q}$
4. $\beta_P(s)$ is stable

Assumption 3 is not crucial and can be weakened to $0 < |g_P| \leq \bar{q}$ if $\text{sign}(g_P)$ is assumed known, without causing much of a problem. What really distinguishes the preceding from the well-known process model assumptions of classical model reference adaptive control {cf. [13]}, is assumption 2 . This assumption is less restrictive than its classical counterpart in that relative degree $t_P(s)$ is no longer assumed to be known exactly. Earlier work dealing with identifier-based adaptive controllers for process model classes containing models of different relative degrees, can be found in [18, 19, 20]. The process model assumptions made here are less restrictive than those made in [18, 19], but more so than those made in [20].

### 4.2 Parameterized Controller $\Sigma_C$

Our aim here is to construct an identifier-based, model reference parameterized controller $\Sigma_C$ suitable for high-order tuning. The subsystems of such a controller consist of a reference model $\Sigma_r$ whose output $y$ is supposed to follow, an identifier $\Sigma_I$ whose function is to generate a tuning error $e$ and an internal regulator $\Sigma_R$ whose output serves as the closed-loop control to the process.

**Reference Model $\Sigma_r$:**

As the first step in the construction of $\Sigma_C$ , we select a suitable reference model $\Sigma_r$. For this, choose a positive number $\lambda$, a monic stable polynomial $\alpha_*(s)$ of degree $n^*$, and a controllable, observable realization $(A_*, b_*, c_*)$ of $\frac{1}{\alpha_*(s)}$. $\Sigma_r$ is then defined by the equations

$$\begin{aligned} \dot{x}_r &= A_* x_r + b_* r \\ \dot{y}_r &= -\lambda y_r + c_* x_r \qquad (44) \\ e_t &= y - y_r \qquad (45) \end{aligned}$$

where $e_t$ is the *tracking error* which is to be driven to zero.

**Identifier $\Sigma_I$ :**

As the next step in the construction of $\Sigma_C$ , we select a suitable parameterized design model $\Sigma_D$ , similar to that used in §3. For this, choose a $n$-dimensional, single-output, observable pair $(c, A)$ with $A$ stable. This and the preceding selections of $\lambda$ and $(A_*, b_*, c_*)$ determine the direct control, design model $\Sigma_D(p)$, parameterized by the vector $p = [p_0 \quad p_1' \quad p_2' \quad p_*]' \in \mathcal{P} = \{\Re \oplus \Re^n \oplus \Re\} \times \{0, 1\}$ and described by the equations

$$\begin{aligned} \dot{x}_{D1} &= A x_{D1} + p_1 y_D + p_2 u_D \\ \dot{x}_{D2} &= A_* x_{D2} + b_*(u_D + c x_{D1}) \\ \sigma_D &= c_* x_{D2} \\ p_* \dot{y}_D &= -\lambda y_D + p_0 \sigma_D \qquad (46) \end{aligned}$$

It is easy to verify that at each point $p \in \mathcal{P}$, $\Sigma_D(p)$ is detectable {through $y_D$} and output stabilized by $u_D = -c x_{D1}$. In addition, for each $\Sigma_P \in \mathcal{C}_P$, there exists a point[8] $q \in \mathcal{P}$ at which $\Sigma_D$ has the same transfer function as $\Sigma_P$ .

Because of matching and the structure of $\Sigma_D$ in (46), it is possible to write

$$q_* \dot{y} = -\lambda y + q_0([\ q_1' \quad q_2'\ ]\, w_I + c_* x_*) + \epsilon_P \tag{47}$$

where $w_I$ is a weighting vector generated by the identifier equations

$$\begin{aligned} w_I' &= c_* H \\ \dot{H} &= A_* H + b_* z' \\ \dot{x}_* &= A_* x_* + b_* u \\ \dot{z} &= \begin{bmatrix} A' & 0 \\ 0 & A' \end{bmatrix} z + \begin{bmatrix} c' & 0 \\ 0 & c' \end{bmatrix} \begin{bmatrix} y \\ u \end{bmatrix} \end{aligned} \tag{48}$$

Here $\epsilon_P$ is a linear combination of decaying exponentials depending on the eigenvalues of $A$ and $A_*$. In view of (47) and(44), it is easy to see that tracking error $e_t$ must satisfy

$$q_* \dot{e}_t = -\lambda e_t + q_0(q_P' w + c_* x_*) + \epsilon_P \tag{49}$$

where

$$w' = [\, w_I' \quad c_* x_r \quad y_r\, ] \tag{50}$$

and $q_P = \left[ q_1' \quad q_2' \quad \frac{q_*}{q_0} \quad \frac{\lambda(1-q_*)}{q_0} \right]'$. Therefore, for $q_* = 1$ it makes sense to use as an identification error

$$e = \widehat{e}_t - e_t \tag{51}$$

where

$$\dot{\widehat{e}}_t = -\lambda \widehat{e}_t + k_0 \widehat{\sigma} \tag{52}$$

$$\widehat{\sigma} = k' w + c_* x_* \tag{53}$$

since this leads to the familiar error equation

$$\dot{e} + \lambda e = (k_0 - q_0)\widehat{\sigma} + q_0(k - q_P)' w - \epsilon_P \tag{54}$$

Here $k_0 \in \Re$, and $k = [\, k_1' \quad k_2' \quad k_3 \quad k_4\, ]'$ where $k_1$ and $k_2$ are in $\Re^n$ and $k_3$ and $k_4$ are in $\Re$. For reasons to become clear in the sequel we shall also take (51) as the definition of $e$ for the case $q_* = 0$, even though the resulting error equation for $e$, namely

$$e = \frac{(k_0 - q_0)\widehat{\sigma} + q_0(k - q_P)' w - \epsilon_P + \lambda \widehat{e}_t - \widehat{\sigma}}{\lambda} \tag{55}$$

---

[8] To calculate such a value $q = [\ q_0 \quad q_1' \quad q_2' \quad q_*\ ]'$, write $\alpha(s)$ for the characteristic polynomial of $A$, set $q_* =$ relative degree $t_P(s)\ - n^*$, and let $\gamma(s)$ and $\rho(s)$ denote the unique quotient and remainder respectively of $(s+\lambda)^{q_*}\alpha(s)\alpha_*(s)$ divided by $\alpha_P(s)$. It is easy to verify that with $q_0 = g_P \lambda^{(1-q_*)}$ and $q_1$ and $q_2$ defined so that $(A, b, q_1')$ and $(A, b, q_2')$ realize $\frac{\beta_P(s)\gamma(s) - \alpha(s)}{\alpha(s)}$ and $\frac{\rho(s)}{g_P \alpha(s)}$ respectively, $q$ will have what's required.

is in a somewhat nonstandard form. $\Sigma_I$ is thus completely described by (48) and (50) to (53).

**Internal Regulator $\Sigma_R$ :**

As in §3, we shall base our selection of $u$ on an appropriately defined surrogate model $\widehat{\Sigma}(k_0, k)$. Since in a model reference context it is the tracking error $e_t$ rather than the process output $y$ which is to be driven to zero, it is natural to define $\widehat{\Sigma}$ so that its output is $\widehat{e}_t$. In view of (44) and (45) which define $\Sigma_R$ and (48), (50)-(53) which define $\Sigma_I$ , it thus makes sense to take as a model for $\widehat{\Sigma}$ the equations

$$\begin{aligned}
\dot{x}_r &= A_* x_r + b_* r \\
\dot{y}_r &= -\lambda y_r + c_* x_r \\
w' &= [c_* H \quad c_* x_r \quad y_r] \\
\dot{H} &= A_* H + b_* z' \\
\dot{x}_* &= A_* x_* + b_* u \\
\dot{z} &= \begin{bmatrix} A' & 0 \\ 0 & A' \end{bmatrix} z + \begin{bmatrix} c' & 0 \\ 0 & c' \end{bmatrix} \begin{bmatrix} \widehat{e}_t + y_r - e \\ u \end{bmatrix} \\
\dot{\widehat{e}}_t &= -\lambda \widehat{e}_t + k_0 \widehat{\sigma} \\
\widehat{\sigma} &= k' w + c_* x_*
\end{aligned} \tag{56}$$

$\widehat{\Sigma}$ may thus be considered to be a time-varying linear system with inputs $e, u,$ and $r$, state $\widehat{x} \sim \{x_r, y_r, H, x_*, z, \widehat{e}_t\}$ and output $\widehat{e}_t$.

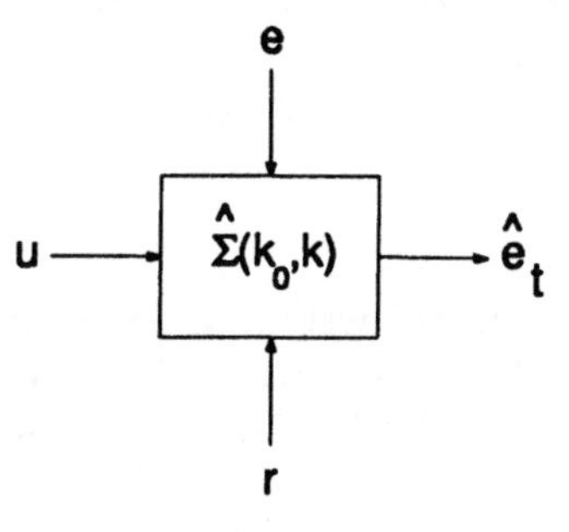


Figure 4: Surrogate Model $\widehat{\Sigma}(k_0, k)$

In analogy with what was done in §3, and for essentially the same reasons, $u$ will be chosen to make $\widehat{e}_t$ go to zero. Since $(A_*, b_*, c_*)$ realizes $\frac{1}{\alpha_*(s)}$, (56) implies that

$$\alpha_*(\delta)\widehat{\sigma} = u + \alpha_*(\delta)(k'w) \tag{57}$$

Thus if $u$ is defined so that

$$u = -\alpha_*(\delta)(k'w) \tag{58}$$

then

$$\alpha_*(\delta)\widehat{\sigma} = 0 \tag{59}$$

and

$$(\delta + \lambda)\alpha_*(\delta)\widehat{e}_t = 0 \tag{60}$$

so both $\widehat{e}_t$ and $\widehat{\sigma}$ go to zero exponentially fast.

Turning to the realizability of the control law given by (58), recall that $\alpha_*(s)$ is of degree $n^*$. Thus (58) can be realized without time function differentiation, provided the first $n^*$ derivatives of both $k$ and $w$ are available signals. In view of (56), $w$ clearly has this property. Therefore, if things can be arranged so that error equations (54) and (55) take on the form of (17) for $q_* = 1$ and $q_* = 0$ respectively, then a tuner of order $n^*$ can be constructed to endow $k$ with the required properties. To get (54) and (55) into this form, its enough to simply set $k_0$ equal to any bounded time function {eg. $k_0 = 0$}, since if this is done (54) and (55) satisfy

$$q_*\dot{e} = -\lambda e + q_0(k - q_P)'w + \epsilon \tag{61}$$

for $q_* = 1$ and $q_* = 0$ respectively, where

$$\epsilon = (k_0 - q_0)\widehat{\sigma} + (1 - q_*)(\lambda\widehat{e}_t - \widehat{\sigma}) - \epsilon_P \tag{62}$$

In view of (59) and (60), $\epsilon$ must go to zero exponentially fast. Henceforth, therefore we shall take (58) as the definition of $\Sigma_R$ . We shall assume that $\mu$ is chosen to satisfy (9) and that $\Sigma_T$ is a tuner of order $\bar{n} = n^*$ defined by (11) to (14). Since there is nothing to prevent us from setting both $k_0$ and $\widehat{e}_t$ equal to zero, for simplicity, we do. As a consequence, the definition of $e$ reduces to

$$e = -e_t \tag{63}$$

### 4.3 Stability Analysis

The overall adaptive control system to be examined consists of the simplified high-order tuner $\Sigma_T$ described by (11) to (14), process model $\Sigma_P$ defined by (43), reference model $\Sigma_R$ defined by (44) and (45) , identifier $\Sigma_I$ modeled by (48) and (50) to (53), and internal regulator $\Sigma_R$ given by (58). The steps involved in establishing global boundedness and model following are as follows:

1. Using (48) and (50) it is easy to see that the component subvectors $w_1, w_2, w_3$, and $w_4$ of $w$ satisfy the differential equations

$$\alpha_*(\delta)(\delta I - A')w_1 = c'(e_t + w_4) \tag{64}$$
$$g_P\beta_P(\delta)\alpha_*(\delta)(\delta I - A')w_2 = c'\alpha_P(\delta)'(e_t + w_4) \tag{65}$$
$$\alpha_*(\delta)w_3 = r \tag{66}$$
$$(\delta + \lambda)w_4 = w_3 \tag{67}$$

The process model assumptions imply that the transfer matrix

$$\frac{\alpha_P(s)}{g_P\beta_P(s)\alpha_*(s)}(sI - A')^{-1}c'$$

is stable and proper[9]. In the light of this, (64) and (65), $w$ can be viewed as the output of a proper, stable linear system $\Sigma_w$ with state $x_w$ and inputs $e_t$ and $w_3$,. The interconnection of $\Sigma_w$ with $\Sigma_T$ determined by (54) and (63), is thus a dynamical system $\Sigma$ of the form

$$\begin{aligned} \dot{x}_\Sigma &= f(x_\Sigma, w_3, \epsilon) \\ w &= g(x_\Sigma, w_3, \epsilon) \end{aligned} \tag{68}$$

where $x_\Sigma = \{q_* e_t, h, X, x_w\}$ and $f$ and $g$ are analytic function.

2. Now invoke Theorem 2 and (37), thereby establishing the boundedness of the $\mathcal{L}^2$ norm of $e_t$ and all components of $x_\Sigma$ , except for $x_w$.

3. From (34) and the definition of $\epsilon$ in (36), it follows that $\epsilon$ is in $\mathcal{D}^{n^*}$, the space of all vector-valued, bounded functions with bounded time derivatives up to and including those of degree $n^*$. From the boundedness of $r$ and (66) it is clear that $w_3$ is also in $\mathcal{D}^{n^*}$.

4. If $q_* = 1$, $x_w$ must be bounded since $\Sigma_w$ is a proper, stable linear system with bounded inputs $r$ and $e_t$. On the other hand, if $q_* = 0$ $x_w$ still must be bounded since in this case $\Sigma_w$ is a strictly proper linear system with two inputs, one which is bounded {i.e., $r$}, the other {i.e., $e_t$} with a finite $\mathcal{L}^2$ norm.

5. The preceding characterizes $\Sigma$ as an analytic dynamical system with a bounded state and an input $\{\epsilon, w_3\}$ in $\mathcal{D}^{n^*}$. From this it follows that $w$ must be in $\mathcal{D}^{n^*}$. The equations defining $\Sigma_T$ show that $k$ must be in $\mathcal{D}^{n^*}$ as well. From these observations, and (58) it follows that $u$ is bounded.

6. Since $r$ is bounded and $\Sigma_R$ is stable, both $x_r$ and $y_r$ must be bounded.

7. To show that $x_P$ is bounded, let $d$ be and column vector which stabilizes $A_P + dc_P$; the existence of such a vector is assured by the assumed detectability of $(c_P, A_P)$. Next rewrite the differential equation for $x_P$ in (43) as $\dot{x}_P = (A_P + dc_P)x_P + b_P u - d(e_t + y_r)$. The boundedness of $u$ and $y_r$ has already been established as has the boundedness of the $\mathcal{L}^2$ norm of $e_t$. From this and the stability of $A_P + dc_P$ it follows that $x_P$ is bounded. The boundedness of $u$ and $x_P$ imply that the identifier state components $H, x_*$ and $z$ defined by (48) are bounded as well.

[9] If $q_* = 0$, this transfer matrix is actually *strictly* proper.

8. The preceding establishes the boundedness of complete adaptive system state $\{h, X, x_P, x_r, H, x_*, z\}$ wherever it exists. From this and the smoothness of the overall adaptive system it follows that its state must exist and be bounded on $[0, \infty)$. It also follows from the boundedness of $\dot{e}_t$ and the finiteness of the $\mathcal{L}^2$-norm of $e_t$, that $e_t$ goes to zero as $t \rightarrow \infty$ [17].

## Concluding Remarks

The aim of this paper has been to introduce two high-order tuning algorithms and to discuss their possible use in the synthesis of adaptive controllers for special families of linear and nonlinear systems. While the resulting controllers, like their predecessors in [5]-[9],[13] are a good deal more complicated than the standard certainty equivalence controllers discussed in [4], the former seem to have several clear advantages over the latter. Among these are their ability to stabilize

- certain type of siso systems with nonlinearities which are not globally Lipschitz.
- siso minimum phase linear systems whose relative degrees are known to be either $n^*$ or $n^* + 1$, $n^*$ being a known positive integer.

It is interesting to note that because the overall adaptive systems under consideration here have been developed without the presumption of slowly tuned parameters, the stability analysis differs somewhat from that used in the study of more familiar adaptive systems {eg., see [12, 4]}. Whereas the latter typically make use of small signal perturbation analysis, the Bellman Gronwall Lemma, and the like, the reasoning used here does not and consequently is somewhat less technical.

The identifier-based algorithms developed in this paper can of course be applied to more familiar process models such as those considered in [4]. Whether or not the departure from certainty equivalence control this paper suggests can improve performance in such applications, remains to be seen.

## Appendix

**Proof of Theorem 1:** For $\bar{n} = 1$ the result is well known and can easily be obtained by evaluating derivative of the "partial Lyapunov function" $e^2 + |q_0|||k - q_P||^2$ along a solution to (1) and (2); this yields

$$\begin{aligned} \frac{d}{dt}(e^2 + |q_0|||k - q_P||^2) &= -2\lambda e^2 + 2\epsilon e \\ &\leq -\lambda e^2 + \frac{\epsilon^2}{\lambda} \end{aligned}$$

Thus by integrating and rearranging terms one gets

$$e^2 + |q_0|||k - q_P||^2 + \lambda \int_0^T e^2 dt \leq C$$

where

$$C = e^2(0) + |q_0|||k(0) - q_P||^2 + \int_0^\infty \frac{\epsilon^2}{\lambda} dt < \infty$$

Hence $e$ and $k$ possess the required properties.

Assume $\bar{n} > 1$. Note from (7) and (8) that

$$\dot{x}_i = (\bar{A}x_i + \bar{b}h_i)(1 + w_i^2), \quad i \in \mathbf{m} \tag{69}$$

$$k_i = \bar{c}x_i, \quad i \in \mathbf{m} \tag{70}$$

where $x_i$ is the ith column of $X$, $h_i$ and $k_i$ are the $i$th components of $h$ and $k$ respectively, and $\mathbf{m} = \{1, 2, \cdots, m\}$. Set

$$z_i = x_i + \bar{A}^{-1}\bar{b}h_i, \quad i \in \mathbf{m} \tag{71}$$

Since $(\bar{c}, \bar{A}, \bar{b})$ realizes $\bar{\alpha}(0)/\bar{\alpha}(s)$, clearly $\bar{c}\bar{A}^{-1}\bar{b} = -1$; thus from (70) and (71)

$$k_i - h_i = \bar{c}z_i, \quad i \in \mathbf{m} \tag{72}$$

In addition, from (6), (69) and (71)

$$\dot{z}_i = \bar{A}z_i(1 + w_i^2) - \text{sign}(q_0)\bar{A}^{-1}\bar{b}w_i\bar{e}, \quad i \in \mathbf{m} \tag{73}$$

Next use (1), (3) and (4) to write

$$\dot{\bar{e}} = -\lambda\bar{e} + q_0(h - k_P)'w - (h_0 - q_0)(k - h)'w + \epsilon \tag{74}$$

Note also from (5) and (6) that

$$\frac{d}{dt}(h_0 - q_0) = (k - h)'w\bar{e} \tag{75}$$

$$\frac{d}{dt}(h - q_P) = -\text{sign}(q_0)w\bar{e} \tag{76}$$

Let $P$ denote the unique positive definite solution to the Lyapunov equation (10) and define

$$\delta = \frac{\lambda}{2m||P\bar{A}^{-1}\bar{b}||^2}$$

$$Q = I - \frac{1}{2||P\bar{A}^{-1}\bar{b}||^2} P\bar{A}^{-1}\bar{b}\bar{b}'(\bar{A}^{-1})'P$$

Note that

$$Q \geq \frac{1}{2}I$$

Evaluation of the rate of change of the partial Lyapunov function

$$V = \bar{e}^2 + (h_0 - q_0)^2 + |q_0|||h - q_P||^2 + \delta \sum_{i=1}^{m} z_i' P z_i$$

along a solution to (73) to (76) yields

$$\begin{aligned}
\dot{V} &= -2\lambda\bar{e}^2 + 2\bar{e}\epsilon - \delta\sum_{i=1}^{m}((1+w_i^2)||z_i||^2 + 2\mathrm{sign}(q_0)z_i'P\bar{A}^{-1}\bar{b}w_i\bar{e}) \\
&\leq -2\lambda\bar{e}^2 + 2\bar{e}\epsilon - \delta\sum_{i=1}^{m}(||z_i||^2 + w_i^2 z_i' Q z_i - 2||P\bar{A}^{-1}\bar{b}||^2\bar{e}^2) \\
&= -\lambda\bar{e}^2 + 2\bar{e}\epsilon - \delta\sum_{i=1}^{m}(||z_i||^2 + w_i^2 z_i' Q z_i) \\
&\leq -\frac{\lambda}{2}\bar{e}^2 + \frac{2}{\lambda}\epsilon^2 - \delta\sum_{i=1}^{m}(1+\frac{w_i^2}{2})||z_i||^2
\end{aligned}$$

Thus by integrating and rearranging terms we get

$$\bar{e}^2+(h_0-q_0)^2+|q_0|||h-q_P||^2+\delta\sum_{i=1}^{m} z_i'Pz_i+\int_0^T(\frac{\lambda}{2}\bar{e}^2+\delta\sum_{i=1}^{m}(1+\frac{w_i^2}{2})||z_i||^2)dt \leq C$$

where

$$C = \bar{e}^2(0)+(h_0(0)-q_0)^2+|q_0|||h(0)-q_P||^2+\delta\sum_{i=1}^{m} z_i'(0)Pz_i(0)+\frac{2}{\lambda}\int_0^\infty \epsilon^2 dt$$

From this it follows that $\bar{e}, h_0, h$ and the $z_i$, $i \in \mathbf{m}$, are bounded on $[0,T)$ and that $\bar{e}$ and the $w_i z_i$, $i \in \mathbf{m}$, have finite $\mathcal{L}^2$ norms. Therefore from (71), $X$ is bounded as well.

Since each $w_i z_i$, $i \in \mathbf{m}$, is $\mathcal{L}^2$-bounded, if follows from (72) that $(k - h)'w$ also is. This together with the boundedness of $h_0$ and (4) imply that $v$ is bounded on $[0,T)$ and has a finite $\mathcal{L}^2$ norm. In view of (3) the same must therefore be true of $e$. ■

**Proof of Theorem 2:** For $\bar{n} = 1$, evaluation of the derivative of the partial Lyapunov function $q_* e^2 + |q_0|||k - q_P||^2$ along a solution to (11) and (17) yields

$$\begin{aligned}
\frac{d}{dt}(q_* e^2 + |q_0|||k - q_P||^2) &= -2\lambda e^2 + 2\epsilon e \\
&\leq -\lambda e^2 + \frac{\epsilon^2}{\lambda}
\end{aligned}$$

Thus by integrating and rearranging terms one gets

$$q_*e^2 + |q_0|||k - q_P||^2 + \lambda \int_0^T e^2 dt \leq C$$

where

$$C = q_*e^2(0) + |q_0|||k(0) - q_P||^2 + \int_0^\infty \frac{\epsilon^2}{\lambda} dt < \infty$$

Hence $e$ and $k$ possess the required properties.

Assume $\bar{n} > 1$. As in the proof of the preceding theorem, set

$$z_i = x_i + \bar{A}^{-1}\bar{b}h_i, \quad i \in \mathbf{m} \tag{77}$$

where $x_i$ is the ith column of $X$, and $h_i$ and $k_i$ are the $i$th components of $h$ and $k$ respectively. The definitions imply that

$$q_*\dot{e} = -\lambda e + (h - q_P)'w + q_0 \sum_{i=1}^m w_i \bar{c} z_i + \epsilon \tag{78}$$

$$\dot{z}_i = \bar{A}z_i(1 + w_i^2) - \text{sign}(q_0)\bar{A}^{-1}\bar{b}w_i e, \quad i \in \mathbf{m} \tag{79}$$

$$k_i - h_i = \bar{c}z_i, \quad i \in \mathbf{m} \tag{80}$$

In addition, from (12)

$$\frac{d}{dt}(h - q_P) = -\text{sign}(q_0)we \tag{81}$$

Define

$$\delta = \frac{\bar{q}||\bar{c}'||}{||P\bar{A}^{-1}\bar{b}||}$$

$$\lambda^* = \lambda - \frac{2m\bar{q}||\bar{c}'||||P\bar{A}^{-1}\bar{b}||}{\mu}$$

and note from (9) that $\lambda^* > 0$. The definitions of $\delta$ and $\lambda^*$ imply that

$$\frac{(\bar{q})^2||\bar{c}'||^2 + \delta^2||P\bar{A}\bar{b}||^2}{\delta\mu} = \frac{\lambda - \lambda^*}{m} \tag{82}$$

Let $V$ denote the partial Lyapunov function

$$V = q_*e^2 + |q_0|||h - q_P||^2 + \delta \sum_{i=1}^m z_i' P z_i$$

Evaluation of its rate of change along a solution to (78) to (81) yields

$$\dot{V} = -2\lambda e^2 + 2e\epsilon - \delta \sum_{i=1}^m (1 + \mu w_i^2)||z_i||^2 + 2eq_0 \sum_{i=1}^m \bar{c} z_i w_i$$

$$
\begin{aligned}
& - 2\delta(\mathrm{sign}(q_0))\sum_{i=1}^{m} z_i' P\bar{A}^{-1}\bar{b}w_i e \\
= \;& -2\lambda e^2 + 2e\epsilon - \sum_{i=1}^{m}(\delta||z_i||^2 + ||\sqrt{\delta\mu}(w_i z_i) \\
& - \frac{1}{\sqrt{\delta\mu}}(q_0\bar{c}' - \delta(\mathrm{sign}(q_0))P\bar{A}^{-1}\bar{b})e||^2) \\
& + \sum_{i=1}^{m}\frac{1}{\delta\mu}||(q_0\bar{c}' - \delta(\mathrm{sign}(q_0))P\bar{A}^{-1}\bar{b})e||^2 \\
\leq \;& -2\lambda e^2 + 2e\epsilon + \sum_{i=1}^{m}\frac{1}{\delta\mu}||(q_0\bar{c}' - \delta(\mathrm{sign}(q_0))P\bar{A}^{-1}\bar{b})e||^2 \\
\leq \;& -2\lambda e^2 + 2e\epsilon + 2e^2\sum_{i=1}^{m}\frac{(\bar{q})^2||\bar{c}'||^2 + \delta^2||P\bar{A}^{-1}\bar{b})||^2}{\delta\mu} \\
= \;& -2\lambda^* e^2 + 2e\epsilon \\
\leq \;& -\lambda^* e^2 + \frac{1}{\lambda^*}\epsilon^2
\end{aligned}
$$

Thus by integrating and rearranging terms we get

$$
q_* e^2 + |q_0|||h - q_P||^2 + \delta\sum_{i=1}^{m} z_i' P z_i + \lambda^*\int_0^T e^2 dt \leq C
$$

where

$$
C = q_* e^2(0) + |q_0|||h(0) - q_P||^2 + \delta\sum_{i=1}^{m} z_i'(0)Pz_i(0) + \frac{1}{\lambda^*}\int_0^\infty \epsilon^2 dt
$$

From this it follows that $q_* e, h$ and the $z_i$, $i \in \mathbf{m}$, are bounded on $[0, T)$ and that $e$ has a finite $\mathcal{L}^2$ norm. Therefore from (77), $X$ is bounded as well. ■

# References


[1] S. Sastry and A. Isidori, "Adaptive Control of Linearizable Systems," *IEEE Trans. Auto. Control*, AC-34, Nov. 1989, pp 1123-1131.

[2] R. Marino, I. Kanellakopoulos and P. V. Kokotovic, "Adaptive Tracking for Feedback Linearizable Systems," *Proc. 28th CDC*, Tampa, 1989, pp. 1002-1007.

[3] L. Praly, G. Bastin, J.-B Pomet and Z. P. Jiang, "Adaptive Stabilization of Nonlinear Systems", Ecole Nationale Superieure des Mines de Paris Technical Report No. A236, October, 1990

[4] S. Sastry and M. Bodson, *Adaptive Control: Stability, Convergence and Robustness*, Prentice-Hall, Inc., Englewood Cliffs, NJ, 1989.

[5] I. Kanellakopoulos, P. V. Kokotovic and A. S. Morse, "Systematic Design of Adaptive Controllers for Feedback Linearizable Systems," *IEEE Trans. Auto. Control*, to appear.

[6] I. Kanellakopoulos, P. V. Kokotovic and A. S. Morse, "Adaptive Output-Feedback Control of Systems with Output Nonlinearities," *Foundations of Adaptive Control*, Springer-Verlag, Berlin, 1991.

[7] R. Marino and P. Tomei, "Global Adaptive Observers and Output-feedback Stabilization for a Class of Nonlinear Systems, *Foundations of Adaptive Control*, Springer-Verlag, Berlin, 1991.

[8] Z. P. Jiang and L. Praly, "Iterative Designs of Adaptive Controllers for Systems with Nonlinear Integrators," *Proc. 1991 IEEE CDC*, submitted.

[9] I. Kanellakopoulos, P. V. Kokotovic and A. S. Morse, "Adaptive Output-Feedback Control of a Class of Nonlinear Systems," Coordinated Science Laboratory Technical Report DC-131, submitted for publication.

[10] D. R. Mudgett, *Problems in Parameter Adaptive Control*, Yale University Doctoral Dissertation, 1988.

[11] D. R. Mudgett and A. S. Morse, "High-Order Parameter Adjustment Laws for Adaptive Stabilization," *Proc. 1987 Conference on Information Sciences and Systems*, March, 1987.

[12] A. S. Morse, "Towards a Unified Theory of Parameter Adaptive Control - Part 2: Certainty Equivalence and Implicit Tuning," *IEEE Trans. Auto. Control*, to appear.

[13] A. Feuer and A. S. Morse, "Adaptive Control of Single-Input, Single-Output Linear Systems," *IEEE Trans. Auto. Control*, AC-23, August, 1978, pp. 557-569.

[14] D. R. Mudgett and A. S. Morse, "Adaptive Stabilization of Linear Systems with Unknown High-Frequency Gains," *IEEE Trans. Auto Control*, AC-30, June, 1985, pp.549-554.

[15] A. S. Morse, "A Comparative Study of Normalized and Unnormalized Tuning Errors in Parameter-Adaptive Control," *Proc. 1991 IEEE Conference on Decision and Control*, Brighton, England, to appear.

[16] A. S. Morse, "High-Order Parameter Tuners for the Adaptive Control of Nonlinear Systems" *Proc. of Workshop on Robust Control*" Tokyo, June, 1991, to appear.

[17] M. A. Aizerman and F. G. Gantmacher, *Absolute Stability of Regulator Systems*, Holden-Day, 1964,p.58.

[18] A. S. Morse, "A $4(n+1)$-Dimensional Model Reference Adaptive Stabilizer for any Relative Degree One or Two Minimum Phase System of Dimension $n$ or Less," *Automatica*, v. 23, 1987, pp.123-125.

[19] G. Tao and P. A. Ioannou, "Model Reference Adaptive Control for Plants with Unknown Relative Degree," *Proc. ACC*, 1989, pp. 2297-2302.

[20] A. S. Morse, G. C. Goodwin and D. Q. Mayne, "Applications of Hysteresis Switching in Parameter Adaptive Control," *Proc., 1991 CDC*, to appear.

Department of Electrical Engineering
Yale University
P. O. Box 1968
New Haven, Ct. 06520-1968, USA

# New Differential Geometric Methods in Nonholonomic Path Finding

HÉCTOR J. SUSSMANN +

**Abstract**

We outline three approaches for nonholonomic path finding —nilpotent approximation, highly oscillatory inputs and path deformation— that are based on the use of the techniques of modern geometric optimal control theory, as well as a more classical one —optimal control— where differential geometric methods are also beginning to play a significant role.

## §1. Introduction

In this note we will outline and compare three approaches that have recently been pursued in our work on path finding for nonholonomic systems, and will suggest adding a fourth one to this list. One of the approaches — optimal control— is old, but has recently attracted renewed interest due to the fact that, by making systematic use of "differential geometric methods," it now appears possible to get explicit solutions of problems that could not be handled by means of the techniques developed in the era of "classical" control theory. Two others —nilpotent approximation and highly oscillatory inputs— have been discussed in our recent joint work with G. Lafferriere and W. Liu. The new addition —which is naturally more speculative than the others— is the *path deformation method*, outlined in §6.

It should be emphasized that this paper is *not* intended to be a survey of the whole subject of nonholonomic path finding. This rich area has produced in recent years many ingenious ideas and methods, that cannot possibly be covered in an article such as this one. (Cf. the work of R. Brockett, e.g. [3], J.-P. Laumond and coworkers, e.g. [5], [9], [10], S. Sastry

+ This work was supported in part by the National Science Foundation under NSF Grant DMS-8902994

and R. Murray, [12], [13], [14], Fernandes, Gurvits and Li, [4].) We will instead limit our discussion to the four approaches in whose development we have been directly involved. A fuller account, describing these approaches in full generality, with precise statements of technical conditions and rigorous convergence theorems, is given in other technical reports, conference proceedings, and journal articles now in preparation.

## §2. Nonholonomic path finding

We consider systems of the form

$$(2.1) \qquad \Sigma: \quad \dot{x} = f_0(x) + u_1 f_m(x) + \ldots + u_m f_m(x) \ , \quad x \in M \ , \quad u \in U \ ,$$

where

1. the *state space* $M$ is a smooth (i.e. $C^\infty$) connected manifold,
2. $f_0, \ldots, f_m$ are smooth vector fields on $M$,
3. the set $U$ of *admissible control values* is a closed, convex subset of $\mathbb{R}^m$ with nonempty interior. (The most common cases are $U = \mathbb{R}^m$, i.e. a totally unconstrained control, $U = B(m)$, where $B(m) = \{u \in \mathbb{R}^m : u_1^2 + \ldots + u_m^2 \leq 1\}$ is the unit ball in $\mathbb{R}^m$, and $U = C(m)$, where $C(m) = \{u \in \mathbb{R}^m : |u_1| \leq 1, \ldots, |u_m| \leq 1\}$ is the unit cube in $\mathbb{R}^m$.)

By a *control* we mean a Lebesgue integrable $U$-valued function defined on a compact interval $[a, b]$ of the real line. If $\eta : [a, b] \longrightarrow \mathbb{R}^m$ is a control, and $t \in [a, b]$, we will use $\eta^t$ to denote the restriction of $\eta$ to the interval $[a, t]$. Usually, we are interested in controls that are "nice" in some sense, e.g. smooth, or piecewise smooth, or piecewise constant, so more general highly pathological controls are to be avoided whenever possible. However, there are cases where these pathological controls cannot be avoided, so we choose to allow them in our definition. As usual, a *trajectory* for a control $\eta = (\eta_1, \ldots, \eta_m) : [a, b] \longrightarrow \mathbb{R}^m$ is an absolutely continuous curve $\gamma : [a, b] \longrightarrow M$ that satisfies the differential equation

$$(2.2) \qquad \dot{\gamma}(t) = f_0(\gamma(t)) + \eta_1(t) f_1(\gamma(t)) + \ldots + \eta_m(t) f_m(\gamma(t))$$

almost everywhere. Given any initial condition $\gamma(a) = x$, and a control $\eta : [a, b] \longrightarrow \mathbb{R}^m$, the trajectory $\gamma$ that corresponds to $\eta$ and satisfies the given initial condition may fail to exist, but if it exists then it is unique, and in that case we will denote it by $\gamma_x^\eta$. Moreover, in any case there is a $t \in [a, b]$ such that $t > a$ and $\gamma_x^{\eta^t}$ exists. (More precisely, given any $x$ there exists an $\alpha(x) > 0$ such that $\gamma_x^\eta$ exists for every $a, b, \eta$ such that

$\eta \in L^1([a,b], \mathbb{R}^m)$ and $||\eta||_{L^1} < \alpha(x)$. In addition, if $K \subseteq M$ is compact, then $\alpha(x)$ can be chosen independently of $x$ for all $x \in K$.) Under various growth conditions, e.g. if $M$ is compact, or if $M = \mathbb{R}^n$ and the $f_i$ are globally Lipschitz, the trajectory $\gamma_x^\eta$ exists for all choices of $x$ and $\eta$. If $\eta : [a,b] \to R^m$ is a control, $\gamma_x^\eta$ exists, and $\gamma_x^\eta(b) = y$, then we say that $\eta$ *steers* (or *drives*) $x$ to $y$ in time $b-a$. If there is a control that steers $x$ to $y$ then we say that $y$ is *reachable*, or *accessible* from $x$.

Roughly, the *path finding problem* is the following: given two points $p$ and $q$, find a way to steer $p$ to $q$ as nicely as possible. Somewhat less vaguely, we want an *algorithm* that, for every pair $(p,q)$ of points of $M$ (or, at least, for every pair $(p,q)$ in some large subset $S$), will produce a control $\eta_{p,q}$ that steers $p$ to $q$. Naturally, the largest set for which such an algorithm might possibly exist is the "reachability set" $\mathcal{R}(\Sigma)$, i.e. the set of all pairs $(p,q)$ such that $q$ is reachable from $p$. If $\mathcal{R}(\Sigma) = M \times M$, then $\Sigma$ is said to be *completely controllable* (CC). This is obviously the only case where we can expect to be able to solve the path finding problem by finding algorithms that work for *all* $(p,q)$. Another property, much weaker than controllability but very important in the theory, is *accessibility*: $\Sigma$ has the *accessibility property* (AP) if for every $x$ the set of points that can be reached from $x$ has a nonempty interior. (This terminology is due to D. Elliott.) For real-analytic systems, accessibility is equivalent to the *Lie algebra rank condition* (LARC), i.e. the condition that $L(\Sigma)(x) = T_xM$ for every $x \in M$, where (i) $T_xM$ denotes the tangent space of $M$ at $x$, (ii) $L(\Sigma)$ —known as the *accessibility Lie algebra* of $\Sigma$— is the Lie algebra of vector fields generated by $f_0, \ldots, f_m$, (iii) $L(\Sigma)(x) = \{X(x) : X \in L(\Sigma)\}$.

The vector field $f_0$ is the *drift*. When $\Sigma$ has no drift, i.e. $f_0 = 0$, then the AP and complete controllability are equivalent. If, in addition, $\Sigma$ is real analytic, then it is CC if and only if it has the LARC. The equivalence is also true in certain cases with drift, e.g. if $M$ is a compact Lie group and $\Sigma$ is a translation-invariant system, or if the flow of $f_0$ is periodic. But, in general, the CC property is hard to characterize by means of "algebraic" conditions such as the LARC.

## §3. Exact solutions for nilpotent systems

The *nilpotent approximation method* works for systems *without drift*. It is based on a procedure that produces an exact solution for nilpotent systems. In this section we discuss this exact solution, and in the next one we explain how to use it for more general systems.

The procedure for finding exact solutions for nilpotent systems is based on two observations. First, for a *complete nilpotent system*, the trajectory can be computed explicitly in terms of a finite number of iterated integrals of the control. (Recall that $\Sigma$ is said to be *nilpotent* if the Lie algebra $L(\Sigma)$ is nilpotent. A Lie algebra $L$ is *nilpotent* if there exists a positive integer $k$ such that every bracket $[X_1,[X_2,\ldots,[X_{k-1},X_k]]]$ of $k$ elements of $L$ vanishes. A system is *complete* if for every control $\eta$ and every initial condition $x$ the trajectory $\gamma_x^\eta$ exists. For driftless nilpotent systems completeness is equivalent to the condition that the vector fields $f_1,\ldots,f_m$ are themselves complete, i.e. all their trajectories are defined for all times $t \in \mathbb{R}$.)

The second observation is that, once the formula for the solution has been expressed in terms of iterated integrals, then it is possible in various ways to invert it and obtain, for any given initial and terminal points $p$, $q$, a control for which the iterated integrals have the values needed to steer $p$ to $q$.

We now explain these ideas in more detail, and begin by developing a formalism (similar to the "chronological calculus" of Agrachev and Gamkrelidze, cf. [1]) that makes the calculations simple. If $f$ is a smooth vector field, let us use exponential notation for the flow of $f$, and let us write the flow on the right, so that $t \rightarrow xe^{tf}$ is the integral curve of $f$ that goes through $x$ at time $t = 0$. Consistently with this, let us also write vector fields on the right, so that $xf$ is an alternative notation for $f(x)$. More generally, use the notation $x\Phi$ as an alternative to $\Phi(x)$ for any map $\Phi$, whether it is a diffeomorphism such as $e^{tf}$ or a vector field such as $f$. If $\Phi : M \rightarrow N$ is a smooth map, $v$ is a tangent vector at $x$, and $y = \Phi(x)$, then $v\Phi$ will denote the tangent vector at $y$ that is usually denoted by $\Phi_* v$, or $d\Phi(x)(v)$, i.e. the value at $v$ of the differential $d\Phi(x) : T_xM \rightarrow T_yN$ of $\Phi$ at $x$. With these notations, it is not hard to prove the formula

$$\frac{d}{dt}(xe^{tg}fe^{-tg}) = xe^{tg}[g,f]e^{-tg} \ , \tag{3.1}$$

where $[g,f]$ is the Lie bracket of $f$ and $g$. (Notice that, to begin with, the formula makes sense because the function $t \rightarrow xe^{tg}fe^{-tg}$ takes values in a fixed linear space, namely $T_xM$, so it is at least meaningful to differentiate with respect to $t$. Moreover, differentiating formally we get

$$\frac{d}{dt}(xe^{tg}fe^{-tg}) = xe^{tg}gfe^{-tg} - xe^{tg}fge^{-tg} \ , \tag{3.2}$$

since $\frac{d}{dt}e^{tg} = e^{tg}g$. Since $[g,f] = gf - fg$, the desired identity follows, at least on the formal level. A complete justification requires some more work, in order to assign a precise meaning to terms such as $x\Phi fg\Psi$ when $\Phi$ and $\Psi$ are diffeomorphisms and $f$, $g$ are vector fields, and then justify all the steps in the calculation, but the formalism in which all this works is easy to construct and then all the steps become rigorous.) We remark in particular that, if $[f, g] = 0$, then Equation (3.1) implies that $\frac{d}{dt}(xe^{tg}fe^{-tg}) = 0$, so that $xe^{tg}fe^{-tg} \equiv \text{constant} \equiv xf$, and therefore $xe^{tg}f = xfe^{tg}$.

Using this, let us show how to solve explicitly the equations for a nilpotent system. For simplicity, we will just consider the case of a system

$$\dot{x} = u_1 f_1(x) + u_2 f_2(x) \ , \ x \in \mathbb{R}^n \ , \tag{3.3}$$

where $f_1$ and $f_2$ are smooth vector fields such that

$$[f_1, [f_1, f_2]] \equiv [f_2, [f_1, f_2]] \equiv 0 \ . \tag{3.4}$$

(The case of a general nilpotent system is of course more complicated, but the idea is the same.) Assume we are given a control $t \rightarrow (u_1(t), u_2(t))$, $0 \leq t \leq T$, and an initial condition $x(0) = \bar{x}$. We rewrite the equation as $\dot{x} = x(u_1 f_1 + u_2 f_2)$, and try a solution of the form

$$x(t) = \bar{x}e^{a_1(t)f_1}e^{a_2(t)f_2}e^{a_3(t)f_3} \ , \tag{3.5}$$

where the $a_i(t)$ are functions to be chosen, and $f_3$ is defined to be $[f_1, f_2]$. Since we want $x(0)$ to be equal to $\bar{x}$, we stipulate that $a_1(0) = a_2(0) = a_3(0)$. Then we can differentiate with respect to $t$ and find that

$$\begin{aligned} \dot{x}(t) =& \frac{d}{dt}\bar{x}e^{a_1 f_1}e^{a_2 f_2}e^{a_3 f_3} \\ =& \quad \bar{x}e^{a_1 f_1}\dot{a}_1 f_1 e^{a_2 f_2}e^{a_3 f_3} \\ & + \bar{x}e^{a_1 f_1}e^{a_2 f_2}\dot{a}_2 f_2 e^{a_3 f_3} \\ & + \bar{x}e^{a_1 f_1}e^{a_2 f_2}e^{a_3 f_3}\dot{a}_3 f_3 \ . \end{aligned} \tag{3.6}$$

Since $[f_2, f_3] = 0$, we have $yf_2e^{a_3 f_3} = ye^{a_3 f_3}f_2$ for every $y$, so in the second term of the above sum we can change the order of the factors $\dot{a}_2 f_2$ and $e^{a_3 f_3}$. Changing the order of $\dot{a}_1 f_1$ and $e^{a_2 f_2}$ is a bit more difficult. We have

$$yf_1e^{a_2 f_2} = ye^{a_2 f_2}e^{-a_2 f_2}f_1e^{a_2 f_2} = ze^{-a_2 f_2}f_1e^{a_2 f_2} \ , \tag{3.7}$$

if we let $z = ye^{a_2 f_2}$. For any fixed $z$, we have

$$\frac{d}{dt}ze^{-a_2 f_2}f_1e^{a_2 f_2} = -ze^{-a_2 f_2}\dot{a}_2[f_2, f_1]e^{a_2 f_2} \ . \tag{3.8}$$

Since $[f_2, f_3] = 0$, we can commute $\dot{a}_2[f_2, f_1]$ past $e^{a_2 f_2}$, and end up with $\frac{d}{dt}ze^{-a_2 f_2}f_1e^{a_2 f_2} = -z\dot{a}_2[f_2, f_1]$. But this then gives

$$ze^{-a_2 f_2}f_1e^{a_2 f_2} = z(f_1 - a_2[f_2, f_1]) \ . \tag{3.9}$$

Letting $z = ye^{a_2 f_2}$ and $y = \bar{x}e^{a_1 f_1}$, we get

$$\bar{x}e^{a_1 f_1}\dot{a}_1 f_1 e^{a_2 f_2}e^{a_3 f_3} = \bar{x}e^{a_1 f_1}e^{a_2 f_2}\dot{a}_1(f_1 - a_2[f_2, f_1])e^{a_3 f_3} . \tag{3.10}$$

Since both $f_1$ and $[f_2, f_1]$ commute with $f_3$, we conclude that

$$\bar{x}e^{a_1 f_1}\dot{a}_1 f_1 e^{a_2 f_2}e^{a_3 f_3} = \bar{x}e^{a_1 f_1}e^{a_2 f_2}e^{a_3 f_3}\dot{a}_1(f_1 - a_2[f_2, f_1]) . \tag{3.11}$$

If we combine all these results we get

$$\dot{x}(t) = x(t)\Big(\dot{a}_1(t)f_1 - \dot{a}_1(t)a_2(t)[f_2, f_1] + \dot{a}_2(t)f_2 + \dot{a}_3(t)f_3\Big) . \tag{3.12}$$

To satisfy the equation $\dot{x}(t) = x(t)(u_1(t)f_1 + u_2(t)f_2)$, since $f_3 = [f_1, f_2]$, we must let

$$\dot{a}_1 = u_1 \ , \quad \dot{a}_2 = u_2 \ , \quad \dot{a}_3 = -u_1 a_2 . \tag{3.13}$$

So, if we let

$$\begin{aligned} U_1(t) &= \int_0^t u_1(s)ds \ , \\ U_2(t) &= \int_0^t u_2(s)ds \ , \\ U_3(t) &= -\int_0^t u_1(s)U_2(s)ds \ , \end{aligned} \tag{3.14}$$

we see that we need to choose $a_i(t) = U_i(t)$.

In other words, we have expressed our solution $x(t)$ as

$$x(t) = \bar{x}e^{U_1(t)f_1}e^{U_2(t)f_2}e^{U_3(t)f_3} , \tag{3.15}$$

where the $U_i$ are defined by (3.14). We now show how this formula can be "inverted" so that, if we are given $p = \bar{x}$, a time $T$, and a desired terminal point $q$, we can find a control that steers $p$ to $q$ in time $t$. Assume that the system (3.3) satisfies the LARC. This means that at every point $x$ the three vectors $f_i(x)$, $i = 1, 2, 3$, span $\mathbb{R}^n$. In particular, $n \leq 3$. Let $F(x) \in \mathbb{R}^{n\times 3}$ be the matrix whose columns are $f_1(x)$, $f_2(x)$ and $f_3(x)$, so $F(x)$ has rank $n$ for every $x$. Let $M(x) = F(x)^\dagger(F(x)F(x)^\dagger)^{-1}$, where "$\dagger$" stands for transpose. (So $M(x)$ is a pseudoinverse of $F(x)$. In particular, $F(x)M(x)$ is the $n \times n$ identity matrix.)

Now consider the "extended system"

$$\dot{x} = v_1 f_1(x) + v_2 f_2(x) + v_3 f_3(x) \ . \tag{3.16}$$

For this system it is very easy to find a trajectory from $p$ to $q$. We just find one path in $\mathbb{R}^n$ that goes from $p$ to $q$ in time $T$, for instance the segment given by $\sigma(t) = p + \frac{t}{T}(q-p)$. We then realize this path $\sigma$ as a trajectory of (3.16) by writing $\dot{\sigma}(t)$ as a linear combination

$$\dot{\sigma}(t) = v_1(t) f_1(\sigma(t)) + v_2(t) f_2(\sigma(t)) + v_3(t) f_3(\sigma(t)) \ , \tag{3.17}$$

which is possible because the three vectors $f_i(\sigma(t))$ span $\mathbb{R}^n$. For instance, we can choose $v(t) = M(\sigma(t))\dot{\sigma}(t)$. For this particular choice of $v$, we now solve Equation (3.16) explicitly in terms of iterated integrals exactly as we did before for Equation (3.3). The only difference is that, in this case, we want $x(t)$ to satisfy $\dot{x}(t) = x(t)(v_1(t)f_1 + v_2(t)f_2 + v_3(t)f_3)$. So we must have

$$\dot{a}_1 = v_1 \ , \quad \dot{a}_2 = v_2 \ , \quad \dot{a}_3 = -v_1 a_2 + v_3 \ . \tag{3.18}$$

We then let

$$\begin{aligned} V_1(t) &= \int_0^t v_1(s)ds \ , \\ V_2(t) &= \int_0^t v_2(s)ds \ , \\ V_3(t) &= \int_0^t (-v_1(s)V_2(s) + v_3(s))ds \ , \end{aligned} \tag{3.19}$$

and choose $a_i(t) = V_i(t)$. In particular, we have shown that

$$q = pe^{V_1(T)f_1} e^{V_2(T)f_2} e^{V_3(T)f_3} \ . \tag{3.20}$$

The above computations establish that, given any $p$, $q$, one can explicitly find numbers $\alpha_i$, $i = 1, 2, 3$, such that $q = pe^{\alpha_1 f_1} e^{\alpha_2 f_2} e^{\alpha_3 f_3}$. The last step is then to show that one can explicitly find, for each triple $(\alpha_1, \alpha_2, \alpha_3)$, a control $u = (u_1, u_2)$, defined on some interval $[0, T]$, for which the three iterated integrals $U_i$ defined above satisfy $U_i(T) = \alpha_i$. This can be done in a number of ways. For instance, we may choose

$$\begin{aligned} u_1(t) &= \frac{\alpha_1}{T} + \beta \sin \frac{2\pi t}{T} \ , \\ u_2(t) &= \frac{\alpha_2}{T} + \gamma \cos \frac{2\pi t}{T} \ , \end{aligned} \tag{3.21}$$

where

$$\beta = \sqrt{\left|\alpha_3 - \frac{\alpha_1\alpha_2}{2}\right|} \quad \text{and} \quad \gamma = \beta \operatorname{sign}\left(\alpha_3 - \frac{\alpha_1\alpha_2}{2}\right). \tag{3.22}$$

Although the algebra becomes much more cumbersome, the idea of the above example generalizes to arbitrary complete nilpotent systems. A more detailed explanation of this is given in [6], [7], [8]. The appropriate mathematical formalism turns out to be the use of *noncommutative power series.* For a general nilpotent system, one can always write the solution explicitly in the form

$$x(t) = x(0)e^{U_1(t)B_1(f)}e^{U_2(t)B_2(f)}\ldots e^{U_r(t)B_r(f)}, \tag{3.23}$$

where

a. $B_1, B_2, \ldots, B_r$ are a basis of $L_k(X_1, X_2, \ldots, X_m)$, the free Lie algebra generated by $m$ noncommuting indeterminates $X_1, X_2, \ldots, X_m$,

b. $k$ is chosen so that all the brackets of the $f$'s of degree $> k$ vanish,

c. for a formal bracket $B$, $B(f)$ is the result of plugging in the $f_i$ for the $X_i$ (so that, for instance, if $B = [X_1, X_2]$, then $B(f) = [f_1, f_2]$),

d. the $U_i(t)$ are certain functionals of the control $u$.

It turns out that, if the basis $\mathcal{B} = (B_1, B_2, \ldots, B_r)$ is chosen to be a *P. Hall basis* of $L_k(X_1, X_2, \ldots, X_m)$, then the $U_i$ are given by very simple formulae as *iterated integrals* (cf. Sussmann [18]).

One then proceeds in general as we did in the simple case considered here. We find a suitably rich Lie bracket extension of the given system, then pick a (not necessarily admissible) path $\gamma : [0, T] \to M$ that joins $p$ to $q$, then find a control $v$ that gives rise to $\gamma$ for the extended system, then express $x(T)$ as in [3.23]. (This turns out to be possible in a simple way even for extended controls.) The last step is then to choose the $u_i$ so that the numbers $U_i(T)$ will be equal to the $V_i(T)$ found by the above procedure. This can be done in several different ways. One such way, involving bang-bang controls, has been proposed in [7] and [8]. Other methods, e.g. involving sinusoids, are proposed by Murray and Sastry in [14] for special classes of nilpotent systems, and can be extended to general nilpotent systems using the formalism of our work with W. Liu. cf. [19], [20], [21].

### §4. Nilpotent approximation of general systems

So far, we have only discussed nilpotent systems. It turns out, however, that the method can be used to obtain an algorithm that works for more general systems. Let us explain this by referring once again to the system (3.3), where $f_1$, $f_2$ are no longer assumed to satisfy (3.4).

Assume that the three brackets $f_1(x)$, $f_2(x)$, $f_3(x)$ span $\mathbb{R}^n$ for every $x$. Then we can still join any two points $p$, $q$ by a path $\sigma$, and then compute $v$ from $\sigma$ as we did above. We can then compute $u$ from $v$ exactly as above. (That is, we define the $V_i$ by (3.19), then we let $\alpha_i = V_i(T)$, and finally choose the $u_i$ according to (3.21).) The only difference, of course, is that now Equations (3.15) and (3.20) do not follow, so we are no longer entitled to conclude that $u$ steers $p$ to $q$. Let $\tilde{q}$ be the point to which $u$ steers $p$. One can then prove an estimate $||\tilde{q} - q|| \leq C||p - q||^{3/2}$. (Precisely: given any compact set $K$ there exists a $C > 0$ such that the estimate holds whenever $p$ and $q$ are in $K$.) In particular, this implies that for every compact $K$ there is a constant $\Delta$ such that the bound $||\tilde{q} - q|| \leq \frac{1}{2}||p - q||$ holds if $p \in K$, $q \in K$, and $||p-q|| \leq \Delta$. (The number $\Delta$ was called the "'critical distance" in [7] and [8].) If we want to join $p$ to $q$ and $||p - q||$ is already $\leq \Delta$, then we may apply the "nilpotent" algorithm to produce a control that steers $p$ to a point $q_1$, then apply it again starting at $q_1$ rather than $p$ (but always with target $q$) to get a point $q_2$, and so on. Each time the distance will go down by a factor of at least $1/2$, so that convergence will be very fast.

In the general case, if $p$ and $q$ are far, it suffices to choose a $\hat{\Delta}$ such that $\hat{\Delta} \leq \Delta$, and apply the nilpotent algorithm starting at $p$ but with target $q_1$ rather than $q$, where $q_1$ lies in the straight line segment from $p$ to $q$, but at a distance $\hat{\Delta}$ from $p$. The resulting control will steer $p$ to a point $p_1$ such that $||p_1 - q_1|| \leq \frac{1}{2}\hat{\Delta}$. On the other hand, since

$$||q_1 - q|| = ||p - q|| - \hat{\Delta} \ , \tag{4.1}$$

the triangle inequality implies that

$$||p_1 - q|| \leq ||p - q|| - \frac{1}{2}\hat{\Delta} \ . \tag{4.2}$$

We can then repeat the whole procedure starting at $p_1$, to produce a point $p_2$, and then continue to iterate. Each step will decrease the distance to $q$ by at least $\frac{1}{2}\hat{\Delta}$, until a point $p_k$ is attained whose distance to $q$ is $\leq \hat{\Delta}$. From then on we just continue to apply the algorithm with target $q$, and each step will bring down the distance by a factor $\frac{1}{2}$, as before.

It then follows that, for any prescribed error margin $\varepsilon$, the algorithm described above will produce a point $\hat{q}$ such that $||\hat{q}-q|| \le \varepsilon$ in at most $N$ steps, where

$$N = 2\left[\frac{||q-p||}{\hat{\Delta}}\right] + \log_2\frac{\hat{\Delta}}{\varepsilon}\,, \tag{4.3}$$

where, as usual, $[x]$ is the largest integer $\nu$ such that $\nu \le x$.

The procedure described above generalizes to arbitrary real analytic completely controllable driftless systems. Naturally, in this more general setting one has to consider a richer "Lie bracket extension" of the original system, and add enough brackets to the right-hand side to span the tangent space at each point. Such an extension always exists locally because of the LARC, and therefore given any compact set $K$ one can find an extension that works for $K$. In many cases, e.g. for systems in $\mathbb{R}^n$ with polynomial vector fields, there is an extension that works globally. The main difference is that the estimate for the general case is $||\tilde{q}-q|| \le C||p-q||^{1+\frac{1}{k}}$, where $k$ is the smallest degree such that the Lie brackets up to order $k$ span the whole space. (Cf. [7] and [8] for more details.) If $k$ is large, the exponent is very close to 1, which means that convergence is very slow and $\Delta$ very small.

The main drawback of the above method is that it requires knowing a lower bound for $\Delta$. In practice, one can simply try a value for $\hat{\Delta}$ and see if it produces an improvement. If it does not, one may try a smaller value and keep lowering until the algorithm starts producing improvements.

## §5. Highly oscillatory inputs

Here again we deal with completely controllable real analytic systems without drift that satisfy the LARC. (The results also apply in some cases where a drift is present, e.g. if the associated driftless system obtained by setting $f_0 = 0$ still satisfies the LARC. )

The idea of this method is to pick a path $\gamma : [0,T] \to M$ such that $\gamma(0) = p$, $\gamma(T) = q$, that need not be admissible, and try to approximate the whole path by a sequence $\{\gamma_j\}$ of admissible paths, also defined on $[0,T]$, and also satisfying $\gamma_j(0) = p$. Exactly as in the previous section, one chooses a Lie bracket extension of the original system, and then expresses the path $\gamma$ as a trajectory of the extended system corresponding to some "extended input" $v$. One then tries to find a sequence of "ordinary inputs" $u^j$ whose trajectories converge to those of the "extended input" $v$. At first it may appear strange that such a convergence is possible, but we will illustrate how this can happen by referring again to the special case of System (3.3) and its extension (3.16). Let $\eta_1$, $\eta_2$, $\eta_3$, $\eta_4$ be smooth real-valued functions on the interval $[0,T]$, and consider the controls $u^j$ defined by $u^j = (u_1^j, u_2^j)$, where $u_1^j(t) = \eta_1(t) + \sqrt{j}\,\eta_3(t)\sin jt$, $u_2^j(t) = \eta_2(t) + \sqrt{j}\,\eta_4(t)\sin jt$. Fix an initial point $p$, and let $x^j$ be the resulting trajectory for the input $u^j$. (For simplicity, assume that the $f_i$ and all their derivatives are bounded, so there are no explosions for (3.3) or any of its Lie bracket extensions.) We can then write

$$x^j(t) = \xi^j(t) + \zeta_1^j(t) + \zeta_2^j(t) \ , \tag{5.1}$$

where

$$\xi^j(t) = p + \int_0^t \eta_1(s) f_1(x^j(s))ds + \int_0^t \eta_2(s) f_2(x^j(s))ds \ , \tag{5.2}$$

and

$$\zeta_1^j(t) = \sqrt{j}\int_0^t \eta_3(s)\sin js f_1(x^j(s))ds \ , \tag{5.3}$$

$$\zeta_2^j(t) = \sqrt{j}\int_0^t \eta_4(s)\cos js f_2(x^j(s))ds \ , \tag{5.4}$$

Integration by parts yields

$$\begin{aligned}
\zeta_1^j(t) &= \theta_1^j(t) f_1(x^j(t)) \\
&\quad - \int_0^t \theta_1^j(s)\cdot Df_1(x^j(s))\cdot\Big(u_1^j(s) f_1(x^j(s)) + u_2^j(s) f_2(x^j(s))\Big)ds \ , \\
\zeta_2^j(t) &= \theta_2^j(t) f_2(x^j(t)) \\
&\quad - \int_0^t \theta_2^j(s)\cdot Df_2(x^j(s))\cdot\Big(u_1^j(s) f_1(x^j(s)) + u_2^j(s) f_2(x^j(s))\Big)ds \ ,
\end{aligned}$$

where

$$\theta_1^j(t) = \sqrt{j}\int_0^t \eta_3(s)\sin js\, ds \ , \tag{5.5}$$

and

$$\theta_2^j(t) = \sqrt{j}\int_0^t \eta_4(s)\cos js\, ds \ . \tag{5.6}$$

Another integration by parts shows that

$$\theta_1(t) = \frac{1}{\sqrt{j}}(\eta_3(0) - \eta_3(t)\cos jt) + \frac{1}{\sqrt{j}}\int_0^t \eta_3'(s)\cos js\, ds \ . \tag{5.7}$$

This implies in particular that $\theta_1^j(t) = O(\frac{1}{\sqrt{j}})$ uniformly in $t$, so in particular $\theta_1^j \to 0$ uniformly as $j \to \infty$. Moreover, the integral $\int_0^t \eta_3'(s)\cos js\,ds$ goes to 0 uniformly by the Riemann-Lebesgue Lemma, which implies that

$$\theta_1(t) = \frac{1}{\sqrt{j}}(\eta_3(0) - \eta_3(t)\cos jt) + \frac{1}{\sqrt{j}}\nu_1^j(t) \ , \tag{5.8}$$

where $\nu_1^j(t) \to 0$ uniformly. We can then write

$$\begin{aligned}\zeta_1^j(t) = &\int_0^t \Big(\eta_3(s)\cos js - \eta_3(0)\Big)\eta_3(s)\sin js(Df_1\cdot f_1)(x^j(s))ds \\ &+ \int_0^t \Big(\eta_3(s)\cos js - \eta_3(0)\Big)\eta_4(s)\cos js(Df_1\cdot f_2)(x^j(s))ds + \mu_1^j(t)\end{aligned} \tag{5.9}$$

where the $\mu_1^j$ go to 0 uniformly. A similar calculation yields

$$\begin{aligned}\zeta_2^j(t) = &-\int_0^t \eta_4(s)\eta_3(s)\sin^2 js\,(Df_2\cdot f_1)(x^j(s))ds \\ &- \int_0^t \eta_4(s)\sin js\ \cos js\,\eta_4(s)\,(Df_2\cdot f_2)(x^j(s))ds + \mu_2^j(t)\end{aligned} \tag{5.10}$$

where the $\mu_2^j$ also go to 0 uniformly. In particular, the integrands that appear in the integrals of the formulas for $\zeta_1^j$ and $\zeta_2^j$ are uniformly bounded, so the integrals are equicontinuous. The $\mu_i^j$ are also equicontinuous, because they converge to 0, and the $\xi^j$ are equicontinuous as well. Therefore the $x^j$ are equicontinuous. By passing to a subsequence, we may assume that they converge uniformly to a limit $x^\infty$. If we let

$$\rho_1^j(s) = \Big(\eta_3(s)\cos js - \eta_3(0)\Big)\eta_3(s)\sin js \ , \tag{5.11}$$

$$\rho_2^j(s) = \Big(\eta_3(s)\cos js - \eta_3(0)\Big)\eta_4(s)\cos js \ , \tag{5.12}$$

$$\rho_3^j(s) = \eta_4(s)\eta_3(s)\sin^2 js \ , \tag{5.13}$$

$$\rho_4^j(s) = \eta_4(s)\eta_4(s)\sin js\ \cos js\,, \tag{5.14}$$

then the four sequences of functions $\{\rho_1^j\}$, $\{\rho_2^j\}$, $\{\rho_3^j\}$, $\{\rho_4^j\}$ are uniformly bounded and converge weakly to 0, $\frac{1}{2}\eta_3(s)\eta_4(s)$, 0, and $\frac{1}{2}\eta_3(s)\eta_4(s)$, respectively. This means that $x^\infty$ satisfies

$$
\begin{aligned}
x^\infty(t) =& p + \int_0^t \eta_1(s) f_1(x^\infty(s))ds + \int_0^t \eta_2(s) f_2(x^\infty(s))ds \\
&+ \frac{1}{2}\int_0^t \eta_3(s)\eta_4(s)(Df_1 \cdot f_2 - Df_2 \cdot f_1)(x^\infty(s))ds ,
\end{aligned} \tag{5.15}
$$

i.e.

$$
\begin{aligned}
x^\infty(t) =& p + \int_0^t \eta_1(s) f_1(x^\infty(s))ds + \int_0^t \eta_2(s) f_2(x^\infty(s))ds \\
&+ \frac{1}{2}\int_0^t \eta_3(s)\eta_4(s)[f_2, f_1](x^\infty(s))ds .
\end{aligned} \tag{5.16}
$$

In other words, $x^\infty$ is a solution of the initial value problem

$$
\begin{aligned}
\dot{x}(t) = \eta_1(t) f_1(x(t)) + \eta_2(t) f_2(x(t)) + \frac{1}{2}\eta_3(t)\eta_4(t)[f_2, f_1](x(t)) ,\\
x(0) = p .
\end{aligned} \tag{5.17}
$$

Although so far we have only asserted this for a subsequence, the argument actually proves that every subsequence of our original sequence has a subsequence that converges to a limit that is a solution of the above initial value problem. Since the initial value problem has a unique solution, it follows that the original sequence $\{x^j\}$ converges to this solution.

It is now easy to find a sequence $\{u^j\}$ of inputs whose trajectories converge to those of a given "extended input" $v$. It suffices to choose the $u^j$ as above, with $\eta_1 \equiv v_1$, $\eta_2 \equiv v_2$, $\eta_3\eta_4 \equiv 2v_3$. Notice that the inputs $u^j$ are *not* uniformly bounded, because of the factor $\sqrt{j}$. This is no accident: for a uniformly bounded (or even $L^1$-bounded) sequence $\{u^j\}$, if the trajectories converge to those of an "extended input" $v$ as above for every choice of smooth $f_1$, $f_2$, then it follows that $v_3 \equiv 0$.

The above procedure generalizes to arbitrary systems that satisfy the LARC. One finds a suitable Lie bracket extension, as before, and then constructs a sequence of "highly oscillatory inputs" given as combinations of complex exponentials $e^{ij\omega t}$ for various frequencies $\omega$ with time-varying coefficients $\eta_k$ to be chosen later, but one takes care to choose the frequencies and the coefficients so that the resulting controls are real. One then computes the limit, and expresses it in terms of the $\eta_k$. It then becomes apparent that the $\eta_k$ can *always* be chosen so as to match any desired limit. However, to do all this it becomes necessary to introduce a considerable amount of algebraic machinery. Some of the details are outlined in [19] and [20], and a complete presentation of the results and proofs will be given in [21].

The main drawback of the method is that it requires the use of very large highly oscillatory controls. On the other hand, the method can be applied to obtain a rough approximation of a given path, to be then supplemented by the use of other methods to get much closer to the desired target point. This is likely to be particularly useful for problems with obstacles, in which case the primary purpose of the approximation is to produce an admissible path that avoids the obstacles.

## §6. Optimal Control

Optimal control is a natural method for path finding problems. In many cases there is a function that one would like to minimize or nearly minimize. And, even if there is no obvious choice of such a function, artificially transforming the path finding problem into one of optimization is likely to make the problem much less underdetermined. Moreover, optimal control leads to *feedback* rather than just families of open-loop controls. On the other hand, the feedback typically cannot be smooth. For example, if a problem (2.1) has no drift and the number $m$ of controls is less than the state space dimension, then the necessary condition for smooth local stablizability discovered by Brockett in [2] will necessarily be violated at every point $x$ where the rank function $y \rightarrow \text{rank}\{f_1(y), \ldots, f_m(y)\}$ has a local maximum. (In particular, it will be violated on an open dense set.)

The main drawback of optimal control methods is that, except in some very special cases, the optimal controls are hard to compute. However, using recently developed differential geometric methods, some progress has been made for a number of systems that could not be dealt with by classical optimal control methods. In several cases, it turns out that, although one cannot actually compute the optimal controls exactly, one can determine a rather small *sufficient family of trajectories*, i.e. a family $\mathcal{F}$ of trajectories such that for every pair $(p, q)$ of points there is an optimal trajectory in $\mathcal{F}$ that goes from $p$ to $q$.

As an example of this, we consider the problem $\dot{x}_1 = u\cos\theta$, $\dot{x}_2 = u\sin\theta$, $\dot{\theta} = v$, where the controls $u$, $v$ are subject to the constraints $|u| \leq 1$, $|v| \leq 1$. This problem was studied by Reeds and Shepp in [15]. They proved that a sufficient family is obtained by considering the set of all concatenations of at most five pieces, each one of which is either a circle (i.e. $(u, v) \equiv (1, 1)$ or $(u, v) \equiv (1, -1)$, or $(u, v) \equiv (-1, 1)$, or $(u, v) \equiv (-1, -1)$) or a straight line, with no more than one straight line and no more than four circles. (They actually proved a number of extra restrictions as well, cf. [15]. They used a different formulation, in which the restriction on $u$ is $|u| = 1$ rather than

$|u| \leq 1$. Since this defines a nonconvex set, existence of optimal trajectories is not guaranteed by the usual existence theorems of control theory, so they could not apply the usual necessary conditions for optimality.)

Recently, the Reeds-Shepp car and a number of other related problems have been investigated from an optimal control theory point of view. It turns out that the tools of classical optimal control, combined with the newer methods of differential geometric nonlinear control, make it possible to analyze the trajectories in detail, and to get a lot of information about the optimal feedback (cf. [22]). So there is new hope that optimal control methods, properly enriched by the addition of the more recently developed techniques, may after all turn out to be more useful than they are currently believed to be.

## §7. Path deformation

The *path deformation approach* is based on extending to the path finding problem the well known continuation method used to solve nonlinear equations. Let us first recall how this method works.

Suppose $M$ and $N$ are smooth manifolds and $\Phi : M \to N$ is a smooth map. Assume $N$ is connected. Suppose we want to solve an equation $\Phi(x) = y$, for which we are given $y \in N$ and want to find $x \in M$. One possible approach is to start with some values $\bar{x}$, $\bar{y}$ such that $\Phi(\bar{x}) = \bar{y}$, find a smooth path $\pi : [0,1] \to N$ such that $\pi(0) = \bar{y}$ and $\pi(1) = y$, and then try to "lift" this path to $M$, i.e. to find a path $\Pi : [0,1] \to M$ such that $\Phi \circ \Pi = \pi$ and $\Pi(0) = \bar{x}$. Clearly, if we find such a path, then $x = \Pi(1)$ will be a solution of our problem.

The reason why this is useful is that, if certain nondegeneracy and nonexplosion conditions are satisfied, then $\Pi$ can be found by solving an ordinary differential equation, which lends itself easily to numerical implementation. To see this, let us assume from now on, for simplicity, that $M$ and $N$ actually are open subsets of Euclidean spaces $\mathbb{R}^m$, $\mathbb{R}^n$. In view of the initial condition $\Pi(0) = \bar{x}$, the identity $\Phi \circ \Pi = \pi$, i.e. $\Phi(\Pi(t)) = \pi(t)$, is equivalent to the equation

$$D\Phi(\Pi(t)) \cdot \dot{\Pi}(t) = \dot{\pi}(t) \,. \tag{7.1}$$

If $D\Phi(\Pi(t))$ has rank $n$, then (7.1) can be solved for $\dot{\Pi}(t)$. Precisely, then, the assumption that we need is that $\Phi$ *is a submersion*, i.e. that $D\Phi(x)$ has rank $n$ for all $x$. If this condition holds, then we can define the pseudoinverse

$$P(x) = D\Phi(x)^{\dagger} \left( D\Phi(x) D\Phi(x)^{\dagger} \right)^{-1} , \tag{7.2}$$

which is a smooth $\mathbb{R}^{m\times n}$-valued function on $M$, and then consider the *path lifting equation*

$$\dot{x}(t) = P(x(t))\dot{\pi}(t) \ . \tag{7.3}$$

If we can find a solution of (7.3) with initial condition $x(0) = \bar{x}$, *which is defined on the whole interval* $[0,1]$, then $x(1)$ will solve our original equation.

The problem now is to make sure that the solutions of (7.3) exist globally. This, of course, requires some other condition, e.g. that the map $\Phi$ be proper. In applications, the map $\Phi$ usually is not a submersion, but the "singular set" $S = \{x \in M : \text{rank}\, D\Phi(x) < n\}$ is small. What one does then is work on the complement $\tilde{M} = M - S$ of this set. Then by definition the restriction $\tilde{\Phi}$ of $\Phi$ to this set is a (not necessarily surjective) submersion. So there is no problem with the local existence and uniqueness of the solutions of (7.3) on $\tilde{M}$. But now the main issue becomes that of explosions: a solution of (7.3) can blow up either because it "goes to infinity" (i.e. does not remain in any compact set) in $M$, *or* because it approaches the singular set. If the latter possibility actually occurs, then extra work is needed to avoid or bypass the singularities. Naturally, this is not always possible, as can be shown by simple examples. (Consider, e.g., the map $\Phi : \mathbb{R} \to \mathbb{R}$ given by $\Phi(x) = x^3 - 3x$. The singular set $S$ consists of 1 and $-1$. The map $\Phi$ is a surjective submersion on $\mathbb{R} - S$, but the path $\pi : [0,1] \to \mathbb{R}$ given by $\pi(t) = -3 + 6t$ cannot be lifted.)

To see how these ideas apply to the path finding problem, we consider once again the simple system (3.3) introduced above. As before we assume, for simplicity, that the $f_i$ and all their derivatives are bounded.

In this case the object that plays the role of $M$ should be the space of all controls, or some sufficiently rich subset. It will be convenient to normalize so that all the controls are defined on $[0,1]$, and to use for $M$ the space $L^2([0,1],\mathbb{R}^2)$ of *square integrable* $\mathbb{R}^2$-valued controls. Then $N$ will be $\mathbb{R}^n$. Suppose we fix $p$, $q$ in $N$ and want to find a control that steers $p$ to $q$. Given any control $\eta \in M$, we can consider the corresponding trajectory $\gamma_p^\eta$, and we let $\Phi(\eta) = \gamma_p^\eta(1)$. With this formulation, our problem now becomes that of solving the equation $\Phi(\eta) = q$ for a given $q$. Clearly, we are exactly in the situation described before, except that now $M$ is an infinite-dimensional Hilbert space.

The singularities of the map $\Phi$ are well understood. It is a general fact that for problems of the kind considered here the singularities of $\Phi$ are precisely the *abnormal extremals*, i.e. the controls that satisfy the necessary conditions of the Pontryagin Maximum Principle with $\lambda_0 = 0$. (Here $\lambda_0$ is the multiplier that is associated to the Lagrangian in the Maximum Principle. Since we are interested here only in the case $\lambda_0 = 0$, the particular choice of a Lagrangian is immaterial.) The zero control is an abnormal extremal. And, unfortunately, for general nonholonomic problems abnormal extremals other than zero can occur, as shown by Montgomery in [11]. However, for our simple system (3.3) they do not. (This follows from a much more general fact, namely, that nontrivial abnormal extremals do not occur for "fat sub-Riemannian metrics," cf. e.g. Strichartz [16], [17]) but can be proved easily in our case, as will be indicated below.)

So for (3.3), if we take $\tilde{M} = M - \{0\}$, then the restriction $\tilde{\Phi}$ of $\Phi$ to $\tilde{M}$ is a surjective submersion. (Surjectivity is trivial, because (3.3) is CC, and $p$ can be reached from $p$ by a nontrivial loop.) To make sure that the solutions of (7.3) do not approach the singular point 0 of $M$, it suffices to use a starting control $\bar{\eta}$ which is not 0, which will then give rise to a starting point $\bar{q} = \gamma_p^{\bar{\eta}}$, and then take the precaution of choosing a path $\pi : [0,1] \to N$ that does not go through $p$ and satisfies $\pi(0) = \bar{q}$, $\pi(1) = q$. (Such a path always exists if $n > 1$, and can always be chosen of a simple kind, e.g. a segment or a circular arc.) The only remaining problem is to show that the solutions do not explode. This can be done, but requires some more delicate estimates, that will be omitted here.

We conclude by writing out in detail the differential equation (7.3) for the evolution of a path for our problem (3.3). If $\eta : [0,1] \to \mathbb{R}^n$ is a control, and $x(t) = \gamma_p^\eta(t)$, then we can consider the *variational equation*

$$\begin{aligned}\dot{y}(t) = &\Big(\eta_1(t)Df_1(x(t)) + \eta_2(t)Df_2(x(t))\Big)y(t) \\ &+ \zeta_1(t)f_1(x(t)) + \zeta_2(t)f_2(x(t)) \ ,\end{aligned} \tag{7.4}$$

corresponding to a control variation $\zeta$. (Precisely, $y(t)$ is the derivative with respect to $s$ at $s = 0$ of $\gamma_p^{\eta+s\zeta}$. The fact that $y$ satisfies (7.4) is a standard result in the theory of ordinary differential equations. Moreover, it is clear that $y(0) = 0$.) Clearly, (7.4) is, for fixed $p$ and $\eta$, a linear time-varying system. The map $\mu$ that assigns to each $\zeta$ the value $y(1)$ is precisely the differential of $\Phi$ at $\eta$. This map $\mu$ is going to be onto if and only if the controllability Gramian $G$ of the linear system (7.4) is nonsingular. If $(t,s) \to M(t,s)$ denotes the fundamental solution of (7.4) (that is, $M(t,s)$ is, as a function of $t$, an $n \times n$ matrix-valued solution of

$$\dot{Y}(t) = \Big(\eta_1(t)Df_1(x(t)) + \eta_2(t)Df_2(x(t))\Big)Y(t) \ , \tag{7.5}$$

and $M(s,s) = \text{identity}$), and $v$ is a vector in $\mathbb{R}^n$, then $v^\dagger G v$ is given by

$$v^\dagger G v = \int_0^1 ||v^\dagger M(1,t)\mathrm{f}(t)||^2 dt \ , \tag{7.6}$$

where $\mathbf{f}(t)$ is the $n \times 2$ matrix whose columns are $f_1(x(t))$ and $f_2(x(t))$. So $G$ will be nonsingular if and only if the function $\varphi_v : t \to v^\dagger M(1,t)\mathbf{f}(t)$ does not vanish identically on $[0,1]$ for any nonzero vector $v$. Assume that $\varphi_v(t) \equiv 0$. Then the derivative $\dot{\varphi}_v$ will also vanish. To compute this derivative, we observe that the derivative of $M(1,t)$ with respect to $t$ is easily seen to be equal to $-M(1,t)(\eta_1(t)Df_1(x(t)) + \eta_2(t)Df_2(x(t)))$. Therefore

$$\begin{aligned}&\frac{d}{dt}M(1,t)f_1(x(t)) = \\ &\quad M(1,t)\cdot Df_1(x(t))\cdot\Big(\eta_1(t)f_1(x(t)) + \eta_2(t)f_2(x(t))\Big) \\ &\quad - M(1,t)\Big(\eta_1(t)Df_1(x(t)) + \eta_2(t)Df_2(x(t))\Big)f_1(x(t)),\end{aligned} \tag{7.7}$$

so that

$$\begin{aligned}&\frac{d}{dt}M(1,t)f_1(x(t)) = \\ &\quad \eta_2(t)M(1,t)\cdot\Big(Df_1(x(t))f_2(x(t)) - Df_2(x(t))f_1(x(t))\Big)\ ,\end{aligned} \tag{7.8}$$

i.e.

$$\frac{d}{dt}M(1,t)f_1(x(t)) = \eta_2(t)M(1,t)[f_2,f_1](x(t))\ . \tag{7.9}$$

Similarly, we have

$$\frac{d}{dt}M(1,t)f_2(x(t)) = \eta_1(t)M(1,t)[f_1,f_2](x(t))\ . \tag{7.10}$$

The above equalities are, of course, valid almost everywhere. If $\eta$ does not vanish almost everywhere, then there exists a $t$ such that $\eta(t) \neq 0$ and both (7.9) and (7.10) hold there. It then follows that

$$v^\dagger M(1,t)[f_1,f_2](x(t)) = 0\ . \tag{7.11}$$

On the other hand, the hypothesis that $\varphi_v$ vanishes identically implies that

$$v^\dagger M(1,t)f_1(x(t)) = v^\dagger M(1,t)f_2(x(t)) = 0\ . \tag{7.12}$$

Since we are assuming that $f_1$, $f_2$ and $[f_1,f_2]$ are independent at every point, it follows that

$$v^\dagger M(1,t) = 0\ . \tag{7.13}$$

Since $M(1,t)$ is nonsingular, this implies that $v = 0$ as desired.

## REFERENCES

[1] Agrachev, A.A., and R.V. Gamkrelidze, "The exponential representation of flows and chronological calculus," *Matem. Sbornik* **109**(149) (1978), pp. 467-532.

[2] Brockett, R.W., "Asymptotic stability and feedback stabilization," in *Differential Geometric Control Theory*, R.W. Brockett R.S. Millman and H.J. Sussmann Eds., Birkhäuser Boston Inc. (1983), pp. 1-115.

[3] Brockett, R.W., and L. Dai, "Non-holonomic kinematics and the role of elliptic functions in constructive controllability," in *Proc. IEEE Int. Conf. Robotics and Automation*, Sacramento, CA, April, 1991.

[4] Fernandes, C., L. Gurvits and Z.X. Li, "Foundations of nonholonomic motion planning," in *Proc. Int. Conf. Robotics and Automation,* Sacramento, CA, April, 1991.

[5] Jacobs, P., J.-P. Laumond, and M. Taix, "A complete iterative motion planner for a car-like robot," In *Journées Géometrie Algorithmique*, INRIA, 1990.

[6] Lafferriere, G., "A general strategy for computing steering controls of systems without drift," to appear in the Proceedings of the 1991 CDC.

[7] Lafferriere, G., and H.J. Sussmann, "Motion planning for controllable systems without drift: A preliminary report," Rutgers University Systems and Control Center Report SYCON-90-04, 1990.

[8] Lafferriere, G., and H.J. Sussmann, "Motion planning for controllable systems without drift," in *Proc. Int. Conf. Robotics and Automation*, Sacramento, CA, April, 1991, pp. 1148-1153.

[9] Laumond, J.P., "Nonholonomic motion planning versus controllability via the multibody car system example," Stanford University Technical Report STAN-CS-90-1345, December 1990.

[10] Laumond, J.P., and T. Siméon, "Motion planning for a two degrees of freedom mobile robot with towing,"In *IEEE International Conference on Control and Applications*, 1989.

[11] R. Montgomery, "Geodesics which do not Satisfy the Geodesic Equations," preprint, 1991.

[12] Murray, R.M., and S.S. Sastry, "Grasping and manipulations using multifingered robot hands," Memorandum No. UCB/ERL M90/24, Electronic Research Laboratory, Univ. of California at Berkeley, March, 1990.

[13] Murray, R.M., and S.S. Sastry, "Steering controllable systems," in *Proc. 29th IEEE Conf. Dec. and Control*, Honolulu, Hawaii, December, 1990.

[14] Murray, R.M., and S.S. Sastry, "Steering nonholonomic systems using sinusoids," *Proc. IEEE Conference on Decision and Control*, 1990.

[15] Reeds, J.A. and L.A. Shepp, "Optimal paths for a car that goes both forwards and backwards," *Pacific J. Math.*, 145(1990), pp 367-393.

[16] R. Strichartz, "Sub-Riemannian Geometry," *J. Diff. Geom.* 24, 221-263, (1983).

[17] R. Strichartz, "Corrections to 'Sub-Riemannian Geometry'," *J. Diff. Geom.* 30, no. 2, 595-596, (1989).

[18] Sussmann, H.J., "A product expansion for the Chen-series," in *Theory and Applications of Nonlinear Control Systems,* C.I. Byrnes, and A. Lindquist Eds., North-Holland, 1986, pp. 323-335.

[19] Sussmann, H.J., and W. Liu, "Motion planning and approximate tracking for controllable systems without drift," Proc. 25th Annual Conf. on Inf. Sciences and Systems, Johns Hopkins University, March 20-22, 1991, pp. 547-551.

[20] Sussmann, H.J., and W. Liu, "Limits of highly oscillatory controls and the approximation of general paths by admissible trajectories," Rutgers Center for Systems and Control Technical Report 91-02, February 1991. To appear, in a slightly abridged form, in the Proceedings of the 30th IEEE Conf. Decision and Control, to be held in Brighton, UK, Dec. 1991.

[21] Sussmann, H.J., and W. Liu, "Limiting behavior of highly oscillatory inputs, noncommutative formal power series, and the approximation of general paths by admissible trajectories," in preparation.

[22] Sussmann, H.J., and G. Tang, "Shortest paths for the Reeds-Shepp car: a worked out example of the use of geometric techniques in nonlinear optimal control," Rutgers Center for Systems and Control Technical Report 91-10, September 1991.

SYCON - Rutgers Center for Systems and Control, October 1991
Department of Mathematics
Rutgers University
New Brunswick, NJ 08903

*E-mail:* sussmann@hamilton.rutgers.edu

# Stability Regions for Differential-Algebraic Systems

V.VENKATASUBRAMANIAN,
H.SCHÄTTLER and J.ZABORSZKY

*This paper is dedicated to Antonio Ruberti, a true pioneer in leading and facilitating the emergence of precise Systems Theory.*

## 1. Introduction

Many problems in the theory of power systems such as voltage stability can be modeled by a parameter dependent differential-algebraic system of the form

$$\dot{x} = f(x,y,p) \quad , \quad f : \mathbb{R}^{n+m+p} \to \mathbb{R}^n \tag{1}$$

$$0 = g(x,y,p) \quad , \quad g : \mathbb{R}^{n+m+p} \to \mathbb{R}^m \tag{2}$$

$$x \in X \subset \mathbb{R}^n, \;\; y \in Y \subset \mathbb{R}^m, \;\; p \in P \subset \mathbb{R}^p$$

In the state space $X \times Y$ dynamic state variables $x$ and instantaneous state variables $y$ are distinguished. Typical dynamic state variables are the time dependent values of generator voltages and rotor phases, instantaneous variables are bus voltages and other load flow variables. The parameter space $P$ is composed of system parameters (which describe the system topography, i.e. which lines, buses etc. are energized, and equipment constants such as inductances, transformation ratios etc.), and operating parameters (such as loads, generation, voltage setpoints etc.) Both of these spaces have distinct structural features, sometimes intertwined. Understanding these structures is essential in analysing the voltage dynamics of the power system and of great practical value in the operation [10,12]. The special nature of the dynamics of differential-algebraic equations was recognized and briefly explored by Takens [6,7]. In this paper we make a first attempt at analyzing the structural features of such systems in the state space giving a definition of region of stability for a differential-algebraic system and describing the structure of the stability boundary. Our definition is strongly motivated by qualitative features of power system models for voltage stability, but the results are applicable to any differential-algebraic system with similar features.

The state variables describe the time behavior of the system, both the stationary one and the transients which may occur around it. The load flow constraint $g = 0$ limits the state to a load flow set $L \subset X \times Y$ at all times. This algebraic constraint $g = 0$ can sometimes be eliminated by inversion or reasonable approximations, but often this is not possible in voltage stability problems. Since the equations for the power system are given by real-analytic functions, $L$ is a stratified set, in simple cases a manifold with maximal dimension $n$. Through every point $q \in L$ with the exception of points $q$ in a singular subset $S$, there exists a unique trajectory (i.e. transient) which passes through $q$ and lies in $L$. The singular set $S$ is defined as the set where the conditions of the implicit function theorem for eliminating $y$ from the algebraic constraint are violated. It is a stratified set of maximal dimension $n-1$ embedded in $L$ and $L$ is separated by $S$ into open regions, the components $C$ of the state space. So the state space is confined to the complement of $S$ in $L$. Typically no solution exists within the singular set $S$.

In voltage stability problems, the singular set $S$ has utmost physical significance. It implies the presence of a fast random event, actually the emergence of a random process, when the state reaches $S$. The quasi-stationarity approximation, which is the basis for deriving mathematical models for $f$ and $g$ in equations (1) and (2), becomes grossly violated near $S$. In fact, the entire apparatus of conventional AC dynamics loses validity. Basic concepts such as RMS voltage, phase shift, reactive power and equivalent $\pi$ circuits are no longer applicable near the singularity. A singular perturbation approach near the singularity is not meaningful in the sense that the results cannot faithfully represent the physical dynamics of the power system. Trajectories cannot cross or bypass the singularity in any predictable manner [9, Section IV.1]. That the system behavior cannot be globally described by the differential-algebraic equations has also been noted by Hill, Hiskens and Mareels [2]. Borrowing their nomenclature, in this work the *point of view* is taken that *the singular set $S$ represents an impasse surface* and we do not allow trajectories to cross the singular set. Our analysis simply stops at $S$. If one interprets the singular surface as the onset of unpredictable events, this is indeed a realistic interpretation for the actual behavior in voltage stability problems.

In section 3 a definition of stable and unstable manifolds will be given which reflects this interpretation and the stability boundary of a stable equilibrium will be described. Since all dynamics are constrained to individual open components, $C$, the stability boundaries typically also contain segments of $S$. The results given here form the core of an in-depth study of voltage stability presented in [9]. Our results are general in the sense that no specific form of equations is assumed, but viewing the singular set as an impasse surface need not be a realistic interpretation for every differential-algebraic system. For differential-algebraic equations with an underlying fast dynamics, trajectories may exhibit jumps or discontinuities near the singularity which can actually be calculated from this fast parasitic dynamics. In this case the regions of attractions as defined and analysed in this paper only provide a conservative estimate for the full regions of stability. In many practical situations, however, this may still be the only reasonable estimate (for instance due to reasons of complexity). We emphasize again that for the question which motivates our work, the voltage stability problem in power systems, we obtain a correct rendering of the stability region for the actual system behavior [9].

An analysis of the main structural features of the state-space with emphasis on those aspects which are due to the presence of the algebraic constraint is given in sections 2.1-2.4. The key tool is a singular transformation introduced in section 2.2 which gives rise to a smooth transformed system which has the same phase portrait as the original system except for time rescaling along the trajectories and reversal of flow on some sections. This transformation completely eliminates all singular aspects of the system. All singular points either become regular or what will be called pseudo equilibria (which are equilibria only in the transformed state). Therefore a complete set of techniques from the realm of dynamical systems is directly applicable to the transformed (and smooth) system, such as $\lambda$-lemmas for stable and unstable manifolds in the investigation of regions of attraction [5]. Since the phase portraits of the original and of the transformed system are identical (topologically equivalent, to be more precise), results about the transformed system translate directly to the original system under consideration. However, the singular surface plays no special role in the transformed equations and therefore an extra analysis of the system near the singularity is needed throughout. This is a novel contribution of this work also from a mathematical perspective. The singular transformation restores complete mathematical unity to the state space and firmly establishes the structural relations between the various components of the state space. It facilitates the analysis without altering the physical facts of the state space.

## 2. The Structure of the State Space

A fixed parameter $p_0$ is considered in equations (1) and (2) and for notational simplicity the parameter is dropped. The equations are written in the form

$$\dot{x} = f(x,y) \quad , \quad f : \mathbb{R}^{n+m} \to \mathbb{R}^n \tag{3}$$

$$0 = g(x,y) \quad , \quad g : \mathbb{R}^{n+m} \to \mathbb{R}^m \tag{4}$$

### 2.1 An Overview of the Main Structural Features

The following sets largely determine the structure of the state space:

$$\begin{aligned}
L &= \{(x,y) \in \mathbb{R}^{n+m} : g(x,y) = 0\} \\
M &= \{(x,y) \in L : \operatorname{rank}(D_x g(x,y), D_y g(x,y)) = m\} \\
S &= \{(x,y) \in L : \Delta(x,y) := \det D_y g(x,y) = 0\} \\
\Psi &= \{(x,y) \in S : \kappa(x,y) := adj(D_y g(x,y)) D_x g(x,y) f(x,y) = 0\} \\
\Xi &= \{(x,y) \in S \setminus \Psi : D_y \Delta(x,y) \kappa(x,y) = 0\} \\
R &= S \setminus (\Psi \cup \Xi)
\end{aligned}$$

(Here $D_x f$ etc. denotes the matrix of partial derivatives of the components of $f$ with respect to $x$. In the definition of $\kappa$, $\operatorname{adj}(D_y g)$ denotes the classical matrix adjoint of the Jacobian $D_y g$). $L$ is the load flow surface determined by the algebraic constraint. The system must stay on $L$ and henceforth all topological definitions are relative to $L$. $M$ is a regular subset of $L$ which, by the implicit function theorem, is an $n$-dimensional manifold. $S$ is called the singular surface or the singularity. Typically no load flow solution exists for points in $S$. $\Psi$ is called the *pseudo equilibrium surface*. This typically high dimensional subset displays all the behavior of an equilibrium surface and indeed becomes an equilibrium surface if the vector field is singularly rescaled as follows: Let

$$\begin{aligned}
Z &= (f(x,y), -(D_y g(x,y))^{-1} D_x g(x,y) f(x,y)) \\
Z^T &= (f(x,y)\Delta(x,y), -adj(D_y g(x,y)) D_x g(x,y) f(x,y))
\end{aligned}$$

The flow of the vector field $Z$ describes the dynamics of the system. The vector field $Z^T$ is obtained from $Z$ by rescaling it at every point by multiplying with the determinant $\Delta$. The transformed vector field is smooth and leaves the load flow surface invariant. It will be used in the proofs, but the results as stated are directly applicable to the original system (3) and (4). The pseudo equilibrium surface is a true equilibrium surface for the rescaled vector field $Z^T$. It is useful to distinguish the following subsets of

$\Psi$:

$$\begin{aligned} (5)\quad N_\psi &= \{(x,y) \in \Psi \cap M : \text{rank}(D_y g) = m-1 \text{ and the Jacobian} \\ &\qquad \text{of } Z^T \text{ has exactly two nonzero eigenvalues}\} \\ (6)\quad B_\psi &= \Psi \setminus N_\psi \end{aligned}$$

Assuming that $M$ is dense in $L$, which is the typical situation, the Jacobian of $Z^T$ (restricted to $L$) has at least $(n-2)$ zero eigenvalues at a point in $\Psi$ (see section 2.4). Hence $N_\psi$ corresponds to the subset of $\Psi$ where the most regular type behavior takes place. It will be shown that near every point in $N_\psi$ the integral curves of the system locally behave like for a linear system. Eventually generic assumptions will be imposed to make sure that the complement $B_\psi$ is a truly lower dimensional subset, so that it suffices to consider $N_\psi$. Finally, $\Xi$ corresponds to points where $Z^T$ is tangent to the singularity (see section 2.3). Henceforth points in $\Xi$ will be called *semi-singular*. Again it will be assumed that the most regular type behavior occurs except for truly lower dimensional subsets. Let

$$\begin{aligned} (7)\qquad N_\xi &= \{(x,y) \in \Xi : \text{rank}\begin{pmatrix} D_x g & D_y g \\ D_x \Delta & D_y \Delta \end{pmatrix} = m+1, \\ &\qquad \text{and } D_y\{(D_y\Delta)\kappa\}\kappa \neq 0\} \\ (8)\qquad B_\xi &= \Xi \setminus N_\xi. \end{aligned}$$

The condition on the second derivative in the definition of $N_\xi$ ensures that trajectories do not cross over as they touch the singularity. The properties of these various subdivisions of the state space are summarized in the next theorem which will be proven in sections 2.2-2.4.

**Theorem 1** *If the subsets are not empty, then*

1. *$L \setminus S$ is an $n$-manifold. The solutions of the system (3) and (4) are the integral curves of the vector field $Z$ on $L \setminus S$.*

2. *$R$ is an $(n-1)$-dimensional manifold. At each point in $R$ two trajectories either approach or leave the singularity transversally.*

3. *$N_\psi$ is an $(n-2)$-dimensional manifold. Ignoring orientations, the solution curves are locally equivalent to those of a linear system near the origin.*

4. *$N_\xi$ is an $(n-2)$-dimensional manifold. Trajectories are tangent to the singularity at $N_\xi$, but do not cross over.*

It will also be assumed that the bad sets are truly lower dimensional. More precisely, the following assumption will be made:

**(A0)** The nice sets $N_\psi$ and $N_\xi$ are dense in $\Psi$ and $\Xi$, respectively. Furthermore, at every point $\psi \in \Psi$ or $\xi \in \Xi$, the sets $B_\psi$ and $B_\xi$ have dimension at most $(n-3)$.

Assumption (A0) is generic and describes the least degenerate situation which will be true for a typical system.

### 2.2 A Singular Transformation

**Lemma 1** *Restricted to $L \setminus S$, the dynamics defined by equations (3) and (4) is locally induced by the vector field $Z$ (i.e. described by an ordinary differential equation).*

**Proof:** This follows from the implicit function theorem ($D_y g$ is nonsingular in $L \setminus S$) and the fact that $(D_x g, D_y g)$ defines a normal vector field for the points in $L \setminus S$. Q.E.D.

The vector field $Z^T$ is obtained when the vector field $Z$ is pointwise scaled by the determinant $\Delta(x, y)$. Define

$$\begin{aligned}(L \setminus S)_+ &= \{(x,y) \in L \setminus S : \Delta(x,y) > 0\} \\ (L \setminus S)_- &= \{(x,y) \in L \setminus S : \Delta(x,y) < 0\}\end{aligned}$$

Clearly, $(L \setminus S) = (L \setminus S)_+ \cup (L \setminus S)_-$ as $\Delta(x,y) \neq 0$ in $L \setminus S$.

**Lemma 2** *The vector fields $Z$ and $Z^T$, restricted to $(L \setminus S)_+$, are equivalent. When restricted to $(L \setminus S)_-$, $Z$ and $-Z^T$ are equivalent.*

**Proof:** Let $t$ denote the time along integral curves of the vector field $Z$. Define a new time scale by

$$\frac{d\tau}{dt} = \Delta(x(t), y(t))$$

The rescaled vector field is then $Z^T$. If $\Delta(x,y) > 0$, the time-orientation is preserved and so $Z$ and $Z^T$ are equivalent on $(L/S)_+$. When $\Delta(x,y) < 0$, as the time orientations are reversed, $Z$ and $-Z^T$ are equivalent. Q.E.D.

In particular, if $(x_0, y_0)$ is an equilibrium point for the vector field $Z$, then it is also an equilibrium for the new vector field $Z^T$, and the eigenvalues in the transformed system are simply scaled by $\Delta(x_0, y_0)$. As curves, the trajectories of $Z$ and $Z^T$ are identical in $L \setminus S$. Whereas $Z$ is not defined on $S$, $Z^T$ defines a smooth vector field on all of the load flow surface $L$. The transformed vector field $Z^T$ has an additional set of equilibrium points in the pseudo equilibrium surface $\Psi$. The subset $M$ of $L$ is an $n$-dimensional manifold by implicit function theorem. By definition $L \setminus M \subset S$. Moreover, all such points on the load flow surface are pseudo equilibrium points.

**Lemma 3** $L \setminus M \subset \Psi$

**Proof:** Let $(x_0, y_0) \in L\backslash M$. Then, at $(x_0, y_0)$, $g(x_0, y_0) = 0$ and $(D_x g,\ D_y g)$ is not of full rank. In particular, $D_y g$ has rank less than $m$. There are two possibilities: (a) If rank $(D_y g) < m - 1$, then $\Delta(x_0, y_0) = 0$ and $\mathrm{adj}(D_y g) = 0$. Therefore $(x_0, y_0) \in \Psi$. (b) If rank $(D_y g) = m - 1$, then necessarily also rank $(D_x g,\ D_y g) = m - 1$ and so

$$D_x g \in \mathrm{Image}(D_y g). \tag{9}$$

Furthermore, by the definition of the classical adjoint for matrices,

$$\mathrm{adj}(D_y g)\ D_y g \ = \ D_y g\ \mathrm{adj}(D_y g) = 0$$

i.e.

$$\mathrm{adj}(D_y g) \in \mathrm{kernel}(D_y g) \tag{10}$$

Equations (9) and (10) imply that, at $(x_0, y_0)$, $\mathrm{adj}(D_y g) D_x g = 0$, i.e. $(x_0, y_0) \in \Psi$. Q.E.D.

**Lemma 4** *$R = S\backslash(\Psi \cup \Phi)$ is an $(n-1)$-dimensional embedded submanifold of $M$ and $Z^T$ is transversal to $R$.*

**Proof:** Since $D_y g \cdot \kappa = \Delta(D_x g) f$, we have on $S$ that

$$\begin{pmatrix} D_x g & D_y g \\ D_x \Delta & D_y \Delta \end{pmatrix} Z^T = \begin{pmatrix} 0 \\ -D_y \Delta \cdot \kappa \end{pmatrix}. \tag{11}$$

Let $(x_0, y_0) \in R$. As $(x_0, y_0) \notin \Psi$, $\mathrm{rank}(D_y g) = m-1$ and rank $(D_x g, D_y g) = m$. As also $(x_0, y_0) \notin \Xi$, $D_y \Delta \cdot \kappa \neq 0$ at $(x_0, y_0)$ and so $D_y \Delta$ does not lie in the row span of $D_y g$. But then the Jacobian

$$\begin{pmatrix} D_x g & D_y g \\ D_x \Delta & D_y \Delta \end{pmatrix}$$

has maximal rank ($=m + 1$) everywhere on $R$. By the implicit function theorem, $R$ is therefore an $(n - 1)$-dimensional manifold. Equation (11) also shows that $Z^T$ is transversal to $R$. Q.E.D.

**Lemma 5** *Any integral curve of the vector field $Z$ which approaches the singular surface has a unique limit on the singularity and this limit agrees with the limit of the solution curve in the transformed system.*

**Proof:** This is an immediate consequence of the fact that $Z^T$ defines a smooth system on the load flow surface. Q.E.D.

Lemma 5 says that $Z^T$ provides a canonical extension of the vector field $Z$ to the singularity. It should be noted that the orientation of the trajectories is reversed for the points in $(L \setminus S)_-$, as $\Delta(x,y) < 0$. Since the solution curves are identical, from here onwards all the proofs will be done for the transformed system and will be interpreted for the original system.

## 2.3 Analysis of the Semi-singular Surface

A special role is played by the pseudo equilibria and semi-singular points in the closure of the region of attraction. Stable manifolds in the quasi-stability boundary which connect with the singular set are anchored there. Here semi-singular points are analysed, in the next subsection the analogous results for pseudo equilibria are derived.

**Lemma 6** *If nonempty, then $N_\xi$ is an $(n-2)$-dimensional embedded submanifold of $M$. The vector field $Z^T$ has a nonzero normal component at every point in $N_\xi$.*

**Proof:** For every point in $N_\xi$ there exists a neighborhood so that $N_\xi$ is given as the zeroset of the functions

$$g(x,y) = 0, \ \Delta(x,y) = 0, \ D_y\Delta(x,y)\kappa(x,y) = 0$$

with extra inequality conditions imposed. Since $D_y\{D_y\Delta \cdot \kappa\}\kappa \neq 0$, it follows that the gradients of these functions are linearly independent on $N_\xi$ and that the normal component of $Z^T$ does not vanish. The result therefore follows from the implicit function theorem. Q.E.D.

The points in the nice set anchor the stability boundary much like equilibrium points and periodic orbits do for a Morse-Smale system. Therefore a detailed analysis of the local picture near these points becomes necessary. Let $\xi \in N_\xi$. Since $\xi \not\in \Psi$, rank $(D_y g) = m-1$ and rank $(D_x g, D_y g) = m$. Hence we can split the coordinates $x$ and $y$ into $\{x_1, x_2\}$ and $\{y_1, y_2\}$, so that $(D_{x_2} g, D_{y_2} g)$ is nonsingular at $(0,0)$. Without loss of generality, assume that

$$x_2 = \{\mathrm{x}_1\}, \ x_1 = \{\mathrm{x}_2, \ldots \mathrm{x}_n\}, \ y_1 = \{\mathrm{y}_1\}, \ y_2 = \{\mathrm{y}_2, \ldots \mathrm{y}_m\}.$$

Then, by the implicit function theorem, the equations $g(x,y) = 0$ can be locally solved as

$$x_2 = x_2(x_1, y_1), \ \ y_2 = y_2(x_1, y_1).$$

Using this representation, the transformed system can be reduced to the chart coordinates $(x_1, y_1)$ to give:

$$\begin{aligned} \dot{x_1} &= f_1(x_1, y_1)\, \Delta_1(x_1, y_1) \\ \dot{y_1} &= g_1(x_1, y_1) \end{aligned}$$

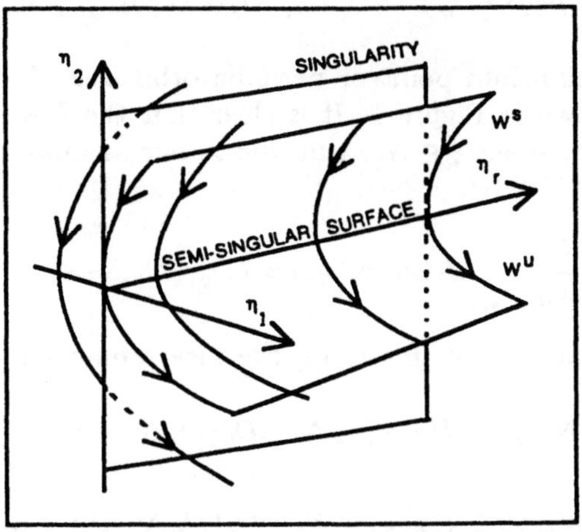


**Fig. 1:** Three-dimensional rendering of the local picture near $N_\xi$

where

$$f_1(x_1,y_1) = (0, I_{n-1})\, f(x_1, x_2(x_1,y_1,), y_1, y_2(x_1,y_1))$$
$$g_1(x_1,y_1) = (1, 0_{m-1})\kappa(x_1, x_2(x_1,y_1), y_1, y_2(x_1,y_1)),$$
$$\Delta_1(x_1,y_1) = \Delta(x_1, x_2(x_1,y_1), y_1, y_2(x_1,y_1))$$

Choose new coordinates $(\eta_1, \ldots, \eta_n)$ which are centered at $\xi$ (i.e. $\eta_i(\xi) = 0$ for all $i$) as follows:

$$\eta_1 = \Delta_1(x,y), \quad \eta_2 = D_y\Delta(x,y)\kappa(x,y)$$

and complete the coordinate chart with an $(n-2)$ vector $\eta_r$ from among the components of $x_1$ to get a diffeomorphism. This can be done since the Jacobian

$$\begin{pmatrix} D_x g & D_y g \\ D_x\Delta & D_y\Delta \\ D_x(D_y\Delta\cdot\kappa) & D_y(D_y\Delta\cdot\kappa) \end{pmatrix}$$

has full rank at $\xi$. In these coordinates the system can be described as

$$\begin{aligned} \dot\eta_1 &= \beta_1(\eta_1,\eta_2,\eta_r), \\ \dot\eta_2 &= \beta_2(\eta_1,\eta_2,\eta_r), \\ \dot\eta_r &= \beta_r(\eta_1,\eta_2,\eta_r). \end{aligned}$$

The singular surface $S$ is locally given as $\{\eta_1 = 0\}$ and $N_\xi$ is the coordinate plane $\{\eta_1 = 0, \eta_2 = 0\}$. Furthermore, by the definition of $N_\xi$,

$$\beta_1(0,0,\eta_r) = (D_{y_1}\Delta)g_{1|N_\xi} = 0,$$

$$\beta_2(0,0,\eta_r) = D_{y_1}(D_y\Delta\cdot\kappa)g_{1|N_\xi} \neq 0.$$

There is no equilibrium point or periodic orbit near $\xi$. Hence the local picture is as shown in Figure 1. It is clear that the flow is tangent to the singularity at any point $\xi \in N_\xi$. But due to our assumption on the second derivative,

$$\ddot{\eta}_1 = \frac{d}{dt}_{|(0,0,\eta_r)}\beta_1(\eta_1,\eta_2,\eta_r) = (D_{y_1}(D_y\Delta\cdot\kappa)g_{1|N_\xi} \neq 0.$$

Depending on the sign of this term, the nice semi-singular points can be split into

$$N_{se.sa} = \{(x,y) \in N_\xi : D_y(D_y\Delta\cdot\kappa)\kappa > 0\}$$

and

$$N_{se.fo} = \{(x,y) \in N_\xi : D_y(D_y\Delta\cdot\kappa)\kappa < 0\}.$$

It is shown in [9] that no semi-singular point can lie in the interior of the closure $\overline{A}$ of the region A of attraction. Hence, if $U$ is a local neighborhood of $\xi \in N_\xi$ in $M$, then only one half of it, namely the intersection $U_C$ of $U$ with the connected component $C$ which contains $A$, can possibly lie in $A$. Without loss of generality assume that $\Delta(x,y) > 0$ in $C$ (otherwise redefine $\Delta$ as $-\Delta$). If $\xi \in N_{se.sa}$, then

$$\frac{d}{dt}_{|\xi}\Delta(x,y) = 0 \text{ and } \frac{d^2}{dt^2}_{|\xi}\Delta(x,y) > 0.$$

In $C$ there exists a unique trajectory which reaches $\xi$ in finite time and immediately leaves the singularity again to the same side. The phaseportrait resembles the phaseportrait near a continuum of saddlepoints along the $\eta_r$ axis, restricted to a half-neighborhood. Hence such a point will be called a *semi-saddle*.

**Lemma 7** *$W^s(N_{se.sa})$ and $W^u(N_{se.sa})$ are $(n-1)$-dimensional manifolds.*

**Proof:** This simply follows from the fact that the flow of $Z^T$ is well-defined and a diffeomorphism near $N_\xi$. Q.E.D.

For points in $N_{se.fo}$ we have

$$\frac{d}{dt}_{|\xi}\Delta(x,y) = 0, \text{ and } \frac{d^2}{dt^2}_{|\xi}\Delta(x,y) < 0.$$

Hence there is no trajectory in $C$ which passes through $\xi$. In particular, $W^s(N_{se.fo})$ and $W^u(N_{se.fo})$ are empty. The local phase portrait in $C$ near such a point is like for a continuum of foci along the $\eta_r$ axis, cut off by the singularity. Hence we call these points *semi-foci*. Because the intersections of the stable and unstable manifolds with $C$ are empty, semi-foci have no significance in stability boundary calculations.

## 2.4 Analysis of the Pseudo equilibrium Surface

The pseudo equilibrium surface $\Psi$ can be split into a nice set $N_\psi$ and a bad set $B_\psi$ as defined in equations (5) and (6). The properties of the set $N_\psi$ will be explored in this section.

**Lemma 8** *If nonempty, then $N_\psi$ is an $(n-2)$-dimensional embedded submanifold of $M$.*

**Proof:** The proof is similar to the proof of Lemma 6. For every point in $N_\psi$, we can find a suitably small neighborhood so that $N_\psi$ can be locally described as the solution of $m+2$ equations,

$$g(x, y) = 0, \ \Delta(x, y) = 0, \ \text{ and } \kappa_1(x, y) = 0,$$

for some component $\kappa_1(x, y)$ of $\kappa$, and with extra inequality conditions imposed in terms of the eigenvalues. These conditions imply that the Jacobian of these functions is of full rank. So $N_\psi$ is an $(n-2)$-dimensional manifold by the implicit function theorem. Q.E.D.

**Lemma 9** *For every point $\psi \in N_\psi$, there exists a neighborhood $U$ of $\psi$ in $M$, and a coordinate chart $(\eta_1, \ldots, \eta_n)$ defined on $U$, such that*

**(o)** *the chart is centered at $\psi$, i.e. $\eta_i(\psi) = 0$ for all $i$*

**(i)** *the dynamics take the form*

$$\dot{\eta}_i = f_i^1 \eta_1 + f_i^2 \eta_2$$

*for some functions $f_i^k : \eta(U) \subset \mathbb{R}^n \longrightarrow \mathbb{R}$, $k = 1, 2$ and $i = 1, \ldots, n$*

**(ii)** $S \cap U = \{\eta_1 = 0\}$.

**Proof:** By definition of $N_\psi$, rank $D_y g = m-1$ and rank $(D_x g, D_y g) = m$. Hence, by a calculation similar to the one given in section 2.3, the flow of $Z^T$ is diffeomorphic to the solutions of ordinary differential equations of the form

$$\begin{aligned} \dot{x}_1 &= f_1(x_1, y_1)\, \Delta_1(x_1, y_1), \\ \dot{y}_1 &= g_1(x_1, y_1) \end{aligned}$$

where dim $x_1 = n-1$ and $y_1$ is scalar. Furthermore, $\Delta_1$ describes $\Delta$ in the new coordinates and $g_1$ is an appropriate component of $\kappa$. Because of the condition on the eigenvalues, the Jacobian

$$J_N := \begin{pmatrix} D_{x_1}\Delta_1 & D_{y_1}\Delta_1 \\ D_{x_1}g_1 & D_{y_1}g_1 \end{pmatrix}$$

has rank 2 at $(0,0)$. The lemma follows by choosing as coordinates

$$\eta_1 = \Delta_1(x_1, y_1) \text{ and } \eta_2 = g_1(x_1, y_1)$$

and then completing the basis from the components of $x_1$. The special form of the dynamics is a consequence of the form of $Z^T$. Q.E.D.

In these coordinates, the origin is the pseudo equilibrium and the Jacobian of the vector field $Z^T$ has $n-2$ zero eigenvalues and 2 nonzero eigenvalues. Hence, the equilibrium points in the set $N_\psi$ are nonhyperbolic. But they are special in the following sense:

**Lemma 10** *The local center manifold for any point in the nice set is $N_\psi$ itself.*

**Proof:** Consider any point $\psi \in N_\psi$. Clearly $N_\psi$ is a center manifold because $N_\psi$ is invariant and satisfies the tangentiality conditions. Suppose $CM$ is some other center manifold. Then it must contain equilibrium points that lie arbitrarily close to $\psi$. Hence $N_\psi$ belongs to $CM$, locally. But $N_\psi$ itself is a $(n-2)$-dimensional surface. Thus, locally, it is the center manifold. Q.E.D.

The set $N_\psi$ can be split into three different subsets, depending on the stability status. Define

$$\begin{aligned} N_{so} &:= \{\psi \in N_\psi : \text{the Jacobian of } Z^T \text{ has two eigenvalues in } \mathbb{C}^+\} \\ N_{sa} &:= \{\psi \in N : \text{the Jacobian of } Z^T \text{ has one eigenvalue in } \mathbb{C}^- \\ &\qquad \text{and one in } \mathbb{C}^+\} \\ N_{si} &:= \{\psi \in N : \text{the Jacobian of } Z^T \text{ has two eigenvalues in } \mathbb{C}^-\} \end{aligned}$$

So $N_\psi = N_{so} \cup N_{sa} \cup N_{si}$. Note that any component of $N_\psi$ entirely lies in one of the three sets. Let us call the points in $N_{so}$, $N_{sa}$ and $N_{si}$ *pseudo-sources*, *pseudo-saddles* and *pseudo-sinks*, respectively. This definition is motivated by the next lemma.

**Lemma 11** *For $\psi \in N_\psi$, the flow of $Z^T$ is locally* equivalent *to a linear system defined as:*

$$\dot{x_u} = x_u, \; \dot{x_s} = -x_s, \; \dot{x_c} = 0, \qquad x_c \in \mathbb{R}^{n-2}.$$

*The dimensions of $x_u$ and $x_s$ depend on the particular set in $N$, i.e.*

$$dim(x_u)=2 \text{ and } dim(x_s)=0, \text{ when } \psi \in N_{so}$$
$$dim(x_u)=1 \text{ and } dim(x_s)=1, \text{ when } \psi \in N_{sa}$$
$$dim(x_u)=0 \text{ and } dim(x_s)=2, \text{ when } \psi \in N_{si}$$

**Proof:** It is known that for nonhyperbolic equilibrium points, the flow can be classified depending on the flow on a center manifold [1]. For $\psi \in N_\psi$, the flow on the center manifold ($= N_\psi$) is trivially the identity map (since $\dot{x_c} = 0$). Hence the result follows. Q.E.D.

The last lemma states that the flow near a nice pseudo equilibrium point looks like that of a linear system. Together with Lemma 8 this proves part 3 of Theorem 1. The remaining statements in Theorem 1 have already been verified ( part 1 is Lemma 1, part 2 is Lemma 4 and part 4 is Lemma 6). For hyperbolic equilibrium points stronger properties are true, for instance, the flow is locally *conjugate* to the flow of a linear system (Hartman-Grobman theorem) [5]. But equivalence is sufficient for our proofs because we are mainly interested in the features of the phase portraits, such as the boundary of the region of attraction. Hence for points in $N_\psi$ there exists a well-defined local "picture" from a dynamical system point-of-view. The local neighborhood of the pseudo saddle surface is summarized in Figure 2.

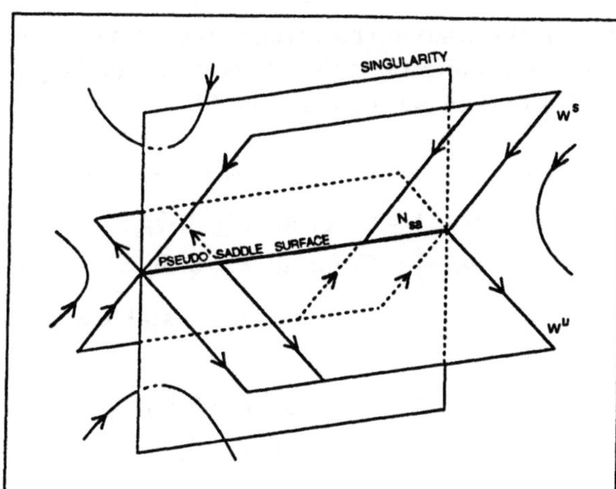


**Fig. 2**: Three-dimensional rendering of the local picture near $N_{sa}$

**Corollary 1** *$W^s(N_\psi)$ and $W^u(N_\psi)$ are manifolds,*

$$\begin{aligned} dim(W^s(N_{so})) &= n-2, & dim(W^u(N_{so})) &= n, \\ dim(W^s(N_{sa})) &= n-1, & dim(W^u(N_{sa})) &= n-1, \\ dim(W^s(N_{si})) &= n, & dim(W^u(N_{si})) &= n-2. \end{aligned}$$

**Proof:** Locally, near any point $\psi \in N_\psi$, the result follows from Lemma 11. For a point $z \in W^s(\psi)$ (or $W^u(\psi)$), a neighborhood of $z$ in $W^s(\psi)$ can be described as the flow in finite time of a local neighborhood near $\psi$. As the flow of the smooth vector field $Z^T$ defines a diffeomorphism, the proof is complete. Q.E.D.

## 3. Stability Boundaries

Equilibria and periodic orbits of $Z$ in $L \setminus S$ are defined as usual, but with the proviso that they do not lie on or intersect the singular set $S$. Denote the coordinates in the state space by $z = (x, y)$ and for a stable equilibrium $z_s$, let $C$ denote the connected component of $L \setminus S$ which contains $z_s$. Recall that it is assumed that the determinant $\Delta(x, y)$ is positive in $C$. For reasons mentioned briefly in the Introduction and explained in detail in [9], if the flow $\Phi_t$ of $Z$ meets the singular set, $\Phi_{t_0}(z) \in S$ for some $t_0 > 0$, then $z$ will not be considered as a point in a stable manifold of an equilibrium or a periodic orbit. The *region of attraction* of a stable equilibrium $z_s$ is defined as

$$A = \{z \in C : \Phi_t(z) \in C\ \forall t \geq 0,\ \Phi_t(z) \to z_s \text{ as } t \to \infty\}.$$

The results given in this section characterize the boundary of the region of attraction of a stable equilibrium point. Proofs are too long to be included here, but a detailed outline can be found in [9].

For any equilibrium $z_0 \in C$ define the *stable* and *unstable manifolds* as

$$\begin{aligned} W^s(z_0) &= \{z \in C : \Phi_t(z) \in C \text{ for all } t \geq 0, \Phi_t(z) \to z_0 \text{ as } t \to \infty\} \\ W^u(z_0) &= \{z \in C : \Phi_t(z) \in C \text{ for all } t \leq 0, \Phi_t(z) \to z_0 \text{ as } t \to -\infty\}. \end{aligned}$$

Similarly, for a periodic orbit $\gamma_0 \in C$ define the stable and unstable manifolds as

$$\begin{aligned} W^s(\gamma_0) &= \{z \in C : \Phi_t(z) \in C \text{ for all } t \geq 0, \Phi_t(z) \to \gamma_0 \text{ as } t \to \infty\} \\ W^u(\gamma_0) &= \{z \in C : \Phi_t(z) \in C \text{ for all } t \leq 0, \Phi_t(z) \to \gamma_0 \text{ as } t \to -\infty\}. \end{aligned}$$

Finally, for a pseudo equilibrium $\psi \in \Psi$ or a semi-singular point $\xi \in \Xi$, these definitions are as follows

$$\begin{aligned} W^s(\psi) &= \{z \in C : \exists t_0 > 0 \text{ such that } \Phi_t(z) \in C \text{ for } 0 \leq t < t_0 \\ &\qquad \text{and } \Phi_t(z) \to \psi \text{ as } t \to t_0\} \\ W^u(\psi) &= \{z \in C : \exists t_0 < 0 \text{ such that } \Phi_t(z) \in C \text{ for } 0 \geq t > t_0 \\ &\qquad \text{and } \Phi_t(z) \to \psi \text{ as } t \to t_0\}. \end{aligned}$$

and verbatim the same for $\xi$.

**Lemma 12** *Let $\psi \in N_\psi$ and suppose the Jacobian of $Z^T$ at $\psi$ has $n_s$ stable eigenvalues. Then $W^s(\psi)$ is an $n_s$-dimensional and $W^u(\psi)$ is a $(2-n_s)$-dimensional manifold. If $\delta(\psi)$ denotes a $\delta$-sized neighborhood of $\psi$ in $N_\psi$, then the stable and unstable manifolds of $\delta(\psi)$, $W^s(\delta(\psi))$ and $W^u(\delta(\psi))$, are $(n-2+n_s)$- and $(n-n_s)$-dimensional manifolds, respectively.*

If $N_C$ denotes a connected component of the nice set $N_\psi$, then the Lemma implies that $W^s(N_C)$ and $W^u(N_C)$ are manifolds. Moreover, all pseudo equilibria in $N_C$ are of the same type. A pseudo-saddle is called *transverse* if its stable and unstable manifolds are not tangent to the singularity. In this case, the singularity divides a neighborhood of $\psi$ into two parts, each containing one stable and one unstable trajectory. Again this is the most regular structure possible and it will be assumed that it persists on the component except for a truly lower dimensional subset. The analogous Lemma for semi-singular points is trivial since convergence occurs in finite time [9]. The local behavior of the system in a half-neighborhood near a semi-singular point is very much like for a transverse pseudo-saddle, except that the trajectory hits tangentially and in finite time (see Figures 1 and 2).

In the statements of the results in this section, in addition to assumption (A0), (which, however, can be relaxed to requiring the conditions only at points in the closure of the region of attraction), the following **assumptions** are made:

**(A1)** (a) Equilibria $z_0$ and periodic orbits $\gamma_0$ in the boundary of the region of attraction, $\partial A$, are hyperbolic.
(b) Except for sets of dimension at most $(n-3)$, all pseudo-saddles in $\partial A$ are transverse.

**(A2)** Stable manifolds from the set $\{W^s(z_0), W^s(\gamma_0), \overline{W^s(N_C)}\}$ intersect transversally with unstable manifolds from the set $\{W^u(z_0), W^u(\gamma_0), W^u(N_C)\}$. Here $N_C$ stands for a component of either $N_\psi$ or $N_\xi$.

**(A3)** All trajectories on the boundary of the region of attraction converge to an equilibrium point, a periodic orbit, a pseudo equilibrium point or a semi-singular point.

It is sufficient, of course, if all these conditions are fulfilled for the objects which lie in the closure of the region of attraction. Assumptions (A1) and (A2) are satisfied generically. In condition (A2) note that the closure of the stable manifold $\overline{W^s(N_C)}$ for the entire component $N_C$ is considered. It is allowed that $\overline{W^s(N_C)}$ intersects nontransversally with a stable manifold $W^s(z_0)$ of an equilibrium, for instance. But intersections with unstable manifolds must be transversal. Assumption (A3) will be satisfied if there exists an energy function as described in the next Lemma (see also [3]).

**Lemma 13** *If there exists a (Lyapunov-) function $V = V(x, y)$ such that*

1. $D_x V f - D_y V (D_y g)^{-1} D_x g f \leq 0$ *in* $A$
2. $V$ *is proper in the closure of* $A$
3. *the* $\omega$*-limit set in* $\overline{A}$ *consists only of equilibria and periodic orbits*

*then assumption (A3) is satisfied: all trajectories in* $\partial A$ *converge to equilibria, periodic orbits, pseudo equilibria or semi-singular points.*

The **main result** of this section follows. Recall that the order of an equilibrium point $z_0$ or a periodic orbit $\gamma_0$ is the dimension of its unstable subspace and it is denoted by a superscript.

**Theorem 2 (Quasi-Stability Boundary Theorem)**
*Under assumptions (A) the boundary of the closure of the region of attraction* $\partial\overline{A}$ *(called quasi-stability boundary in [11]) is composed of the following four principal elements;*

1. *stable manifolds of order 1 equilibria* $z_0^1$
2. *stable manifolds of order 2 periodic orbits* $\gamma_0^2$
3. *stable manifolds of pseudo-saddles or semi-saddles*
4. *singular boundary pieces.*

*More precisely*

$$(12) \quad \partial\overline{A} = \overline{\cup_{z_0^1 \in \partial\overline{A}} W^s(z_0^1) \cup_{\gamma_0^2 \in \partial\overline{A}} W^s(\gamma_0^2) \cup W^s(N_{n-1}) \cup (S \cap \partial\overline{A})}$$

*where* $N_{n-1} = N_{n-1}^{\psi} \cup N_{n-1}^{\xi}$ *and*

$$(13) \quad N_{n-1}^{\psi} = \{\psi \in N_\psi \cap \partial\overline{A} : \text{there exists a } \delta > 0 \text{ such that the } \delta\text{-sized neighborhood } \delta(\psi) \text{ of } \psi \text{ in } N_\psi \text{ satisfies } \delta(\psi) \subset \partial\overline{A} \text{ and } n_s = 1\}$$

$$(14) \quad N_{n-1}^{\xi} = \{\xi \in N_\xi \cap \partial\overline{A} : \text{there exists a } \delta > 0 \text{ such that the } \delta\text{-sized neighborhood } \delta(\xi) \text{ of } \xi \text{ in } N_\xi \text{ satisfies } \delta(\xi) \subset \partial\overline{A}\}$$

The principal structure of the full boundary is characterized next.

**Theorem 3** *Under assumptions (A) the set*

$$\partial A \;/\; \overline{\cup_{z_0^1 \in \partial A} W^s(z_0^1) \cup_{\gamma_0^2 \in \partial A} W^s(\gamma_0^2) \cup W^s(N_{n-1}) \cup (S \cap \partial A)}$$

*has maximal dimension* $n - 2$ *at every point.*

Thus, modulo a set of dimension strictly less than $n-1$, the boundary $\partial A$ consists of the $(n-1)$-dimensional stable manifolds of equilibria, periodic orbits, pseudo-saddles and semi-saddles, and of a singular segment. A complete description of the full boundary is much more difficult because there may exist components of $\partial A$ of dimension strictly less than $n-1$ which lie in the interior of $\partial\overline{A}$. Moreover, these components may consist only of part of a stable manifold. The quasi-stability boundary was introduced in [11] to eliminate possible internal segments of the stability boundary from consideration since these are not significant from a practical viewpoint.

For the power system model, only polynomials and polynomial functions of sinussoidal functions are present. If the angles are restricted to a variation of $2\pi$, it follows from algebraic stratification theory (essentially, from the Tarski-Seidenberg Theorem [4]), that the numbers of equilibria and the numbers of connected components of $N$ and $S$ are finite. Therefore

**Corollary 2** *Suppose assumption (A) holds and there exist only finitely many periodic orbits. Then for a power system model there are only finitely many terms in the union*

$$\partial\overline{A} = \cup_{z_0^1\in\partial\overline{A}}\overline{W^s(z_0^1)} \cup_{\gamma_0^2\in\partial\overline{A}} \overline{W^s(\gamma_0^2)} \cup \overline{W^s(N_{n-1})} \cup (S\cap\partial\overline{A})$$

## REFERENCES


[1] D.K. Arrowsmith and C.M. Place, An introduction to Dynamical Systems, Cambridge University Press, 1990

[2] D. J. Hill, I. A. Hiskens and I. M. Y. Mareels, *Stability theory of differential algebraic models of power systems*, Proceedings of the 11th IFAC World Congress, Tallinn, USSR, Aug. 1990, Volume 12, pp. 19-24.

[3] I. A. Hiskens and D. J. Hill, *Energy function, transient stability and voltage behavior in power systems with nonlinear loads*, IEEE Transactions on Power Systems, Vol. 4, No. 4, October 1989, pp. 1525-1533.

[4] N. Jacobson, Basic Algebra I, Freeman, 1974.

[5] J. Palis, Jr. and W. de Melo, Geometric theory of dynamical systems, an introduction, Springer-Verlag, New York, 1982.

[6] F. Takens, *Constrained equations: a study of implicit differential equations and their discontinuous solutions*, in: Structural stability, the theory of catastrophes and the application in the sciences, Lecture Notes in Mathematics, 525, Springer-Verlag, Berlin and New York, 1976, pp. 143-234.

[7] F. Takens, *Implicit differential equations: some open problems*, in: Singularites d'Applications Differentiables, Plans-sur-Bex, 1975, Lecture Notes in Mathematics, vol. 535, Springer Verlag, 1976, pp.237-253.

[8] V. Venkatasubramanian, H. Schättler and J. Zaborszky, *Global voltage dynamics: study of a generator with voltage control, transmission and matched MW load*, Proceedings of the 29th IEEE Conference on Decision and Control, Hawaii, December 1990, pp. 3045-3056.

[9] V. Venkatasubramanian, H. Schättler and J. Zaborszky, *A Taxonomy of the Dynamics of the Large Power System with Emphasis on its Voltage Stability*, Proceedings of a Workshop on Bulk Power System Voltage Phenomena, Deep Creek Lake, McHenry, Maryland, August 1991, to appear

[10] J. Zaborszky *Some basic issues in voltage stability and viability*, Proceedings of Bulk-Power Voltage Phenomena – Voltage Stability and Security, Potosi, MO September, 1988, pp. 1.17-1.60.

[11] J. Zaborszky, G. Huang, B. Zheng and T. C. Leung, *On the phase portraits of a class of large nonlinear dynamic systems such as the power system*, IEEE Transactions on Automatic Control, Vol. 33, No. 1, January, 1988, pp. 4-15.

[12] J. Zaborszky and B. Zheng, *Structure features of the dynamic state space for studying voltage-reactive control*, Proceedings of The Power Systems Computation Conference, Graz, Austria, August 1990, pp. 319-326.

DEPARTMENT OF SYSTEMS SCIENCE AND MATHEMATICS,
CAMPUS BOX 1040, WASHINGTON UNIVERSITY,
ONE BROOKINGS DRIVE, ST.LOUIS, MISSOURI, 63130-4899

# Progress in Systems and Control Theory

*Series Editor*

Christopher I. Byrnes
Department of Systems Science and Mathematics
Washington University
Campus P.O. 1040
One Brookings Drive
St. Louis, MO 63130-4899

*Progress in Systems and Control Theory* is designed for the publication of workshops and conference proceedings, sponsored by various research centers in all areas of systems and control theory, and lecture notes arising from ongoing research in theory and applications control.

We encourage preparation of manuscripts in such forms as LATEX or AMS TEX for delivery in camera-ready copy which leads to rapid publication, or in electronic form for interfacing with laser printers.

Proposals should be sent directly to the editor or to: Birkhäuser Boston, 675 Massachusetts Avenue, Cambridge, MA 02139, U.S.A.

PSCT1 Computation and Control: Proceedings of the Bozeman Conference, Bozeman, Montana, 1988
*K. Bowers and J. Lund*

PSCT2 Perspectives in Control Theory: Proceedings of the Sielpia Conference, Sielpia, Poland, 1988
*B. Jakubczyk, K. Malanowski, and W. Respondek*

PSCT3 Realization and Modelling in System Theory: Proceedings of the International Symposium MTNS-89, Volume 1
*M. A. Kaashoek, J. H. van Schuppen, and A. C. M. Ran*

PSCT4 Robust Control of Linear Systems and Nonlinear Control: Proceedings of the International Symposium MTNS, Volume II
*M. A. Kaashoek, J. H. van Schuppen, and A. C. M. Ran*

PSCT5 Signal Processing, Scattering and Operator Theory, and Numerical Methods: Proceedings of the International Symposium MTNS-89, Volume III
*M. A. Kaashoek, J. H. van Schuppen, and A. C. M. Ran*

PSCT6 Control of Uncertain Systems: Proceedings of an International Workshop, Bremen, West Germany, June 1989
*D. Hinrichsen and B. Martensson*

PSCT7 New Trends in Systems Theory: Proceedings of the Università di Genova - The Ohio State University Joint Conference, July 9-11, 1990
*G. Conte, A. M. Perdon, and B. Wyman*

PSCT8 Analysis of Controlled Dynamical Systems: Proceedings of a Conference in Lyon, France, July 1990
*B. Bonnard, B. Bride, J. P. Gauthier and I. Kupka*

PSCT9 Nonlinear Synthesis: Proceedings of a IIASA Workshop held in Sopron, Hungary, June 1989
*Christopher I. Byrnes and Alexander Kurzhansky*

PSCT10 Modeling, Estimation and Control of Systems with Uncertainty: Proceedings of a Conference held in Sopron, Hungary, September 1990
*Giovanni B. Di Masi, Andrea Gombani and Alexander B. Kurzhansky*

PSCT 11 Computation and Control II: Proceedings of the Second Conference at Bozeman, Montana, August 1990
*Kenneth Bowers and John Lund*

PSCT 12 Systems, Models, and Feedback: Theory and Applications. Proceedings of a U.S.-Italy Workshop in honor of Professor Antonio Ruberti, held in Capri, Italy, 15-17, June 1992
*A. Isidori and T.J. Tarn*

GPSR Compliance

*The European Union's (EU) General Product Safety Regulation (GPSR) is a set of rules that requires consumer products to be safe and our obligations to ensure this.*

*If you have any concerns about our products, you can contact us on ProductSafety@springernature.com*

In case Publisher is established outside the EU, the EU authorized representative is:

Springer Nature Customer Service Center GmbH
Europaplatz 3
69115 Heidelberg, Germany

**Batch number: 10577878**

Printed by Printforce, the Netherlands